Essentials of Toxicology

Edited by Braylon Holden

hayle
medical

New York

Hayle Medical,
750 Third Avenue, 9ᵗʰ Floor,
New York, NY 10017, USA

Visit us on the World Wide Web at:
www.haylemedical.com

ISBN: 978-1-63241-471-7

Cataloging-in-Publication Data

Essentials of toxicology / edited by Braylon Holden.
   p. cm.
Includes bibliographical references and index.
ISBN 978-1-63241-471-7
1. Toxicology. 2. Toxicity testing. 3. Medicine. I. Holden, Braylon.
RA1211 .E87 2017
615.9--dc23

Table of Contents

Permissions

List of Contributors

Index

Preface

Toxicology studies the damaging effects caused by a chemical substance to living organisms. This book on toxicology serves as an important source of information about toxins and toxicity levels. Toxicology studies fundamental determinants of poisoning such as level of dosage, route of entry and the vital statistics of the person exposed to the substance. Contents in the text have been compiled keeping in mind the high level of research that takes place in this field. This book includes contributions of experts and scientists which will provide innovative insights into this field. It attempts to assist those with a goal of delving into the field of toxicology.

This book is a comprehensive compilation of works of different researchers from varied parts of the world. It includes valuable experiences of the researchers with the sole objective of providing the readers (learners) with a proper knowledge of the concerned field. This book will be beneficial in evoking inspiration and enhancing the knowledge of the interested readers.

In the end, I would like to extend my heartiest thanks to the authors who worked with great determination on their chapters. I also appreciate the publisher's support in the course of the book. I would also like to deeply acknowledge my family who stood by me as a source of inspiration during the project.

Editor

Contribution for the Derivation of a Soil Screening Value (SSV) for Uranium, Using a Natural Reference Soil

Ana Luisa Caetano[1,2]*, Catarina R. Marques[1,2], Ana Gavina[6,2], Fernando Carvalho[3], Fernando Gonçalves[1,2], Eduardo Ferreira da Silva[4], Ruth Pereira[5,6]

1 Department of Biology, University of Aveiro, Campus Universitário de Santiago, Aveiro, Portugal, 2 CESAM, University of Aveiro, Campus Universitário de Santiago, Aveiro, Portugal, 3 Nuclear and Technological Institute (ITN) Department of Radiological Protection and Nuclear Safety, Sacavém, Portugal, 4 Department of Geosciences, University of Aveiro, GeoBioTec Research Center, Campus Universitário de Santiago, Aveiro, Portugal, 5 Department of Biology, Faculty of Sciences of the University of Porto, Porto, Portugal, 6 Interdisciplinary Centre of Marine and Environmental Research (CIIMAR/CIMAR), University of Porto, Porto, Portugal

Abstract

In order to regulate the management of contaminated land, many countries have been deriving soil screening values (SSV). However, the ecotoxicological data available for uranium is still insufficient and incapable to generate SSVs for European soils. In this sense, and so as to make up for this shortcoming, a battery of ecotoxicological assays focusing on soil functions and organisms, and a wide range of endpoints was carried out, using a natural soil artificially spiked with uranium. In terrestrial ecotoxicology, it is widely recognized that soils have different properties that can influence the bioavailability and the toxicity of chemicals. In this context, SSVs derived for artificial soils or for other types of natural soils, may lead to unfeasible environmental risk assessment. Hence, the use of natural regional representative soils is of great importance in the derivation of SSVs. A Portuguese natural reference soil PTRS1, from a granitic region, was thereby applied as test substrate. This study allowed the determination of NOEC, LOEC, EC_{20} and EC_{50} values for uranium. Dehydrogenase and urease enzymes displayed the lowest values (34.9 and <134.5 mg U Kg, respectively). *Eisenia andrei* and *Enchytraeus crypticus* revealed to be more sensitive to uranium than *Folsomia candida*. EC_{50} values of 631.00, 518.65 and 851.64 mg U Kg were recorded for the three species, respectively. Concerning plants, only *Lactuca sativa* was affected by U at concentrations up to 1000 mg U kg^1. The outcomes of the study may in part be constrained by physical and chemical characteristics of soils, hence contributing to the discrepancy between the toxicity data generated in this study and that available in the literature. Following the assessment factor method, a predicted no effect concentration (PNEC) value of 15.5 mg kg$^{-1}_{dw}$ was obtained for U. This PNEC value is proposed as a SSV for soils similar to the PTRS1.

Editor: Stephen J. Johnson, University of Kansas, United States of America

Funding: ALC was supported by a PhD grant from Fundação para a Ciência e Tecnologia (FCT) (http://www.fct.pt/). The funders had no role in study design, data collection and analysis, decision to publish, or preparation of the manuscript.

Competing Interests: The authors have declared that no competing interests exist.

* Email: ana.caetano@ua.pt

Introduction

Uranium (U) is a natural soil component, being originated from rocks in the Earth's crust, where it mainly occurs in the form of oxides. Natural processes acting on rocks and soils, such as wind, water erosion, dissolution, precipitation and volcanic activity contribute for U dispersal in the environment [1]. The use of U as fuel in nuclear power plants has driven to its large-scale exploration worldwide. The U exploration became significantly important in the world during the Second World War, and later on during the Cold War, in both cases to supply military needs of the greatest potencies. Recently, the World Nuclear Association estimated worldwide reserves of U at 5.4 million tons in 2009, of which Australia had about 31%, followed by Kazakhstan (12%), Canada and Russia with 9% (http://www.world-nuclear.org/info/inf75.html). The remarkable energy crisis that is currently faced worldwide due to the exhaustion of carbon based energy resources is demanding further extraction of U, as nuclear energy arises as a potential solution. Hence, it is expected that the mining and milling of U will increase in the next decades, contributing for its widespread in the environment [2].

During the last century, Portugal has actively explored radioactive ores and was for some time ranked as one of the main U producers. The extraction of U ore in Portugal started in 1908, first driven by the interest in radium (being U a by-product) and then by the interest in its military applications, till 2001 [3,4]. Most of the old U mines were located in the granitic regions of the Iberian Meseta, in the centre-north of Portugal (Beiras), [5]. Nowadays, although the mining activities ceased, like in several other places in the world, the old U mines represent a serious environmental problem, due to waste accumulation (mainly tailings and sludge) and improper disposal of radioactive material, composed by U and its daughter radionuclides [1,5–16]. Soils and water are the two major environmental matrices affected by U contamination.

U has a long half-life, persisting in nature as different isotopes, with different chemical and radiological characteristics [17]. The toxic effects induced by this metal are caused by both properties.

However, since U isotopes mainly emit alpha particles, with little penetration capacity, the main radiation hazards only occur after ingestion or inhalation of these isotopes and daughter radionuclides [17]. Once in the soil, U interacts with all the components of this matrix, such as clay minerals, aluminum and iron oxides, organic matter and microorganism, in a very complex system, where pH and organic matter seem to have the major role in controlling U mobility (pH 6) and leaching (pH<6) [18]. The high mobility/availability of U will in turn increase the ecological risks posed to soil and water compartments [19–27].

The soil has been recognized as an important compartment that provides crucial ecosystem services (e.g., filtering of contaminants, reservoir of carbon and a bank of genes) and is the support of agro-sylvo-pastoral production [28,29] and of several other human activities. The soil compartment offers raw materials (e.g., peat, clay, ore) and contributes for climate regulation and biodiversity conservation, as well as other cultural services [30,31]. The recognition of the importance of maintaining the provision of such services has increased the necessity to create appropriate legal tools to correctly and effectively protect and manage this resource. In this sense, the Soil Framework Directive proposed by the Commission of the European Communities (CEC), aims to establish a common strategy for the protection and sustainable use of soils [32]. For that end, this proposal defines measures for the identification of the main problems faced by soils, the adoption of strategies to prevent their degradation, as well as for the rehabilitation of contaminated or degraded soils [33]. The Soil Framework Directive will fill in the gap regarding soil protection, since this compartment has never been a target of specific protection policies at the European Community level [32]. Many countries, committed in regulating the management of contaminated land, have adopted generic quality standards, the soil screening values (SSVs) [34]. SSVs are concentration thresholds above which, more site-specific evaluations are required to assess the risks posed by soil contamination [35]. The SSVs should provide a level of protection to terrestrial species and ecological functions of the soil [35–37]. SSVs are particularly useful for the first tier of Ecological Risk Assessment (ERA) processes applied to contaminated sites, supporting the decision-making at this initial stage of assessment [38], which at the end is aimed in setting priorities for remediation and risk reduction measures [39]. In the case of Portugal, SSVs for soils have never been established for metals or organics. Only threshold concentrations of metals on sewage sludge were legally established to regulate the application of this solid waste on agricultural soils [40]. However, they are not appropriate for soil ERA purposes.

The use of natural reference soils in ecotoxicological tests has been recommended by several authors [41–43]. This is because the properties of the OECD artificial soil are not representative of the great majority of natural soils [44]. Different levels of toxicity, for each contaminant, can be expected in soils with different properties [45–48], hence it is important each country derives their own SSVs using natural reference soils representing the main types of soils within their territories. In this context, the main aim of this work was to obtain ecotoxicological data for U, performing soil enzymes activity tests, invertebrates and plant tests, using for that a Portuguese natural reference soil (PTRS1), that represents one of the dominant types of soil from a granitic region (cambisol) of the country [49]. As a result, enough data are gathered as to make the first proposal of a SSV for this metal.

Materials and Methods

The present study used a natural soil that was collected in a non-protected area, requiring no specific permission for its collection. Further, no work with endangered species was performed, and no vertebrate species were used in the ecotoxicological assays. Only tests with invertebrates and plants were performed. The invertebrates were obtained from laboratorial cultures maintained by the authors of this manuscript and plant seeds were obtained from a local supplier.

1. Test soil

The natural soil (PTRS1) used as test substrate in this study was collected in Ervas Tenras [Pinhel, Guarda, Portugal center; geographical coordinates: 40°44'4.27"N and 7°10'54.3"W), at 655 m altitude, in a granitic region.

A composite soil sample was collected and immediately brought to the laboratory where it was air dried. Another portion of the soil, was immediately sieved through a 2 mm mesh size and the sieved fraction (<2 mm) was stored in polyethylene bags, at − 20°C, until further analysis of soil microbial parameters, which were performed within the period of one month. For the tests with soil organisms and plants, the soil was passed through a 4 mm mesh sieve and the sieved fraction (<4 mm) was defaunated through two freeze–thawing cycles (48 h at −20°C followed by 48 h at 25°C), before the beginning of the assays.

The physical and chemical properties (including total metal contents) of the PTRS1 soil were presented in a preliminary study by Caetano et al. [49], aimed in characterizing this soil as a reference substrate for ecotoxicological purposes. The main properties of the PTRS1 are also described in Tables 1 and 2. Briefly, soil-KCl 1 M and soil-deionized water suspensions (1:5 m/v) were used for pH (KCl, 1 M) and pH-H_2O measurements, respectively, according to ISO 17512–1 [50]- After 15 min of magnetic stirring and 1 h resting period, the pH of the suspension was measured using a WTW 330/SET-2 pH meter. A soil water suspension (1:5 w/v) was used for the measurement of soil conductivity [51] Ten grams of PTRS1 were mechanically shaken in polypropylene flasks with 50 ml with deionized water filtered in a Milli-Q equipment (hereinafter referred as deionized water), water for 15 min. The mixture was left to rest overnight for soil bulk settling [51]. The conductivity of the resulting suspension was measured using an LF 330/SET conductivity meter. Soil water content was determined from the loss of weight after drying at 105°C, for 24 h. Organic matter (OM) content was determined by loss of ignition of dried soil samples at 450°C during 8 h [52]. For determination of water holding capacity (WHC) polypropylene flasks were prepared with a filter paper-replaced bottom, which after being filled up with soil samples, were immersed in water for 3 h. After this period, samples were left for water drainage during 2 h and the WHC was determined accounting to the loss of weight after drying at 105°C until weight stabilization [50].

2. Test substance

For all the test organisms, the natural soil was spiked with a stock solution of uranyl nitrate 6-hydrate, $UO_2(NO_3)_26H_2O$ (98%, PANREAC) prepared with deionized water in order to obtain a range of concentrations, which were ascertained by range finding tests performed with the different test species.

For soil enzyme tests, the PTRS1 soil was spiked with the following concentrations: 0.0, 134.6, 161.5, 193.8, 232.5, 279.0, 334.8, 401.8, 482.2, 578.7, 694.4, 833.3, 1000 mg U kg^{-1}_{dw}. To obtain these concentrations, the stock solution of uranyl nitrate

Table 1. Physical and chemical properties of PTRS1 soil (retrieved from Caetano et al. [49]).

	pH (H$_2$O)	pH (KCl, 1 M)	Conductivity (mS cm^{-1})	OM (%)	WHC (%)	Particle-size distribution (%)				[U] (mg Cu kg^{-1} soil$_{dw}$)
						Clay (<4 µm)	Silt (4-63 µm)	Sand (63 µm-2 mm)	Gravel (>2 mm)	
PTRS1	5.9±0.09	4.3±0.02	4.8±0.23	6.5±0.004	23.9±1.84	3.3	22.8	46.9	23.9	9.0

pH (H$_2$O), pH (KCl, 1 M), OM (organic matter), and WHC (water holding capacity) are represented as average ± STDEV.

was diluted in the volume of deionized water required to adjust the soil moisture at 80% of its maximum water holding capacity (WHC$_{max}$).

The following U concentrations were used to expose the earthworms in the reproduction tests: 0.0, 113.1, 124.4, 136.9, 150.5, 165.6, 231.9, 324.6, 454.5, 500.0, 550.0, 605.0, 665.5 mg U kg^{-1}$_{dw}$. For potworms, collembolans and terrestrial plant assays the same range of concentrations was tested: 0.0, 167.4, 192.5, 221.4, 254.6, 292.7, 336.6, 420.8, 526.0, 657.5, 756.1, 869.6, 1000 mg U kg^{-1}$_{dw}$.

The volume of deionized water required to adjust the WHC of the soil to a given percentage of its maximum value was used to dilute the stock solution for these tests. After spiking the soil was left to rest for equilibration for 48 h before testing.

3. Ecotoxicological assessment

3.1 Soil microbial activity. For testing the effect of increasing concentrations of U on soil microbial parameters, a 30-day exposure was firstly conducted. Ten grams of sieved PTRS1 soil per replicate and concentration were spiked with different U concentrations, a total of three replicates were used per treatment. Six replicates with the same amount of soil only moistened with deionized water were also prepared for the control. The soil was incubated at 20±2°C and a photoperiod of 16 h^L:8 h^D. During the incubation period, the soil moisture was weekly monitored by weighing the pots, and whenever needed it was adjusted to 80% of its WHC$_{max}$ by adding deionized water. At the end, 1 g of each replicate from the control and concentrations tested was stored in individual falcon tubes at -20°C for approximately one month. Thereby, a total of 9 sub-replicates were made for each concentration. The soil was thawed at 4°C before analysis.

The biochemical parameters analyzed were: the activity of arylsulphatase, dehydrogenase, urease, and cellulase enzymes and changes in the nitrogen mineralization (N mineralization) and potential nitrification.

For the determination of arylsulphatase activity, the method proposed by Tabatabai and Bremner [53] and Schinner et al. [54] was followed. After addition of 1 mL of p-nitrophenylsulfate (0.02 M), soil sub-samples were incubated for one hour, at 37°C. The nitrophenyl liberated by the activity of arylsulphatase was extracted and colored with a 4 mL of sodium hydroxide (0.5 M) and determined photometrically at 420 nm. The results were expressed as µg p-nitrophenylsulfate (p-NP) g^{-1} soil$_{dw}$ h^{-1}.

The method proposed by Öhlinger [55] was used to assess the dehydrogenase activity. The samples were suspended in 1 mL of trifeniltetrazol chloride (TTC) (3.5 g L^{-1}) and incubated at 40°C for 24 h. The triphenylformazan (TPF) produced was extracted with acetone and measured spectrophotometrically at 546 nm. The results were expressed as µg TPF g^{-1} soil$_{dw}$ h^{-1}.

The cellulase activity was tested according to the method proposed by Schinner et al. [54] and Schinner and von Mersi [56]. The reducing sugars produced during the incubation period, after addition of 1.5 mL of acetate buffer (2 M), caused the reduction of hexacyanoferrate (III) potassium to hexacyanoferrate (II) potassium in an alkaline solution. This last compound reacts with ferric ammonium sulfate in acid solution to form a ferric complex of hexacyanoferrate (II), of blue colour, which is colorimetrically measured at 690 nm and expressed as µg glucose g^{-1} soil$_{dw}$ 24 h^{-1}.

N mineralization activity was measured according to Schinner et al. [54]. For this purpose the soil samples were incubated for 7 days at 40°C. During this period, the organic forms of N were converted to inorganic forms (mainly ammonium ion, NH$_4^+$),

Table 2. Pseudo-total concentrations (mg/kg) of metals recorded in PTRS1 soil (average ± standard deviation) extracted with aqua régia, (retrieved from Caetano et al.[49]).

Metal	PTRS1
Ag	0.1±0.0
Al	25628.5±5130.0
B	2.2±0.8
Ba	45.8±8.0
Be	1.2±0.2
Cd	0.1±0.1
Co	5.6±1.1
Cr	10.8±2.1
Cu	9.0±1.8
Fe	24921.4±4534.4
Li	124.4±22.9
Hg	5253.5±1025.5
Mn	386.8±77.9
Mo	0.9±0.2
Na	78.1±14.9
Ni	4.6±0.9
Pb	12.5±2.2
Sb	0.2±0.0
Sn	10.4±1.9
U	7.8±1.7
V	37.8±14.1
Zn	57.1±8.9

which were determined by a modification of the Berthelot reaction, after extraction with 3 mL of potassium chloride (2 M). The reaction of ammonia with sodium salicylate in the presence of sodium dichloroisocyanurate formed a green colored complex in alkaline pH that was measured at 690 nm and expressed as μg N g^{-1} $soil_{dw}$ d^{-1}.

The urease activity was assayed according to the method proposed by Kandeler and Gerber [57] and, Schinner et al. [54]. The samples were incubated for 2 h at 37°C after the addition of 4 mL of a buffered urea solution (720 mM). The ammonia released was extracted with 6 mL of potassium chloride (2 M) and determined by the modified Berthelot reaction. The quantification was based on the reaction of sodium salicylate with ammonia in the presence of chlorinated water. UR was detected at 690 nm and expressed as μg N g^{-1} $soil_{dw}$ 2 h^{-1}.

The quantification of potential nitrification was determined by the method of Kandeler [58], which is a modification of the technique proposed by Berg and Rosswall [59]. The ammonium sulphate (4 mL, 10 mM) was used as substrate, and soil samples were incubated for 5 h, at 25°C. Nitrate released during the incubation period was extracted with 1 mL of potassium chloride (2 mM) and determined colorimetrically at 520 nm. This reaction was expressed as μg nitrite (N) g^{-1} $soil_{dw}$ h^{-1}.

3.2. Invertebrate and plant tests. Test organisms and culture conditions: The earthworm *Eisenia andrei* (Oligochaeta: Lumbricidae), the potworm *Enchytraeus crypticus* (Oligochaeta: Enchytraeidae) and the springtail *Folsomia candida* (Collembola: Isotomidae) were used as invertebrate test organisms.

All organisms were obtained from laboratorial cultures, kept under controlled environmental conditions (temperature: 20±2°C; photoperiod: 16 h^L:8 h^D). The earthworms (*E. andrei*) are maintained in plastic boxes (10 to 50 L) containing a substrate composed by peat, dry and defaunated horse manure (through two freeze–thawing cycles (48 h at −20°C followed by 48 h at 65°C), and deionized water. The pH of the culture medium is adjusted to 6.0–7.0 with $CaCO_3$. The organisms are fed, every 2 weeks, with six table spoons of oatmeal previously hydrated with deionized water and cooked for 5 min. The potworms (*E. crypticus*) are cultured in plastic containers (25.5 cm length; 17.4 cm width; 6.5 cm height), which are filled with pot soil moistened to the nearest 60% of its WHC_{max} and with pH adjusted to 6.0±0.5. The organisms are fed twice a week with a tea spoon of macerated oat. The collembolans (*F. candida*) are maintained in plastic containers filled with culture medium composed by moistened Plaster of Paris mixed with activated charcoal 8:1 (w:w). They are fed with half of a tea spoon of granulated dry yeast, twice a week. The food is added in small amounts to avoid spoilage by fungi.

Seeds from four plant species (two dicotyledonous and two monocotyledonous), purchased from a local supplier, were used for seed germination and growth tests: *Avena sativa*, *Zea mays*, *Lacuta sativa* and *Lycopersicon esculentum*.

Reproduction tests with invertebrates: Previous studies from our team, at least with earthworms from the same laboratorial cultures, have proved that these organisms were not exposed to meaningful levels of metals (especially U, in laboratorial culture conditions) [60]. The accomplishment of validity criteria, by all the controls of the assays (herein described) with the three invertebrate species, also confirmed that the test animals were not previously exposed to toxic levels of metals through test containers, substrates or food. The reproduction tests with *E. andrei*, *E. albidus* and *F. candida* were carried out according to the ISO guidelines 11268-2 [61], 16387 [62] and 11267 [63], respectively. Each replicate of the invertebrate tests contained 10 individuals in a certain developmental stage: the earthworms had a fully developed clitellum and an individual fresh weight between 250 and 600 mg; the potworms were 12-mm size; and the springtails were 10–12 days old. Five hundred grams of dry soil were weighted per test vessel for earthworms. For the tests with potworms and collembolans 20 g and 30 g of soil were weighted per replicate, respectively. Following an ECx sampling design, which considers more concentrations and less number of replicates, two replicates per concentration and five replicates for the control were prepared in the reproduction tests with *E. andrei*. Adult earthworms were removed from the test containers after 28 days. The produced cocoons persisted in the soil until 56 days have been completed. After this period, the juveniles from each test container were counted. During the test, organisms were fed once a week, with 5 g per box of defaunated horse manure (using the same procedure above described), and the soil moisture content was weekly monitored (following the procedures outlined in ISO guideline 11268-2 [61]).

The *E. albidus* reproduction test was held for 28 days and the adults were left in the vessels until the end of the test. About 2 mg of rolled oats were placed on the soil surface, weekly to feed the animals. At the end of the test, the potworms were killed with alcohol, colored with Bengal red and counted according to the Ludox Flotation Method, as described in ISO 16387 [62]. The reproduction tests with *F. candida* took four weeks to be completed. The collembolans were fed with granulated dry yeast, obtained from a commercial supplier, being weekly added (about 2 mg of yeast per test vessel) to the soil surface. At the end of the test, the containers were filled with water and the juveniles were

Table 3. Toxicity data obtained for copper (mg U kg^{-1} soil$_{dw}$) in PTRS1 soil on soil microbial processes, invertebrates and plants.

Biota	Endpoint	NOEC	LOEC	EC$_{20}$	EC$_{50}$
Microbial parameters					
Arylsulphatase		232.5	279	155.3 (84.76–255.87)	295.6 (216.09–375.17)
Dehydrogenase		<134.5	≤134.5	34.9 (20.52–59.35)	110.3 (83.25–137.47)
Nitrogen mineralization	Enzyme activity	694.4	833.3	152.2 (46.66–257.79)	347.0 (211.25–482.91)
Celulase		≤134.5	≥134.5	n.d.	n.d.
Urease		<134.5	≤134.5	<134.5	<134.5
Potencial nitrification		<134.5	≤134.5	429.5 (229.53–629.46)	610.0 (459.07–761.11)
Invertebrates					
Eisenia andrei	Rep. (56 days)	500.0	550.0	474.8 (391.47–558.04)	631.0 (532.78–699.21)
Enchytraeus crypticus	Rep. (28 days)	420.8	526.0	469.7 (355.47–584.04)	518.6 (480.40–556.90)
Folsomia candida	Rep. (28 days)	675.5	756.1	343.4 (172.23–514.60)	851.64 (606.10–1097.18)
Plants					
Avena sativa	Germination	≥1000	>1000	n.d.	n.d.
Zea mays	Germination	≥1000	>1000	n.d.	n.d.
Lactuca sativa	Germination	≥1000	>1000	n.d.	n.d.
Lycopersicon esculentum	Germination	≥1000	>1000	n.d.	n.d.
Avena sativa	Fresh mass	≥1000	>1000	n.d	n.d.
Zea mays	Fresh mass	≥1000	>1000	n.d	n.d.
Lactuca sativa	Fresh mass	≥1000	>1000	n.d	n.d.
Lycopersicon esculentum	Fresh mass	≥1000	>1000	n.d	n.d.
Avena sativa	Dry mass	≥1000	>1000	n.d.	n.d.
Zea mays	Dry mass	≥1000	>1000	n.d.	n.d.
Lactuca sativa	Dry mass	<167.4	≤167.4	n.d.	n.d.
Lycopersicon esculentum	Dry mass	≥1000	>1000	n.d.	n.d.

For ECx point estimates the 95% confidence limits are presented in brackets. n.d.- not determined; Rep. – reproduction.

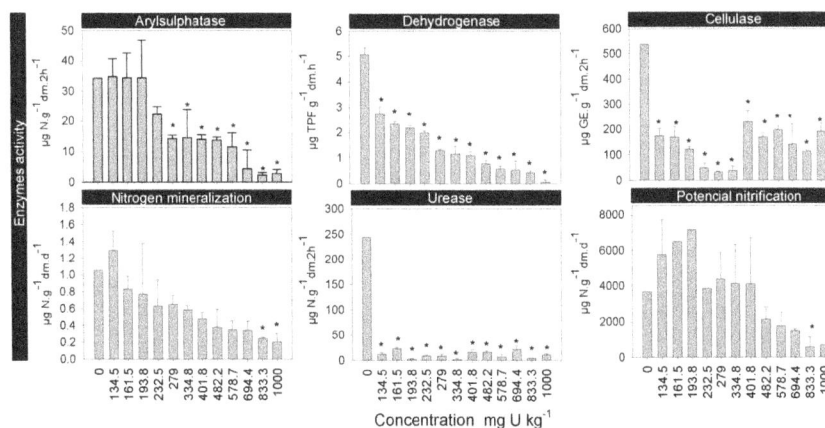

Figure 1. Soil enzyme activities, N mineralization and potential nitrification. Response of the arylsulphatase, dehydrogenase, cellulase urease, activity, N mineralization and potential nitrification to soils spiked with a range of uranium concentrations. The error bars indicate the standard deviation. The asterisks point out significantly differences from the control (P<0.05).

Table 4. Toxicity of copper (mg U kg^{-1} soil$_{dw}$) reported in the literature for the reproduction of soil invertebrates using different soil types with different physical and chemical characteristics.

Species	Endpoint	Soil type	Physical-chemical parameters			Point estimates (mg U kg^{-1} soil$_{dw}$)				Reference
			pH	OM (%)	Clay (%)	NOEC	LOEC	EC$_{20}$	EC$_{50}$	
		Canadian soil	6.2	1.0	2.0	n.d.	n.d.	>1000	n.d.	Sheppard and Stephenson [86]
Eisenia fetida	Rep. (56 days)	Canadian soil	6.2	1.0	2.0	n.d.	n.d.	>1120	n.d.	Sheppard and Stephenson [86]
		Canadian soil	7.5	2.2		>838	n.d.	n.d.	n.d.	Sheppard and sheppard [98]
		Canadian soil	7.5	18.4		>994	n.d.	n.d.	n.d.	Sheppard and sheppard [98]
Folsomia Candida	Rep. (28 days)	Canadian soil	7.5	2.2	24.0	n.d.	n.d.	840.0	n.d.	Sheppard and Stephenson [86]
		Canadian soil	7.5	n.d.	n.d.	n.d.	n.d.	>720	n.d.	Sheppard and Sheppard [98]
Elymus lanceolatus	Germination	Canadian soil	6.2	1	2	n.d.	>1000	n.d.	n.d.	Sheppard and Stephenson [86]
Elymus lanceolatus	Germination	Canadian soil	7.5	2.2	24	n.d.	>1001	n.d.	n.d.	Sheppard and Stephenson [86]
Zea mays	Dry mass	European soil	5.2	2.5	n.d.	n.d.	>100	n.d.	n.d.	Stojanović et al., [108]

OM - organic matter, Rep. - reproduction, n.d. - not determined., germ.- germination.

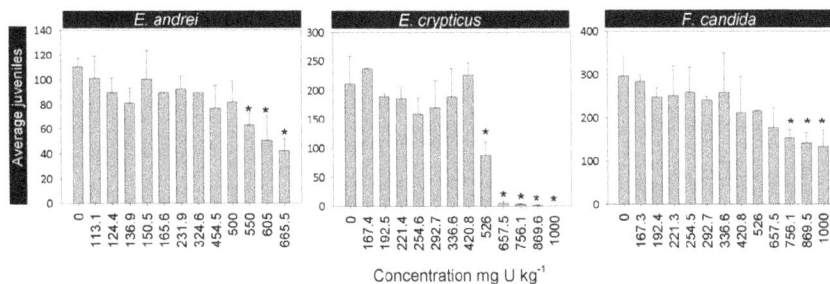

Figure 2. Reproduction of invertebrates. Results obtained exposing *Eisena andrei*, *Enchytraeus crypticus* and *Folsomia candida*, to natural PTRS1 soil, contaminated with different concentrations of U. The error bars indicate the standard deviation. The asterisks point out significantly differences from the control ($P < 0.05$).

counted after flotation. The addition of a few dark ink drops provided a higher contrast between the white individuals and the black background. The organisms were then counted through the use of the ImageJ software (online available: http://rsb.info.nih. gov/ij/download.html). The exposure was carried out at $20 \pm 2°C$ and a photoperiod of $16^{L}:8^{D}$. For both species five replicates of uncontaminated natural PTRS1 soil were prepared for the control. The same ECx sampling design applied for earthworms was followed. However, in order to reduce the variability of the results, three replicates were prepared per test concentration (instead of two for the earthworms).

Seed germination and plant growth tests: Germination and growth tests with terrestrial plants were performed following standard procedures described by the ISO guideline 11269-2 [64]. For this purpose, 200 g_{dw} of the spiked soil with the concentrations

described above was placed in plastic pots (11.7 cm diameter, 6.2 cm height) and tested. In this case, the amount of water required to adjust the WHC_{max} of the soil to 45% was used to dilute the stock solution and to moist the soil at the beginning of the test. The soil was placed in the plastic pots (11.7 cm diameter, 6.2 cm height). In the bottom of each plastic pot a hole was previously made to let a rope passing through, hence allowing communication with the pot below that was filled with deionized water. After soil spiking and soil saturation with water twenty seeds were added to each pot and gently covered with the spiked soil. The level of water in the lower recipient was adjusted whenever needed, as to guarantee the necessary conditions of moisture according to, the recommendations specified in [64]. Five replicates of uncontaminated natural PTRS1 soil were prepared for the control, while three replicates were tested per

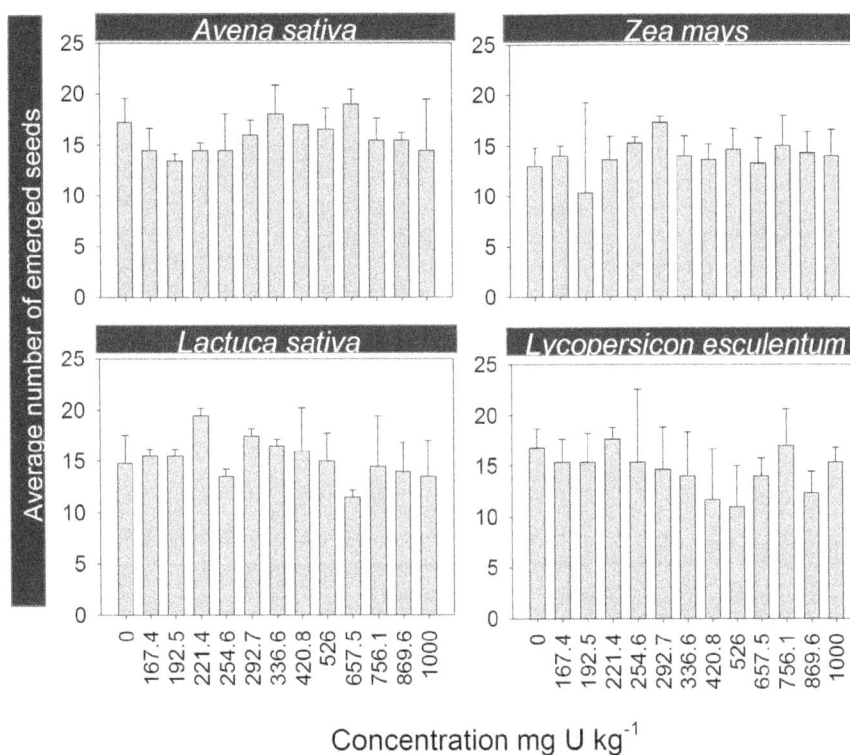

Figure 3. Seed germination of plants. Average number of emerged seeds in monocotyledonous, *Avena sativa and Zea mays* and in dicotyledonous species, *Lycopersicon esculentum and Lactuca sativa,* grown in PTRS1 soil contaminated with U. The error bars indicate the standard deviation.

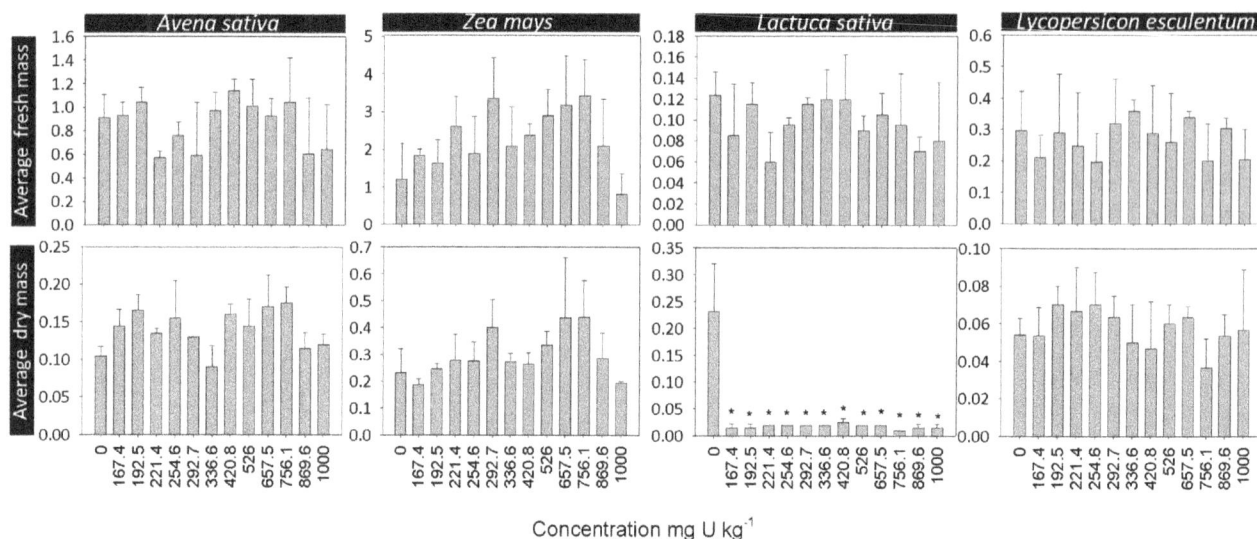

Figure 4. Growth of plants. Average values of fresh mass and dry mass in monocotyledonous, *Avena sativa and Zea mays* and in dicotyledonous species, *Lycopersicon esculentum and Lactuca sativa* grown in PTRS1 soil, contaminated with U. The error bars indicate the standard deviation. The asterisks point out significantly differences from the control (P<0.05).

concentration, in order to minimize the variability of the results, and to follow the ECx sampling design, similarly used for the invertebrate tests.

At the beginning of the test, nutrients (Substral - Plants fertilizer using 1 bottle cap for 2 L of water proportion according to the manufacturer recommendation; Fertilizer NPK: 6-3-6; nitrogen (N): 6%; phosphate (P_2O_5): 3%; potassium (K_2O): 6%; iron (Fe): 0,03%; trace elements: Cu, Mn, Mo and Zn), were added in each lower recipient containing the water. Pots were maintained at constant conditions of temperature ($20\pm2°C$), photoperiod (16 hL: 8 hD) and light intensity (25.000 lux). Daily observations were carried out to record the number of emerged seeds. Only the first five emerged seeds were left to grow, the remaining ones were counted and harvested. Fourteen days later, the assay was finished and the fresh and dry biomass above soil was assessed for each test species at the end of the exposure period.

The endpoints seed germination, and fresh and dry biomass, above soil, were assessed for each species at the end of the exposures according to the methods outlined in ISO guideline 11269-2 [64].

For this work, a battery of enzymes involved in different biogeochemical cycles S (sulfur cycle), N (Nitrogen cycle), C (Carbon cycle), as well as enzymes more indicative of the good physiological conditions of the whole microbial community (e.g.

dehydrogenases) were selected. The species of invertebrates and plants were selected based on the availability of standard protocols. Since we aimed to obtain data for the derivation of SSVs, for regulatory purposes, this procedure is recommended.

Statistical Analysis

A one-way analysis of variance (one-way ANOVA) was performed to test significant differences between the uranium concentrations tested for each endpoint analyzed: the activity of enzymes, the number of juveniles produced by potworms and collembolans, the number of emerged seeds, and the fresh and dry mass of the plants. The Kolmogorov-Smirnov test was applied to check data normality, whereas homoscedasticity of variances was checked by the Levene's test. When these two assumptions of the one-way ANOVAs were not met, a Kruskal-Wallis analysis was performed. The statistical analysis was run in the SigmaPlot 11.0 software for Windows. When statistical significant differences were recorded, the Dunnett's (for parametric one-way ANOVA) or the Dunn's test (for non-parametric ANOVA) was carried out to perceive which concentrations were significantly different from the respective control. Based on the outcomes of the multiple comparison tests the NOEC (no-observed-effect-concentration) and LOEC (low-observed-effect-concentration) values were deter-

Table 5. Soil quality guideline values derived for copper in Portugal, USA and Canada (mg U Kg^{-1} soildw).

Portugal				Canada	Other reference
Backgound concentrations	PNEC		Proposed SSV[b]	SQG$_E$[c]	
	NOEC	EC$_{20}$			
7.8[a]	23.3	15.5	15.5	23	100[d]

[a]Caetano et al.[49];
[b]SSV - soil screening value;
[c]Canadian Soil Quality Guidelines for environmental health (SQGe), Scott-Fordsmand and Pedersen [116].;
[d]Sheppard and Sheppard [98].

mined. The EC_{20} and EC_{50} values for each endpoint were calculated whenever possible, after fitting the data to a log-logistic model using the STATISTICA 7.0 software.

PNEC-Based SSV Derivation

Following the approach suggested by the Technical Guidance Document published by the European Commission [65], a predicted no effect concentration (PNEC) for U in the PTRS1 soil was determined, based on the assessment factor method For that, it was by used the lowest point estimate (i.e., NOEC and EC_{20} values) and applied the appropriate assessment factor based on the criteria of the Guidance Document [65]. The lowest point estimate calculated was for arylsulphatase activity. Considering that more than three NOEC values were obtained in this study, for at least three species, an assessment factor of 10 was applied, The PNEC value was calculated through the application of the following equation:

$$PNEC = \frac{\text{lowest point estimate}}{10}$$

Results and Discussion

1. Soil microbial activity

As far as authors are aware, this study is one of the few studies gathering data regarding the ecotoxicity of spiked soils with U on soil microbial parameters. Only a study from Sheppard et al. [66] has analyzed the effect of U on soil phosphatase activity in eleven different Canadian soils (including an agricultural, a boreal forest and a garden soil). This study recorded a significantly depressed activity only at the highest concentration tested (1000 mg U kg soil$_{dw}^{-1}$) for all the soils. These results suggested that probably, soil phosphatase activity was one of the less sensitive soil microbial parameters to U. In fact, Pereira et al. [7] also reported the low sensitivity of this parameter in mine soils contaminated with metals.

The variation in soil enzyme activities, N mineralization and potential nitrification in the PTRS1 soil, spiked with different U concentrations, is shown in Figure 1, and the Table 3 summarizes the toxicity values obtained for each biochemical parameter.

U had a clear inhibitory effect in almost all functional parameters tested. Overall, dehydrogenase and urease were the most affected soil enzymes by U, being their activity significantly inhibited at concentrations equal or lower than 134.5 mg U kg soil$_{dw}^{-1}$ (Table 3). Dehydrogenases have a relevant role in the oxidation of soil organic matter (SOM), being a good indicator of the active microbial biomass in the soil compartment [67]. As such, U (in the form of uranyl) strongly affected the normal microbial activity in PTRS1 soil. Meyer et al. [68] also observed a significant reduction in respiration rates of a soil exposed to depleted uranium (DU), but only for concentration equal and higher than 500 mg U kg soil$_{dw}^{-1}$. Indeed, the inhibition of urease activities indicates that U had a deleterious effect on soil N-cycle (Figure 1, Table 3). The reduction in the activity of this enzyme may have been caused by a negative effect of U on the overall microbial biomass, which in turn was translated into a reduction in the oxidation rate of organic N into ammonium [58,69]. Arylsulphatase is regularly involved in the S-cycle by catalyzing hydrolysis reactions in the biogeochemical transformation of S [67]. This parameter was significantly affected by U at a LOEC of 279.0 mg U kg soil$_{dw}^{-1}$. On its turn, the cellulase activity was

significantly inhibited at intermediate U concentrations. However in the highest concentrations the tendency was reversed and the activity increased, but not for levels significantly different from the control (Figure 1). Thereby, we can conclude that the C-metabolism associated with the degradation of SOM and catalyzed by these extracellular enzymes [70] was constrained by U. N mineralization and potential nitrification are indicators of the functioning of the N-cycle, hence providing an overview of the activity of specific microbial groups (nitrifying bacteria) directly involved in both processes [71]. The general pattern of response observed for these two parameters corresponded to stimulation at the lower U concentrations and inhibition under the highest ones (Figure 1), leading to EC_{50} values of 347.0 and 610.0 mg U kg soil$_{dw}^{-1}$ (Table 3), respectively. It has been stated that N mineralization is normally less sensitive than potential nitrification, since the former is carried out by a wider diversity of microorganisms [71]. However, our data showed the opposite (Figure 1). Meyer et al [68] did not observe effects on nitrogen mineralization of the test soil for U concentrations up to 25000 mg kg soil$_{dw}^{-1}$, however the form of U tested by these authors (schoepite $UO2(OH_2).H_2O$) was less soluble than the one tested in this soil.

The sensitivity of soil microbial parameters to metals has already been demonstrated by several authors, either in metal-polluted or in artificially spiked soils (e.g.,[4,72–77]). Dehydrogenase and urease had generally been referred as the most affected enzymes for different metals (e.g., Cu, Pb, Zn, Cd, Fe, Cr, Ni), (e.g.,[72,73,78,79]). Arylsulphatase and cellulase, however, have shown contradictory responses in different studies. Some authors observed negative correlations between arylsulphatase and cellulase activities and Zn [75] and Cu concentrations, respectively [80,81]; while others observed positive correlations between arylsulphatase and Cd [81], and no changes on cellulase activities in the presence of metals in urban soils was observed [82]. Usually, potential nitrification is negatively influenced by the presence of metals and metalloids such as Pb, Cu and As [7,81]; and the inhibitory effect of some metals (like Zn, Cd and Pb) on N mineralization was also reported by Dai et al. [83]. Antunes et al [81] found negative correlations (based on the Spearman coefficient) between U levels in soils from an abandoned U mine (presenting a mixture of metals) and the activities of urease and cellulase enzymes. For dehydrogenase, potential nitrification and arylsulphatase no significant correlations were detected. Nevertheless, this study analyzed mine contaminated soils, where the mixture of metals, may cause either synergistic or antagonistic effects, and where a well adapted and functional microbial community was likely established.

The inhibition of soil enzyme activities recorded could have been caused by toxicological effects of metals on soil microorganisms with subsequent decrease in their abundance and/or biomass; and/or by the direct inactivation of extracellular enzymes by metals [84]. Notwithstanding, the levels of metals may be not the sole effect on soil microbial activity. Soil properties (e.g., pH, organic matter content, nutrients and soil texture) may also interfere and modulate the bioavailability and toxicityof metals on soil enzymes [74,85]. Clays can retain and protect extracellular hydrolases, namely urease [73]. But the low clay content of PTRS1 soil (3.32%) (Table 1) might have increased U bioavailability, leading to the impairment of soil microbial community through cytotoxic effects, hence reducing their metabolic activity [81]. Additionally, the low pH of PTRS1 soil (Table 1) might have contributed for U availability and impacts on enzyme processes, potential nitrification and N mineralization, particularly at higher

U concentrations, as previously observed by Coppolecchia et al. [75] for arylsulphatase in the presence of Zn and low pH.

The above results illustrated well the effects of U in the performance of soil enzymes, reinforcing the importance of these parameters as bioindicators of soil quality. Indeed, the EC_{20} values calculated for dehydrogenase (34.9 mg Ukg soil$_{dw}^{-1}$), urease ($<$ 135.5 mg Ukg soil$_{dw}^{-1}$), N mineralization (152.2 mg Ukg soil$_{dw}^{-1}$) and arylsulphatase (155.3 mg Ukg soil$_{dw}^{-1}$) are within the environmental concentrations quantified in soils from an abandoned U mine, following extractions with *aqua regia* or with rainwater [8]. In this sense, the data herein generated represent a great asset for the derivation of SSVs, since they have a great ecological representativeness.

2. Uranium toxicity to the reproduction of soil invertebrates

The reproduction tests with the three invertebrate species revealed that *E. andrei*, *E. crypticus and F. candida* were quite sensitive to U in the PTRS1 soil. Tests fulfilled the validity criteria established by the standard guidelines for control replicates [61–63]. The resulting NOEC, LOEC, EC_{20} and EC_{50} values obtained in this study and toxicity data available in the literature are summarized in the Table 3.

The effects of U in the reproduction of *E. andrei* were evident, since statistical significant differences were found between the control and the highest tested concentrations of U for this organism ($F = 5.218$, d.f. $= 23$, $p = 0.002$) (Figure 2). The tested metal did not significantly affect the reproduction of *E. andrei* at concentrations up to 500.0 mg U kg soil$_{dw}^{-1}$ (NOEC) but compromised this endpoint for concentrations above 550.0 mg U kg soil$_{dw}^{-1}$ (LOEC). EC_{20} and EC_{50} values of U for *E. andrei* reproduction were 474.83 mg U kg soil$_{dw}^{-1}$ and 631.00 mg U kg soil$_{dw}^{-1}$, respectively (Table 3). The results obtained in our study, did not support those of Sheppard and Stephenson [86] (Table 4) that did not record toxic effects for *E. andrei* below 1000 mg U kg soil$_{dw}^{-1}$ (soils (carbonated): pH 7.5, 18% organic matter, 18% clay). However, they found an inhibition of juveniles production in two soils spiked with U, presenting low organic matter (2.2% and 1%) and a pH of 7.5 and 6.2, respectively (Table 4). According to the literature, the adsorption of metals to soil components is dependent on its physical and chemical properties, therefore influencing their toxicity to soil organisms [41,47,48]. Chelinho et al. [87], observed that soils with an organic matter content below 4% reduced or completely inhibited earthworms reproduction. However, the PTRS1 natural soil, had a high organic matter content, 6.2% (according to the classification provided by Murphy et al. [88]). Besides, as previously checked, the intrinsic properties of this soil did not compromise the performance of earthworms [49]. A high organic matter content of soils is usually related with a decrease in the toxicity of the contaminants for the organisms [41,43,89]. However, this was not the case in the study. In fact, Lourenço [60,90] exposed *E. andrei* to a U mine soil with 215.72±8.50 mg U kg soil$_{dw}^{-1}$, a pH of 7.79±0.01, and 7.71±0.60% of organic matter and observed that the bioaccumulation of U and daughter radionuclides was in tandem with loss of DNA integrity of coelomocyte cells, changes in the frequency of cells of immune system and also with histopathological changes (especially of the epidermis and chloragogenous tissue and intestinal epithelium). In fact, some other authors [91] had also suggested that the direct dermal exposure of earthworms to metals in the soil pore water, the ingestion of water, and/or soil particles may strongly favor the bioaccumulation of metals. Since pH is variable in the different compartments of gastrointestinal tract of earthworms, it can increase the mobilization of contaminants from soil after its ingestion [92,93].

Although, other metals were present in the contaminated soil tested by Lourenço et al. [60,90] U likely had a crucial role in the toxic effects observed, because it persisted in the whole body till 56 days. These authors suggested that the changes observed in DNA integrity were likely early warning indicators of effects on the growth and reproduction of the organisms. And in fact, effects on reproduction were observed in our study. Further, Giovanetti et al. [94]. exposed *E. fetida* natural U- and DU-contaminated soil (no information on soil type) for 7 and 28 days. Regarding natural U, no mortality or significant changes in weight were observed for both exposure periods up to 600 mg U kg$^{-1}_{dw}$. The chloragogeneous tissue, the main storage tissue of U, presented meaningful changes after 7 days of exposure for 300 mg U Kg^{-1}, while DNA strand breaks increased in a dose dependent manner above 150 mg U kg soil$_{dw}^{-1}$.

Regarding *E. crypticus* reproduction, it was significantly not reduced above 526.0 mg U kg soil$_{dw}^{-1}$ (LOEC) (Table 3) ($F = 31.05$, d.f. $= 12$, $p<0.05$). The EC_{20} and EC_{50} values estimated were respectively 469.7 and 518.6 mg U kg soil$_{dw}^{-1}$. Although no toxicity values are reported for the lowest concentrations tested, enchytraeids showed considerable sensitivity to U, since the number of juveniles was minimal or no juveniles were produced by *E. crypticus* at concentrations above 657.5 mg U kg soil$_{dw}^{-1}$ (Figure 2). Despite enchytraeids are commonly used in standardized toxicity tests, to the best of our knowledge, no data are available in the literature regarding the effects of U on the reproduction of this species. The available information concerns only the toxic effects caused by other metals or by natural soil properties in the reproduction of this species [43,48,95–97]. Thus, taking into account this literature review, pH and CEC were the most important parameters controlling the high sensitivity of enchytraeids to metals. Additionally, and according to Kuperman et al. [43], adults survival and juveniles production by *E. crypticus* can be maximized in natural soils with properties within the following ranges: 4.4–8.2 pH; 1.2–42% OM; 1–29% clay. The PTRS1 natural soil used as test substrate fell into in these ranges (Table 3), and similarly to *E. andrei*, the reproduction of this species was not compromised during the validation of the PTRS1 natural soil as a reference soil [49], meaning that the soil properties did not limit the performance of *E. crypticus*.

Concerning *F. candida*, U affected the production of juveniles, as shown by a significant decrease of this endpoint along the concentrations tested ($F = 11.6$, d.f. $= 12$, $p<0.05$) (Figure 2). The number of juveniles was not significantly affected up to 675.50 mg U kg soil$_{dw}^{-1}$ (NOEC), but it was significantly decreased for U concentrations equal to or greater than 756.10 mg U Kg^{-1} (LOEC) (Table 3). The EC_{20} value estimated for reproduction was 343.41 mg U kg soil$_{dw}^{-1}$ which is considerably lower than the toxicity data reported by Shepard et al. [98], $EC_{20}>710$ mg U kg soil$_{dw}^{-1}$, in two loam soils with pH 7.5 (Table 4). The low sensitivity of *F. candida* to U was also observed by Sheppard and Stephenson [86] which tested 3 soils amended with a range of U concentrations and aged for 10 years before testing. In this study, the lowest EC_{20} value obtained was 840 mg U kg soil$_{dw}^{-1}$ in a loam soil (pH 7.5, 24% clay, 2.2% OM) (Table 4). Despite this, *F. candida* was more sensitive in the study of Sheppard and Stephenson (since their EC_{20} value was similar to the EC_{50} recorded in our study 851.64 mg U kg soil$_{dw}^{-1}$). When considering the number of juveniles produced, U was less toxic to *F. candida* comparatively to *E. andrei* and *E. crypticus*. The lower sensitivity of *F. candida* is also consistent with other studies, when the effects of other metals in the reproduction of the three species was investigated [97,99], or even when other species of collembolans were analyzed [86]. The exposure of *F. candida* to chemicals in soil is apparently lower than for earthworms, which are exposed both by ingestion of contaminated soil (mineral particles, organic matter and chemicals in the soil solution) and also through direct dermal contact [100]. Despite the widely known influence of soil parameters on the bioavailability of chemicals and their influence on the reproduction of soil organisms, less is known about the intrinsic effects of physico-chemical parameters of the soils in the reproduction of *F. candida*. In general, several authors had reported a high tolerance of *F. candida* reproduction to a wide range of soil textural classes, organic matter contents and soil pH [48,101,102]. Once again the performance of this species was not compromised by the intrinsic properties of the PTRS1 soil. Hence, the effects observed can undoubtedly be attributed to U exposure.

3. Phytotoxicity of uranium

Relatively to terrestrial plants tests, all the validity criteria as described by the standard guidelines were attained [64]. Data obtained showed no significant effects on seeds emergence for all species tested (p>0.05). In fact, it was observed a relatively high rate of germination, either in monocotyledonous and dicotyledonous species (Figure 3). This outcome was somewhat expected, based on previous studies (e.g.,[22]). Seed coats form a barrier which protects embryos from a wide range of contaminants, especially metals. Thus, the germination relies almost exclusively on the seed reserves making it a less sensitive endpoint to the toxicity of soil pollutants [103].

An apparent hormetic effect was recorded for the other endpoints measured for almost all plant species. Such occurrence was recorded by other authors and it was attributed to the use of U as uranyl nitrate, which corresponds to a supplementary dose of N given to plants [98].

With regard to production of fresh- and dry-mass, it was possible to perceive that the tested plants displayed different sensitivities to this metal. However, no significant differences were generally observed comparatively to the control, exception for *L. sativa* dry mass (H = 22.8, d.f. = 12, p = 0.029). Thus, and according to Figure 4, *L. sativa* was the most sensitive terrestrial plant to U. The high sensitivity of *L. sativa* was also found by Hubálek et al. [104] and Soudek [105]. This was probably caused by the high capacity of this species to bioaccumulate high concentrations of metals, including U [22].

The exposure of plants to metals, was already extensively studied, showing that these contaminants can induce biological effects on germination, growth and development, as well as, alterations in the nutrient profile of plants [22,106]. However, only some studies (e.g., [66,107] and others reviewed [98]) have assessed the ecotoxicological effects of U on terrestrial plant species.

Based on our study, once again was proved the diverse ecotoxicological outcomes for U effects on plant species, since no effects were observed, in the range of tested concentrations for the three evaluated endpoints (in three out of four species), in PTRS1 soil. Similar results were obtained by Sheppard and Sheppard [86] in acidic soils (Table 4), when testing the emergence and growth of wheatgrass *Elymus lanceolatus*. Like in our study, these authors did not observe any effect on this species (up to 1000 mg U kg soil$_{dw}$$^{-1}$). In opposition, Sheppard and Sheppard [81] revised data on U toxicity to terrestrial plants and reported EC$_{25}$ values ranging from 300 to 500 mg U kg soil$_{dw}$$^{-1}$, considering only the most reliable studies. Stojanović et al. [108] also reported phytotoxic effects of U on *Zea mays* exposed, in

different soil types, to 250, 500 and 1000 mg U kg soil$_{dw}$$^{-1}$, but especially at the highest concentration tested and in the most acidic soil. However, no statistical analysis of the data was performed in this study.

Soil properties are also the factors that most strongly affect U uptake and phytotoxic effects, [18,109–111]. The bivalent uranyl ion (UO$_2$$^{2+}$) is sorbed to the negatively charged surfaces of clay minerals and organic compounds. In acidic soils subjected to pH increase, more negatively charged binding sites are available on mineral surfaces due to the progressive reduction of protons occupying these sites. However, pH values close to 6, like the one of PTRS1, favors U availability, since the concentrations of carbonates tends to increase, and U is released to the soil solution in the form of U-carbonate complexes [18]. The natural soil PTRS1, besides being acidic, has a lower clay content, which means lower binding sites for the bivalent uranyl ion (UO$_2$$^{2+}$), hence constraining U bioavailability. other soil properties and plant mechanisms may explain the reduced sensitivity of the plants in comparison with soil microbial parameters and invertebrates. Viehweger and Geipel [112] reported an increased U absorption by *Arabidopsis halleri* attributed to Fe deficiency in the medium of hydroponically grown plants. With respect to this metal, in the natural PTRS1 soil, the analyses done by Caetano et al. [49] showed that Fe surpassed the soil benchmark values proposed by two EPA regions (http://rais.ornl.gov/tools/eco_search.php http://rais.ornl.gov/tools/eco_search.php). In this sense, it is hypothesized that the high Fe content of the PTRS1 soil, may have also contributed for reducing the absorption of U by plants. As far as plant mechanisms are considered, in several studies reviewed by Mitchell et al. [113] the transport of U within plants was reduced and higher concentrations were consistently found in the roots. Using X-ray absorption spectroscopy (XAS) and transmission electron microscopy (TEM), Laurette et al. [114] observed that when plants are exposed to U and phosphates, needle-like U-phosphates are formed and precipitate, both outside and inside the cells, or persist in the subsurface of root tissues. The precipitation of U-phosphate complexes acts as a protective mechanism preventing U translocation to the shoots and leaves. This can also occur when the culture medium of the plants has no phosphate, since some plants are able to exudate phosphates. Further, U may be also absorbed like UO$_2$$^{2+}$ and linked to endogenous organophospate groups [114]. In opposition, when translocation occurs within plants, U has mainly formed U-carboxylated complexes. Plants can also exudate organic acids to the rhizosphere environment or UO$_2$$^{2+}$ may form complexes with endogenous compounds like malic, citric, oxalic and acetic acid [114]. In summary, the different resistance mechanisms described above could explain the lack of toxic effects observed for *A. sativa*, *Z. mays* and *L. esculentum*, in opposition to *L. sativa*. Most concerning is the fact that the majority of studies testing the phytotoxicity of U, including those performed by us, were made with the addition of nutrients solution, which increased the availability of phosphates to the test soil, likely decreasing the sensitivity of plants to U. Hence, to enhance the protection level of SSVs derived for plants, more assays with different plant species should be performed and the addition of nutrients should be prevented, or at least the tests may include replicates with and without nutrients.

Derivation of a Soil Screening Value (SSV) for Uranium Applying Assessment Factors

The PNEC values obtained for U were based in EC$_{20}$ and NOEC values varied between 15.5 and 23.3 mg kg soil$_{dw}$$^{-1}$, respectively (Table 5). These values were six to four times lower than the PNEC value suggested by Sheppard and Sheppard [86] (Table 5), which was 100 mg Ukg $soil_{dw}^{-1}$. In opposition, they are close to the lowest Canadian Soil Quality Guideline for both environment and human health (23 mg U kg $soil_{dw}^{-1}$). Thereby, while more ecotoxicological data is being obtained or other methods are being applied to derive soil screening values (SSVs) we prefer to be precautious by proposing a PNEC of 15.5 mg Kg^{-1} $soil_{dw}$ as a SSV for U, in soils similar to the PTRS1. This SSV value is near the background value found in non-contaminated soils [8,48], but not in some areas with naturally occurring U anomalies in soils, where concentrations ranging between 13–724 mg U kg $soil_{dw}^{-1}$ can be found [115].

Conclusion

With the present study it was possible to generate a set of important ecotoxicological data for the derivation of a SSV for U, using a Portuguese natural soil representative of a granitic region, where this type of mine exploration occurred.

Soil enzyme activities were clearly inhibited by U. The obtained results depended not only on the concentrations of U but also on the properties of soil, which were likely responsible for the bioavailability of U and subsequent impairments on soil microbial population and, consequently, in their activity. Dehydrogenase and urease were particularly sensitive to U. Further, and comparatively to the remaining effect concentrations obtained/estimated for invertebrates and plants, the soil microbial parameters were more affected by U contamination[1].

The toxic effects of U in soil invertebrates were also confirmed, but the tested species showed a variable sensitivity to this metal.

References

The increasing order of species sensitivity to U based on EC_{50} values for reproduction was *E. crypticus* > *E. andrei* > *F. candida*. However, if EC_{20} values are considered *F. candida* is the most sensitive invertebrate, since its EC_{20} value was 343.41 mg U kg $soil_{dw}^{-1}$, compared to 474.83 mg U kg $soil_{dw}^{-1}$ and 469.76 mg U kg $soil_{dw}^{-1}$ EC_{20} values estimated for *E. andrei* and *E. crypticus*, respectively. The EC_{20} values estimated were lower than the NOEC values for *E. andrei* and *F. candida*. Thus, the EC_{20} point estimate should be selected for the derivation of more protective SSVs. Relatively to the plants, the tested species showed no adverse effects caused by U in PTRS1, with the exception of *L. sativa* dry mass yield. Considering the results obtained, it was possible to verify a great variability between the EC_x values estimated in this study and those reported in the scientific literature. Multiple factors can contribute to this discordance, but probably at least for some species, soils physical and chemical properties were the main factors responsible for such differences. Although, this reinforces, at least in part, the importance of using natural soils representatives of the main types of soil from each region in ecotoxicological evaluations and for the derivation of SSVs, the data generated suggests that the SSV (15.5 mg Kg^{-1} $soil_{dw}$) derived for U, was six times lower than the PNEC value proposed by other authors. Nevertheless, as mentioned previously, more data should be obtained following standard protocols.

Author Contributions

Conceived and designed the experiments: ALC CRM RP. Performed the experiments: ALC AG. Analyzed the data: ALC CRM RP. Contributed reagents/materials/analysis tools: FC FG EFS. Contributed to the writing of the manuscript: ALC CRM RP.

1. Gavrilescu M, Pavel L, Cretescu I (2009) Characterization and remediation of soils contaminated with uranium. J Hazard Mater 163: 475–510. doi:10.1016/j.jhazmat.2008.07.103.

2. Malyshkina N, Niemeier D (2010) Future sustainability forecasting by exchange markets: basic theory and an application. Environ Sci Technol 44: 9134–9142. Available: http://www.ncbi.nlm.nih.gov/pubmed/21058697.

3. Carvalho F, Oliveira J, Faria I (2009) Alpha emitting radionuclides in drainage from Quinta do Bispo and Cunha Baixa uranium mines (Portugal) and associated radiotoxicological risk. Bull Environ Contam Toxicol 83: 668–673. Available: http://www.ncbi.nlm.nih.gov/pubmed/19590808. Accessed 19 November 2012.

4. Pereira R, Barbosa S, Carvalho FP (2014) Uranium mining in Portugal: a review of the environmental legacies of the largest mines and environmental and human health impacts. Environ Geochem Health 36: 285–301. Available: http://www.ncbi.nlm.nih.gov/pubmed/24030454. Accessed 7 March 2014.

5. Carvalho F, Madruga M, Reis M, Alves J, Oliveira J, et al. (2007) Radioactivity in the environment around past radium and uranium mining sites of Portugal. J Environ Radioact 96: 39–46. Available: http://www.ncbi.nlm.nih.gov/pubmed/17433852. Accessed 25 October 2012.

6. Vandenhove H, Sweeck L, Mallants D, Vanmarcke H, Aitkulov A, et al. (2006) Assessment of radiation exposure in the uranium mining and milling area of Mailuu Suu, Kyrgyzstan. J Environ Radioact 88: 118–139. Available: http://www.ncbi.nlm.nih.gov/pubmed/16581165.

7. Pereira R, Sousa J, Ribeiro R, Gonçalves F (2006) Microbial Indicators in Mine Soils (S. Domingos Mine, Portugal). Soil Sediment Contam 15: 147–167. Available: http://www.informaworld.com/openurl?genre=article&doi=10%2e1080%2f1532038050050681 3&magic =crossref%7c%7cD404A21C5BB053405B1A640AFFD44AE3. Accessed 5 October 2013.

8. Pereira R, Antunes S, Marques S, Gonçalves F (2008) Contribution for tier 1 of the ecological risk assessment of Cunha Baixa uranium mine (Central Portugal): I soil chemical characterization. Sci Total Environ 390: 377–386. Available: http://www.ncbi.nlm.nih.gov/pubmed/17919686. Accessed 6 September 2011.

9. Arogunjo A, Höllriegl V, Giussani A, Leopold K, Gerstmann U, et al. (2009) Uranium and thorium in soils, mineral sands, water and food samples in a tin mining area in Nigeria with elevated activity. J Environ Radioact 100: 232–240. Available: http://www.ncbi.nlm.nih.gov/pubmed/19147259. Accessed 14 October 2013.

10. Momčilović M, Kovačević J, Dragović S (2010) Population doses from terrestrial exposure in the vicinity of abandoned uranium mines in Serbia.

    Radiat Meas 45: 225–230. Available: http://linkinghub.elsevier.com/retrieve/pii/S1350448710000363. Accessed 25 October 2012.

11. Figueiredo M, Silva T, Batista M, Leote J, Ferreira M, et al. (2011) Uranium in surface soils: An easy-and-quick assay combining X-ray diffraction and X-ray fluorescence qualitative data. J Geochemical Explor 109: 134–138. Available: http://linkinghub.elsevier.com/retrieve/pii/S0375674210001366. Accessed 23 October 2012.

12. Patra A, Sumesh C, Mohapatra S, Sahoo S, Tripathi R, et al. (2011) Long-term leaching of uranium from different waste matrices. J Environ Manage 92: 919–925. Available: http://www.ncbi.nlm.nih.gov/pubmed/21084148. Accessed 14 October 2013.

13. Scheele F (2011) Uranium from Africa: Mitigation of Uranium Mining Impacts on Society and Environment by Industry and Governments. SSRN Electron J. Available: http://papers.ssrn.com/abstract=1892775. Accessed 14 October 2013.

14. Niemeyer J, Moreira-Santos M, Nogueira M, Carvalho G, Ribeiro R, et al. (2010) Environmental risk assessment of a metal-contaminated area in the Tropics. Tier I: screening phase. J Soils Sediments 10: 1557–1571. Available: http://www.springerlink.com/index/10.1007/s11368-010-0255-x. Accessed 29 October 2012.

15. Carvalho F (2011) Environmental Radioactive Impact Associated to Uranium Production Nuclear and Technological Institute Department of Radiological Protection and Nuclear Safety,. Journal, Am Sci Environ Publ Sci 7: 547–553.

16. Wang J, Lu A, Ding A (2007) Effect of cadmium alone and in combination with butachlor on soil enzymes. Heal (San Fr: 395–403. doi:10.1007/s10653-007-9084-2.

17. ASTDR (2011) Agency for Toxic Substances & Disease Registry-Toxicological profile for uranium. Draft. U.S. Department of Health and Human Services. Public Health Service and Agency for Toxic Substances and Disease Registry. 452 + annexes.

18. Vandenhove H, Van Hees M, Wouters K, Wannijn J (2007) Can we predict uranium bioavailability based on soil parameters? Part 1: effect of soil parameters on soil solution uranium concentration. Environ Pollut 145: 587–595. Available: http://www.ncbi.nlm.nih.gov/pubmed/16781802. Accessed 19 November 2012.

19. Gongalsky K (2003) Impact of pollution caused by uranium production on soil macrofauna. Environ Monit Assess 89: 197–219. Available: http://www.ncbi.nlm.nih.gov/pubmed/14632090. Accessed 15 October 2013.

20. Geras'kin S, Evseeva T, Belykh E, Majstrenko T, Michalik B, et al. (2007) Effects on non-human species inhabiting areas with enhanced level of natural

radioactivity in the north of Russia: a review. J Environ Radioact 94: 151–182. Available: http://www.ncbi.nlm.nih.gov/pubmed/17360083. Accessed 26 October 2012.

21. Joner E, Munier-Lamy C, Gouget B (2007) Bioavailability and microbial adaptation to elevated levels of uranium in an acid, organic topsoil forming on an old mine spoil. Environ Toxicol Chem 26: 1644–1648. Available: http://www.ncbi.nlm.nih.gov/pubmed/17702337.

22. Pereira R, Marques CR, Ferreira MJS, Neves MFJV, Caetano AL, et al. (2009) Phytotoxicity and genotoxicity of soils from an abandoned uranium mine area. Appl Soil Ecol 42: 209–220. Available: http://linkinghub.elsevier.com/retrieve/pii/S0929139309000778. Accessed 6 September 2011.

23. Kenarova A, Radeva G, Danova I, Boteva S, Dimitrova I (2010) Soil bacterial abundance and diversity of uranium impacted. Second Balkan conference on Biology: 5–9.

24. Islam E, Sar P (2011) Molecular assessment on impact of uranium ore contamination in soil bacterial diversity. Int Biodeterior Biodegradation 65: 1043–1051. Available: http://linkinghub.elsevier.com/retrieve/pii/S0964830511001697. Accessed 29 October 2012.

25. Geng F, Hu N, Zheng J-F, Wang C-L, Chen X, et al. (2011) Evaluation of the toxic effect on zebrafish (*Danio rerio*) exposed to uranium mill tailings leaching solution. J Radioanal Nucl Chem 292: 453–463. Available: http://link.springer.com/10.1007/s10967-011-1451-x. Accessed 15 October 2013.

26. Lourenço J, Pereira R, Silva A, Carvalho F, Oliveira J, et al. (2012) Evaluation of the sensitivity of genotoxicity and cytotoxicity endpoints in earthworms exposed in situ to uranium mining wastes. Ecotoxicol Environ Saf 75: 46–54. Available: http://www.sciencedirect.com/science/article/pii/S014765131100 2685. Accessed 14 October 2013.

27. Islam E, Sar P (2011) Molecular assessment on impact of uranium ore contamination in soil bacterial diversity. Int Biodeterior Biodegradation 65: 1043–1051. Available: http://www.sciencedirect.com/science/article/pii/S0964830511001697. Accessed 25 October 2013.

28. Lavelle P, Decaens T, Aubert M, Barot S, Blouin M, et al. (2006) Soil invertebrates and ecosystem services. Eur J Soil Biol 42: S3–S15. Available: http://linkinghub.elsevier.com/retrieve/pii/S1164556306001038. Accessed 19 July 2011.

29. O'Halloran K (2006) Toxicological Considerations of Contaminants in the Terrestrial Environment for Ecological Risk Assessment. Hum Ecol Risk Assess An Int J 12: 74–83. Available: http://www.tandfonline.com/doi/abs/10.1080/10807030500428603. Accessed 7 September 2011.

30. Barrios E (2007) Soil biota, ecosystem services and land productivity. Ecol Econ 64: 269–285. Available: http://linkinghub.elsevier.com/retrieve/pii/S0921800907001693. Accessed 4 October 2012.

31. Dominati E, Patterson M, Mackay A (2010) A framework for classifying and quantifying the natural capital and ecosystem services of soils. Ecol Econ 69: 1858–1868. Available: http://www.sciencedirect.com/science/article/pii/S0921800910001928. Accessed 14 October 2013.

32. CEC (2006) Comission of the European Cominities, Directive 2004/35/EC. Commissiona of the European Communities, Brussels.

33. Bone J, Head M, Barraclough D, Archer M, Scheib C, et al. (2010) Soil quality assessment under emerging regulatory requirements. Environ Int 36: 609–622. Available: http://www.ncbi.nlm.nih.gov/pubmed/20483160. Accessed 6 October 2012.

34. Jensen J, Mesman M (2006) Ecological Risk Assessment of Contaminated Land. Decision support for site specific investigations. ISBN 90-6960-138-9 978-90-6960-138-0.

35. Fishwick S (2004) Soil screening values for use in UK ecological risk assessment. Soil Quality & Protection, Air Land and Water Group. Science Environment Agency's Project Manager.

36. USEPA (2003) United States Environmental Protection Agency. Guidance for Developing Ecological Soil Screening Levels (Eco-SSLs). Attachment 1–2: Assessment of Whether to Develop Ecological Soil Screening Levels for Microbes and Microbial Processes, OSWER Directi. Environ Prot.

37. Carlon C (2007) Derivation methods of soil screening values in europe. a review and evaluation of national procedures towards harmonisation. European Commission, Joint Research Centre, Ispra, EUR 22805-EN, 306.

38. Provoost J, Reijnders L, Swartjes F, Bronders J, Carlon C, et al. (2008) Parameters causing variation between soil screening values and the effect of harmonization. J Soils Sediments 8: 298–311. Available: http://www.springerlink.com/index/10.1007/s11368-008-0026-0. Accessed 29 October 2012.

39. Van Gestel C (2012) Soil ecotoxicology: state of the art and future directions. Zookeys 296: 275–296. Available: http://www.pubmedcentral.nih.gov/articlerender.fcgi?artid=3335420&tool=pmcentrez&rendertype=abstract. Accessed 29 October 2012.

40. MAOTDR (2006) Ministerio do Ambiente do Ordenamento do Territorio e do Desenvolvimento Regional. Diário da República, 1. a série – N. o 208–27 de Outubro.

41. Römbke J, Jänsch S, Junker T, Pohl B, Scheffczyk A, et al. (2006) Improvement of the applicability of ecotoxicological tests with earthworms, springtails, and plants for the assessment of metals in natural soils. Environ Toxicol Chem 25: 776–787. Available: http://www.ncbi.nlm.nih.gov/pubmed/16566163.

42. Van Assche F, Alonso JL, Kapustka L, Petrie R, Stephenson GL, et al. (2002) Terrestrial plant toxicity tests. In: Fairbrother A., Glazebrock P.W., Van Straalen N.M., Tarazona J.V. (eds) Test methods to determine hazards of

43. Kuperman R, Amorim M, Römbke J, Lanno R, Checkai R, et al. (2006) Adaptation of the enchytraeid toxicity test for use with natural soil types. Eur J Soil Biol 42: S234–S243. Available: http://www.sciencedirect.com/science/article/pii/S1164556306000719. Accessed 14 October 2013.

44. Hofman J, Hovorková I, Machát J (2009) Comparison and Characterization of OECD Artificial Soils. Ecotoxicological Characterization of Waste 2009, 223–229. Available: http://www.rivm.nl/bibliotheek/rapporten/711701047.html.

45. Song J, Zhao F, McGrath S, Luo Y (2006) Influence of soil properties and aging on arsenic phytotoxicity. Environ Toxicol Chem 25: 1663–1670. Available: http://www.ncbi.nlm.nih.gov/pubmed/16764487. Accessed 14 October 2013.

46. Rooney C, Zhao F, McGrath S (2007) Phytotoxicity of nickel in a range of European soils: influence of soil properties, Ni solubility and speciation. Environ Pollut 145: 596–605. Available: http://www.ncbi.nlm.nih.gov/pubmed/16733077. Accessed 14 October 2013.

47. Van Gestel C, Borgman E, Verweij R, Ortiz M (2011) The influence of soil properties on the toxicity of molybdenum to three species of soil invertebrates. Ecotoxicol Environ Saf 74: 1–9. Available: http://www.ncbi.nlm.nih.gov/pubmed/20951431. Accessed 14 October 2013.

48. Domene X, Chelinho S, Campana P, Natal-da-Luz T, Alcañiz JM, et al. (2011) Influence of soil properties on the performance of Folsomia candida: implications for its use in soil ecotoxicology testing. Environ Toxicol Chem 30: 1497–1505. Available: http://www.ncbi.nlm.nih.gov/pubmed/21437938. Accessed 13 March 2013.

49. Caetano A, Gonçalves F, Sousa J, Cachada A, Pereira E, et al. (2012) Characterization and validation of a Portuguese natural reference soil to be used as substrate for ecotoxicological purposes. J Environ Monit 14: 925–936. Available: http://www.ncbi.nlm.nih.gov/pubmed/22297688. Accessed 31 May 2013.

50. ISO 17512-1 (2008) International Organization for Standardization 17512-1: 2008.Soil Quality: Avoidance Test for Testing the Quality of Soils and the Toxicity of Chemicals-Test with Earthworms (Eisenia Fetida). Geneva, Switzerland. International Organization for Standard. Geneva, Switzerland.

51. FAOUN (1984) Food and agriculture organization of the United Nations – physical and chemical methods of soil and water analysis. Soils Bull. 10, 1–275.

52. SPAC (2000) Soil and Plant Analysis Council – Handbook of Reference Methods. CRC Press, Boca Raton, Florida.

53. Tabatabai M, Bremner J (1970) Arylsulfatase activity in soils. Soil Science Society of America, 34: 225–9.

54. Schinner F, Kandeler E, Öhlinger R, Margesin R (1996) Methods in soil biology. Germany: Springer-Verlag.

55. Öhlinger R (1996) Soil sampling and sample preparation. In: Schinner F, Öhlinger R, Kandeler E, Margesin R, editors. Methods in Soil Biology. Springer-Verlag.

56. Schinner F, von Mersi W (1990) Xylanase, CM-cellulase and invertase activity in soil, an improved method. Soil Biol Biochem 22: 511–5.

57. Kandeler E, Gerber H (1988) Short-term assay of soil urease activity using colorimetric determination of ammonium. biol Fert Soils 6: 68–72.

58. Kandeler E (1996) Potential nitrification. In: Schinner F, Öhlinger R, Kandeler E, Margesin R, editors. Methods in soil biology. Germany: Springer-Verlag-Berlin-Heidelberg, 146–9.

59. Berg P, Rosswall T (1985) Ammonium oxidizer numbers, potential and actual oxidation rates en two Swedish arable soils. Biology Fertility Soils 1: 131–40.

60. Lourenço JI, Pereira RO, Silva AC, Morgado JM, Carvalho FP, et al. (2011) a) Genotoxic endpoints in the earthworms sub-lethal assay to evaluate natural soils contaminated by metals and radionuclides. J Hazard Mater 186: 788–795. Available: http://www.ncbi.nlm.nih.gov/pubmed/21146299. Accessed 21 September 2013.

61. ISO 11268-2 (1998) International Organization for Standardization ISO 11268-2. Soil quality: effects of pollutants on earthworms (Eisenia fetida) - Part 2: Determination of effects on reproduction. ISO 11268-2. Geneva, Switzerland: International Organization for Standardiza. Geneva, Switz.

62. ISO 16387 (2004) International Organization for Standardization ISO 16387. Soil quality: effects of pollutants on Enchytraeidae (Enchytraeus sp.)-Determination of effects on reproduction and survival. ISO16387. Geneva, Switzerland: International Organization for Standard. Geneva, Switz.

63. ISO 11267 (1999) International Organization for Standardization ISO 11267. Soil quality: inhibition of reproduction of Collembola (Folsomia candida) by soil pollutants. ISO 11267. Geneva, Switzerland: International Organization for Standardization. Geneva, Switz.

64. ISO 11269-2 (2005) International Organization for Standardization ISO 11269-2. Soil quality: determination of the effects of pollutants on soil flora - Part 2: Effects of chemicals on the emergence and growth of higher plants. ISO 11269-2. Geneve, Switzerlan. International O. Geneva, Switz.

65. EC. Commission European (2003) Commission European. Technical Guidance Document on Risk Assessment. in support of Commission Directive 93/67/EEC on Risk Assessment for new notified substances Commission Regulation (EC) No 1488/94 on Risk Assessment for existing substances Directive 98/.

66. Sheppard S, Evenden W (1992) Bioavailability Indices for Uranium: Effect of Concentration in Eleven Soils. Arch Environ Contain Toxicol 23, 117–124.

67. Taylor J, Wilson B, Mills M, Burns R (2002) Comparison of microbial numbers and enzymatic activities in surface soils and subsoils using various techniques. 34.

68. Meyer MC, Paschke MW, McLendon T, Price D (1998) Decreases in Soil Microbial Function and Functional Diversity in Response to Depleted Uranium. J Environ Qual 27: 1306. Available: https://www.agronomy.org/publications/jeq/abstracts/27/6/JEQ0270061306. Accessed 31 March 2014.

69. Wang M, Markert B, Shen W, Chen W, Peng C, et al. (2011) Microbial biomass carbon and enzyme activities of urban soils in Beijing. Environ Sci Pollut Res Int 18: 958–967. Available: http://www.ncbi.nlm.nih.gov/pubmed/21287285. Accessed 25 September 2013.

70. Alvarenga P, Palma P, Gonçalves AP, Baião N, Fernandes RM, et al. (2008) Assessment of chemical, biochemical and ecotoxicological aspects in a mine soil amended with sludge of either urban or industrial origin. Chemosphere 72: 1774–1781. Available: http://www.ncbi.nlm.nih.gov/pubmed/18547605. Accessed 13 May 2014.

71. Winding A, Hund-Rinke K, Rutgers M (2005) The use of microorganisms in ecological soil classification and assessment concepts. Ecotoxicol Environ Saf 62: 230–248. Available: http://www.ncbi.nlm.nih.gov/pubmed/15925407.

72. Khan S, Cao Q, Hesham AE-L, Xia Y, He J-Z (2007) Soil enzymatic activities and microbial community structure with different application rates of Cd and Pb. J Environ Sci (China) 19: 834–840. Available: http://www.ncbi.nlm.nih.gov/pubmed/17966871. Accessed 15 October 2013.

73. Lee S, Kim E, Hyun S, Kim J (2009) Metal availability in heavy metal-contaminated open burning and open detonation soil: assessment using soil enzymes, earthworms, and chemical extractions. J Hazard Mater 170: 382–388. Available: http://www.ncbi.nlm.nih.gov/pubmed/19540045. Accessed 5 April 2013.

74. Papa S, Bartoli G, Pellegrino A, Fioretto A (2010) Microbial activities and trace element contents in an urban soil. Environ Monit Assess 165: 193–203. Available: http://www.ncbi.nlm.nih.gov/pubmed/19444636. Accessed 13 May 2013.

75. Coppolecchia D, Puglisi E, Vasileiadis S, Suciu N, Hamon R, et al. (2011) Soil Biology & Biochemistry Relative sensitivity of different soil biological properties to zinc. Soil Biol Biochem 43: 1798–1807. Available: http://dx.doi.org/10.1016/j.soilbio.2010.06.018.

76. Lee S-H, Park H, Koo N, Hyun S, Hwang A (2011) Evaluation of the effectiveness of various amendments on trace metals stabilization by chemical and biological methods. J Hazard Mater 188: 44–51. Available: http://www.ncbi.nlm.nih.gov/pubmed/21333442. Accessed 17 August 2013.

77. Hu B, Liang D, Liu J, Xie J (2013) Ecotoxicological effects of copper and selenium combined pollution on soil enzyme activities in planted and unplanted soils. Environ Toxicol Chem 32: 1109–1116. Available: http://www.ncbi.nlm.nih.gov/pubmed/23401089. Accessed 8 October 2013.

78. Gülser F, Erdoğan E (2008) The effects of heavy metal pollution on enzyme activities and basal soil respiration of roadside soils. Environ Monit Assess 145: 127–133. Available: http://www.ncbi.nlm.nih.gov/pubmed/18027096. Accessed 17 August 2013.

79. Thavamani P, Malik S, Beer M, Megharaj M, Naidu R (2012) Microbial activity and diversity in long-term mixed contaminated soils with respect to polyaromatic hydrocarbons and heavy metals. J Environ Manage 99: 10–17. Available: http://dx.doi.org/10.1016/j.jenvman.2011.12.030.

80. Alvarenga P, Palma P, de Varennes A, Cunha-Queda AC (2012) A contribution towards the risk assessment of soils from the São Domingos Mine (Portugal): chemical, microbial and ecotoxicological indicators. Environ Pollut 161: 50–56. Available: http://www.ncbi.nlm.nih.gov/pubmed/22230067. Accessed 5 October 2013.

81. Antunes S, Pereira R, Marques S, Castro B, Gonçalves F (2011) Impaired microbial activity caused by metal pollution A field study in a deactivated. Sci Total Environmen 410–411.

82. Sivakumar S, Nityanandi D, Barathi S, Prabha D, Rajeshwari S, et al. (2012) Selected enzyme activities of urban heavy metal-polluted soils in the presence and absence of an oligochaete, Lampito mauritii (Kinberg). J Hazard Mater 227–228: 179–184. Available: http://www.ncbi.nlm.nih.gov/pubmed/22658212. Accessed 17 August 2013.

83. Dai J, Becquer T, Rouiller JH, Reversat G, Bernhard-Reversat F, et al. (2004) Influence of heavy metals on C and N mineralisation and microbial biomass in Zn-, Pb-, Cu-, and Cd-contaminated soils. Appl Soil Ecol 25: 99–109. Available: http://www.sciencedirect.com/science/article/pii/S0929139303001355. Accessed 15 October 2013.

84. Kızılkaya R, Bayraklı B (2005) Effects of N-enriched sewage sludge on soil enzyme activities. Appl Soil Ecol 30: 192–202. Available: http://www.sciencedirect.com/science/article/pii/S0929139305000594. Accessed 15 October 2013.

85. Turner BL, Hopkins DW, Haygarth PM, Ostle N (2002) β-Glucosidase activity in pasture soils. Available: http://www.sciencedirect.com/science/article/pii/S0929139302000203. Accessed 15 October 2013.

86. Sheppard SC, Stephenson GL (2012) Ecotoxicity of aged uranium in soil using plant, earthworm and microarthropod toxicity tests. Bull Env Contam Toxicol 8843–47: 43–47. doi:10.1007/s00128-011-0442-5.

87. Chelinho S, Domene X, Campana P, Natal-da-Luz T, Scheffczyk A, et al. (2011) Improving ecological risk assessment in the Mediterranean area: selection of reference soils and evaluating the influence of soil properties on avoidance and reproduction of two oligochaete species. Environ Toxicol Chem

88. Murphy S, Giménez D, Muldowney LS, Heckman JR (2012) Soil Organic Matter Level and Interpretation How is the Organic Matter Content of Soils Determined? How Is Organic Matter Level: 1–3.

89. Natal-da-Luz T, Ojeda G, Pratas J, Van Gestel CA, Sousa JP (2011) Toxicity to Eisenia andrei and Folsomia candida of a metal mixture applied to soil directly or via an organic matrix. Ecotoxicol Environ Saf 74: 1715–1720. Available: http://www.ncbi.nlm.nih.gov/pubmed/21683441. Accessed 13 March 2013.

90. Lourenço J, Silva A, Carvalho F, Oliveira J, Malta M, et al. (2011) b) Histopathological changes in the earthworm Eisenia andrei associated with the exposure to metals and radionuclides. Chemosphere 85: 1630–1634. Available: http://www.ncbi.nlm.nih.gov/pubmed/21911243. Accessed 4 October 2013.

91. Hobbelen PHF, Koolhaas JE, van Gestel CAM (2006) Bioaccumulation of heavy metals in the earthworms Lumbricus rubellus and Aporrectodea caliginosa in relation to total and available metal concentrations in field soils. Environ Pollut 144: 639–646. Available: http://www.ncbi.nlm.nih.gov/pubmed/16530310. Accessed 15 October 2013.

92. Li L, Wu J, Tian G, Xu Z (2009) Effect of the transit through the gut of earthworm (Eisenia fetida) on fractionation of Cu and Zn in pig manure. J Hazard Mater 167: 634–640. Available: http://www.ncbi.nlm.nih.gov/pubmed/19232822. Accessed 19 September 2013.

93. Peijnenburg W, Jager T (2003) Monitoring approaches to assess bioaccessibility and bioavailability of metals: matrix issues. Ecotoxicol Environ Saf 56: 63–77. Available: http://www.ncbi.nlm.nih.gov/pubmed/12915141. Accessed 15 October 2013.

94. Giovanetti A, Fesenko S., Cozzella ML, Asencio LD, Sansone U (2010) Bioaccumulation and biological effects in the earthworm Eisenia fetida exposed to natural and depleted uranium. J Environ Radioact 101: 509–516. Available: http://www.ncbi.nlm.nih.gov/pubmed/20362371. Accessed 15 October 2013.

95. Peijnenburg W, Baerselman R, de Groot C, Jager T, Posthuma L, et al. (1999) Relating environmental availability to bioavailability: soil-type-dependent metal accumulation in the oligochaete Eisenia andrei. Ecotoxicol Environ Saf 44: 294–310. Available: http://www.ncbi.nlm.nih.gov/pubmed/10581124.

96. Amorim M, Römbke J, Scheffczyk A, Soares AM (2005) Effect of different soil types on the enchytraeids Enchytraeus albidus and Enchytraeus luxuriosus using the herbicide Phenmedipham. Chemosphere 61: 1102–1114. Available: http://www.ncbi.nlm.nih.gov/pubmed/16263380.

97. Kuperman R (2004) Manganese toxicity in soil for Eisenia fetida, Enchytraeus crypticus (Oligochaeta), and Folsomia candida (Collembola). Ecotoxicol Environ Saf 57: 48–53. Available: http://linkinghub.elsevier.com/retrieve/pii/S0147651303001544. Accessed 22 June 2011.

98. Sheppard S, Sheppard M (2005) Derivation of ecotoxicity thresholds for uranium. J Environ 79: 55–83. Available: http://www.ncbi.nlm.nih.gov/pubmed/15571876. Accessed 19 November 2012.

99. Lock K, Janssen CR (2001) Cadmium Toxicity for Terrestrial Invertebrates: Taking Soil Parameters Affecting Bioavailability into Account. Environ Toxicol: 315–322.

100. Layinka T, Idowu B, Dedeke A, Akinloye A, Ademolu O, et al. (2011) Earthworm as bio-indicator of heavy metal pollution around Lafarge, Wapco Cement Factory, Ewekoro, Nigeria. Proceedings of the Environmental Management Conference, Federal University of Agriculture, Abeokuta, Nigeria: 489–496.

101. Amorim M, Rçmbke J, Scheffczyk A, Nogueira A, Soares A (2005) Effects of Different Soil Types on the Collembolans Folsomia candida and Hypogastrura assimilis using the Herbicide Phenmedipham. 352: 343–352. doi:10.1007/s00244-004-0220-z.

102. Jänsch S, Amorim MJ, Römbke J (2011) Identification of the ecological requirements of important terrestrial ecotoxicological test species. Environmental Reviews, 13(2): 51–83. Available: http://www.nrcresearchpress.com/doi/abs/10.1139/a05-007#.Ul0AxlDENPB. Accessed 15 October 2013.

103. Liu X, Zhang S, Shan X-Q, Christie P (2007) Combined toxicity of cadmium and arsenate to wheat seedlings and plant uptake and antioxidative enzyme responses to cadmium and arsenate co-contamination. Ecotoxicol Environ Saf 68: 305–313. Available: http://www.ncbi.nlm.nih.gov/pubmed/17239437. Accessed 15 October 2013.

104. Hubálek T, Vosáhlová S, Mateju V, Kováčová N, Novotný C (2007) Ecotoxicity monitoring of hydrocarbon-contaminated soil during bioremediation: a case study. Arch Environ Contam Toxicol 52: 1–7. Available: http://www.ncbi.nlm.nih.gov/pubmed/17106791. Accessed 13 August 2013.

105. Soudek P, Petrová S, Benešová D, Dvořáková M, Vaněk T (2011) Uranium uptake by hydroponically cultivated crop plants. J Environ Radioact 102: 598–604. Available: http://www.ncbi.nlm.nih.gov/pubmed/21486682. Accessed 13 September 2013.

106. Gopal R, Rizvi A (2008) Excess lead alters growth, metabolism and translocation of certain nutrients in radish. Chemosphere 70: 1539–1544. Available: http://www.ncbi.nlm.nih.gov/pubmed/17923149. Accessed 15 October 2013.

107. Sheppard S, Evenden W, Anderson A (1992) Multiple assays of uranium toxicity in soil. Environ. Toxicol. Water Qual., 7: 275–294.

108. Stojanović M, Stevanović D, Milojković J, Grubišić M, Ileš D (2009) Phytotoxic Effect of the Uranium on the Growing Up and Development the Plant of Corn.

30: 1050–1058. Available: http://www.ncbi.nlm.nih.gov/pubmed/21305581. Accessed 17 October 2013.

Water, Air, Soil Pollut 209: 401–410. Available: http://www.springerlink.com/index/10.1007/s11270-009-0208-4. Accessed 25 September 2011.

109. Bednar A, Medina V, Ulmer-Scholle D, Frey B, Johnson B, et al. (2007) Effects of organic matter on the distribution of uranium in soil and plant matrices. Chemosphere 70: 237–247. Available: http://www.sciencedirect.com/science/article/pii/S0045653507008168. Accessed 15 October 2013.

110. Tunney H, Stojanovic M, Mrdakovic Popic J, McGrath D, Zhang C (2009) Relationship of soil phosphorus with uranium in grassland mineral soils in Ireland using soils from a long-term phosphorus experiment and a National Soil Database. J Plant Nutr Soil Sci 172: 346–352. Available: http://doi.wiley.com/10.1002/jpln.200800069. Accessed 13 September 2013.

111. Soudek P, Petrová Š, Benešová D, Vaněk T (2011) Uranium uptake and stress responses of in vitro cultivated hairy root culture of Armoracia rusticana. Agrochimical 1: 15–28.

112. Viehweger K, Geipel G (2010) Uranium accumulation and tolerance in *Arabidopsis halleri* under native versus hydroponic conditions. Environ Exp Bot 69: 39–46. Available: http://dx.doi.org/10.1016/j.envexpbot.2010.03.001.

113. Mitchell N, Pérez-Sánchez D, Thorne MC (2013) A review of the behaviour of U-238 series radionuclides in soils and plants. J Radiol Prot 33: R17–48. Available: http://www.ncbi.nlm.nih.gov/pubmed/23612607. Accessed 18 September 2013.

114. Laurette J, Larue C, Llorens I, Jaillard D, Jouneau P-H, et al. (2012) Speciation of uranium in plants upon root accumulation and root-to-shoot translocation: A XAS and TEM study. Environ Exp Bot 77: 87–95. Available: http://www.sciencedirect.com/science/article/pii/S0098847211002814. Accessed 15 October 2013.

115. Pereira A, Neves L (2012) Estimation of the radiological background and dose assessment in areas with naturally occurring uranium geochemical anomalies–a case study in the Iberian Massif (Central Portugal). J Environ Radioact 112: 96–107. Available: http://www.ncbi.nlm.nih.gov/pubmed/22694913. Accessed 14 September 2013.

116. Scott-Fordsmand JJ, Pedersen MB (1995) Soil quality criteria for selected inorganic compounds. Danish Environmental Protection Agency, Working Report No. 48, 200.

2

Role of Key *TYMS* Polymorphisms on Methotrexate Therapeutic Outcome in Portuguese Rheumatoid Arthritis Patients

Aurea Lima[1,2,3]*, **Vítor Seabra**[1], **Miguel Bernardes**[4,5], **Rita Azevedo**[2,4], **Hugo Sousa**[2,4,6], **Rui Medeiros**[2,3,6,7]

1 CESPU, Institute of Research and Advanced Training in Health Sciences and Technologies, Department of Pharmaceutical Sciences, Higher Institute of Health Sciences-North (ISCS-N), Gandra PRD, Portugal, 2 Molecular Oncology Group CI, Portuguese Institute of Oncology of Porto (IPO-Porto), Porto, Portugal, 3 Abel Salazar Institute for the Biomedical Sciences (ICBAS) of University of Porto, Porto, Portugal, 4 Faculty of Medicine of University of Porto (FMUP), Porto, Portugal, 5 Rheumatology Department of São João Hospital Center, Porto, Portugal, 6 Virology Service, Portuguese Institute of Oncology of Porto (IPO-Porto), Porto, Portugal, 7 Research Department-Portuguese League Against Cancer (LPCC-NRNorte), Porto, Portugal

Abstract

Background: Therapeutic outcome of rheumatoid arthritis (RA) patients treated with methotrexate (MTX) can be modulated by thymidylate synthase (TS) levels, which may be altered by genetic polymorphisms in TS gene (*TYMS*). This study aims to elucidate the influence of *TYMS* polymorphisms in MTX therapeutic outcome (regarding both clinical response and toxicity) in Portuguese RA patients.

Methods: Clinicopathological data from 233 Caucasian RA patients treated with MTX were collected, outcomes were defined and patients were genotyped for the following *TYMS* polymorphisms: 1) 28 base pairs (bp) variable number tandem repeat (rs34743033); 2) single nucleotide polymorphism C>G (rs2853542); and 3) 6 bp sequence deletion (1494del6, rs34489327). Chi-square and binary logistic regression analyses were performed, using genotype and haplotype-based approaches.

Results: Considering *TYMS* genotypes, 3R3R ($p = 0.005$, OR = 2.34), 3RC3RG ($p = 0.016$, OR = 3.52) and 6bp− carriers ($p = 0.011$, OR = 1.96) were associated with non-response to MTX. Multivariate analysis confirmed the increased risk for non-response to MTX in 6bp− carriers ($p = 0.016$, OR = 2.74). Data demonstrated that *TYMS* polymorphisms were in linkage disequilibrium ($p < 0.00001$). Haplotype multivariate analysis revealed that haplotypes harboring both 3R and 6bp− alleles were associated with non-response to MTX. Regarding MTX-related toxicity, no statistically significant differences were observed in relation to *TYMS* genotypes and haplotypes.

Conclusion: Our study reveals that *TYMS* polymorphisms could be important to help predicting clinical response to MTX in RA patients. Despite the potential of these findings, translation into clinical practice needs larger studies to confirm these evidences.

Editor: Yue Wang, National Institute for Viral Disease Control and Prevention, CDC, China, China

Funding: Funding support from a Doctoral Grant (SFRH/BD/64441/2009) for Aurea Lima - Fundação para a Ciência e Tecnologia (FCT). The funders had no role in study design, data collection and analysis, decision to publish, or preparation of the manuscript.

Competing Interests: The authors have declared that no competing interests exist.

* Email: aurea.lima@iscsn.cespu.pt

Introduction

Methotrexate (MTX) is the cornerstone for rheumatoid arthritis (RA) treatment and is the most widely used disease-modifying antirheumatic drug (DMARD) in newly diagnosed patients [1,2]. Despite its cost-effectiveness, therapeutic outcome is variable mainly concerning to MTX clinical response and/or development of MTX-related toxicity [3–7]. MTX is an antifolate drug with important anti-inflammatory and antiproliferative effects, partly achieved by the intracellular inhibition of thymidylate synthase (TS) [8–10]. TS is a key protein for the *de novo* pyrimidine synthesis and is responsible for the simultaneous conversion of deoxyuridine monophosphate (dUMP) and 5,10-methylenetetra-

hydrofolate (5,10-MTHF) to deoxythymidine monophosphate (dTMP) and dihydrofolate (DHF). Subsequently, the dTMP is phosphorylated to deoxythymidine triphosphate (dTTP) and used for the deoxyribonucleic acid (DNA) synthesis and repair [6,11,12] (Figure 1A). Since TS levels were found to be predictive of MTX therapeutic outcome [13,14] and genetic polymorphisms in TS gene (*TYMS*) have been associated with TS levels [15,16], pharmacogenomics has raised great interest and, in fact, some studies have attempted to clarify the influence of genetic variations on clinical response to MTX in RA [17]. The most studied polymorphisms (rs34743033, rs2853542 and rs34489327) are represented on Figure 1B. Polymorphism rs34743033 is a 28 base pairs (bp) variable number tandem repeat (VNTR), located on 5' untranslated region (UTR) [18]. Is characterized by exhibiting a putative Enhancer box (E-box) sequence on the first 28 bp repeat of 2R allele and on the two first repeats of 3R allele [15,19]. Therefore, a higher number of repeats should increase the amount of E-box binding sites for the upstream stimulating factors (USF), leading to an increased transcription of *TYMS* and, consequently, to higher TS levels [20]. In addition, a single nucleotide polymorphism (SNP) characterized by a cytosine to guanine (C>G) transition on the twelfth nucleotide of the second repeat of VNTR 3R allele (rs2853542) has been described [15]. In the presence of cytosine (3RC) the E-box seems to be disrupted, reducing the stimulation of transcription in comparison to 3RG, thereby decreasing TS levels [15]. Since this SNP occurs within the *TYMS* 28 bp VNTR polymorphism, several studies have been performed combining the information from both *TYMS* enhancer region (TSER) polymorphisms [6,21]. Another important polymorphism is a 6 bp sequence (TTAAAG) deletion (1494del6, rs34489327) at 3'UTR, which seems to affect a region of TS pre-messenger ribonucleic acid (mRNA) that contains *cis* adenylate-uridylate-rich elements (AREs) [22,23]. These elements bind to a *trans* AU-rich factor 1 (AUF1), preferentially in the presence of deletion allele (6bp−), diminishing mRNA stability and, consequently, decreasing TS levels [16,22,23]. Therefore, the aim of this study was to elucidate the clinical relevance of these *TYMS* polymorphisms, by genotype and haplotype-based approaches, in MTX therapeutic outcome of Portuguese RA patients.

Methods

Patients and study design

A retrospective study was performed between January 2009 and December 2012 at São João Hospital Center (Porto, Portugal) in a cohort of consecutive Caucasian patients (≥18 years) with RA treated with MTX. Patients were excluded from the study if there was history of drug abuse, recent pregnancy or desire to become pregnant. The study was approved by the Ethical Committee of São João Hospital Center (reference 33/2009), procedures were considered to be according to the standards of the Helsinki Declaration and all patients provided an informed written consent.

After diagnosis, patients were classified according the 1987 criteria of the American College of Rheumatology (ACR) and reclassified according the 2010 criteria of the ACR and the European League Against Rheumatism (EULAR) [24]. All patients were initially treated with 10 mg *per os* (PO)/week of MTX in monotherapy. This dose was increased 5 mg at each three weeks if the patients did not meet the EULAR criteria for response, i.e., if presented a Disease Activity Score in 28 joints (DAS28) >3.2. Every 3 months treatment response was evaluated and, on the: 1) first evaluation, if patients have no response or show gastrointestinal toxicity, administration route was changed to subcutaneous (SC); 2) second evaluation, if maximum tolerable dose was used without response, MTX therapy was discontinued or associated with other synthetic DMARD; and 3) third evaluation, in patients without response and other contraindication, therapy was changed by associating a biological DMARD. The occurrence of MTX-related toxicity was registered at each visit and, according to severity, MTX dose was adjusted or discontinued. Folic acid supplementation was prescribed to all patients for the prevention of toxicity occurrence and their regular compliance was registered [7,25,26]. Other concomitant drugs, such as corticosteroids and non-steroidal anti-inflammatories (NSAIDs) were allowed during the study.

Outcome definition

Non-response. MTX clinical response was recorded at time of each visit. Non-response was defined when patients presented a DAS28>3.2, calculated and defined as described by Prevoo *et al*. [27], in two consecutive evaluations. Therefore, non-response to MTX had a minimum period of MTX therapy, at least, of six months.

Toxicity. The occurrence of MTX-related toxicity, defined when patients presented any adverse drug reaction (ADR) related to MTX, was recorded upon each visit. The type of ADR was classified in System Organ Class (SOC) disorders, in accordance with Common Terminology Criteria for Adverse Events (CTCAE) [28].

Samples handling and *TYMS* genotyping

Whole blood samples from each patient were obtained with standard venipuncture technique in ethylenediaminetetraacetic acid (EDTA) containing tubes. Genomic DNA was extracted with QIAamp DNA Blood Mini Kit (QIAGEN, Hilden, Germany) according to manufacturer instructions and total genomic DNA was quantified, and its purity analyzed, using the NanoDrop 1000 Spectrophotometer v3.7 (Thermo Scientific, Wilmington DE, USA).

TSER polymorphisms. 28 bp VNTR polymorphism (rs34743033) and SNP C>G (rs2853542) at the twelfth nucleotide of the second repeat of 3R allele were genotyped as described by Lima *et al*. [29]. For quality control, 10% of the samples were randomly selected for a second analysis and 10% percent of cases were confirmed by automated sequencing in a 3130×l Genetic Analyzer using the Kit BigDye Terminator v3.1 (Life Technologies, Foster City, CA, USA). Results were 100% concordant.

TYMS 1494del6 polymorphism. 1494del6 polymorphism (rs34489327) was genotyped as described by Lima *et al*. [29] with slight modifications. PCR products were purified with USB ExoSAP-IT (Affymetrix, Santa Clara, CA, USA) before cycle sequencing. Sequence reactions were carried out using the sequencing Kit BigDye Terminator v.3.1 (Life Technologies, Foster City, CA, USA) according to manufacturer's specifications. The sequencing profile was 30 cycles at 96°C for 10 seconds, 55°C for 10 seconds and 60° for 60 seconds, followed by an extension cycle at 60°C for 10 minutes. The sequence products were purified with illustra Sephadex G-50 Fine DNA Grade (GE Healthcare, Fairfield, CT, EUA) columns, denatured with Hi-Di Formamide and run in an 3130×l Genetic Analyzer (Life Technologies, Foster City, CA, USA). For quality control, 10% of the samples were randomly selected for a second analysis and results were 100% concordant.

Polymorphisms classification and linkage disequilibrium measure

TSER polymorphisms were classified according to their theoretical TS functional *status* as previously described [6] and grouped by predicted expression levels, as follow: low expression genotypes (2R2R, 2R3RC and 3RC3RC), median expression genotypes (2R3RG and 3RC3RG) and high expression genotype (3RG3RG). Haplotype analysis was performed using a two-stage iterative method named expectation maximization algorithm (SNPStats software) [30]. In order to estimate LD between pairs of alleles at TSER and *TYMS* 1494del6 *loci*, D' coefficients were calculated in Arlequin for Windows, Version 3.11 (University of Berne, Bern, Switzerland) [31] with 100,000 number of steps in Markov chain. The measure was interpretable as the proportion of maximum possible level of association between two *loci*, given the allele frequencies, ranging from 0 (linkage equilibrium) to 1

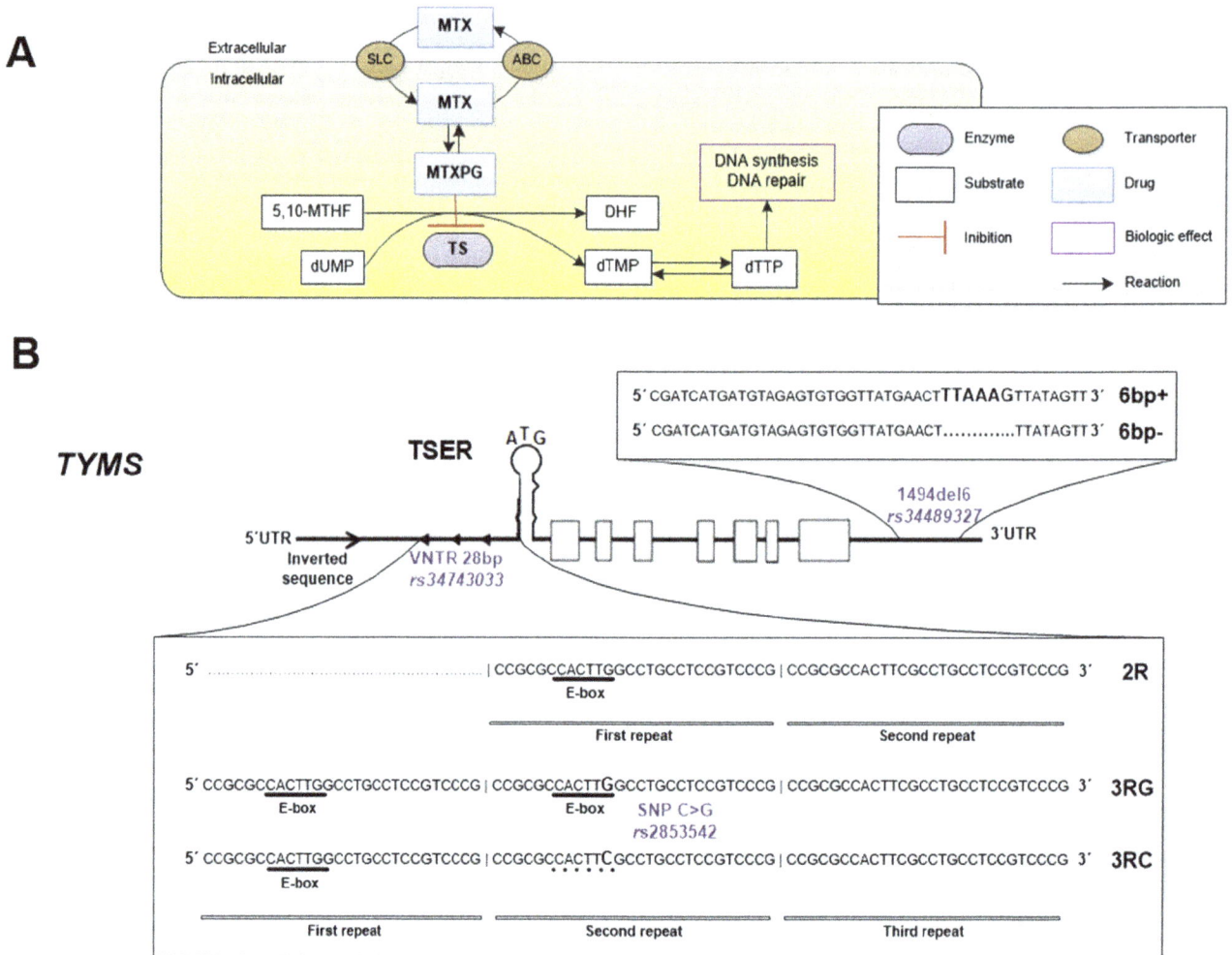

Figure 1. Part of MTX action mechanism in which thymidylate synthase (TS) is involved (A). MTX enters the cell after binding to folate transporters, mainly by solute carriers (SLC), and can be exported by members of the ATP-binding cassette (ABC) transporters family. To prevent MTX rapid efflux from cells and enhance its intracellular retention, MTX is polyglutamated by the enzyme folylpolyglutamyl synthase into MTX polyglutamates (MTXPGs) which inhibit TS activity. TS is a key protein for the *de novo* pyrimidine synthesis and is responsible for the simultaneous conversion of deoxyuridine monophosphate (dUMP) and 5,10-methylenetetrahydrofolate (5,10-MTHF) to deoxythymidine monophosphate (dTMP) and dihydrofolate (DHF). Subsequently, the dTMP is phosphorylated to deoxythymidine triphosphate (dTTP) and used for the DNA synthesis and repair. *TYMS* structure and location of VNTR 28 bp (rs34743033), SNP C>G (rs2853542) and 1494del6 (rs34489327) polymorphisms (B). 5,10-MTHF: 5,10-methylenetetrahydrofolate; A: adenine; ABC: ATP-binding cassette; bp: base pairs; C: cytosine; del: deletion; DHF: dihydrofolate; dTMP: deoxythymidine monophosphate; dTTP: deoxythymidine triphosphate; dUMP: deoxyuridine monophosphate; E-box: enhancer box; G: guanine; MTXPG: methotrexate polyglutamates; R: repeat; SLC: solute carrier; SNP: single nucleotide polymorphism; TS: thymidylate synthase (protein); TSER: thymidylate synthase enhancer region; T: thymine; *TYMS*: thymidylate synthase (gene); UTR: untranslated region; VNTR: variable number tandem repeat.

(complete LD) [32]. Possible haplotypes were tested for association with risk for non-response to MTX and for MTX-related toxicity by taking the most frequent haplotype as reference.

Statistical analysis

Statistical analyses were performed with either IBM SPSS Statistics for Windows, Version 20.0 (IBM Corp, Armonk, NY, USA), OpenEpi for Windows, Version 2.3.1 [33] and SNPStats software [30]. Genotype and allele frequencies were assessed and tested for Hardy-Weinberg equilibrium (HWE). All statistical tests were two-sided and a probability (p) value of 5% or less was considered as statistically significant. The Pearson Chi-square test or Fisher's exact test were used to compare the outcome variables and *TYMS* polymorphisms. The odds ratio (OR) and the

correspondent 95% confidence intervals (CI) were calculated as a measure of the association between the categorical variables. To correct for multiple comparisons, Bonferroni's method was applied in order to control the false positive rate, and a significance level of $\alpha = 0.05/(\text{n comparisons})$ was used [34]. Forest plot was performed using MedCalc software for Windows, Version 13.1.2 [35]. Multivariate analysis with binary logistic regression was used to identify which *TYMS* genotypes or haplotypes could predict the occurrence of non-response to MTX and MTX-related toxicity. This analysis was performed adjusting to potential confounding clinicopathological variables in three steps. In the first step patient-related variables (age, gender and smoking) were considered; in a second step, beyond patient-related variables, disease-related variables (diagnosis age and disease duration) were added; and in a third step, beyond patient and disease-related variables,

Table 1. Thymidylate synthase polymorphisms and methotrexate therapeutic outcome.

	MTX Response				MTX Toxicity			
	Response	Non-Response	p	OR (95% CI)	Non-Toxicity	Toxicity	p	OR (95% CI)
TYMS 28 bp VNTR (rs34743033)#								
2R2R	19 (54.3)	16 (45.7)		Reference	25 (71.4)	10 (28.6)		Reference
2R3R	62 (49.6)	63 (50.4)	0.624	1.21 (0.57–2.56)	78 (62.4)	47 (37.6)	0.324	1.51 (0.67–3.41)
3R3R	21 (30.4)	48 (69.6)	0.018*	2.71 (1.17–6.29)	50 (72.5)	19 (27.5)	0.911	0.95 (0.39–2.35)
2R carriers	81 (50.6)	79 (49.4)		Reference	103 (64.4)	57 (35.6)		Reference
3R3R	21 (30.4)	48 (69.6)	0.005*	2.34 (1.29–4.27)	50 (72.5)	19 (27.5)	0.233	0.69 (0.37–1.28)
2R2R	19 (54.3)	16 (45.7)		Reference	25 (71.4)	10 (28.6)		Reference
3R carriers	83 (42.8)	111 (57.2)	0.208	1.59 (0.77–3.27)	128 (66.0)	66 (34.0)	0.529	1.29 (0.58–2.84)
2R allele	100 (51.3)	95 (48.7)		Reference	128 (65.6)	67 (34.4)		Reference
3R allele	104 (39.5)	159 (60.5)	0.012*	1.61 (1.09–2.38)	178 (67.7)	85 (32.3)	0.647	0.91 (0.60–1.38)
TSER polymorphisms (rs2853542* and rs34743033)								
Functional 2R								
2R2R	19 (54.3)	16 (45.7)		Reference	25 (71.4)	10 (28.6)		Reference
2R3RC	32 (47.1)	36 (52.9)	0.487	1.34 (0.59–3.03)	44 (64.7)	24 (35.3)	0.492	1.36 (0.56–3.31)
3RC3RC	9 (31.0)	20 (69.0)	0.062	2.64 (0.94–7.39)	23 (79.3)	6 (20.7)	0.469	0.65 (0.20–2.08)
Functional 3R								
2R3RG	30 (52.6)	27 (47.4)		Reference	34 (59.6)	23 (40.4)		Reference
3RC3RG	6 (24.0)	19 (76.0)	0.016*	3.52 (1.23–10.10)	16 (64.0)	9 (36.0)	0.710	0.83 (0.31–2.20)
3RG3RG	6 (40.0)	9 (60.0)	0.384	1.67 (0.52–5.30)	11 (73.3)	4 (26.7)	0.384§	0.54 (0.15–1.90)
2R allele	100 (51.3)	95 (48.7)		Reference	128 (65.6)	67 (34.4)		Reference
3RC allele	56 (37.1)	95 (62.9)	0.008*	1.79 (1.13–2.82)	106 (70.2)	45 (29.8)	0.369	0.81 (0.50–1.31)
3RG allele	48 (42.9)	64 (57.1)	0.155	1.40 (0.86–2.30)	72 (64.3)	40 (35.7)	0.810	1.06 (0.63–1.78)
TSER polymorphisms grouped according to theoretically TS expression levels **								
Low expression	60 (45.5)	72 (54.5)		Reference	92 (69.7)	40 (30.3)		Reference
Median expression	36 (43.9)	46 (56.1)	0.824	1.07 (0.61–1.85)	50 (61.0)	32 (39.0)	0.189	1.47 (0.83–2.63)
High expression	6 (40.0)	9 (60.0)	0.687	1.25 (0.42–3.71)	11 (73.3)	4 (26.7)	1.000§	0.84 (0.25–2.79)
Low+Median expression	96 (44.9)	118 (55.1)		Reference	142 (66.4)	72 (33.6)		Reference
High expression	6 (40.0)	9 (60.0)	0.714	1.22 (0.42–3.55)	11 (73.3)	4 (26.7)	0.778§	0.72 (0.22–2.33)
Low expression	60 (45.5)	72 (54.5)		Reference	92 (69.7)	40 (30.3)		Reference
Median+High expression	42 (43.4)	55 (56.7)	0.746	1.09 (0.64–1.85)	61 (62.9)	36 (37.1)	0.279	1.36 (0.78–2.36)
TYMS 1494del6 (rs34489327)								
6bp+6bp+	61 (53.5)	53 (46.5)		Reference	78 (68.4)	36 (31.6)		Reference
6bp+6bp–	38 (39.2)	59 (60.8)	0.038	1.79 (1.03–3.10)	59 (60.8)	38 (39.2)	0.249	1.40 (0.79–2.46)
6bp–6bp–	6 (27.3)	16 (72.7)	0.024*	3.07 (1.12–8.41)	19 (86.4)	3 (13.6)	0.122§	0.34 (0.10–1.23)

Table 1. Cont.

	MTX Response				MTX Toxicity			
	Response	Non-Response	p	OR (95% CI)	Non-Toxicity	Toxicity	p	OR (95% CI)
6bp+6bp+	61 (53.5)	53 (46.5)		Reference	78 (68.4)	36 (31.6)		Reference
6bp− carriers	44 (37.0)	75 (63.0)	**0.011**^∀	1.96 (1.16-3.31)	78 (65.5)	41 (34.5)	0.641	1.14 (0.66-1.97)
6bp+ carriers	99 (46.9)	112 (53.1)		Reference	137 (64.9)	74 (35.1)		Reference
6bp−6bp−	6 (27.3)	16 (72.7)	0.078	2.36 (0.89-6.26)	19 (86.4)	3 (13.6)	0.055^§	0.29 (0.08-1.02)
6bp+ allele	160 (49.2)	165 (50.8)		Reference	215 (66.2)	110 (33.8)		Reference
6bp− allele	50 (35.5)	91 (64.5)	**0.006**^∀	1.76 (1.15-2.71)	97 (68.8)	44 (31.2)	0.578	0.89 (0.57-1.38)

Results are expressed in n (%). p value<0.05 was considered to be of statistical significance (highlighted in bold).
^§Fisher's exact test used when number of cases of one cell was less than 5.
^∀Statistically significant when p values were adjusted for multiple comparisons correction using Bonferroni's method (α =0.05/n comparisons).
^#3R4R genotype (n = 4) was excluded from analyses due to the low frequency.
^*rs2853542 - TYMS SNP C>G on 3R allele.
^**Genotypes theoretically associated with TS expression: a) high: 3RG3RG; b) median: 2R3RG and 3RC3RG; c) low: 2R2R, 2R3RC and 3RC3RC.
bp: base pairs; C: cytosine; del: deletion; G: guanine; OR: odds ratio; R: repeat; SNP: single nucleotide polymorphism; TS: thymidylate synthase (protein); TSER: TYMS enhancer region; TYMS: thymidylate synthase (gene); VNTR: variable number tandem repeat.

treatment-related variables (folic acid supplementation, corticosteroids therapy, use of NSAIDs, other concomitant DMARDs used and MTX administration characteristics - dose, treatment duration and administration route) were also considered.

Results

Population description

This study included follow-up data of 233 patients, 196 (84.1%) females and 37 (15.9%) males, with a mean age of 51 ± 11.6 years old, of which 32 (13.7%) were smokers. Considering the disease-related variables, the mean age at diagnosis was 40.3 ± 13.2 years old and the median disease duration was 7.0 years (0.3–51.0). All 233 (100.0%) patients were treated with MTX with a median dose of 15.0 mg/week (2.5–25.0), 118 (50.6%) complied regularly to folic acid supplementation, 188 (80.7%) were under corticosteroid therapy and 170 (73.0%) used NSAIDs.

Non-response to MTX (DAS28 >3.2 in two consecutive evaluations) was observed in 128 (54.9%) patients. Regarding disease activity, the mean for DAS28 was 4.2 ± 1.3. MTX-related toxicity was registered in 77 (33.0%) patients. The observed ADRs were classified in SOCs disorders as follow: 58 (75.3%) gastrointestinal disorders (abdominal distension, diarrhea, dyspepsia, nauseas, stomach pain and/or vomiting); 9 (11.7%) skin and subcutaneous tissue disorders (alopecia, rash maculo-papular and rheumatoid nodulosis exacerbation); 5 (6.5%) hepatobiliary disorders (determined by transaminases serum elevation); and 5 (6.5%) respiratory, thoracic and mediastinal disorders (hypersensitivity pneumonitis). Since the number of cases in each SOCs disorders were small, the evaluation of *TYMS* polymorphisms with clinical relevance as possible biomarkers of MTX-related toxicity was performed for MTX-related overall toxicity.

TYMS genotype and haplotype analyses

Genotypes distribution of *TYMS* polymorphisms was in HWE ($p>0.050$) in the studied population. Frequencies of 28 bp VNTR alleles and genotypes were: 2R allele 41.8%; 3R allele 57.3%; 4R allele 0.9%; 2R2R 15.0% (n = 35); 2R3R 53.7% (n = 125); 3R3R 29.6% (n = 69); and 3R4R 1.7% (n = 4). Due to the low frequency of 3R4R genotype, it was excluded from the analyses. Considering TSER polymorphisms, genotypes distribution was: 2R allele 42.6%; 3RC allele 33.0%; 3RG allele 24.4%; 2R2R 15.3% (n = 35); 2R3RC 29.7% (n = 68); 2R3RG 24.9% (n = 57); 3RC3RC 12.6% (n = 29); 3RC3RG 10.9% (n = 25); and 3RG3RG 6.6% (n = 15). According to TS theoretical functional *status*, genotypes frequencies were: low expression 57.6% (n = 132); median expression 35.8% (n = 82); and high expression 6.6% (n = 15). Frequencies of 1494del6 alleles and genotypes were: 6pb+ allele 70.0%; 6bp− allele 30.0%; 6bp+6bp+ 48.9% (n = 114); 6bp+6bp− 41.6% (n = 97); and 6bp−6bp− 9.5% (n = 22).

Haplotype analysis revealed that 28 bp VNTR and 1494del6 polymorphisms were in LD ($p<0.00001$). Alleles 2R and 6bp+, and alleles 3R and 6bp− were the most linked ones ($D' = 0.67$ for both). The analysis demonstrated four haplotypes: 2R6bp+ 38.4%; 2R6bp− 4.1%; 3R6bp+ 31.7% and 3R6bp− 25.8%. TSER and 1494del6 polymorphisms were also in LD ($p<0.00001$). Alleles 2R and 6bp+ ($D' = 0.67$) and 3RG and 6bp− ($D' = 0.48$) demonstrated to be the most linked ones. This analysis showed six haplotypes: 2R6bp+ 38.4%; 2R6bp− 4.1%; 3RC6bp+ 22.7%; 3RG6bp+ 9.0%; 3RC6bp− 10.3%; and 3RG6bp− 15.5%.

Table 2. *Thymidylate synthase* haplotypes and methotrexate therapeutic outcome.

TYMS Haplotypes	MTX Response				MTX Toxicity			
	Response	Non-Response	p	OR (95% CI)	Non-Toxicity	Toxicity	p	OR (95% CI)
Based on *TYMS* 28 bp VNTR and *TYMS* 1494del6 polymorphisms								
2R6bp+	43.0	30.0		Reference	36.2	33.9		Reference
2R6bp−	6.0	7.4	0.360	1.70 (0.54–5.32)	5.7	10.2	0.190	2.20 (0.69–7.03)
3R6bp+	33.9	34.5	0.100	1.55 (0.92–2.60)	33.1	37.8	0.490	1.23 (0.69–2.20)
3R6bp−	17.1	28.1	**0.001**	2.54 (1.46–4.43)	25.0	18.1	0.320	0.74 (0.41–1.34)
Based on TSER and *TYMS* 1494del6 polymorphisms								
2R6bp+	43.2	30.2		Reference	36.2	34.8		Reference
2R6bp−	5.8	7.2	0.360	1.70 (0.55–5.24)	5.6	9.3	0.220	2.02 (0.66–6.20)
3RC6bp+	21.2	25.6	**0.041**	1.79 (1.03–3.12)	23.8	23.6	0.820	1.07 (0.59–1.95)
3RC6bp−	6.2	11.8	**0.013**	2.80 (1.25–6.25)	10.9	6.0	0.240	0.55(0.21–1.47)
3RG6bp+	12.5	8.7	0.880	1.06 (0.50–2.24)	9.3	13.3	0.300	1.53 (0.69–3.38)
3RG6bp−	11.1	16.5	**0.009**	2.39 (1.24–4.59)	14.2	13.0	0.810	0.92 (0.46–1.82)

Results are expressed in estimated frequencies (%) under linkage disequilibrium. p value<0.05 was considered to be of statistical significance (highlighted in bold).
bp: base pairs; C: cytosine; del: deletion; G: guanine; OR: odds ratio; R: repeat; TSER: *TYMS* enhancer region; *TYMS*: thymidylate synthase (gene); VNTR: variable number tandem repeat.

Table 3. Multivariate analysis of *thymidylate synthase* polymorphisms and clinical response to methotrexate.

	Patient-related		Patient+Disease-related		Patient+Disease+Treatment-related	
	p	OR (95% CI)	*p*	OR (95% CI)	*p*	OR (95% CI)
TYMS **genotypes**						
TYMS **28 bp VNTR (rs34743033)**						
2R carriers		Reference		Reference		Reference
3R3R	**0.013**	2.23 (1.19–4.17)	**0.013**	2.24 (1.19–4.21)	0.135	1.99 (0.81–4.91)
TSER polymorphisms						
Functional 3R						
2R3RG		Reference		Reference		Reference
3RC3RG	0.069	2.90 (0.92–9.13)	0.071	2.91 (0.91–9.25)	0.203	2.70 (0.59–12.47)
TYMS **1494del6 (rs34489327)**						
6bp+6bp+		Reference		Reference		Reference
6bp− carriers	**0.003**	2.33 (1.32–4.10)	**0.003**	2.38 (1.34–4.23)	**0.016**	2.74 (1.21–6.23)
TYMS **haplotypes**						
Based on *TYMS* **28 bp VNTR and** *TYMS* **1494del6 polymorphisms**						
2R6bp+		Reference		Reference		Reference
3R6bp−	**<0.001**	2.87 (1.59–5.19)	**<0.001**	2.92 (1.60–5.32)	**0.012**	2.68 (1.25–5.75)
Based on TSER and *TYMS* **1494del6 polymorphisms**						
2R6bp+		Reference		Reference		Reference
3RC6bp+	**0.041**	1.81 (1.03–3.20)	**0.035**	1.86 (1.05–3.31)	0.090	1.85 (0.91–3.76)
3RC6bp−	**0.012**	2.97 (1.28–6.93)	**0.018**	2.75 (1.19–6.32)	**0.048**	2.89 (1.01–8.21)
3RG6bp−	**0.004**	2.78 (1.39–5.56)	**0.003**	3.06 (1.49–6.31)	**0.043**	2.60 (1.04–6.49)

P value<0.05 is considered to be of statistical significance (highlighted in bold).
Adjusted variables include: 1) patient-related variables (age, gender and smoking); 2) disease-related variables (diagnosis age and disease duration); and 3) treatment-related variables (folic acid supplementation, corticosteroids, non-steroidal anti-inflammatories, other concomitant disease-modifying antirheumatic drugs and methotrexate administration characteristics - dose, treatment duration and administration route). Genetic variables include: *TYMS* genotypes and *TYMS* haplotypes.
bp: base pairs; C: cytosine; del: deletion; G: guanine; OR: odds ratio; R: repeat; TSER: *TYMS* enhancer region; *TYMS*: thymidylate synthase (gene); VNTR: variable number tandem repeat.

TYMS genotypes and MTX therapeutic outcome

Table 1 reports the relation between *TYMS* polymorphisms and MTX therapeutic outcome both regarding MTX non-response and toxicity.

Non-response. In relation to 28 bp VNTR polymorphism, 3R allele was significantly associated with non-response to MTX when compared to 2R allele ($p = 0.012$, OR = 1.61). In addition, 3R homozygotes were associated with more than 2-fold increased risk for non-response to MTX when compared to 2R homozygotes ($p = 0.018$, OR = 2.71) and 2R carriers ($p = 0.005$, OR = 2.34) and remained significant after corrected for multiple comparisons. For TSER polymorphisms, 3RC allele shown to be associated with non-response to MTX when compared to 2R allele ($p = 0.008$, OR = 1.79). Furthermore, and attending to functional 3R, 3RC3RG was related with more than 3-fold increased risk for non-response to MTX when compared to 2R3RG ($p = 0.016$, OR = 3.52), which remained significant after multiple comparisons correction. Considering the 1494del6 polymorphism, 6bp− allele was significantly associated with non-response to MTX when compared to 6bp+ allele ($p = 0.006$, OR = 1.76). Moreover, and compared to 6bp+ homozygotes, 6bp+6bp− ($p = 0.038$, OR = 1.79), 6bp−6bp− ($p = 0.024$, OR = 3.07) and 6bp− carriers ($p = 0.011$, OR = 1.96) presented a statistically significant increased risk for non-response to MTX and, excepting for 6bp+ 6bp−, continued significant after correcting for multiple comparisons.

Toxicity. No statistically significant differences were observed in relation to *TYMS* genotypes and MTX-related overall toxicity.

TYMS haplotypes and MTX therapeutic outcome

Table 2 represents the relationship between *TYMS* haplotypes and MTX therapeutic outcome both regarding MTX non-response and toxicity.

Non-response. 3R6bp− haplotype was found significantly associated with non-response to MTX when compared to 2R6bp+ haplotype ($p = 0.001$, OR = 2.54). Moreover, 3RC6bp+, 3RC6bp− and 3RG6bp− haplotypes were statistically significant associated with non-response to MTX when compared to 2R6bp+ haplotype ($p = 0.041$, OR = 1.79; $p = 0.013$, OR = 2.80; and $p = 0.009$, OR = 2.39, respectively).

Toxicity. No statistically significant differences were observed in relation to *TYMS* haplotypes and MTX-related overall toxicity.

Multivariate analysis

Multivariate analysis was performed in three steps adjusting to potential confounding variables. Table 3 shows multivariate analysis results of *TYMS* genotypes and haplotypes and clinical response to MTX. Figure 2 resumes the impact of all potential confounding variables in the association of *TYMS* genotypes and haplotypes with clinical response to MTX. Regarding *TYMS* genotypes, results demonstrated that 6bp− carriers were statistically significant associated with more than 2-fold increased risk for

Forest plot for *thymidylate synthase* polymorphisms and clinical response to methotrexate

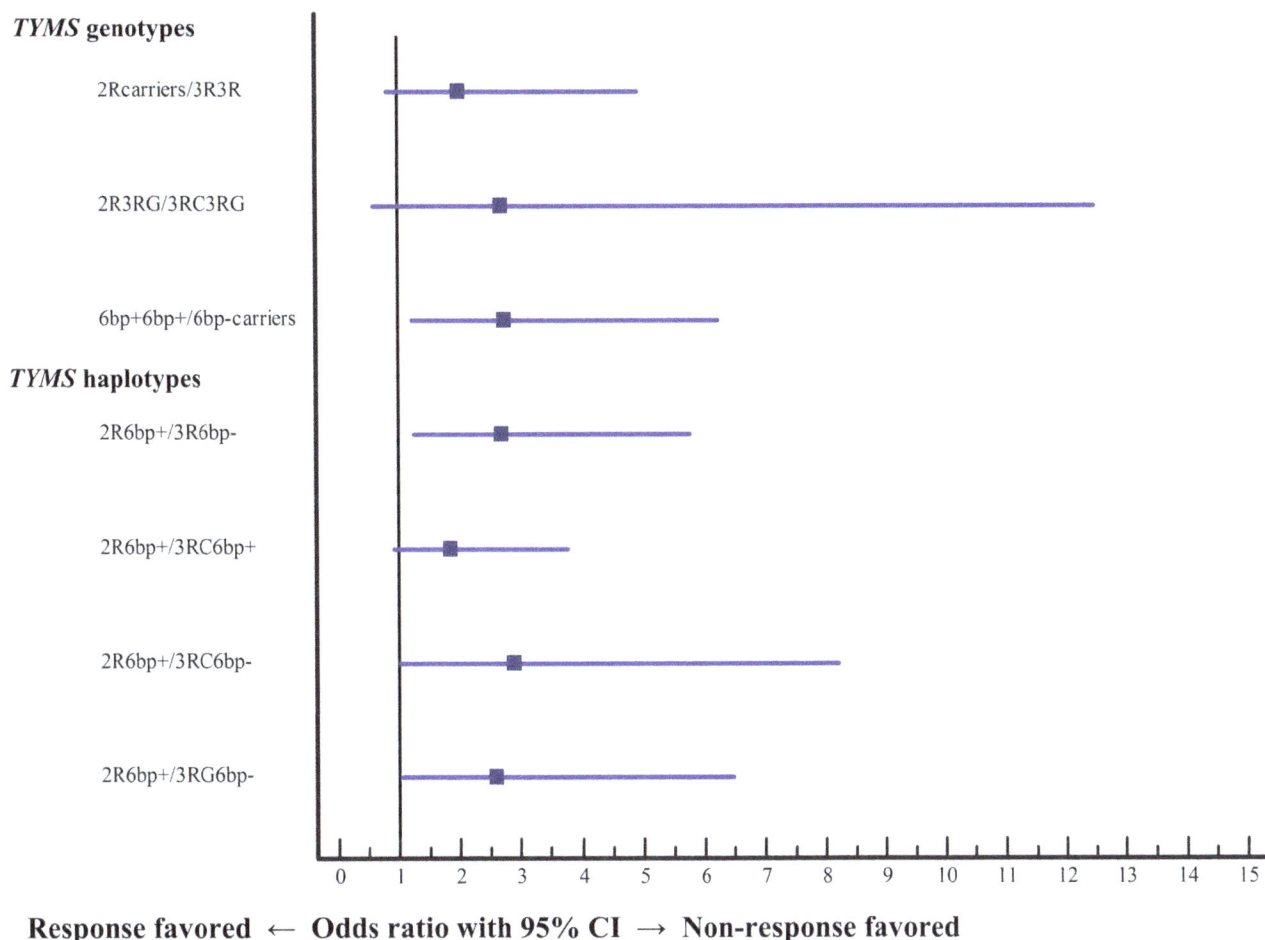

TYMS genotypes

2Rcarriers/3R3R

2R3RG/3RC3RG

6bp+6bp+/6bp-carriers

TYMS haplotypes

2R6bp+/3R6bp-

2R6bp+/3RC6bp+

2R6bp+/3RC6bp-

2R6bp+/3RG6bp-

0 1 2 3 4 5 6 7 8 9 10 11 12 13 14 15

Response favored ← Odds ratio with 95% CI → Non-response favored

Figure 2. Forest plot of multivariate analysis in the association of *thymidylate synthase* genotypes and haplotypes with clinical response to methotrexate. Odds ratio and 95% confidence intervals are reported for clinical response to methotrexate. bp: base pairs; C: cytosine; CI: confidence interval; del: deletion; G: guanine; R: repeat; TYMS: thymidylate synthase (gene).

non-response to MTX when compared to 6bp+ homozygotes ($p = 0.016$, OR = 2.74). According to *TYMS* haplotypes, our results shown that haplotypes harboring simultaneously 3R and 6bp− alleles were statistically significant associated with almost 3-fold increased risk for non-response to MTX when compared to 2R6bp+ haplotype.

Discussion

Thymidylate synthase is a key enzyme for DNA synthesis and repair [11,12] inhibited by MTXPGs and, therefore, contributes for MTX antiproliferative and anti-inflammatory effects [10]. In fact, TS levels were found to be predictive of MTX therapeutic outcome [13,14]. Since genetic polymorphisms in *TYMS* have been associated with TS levels [6], in this study we aimed to elucidate the influence of *TYMS* polymorphisms (28 bp VNTR, SNP C>G and 1494del6) in MTX therapeutic outcome of Portuguese RA patients.

All patients enrolled in this study were recruited within a well-defined geographical area and were of Caucasian ethnicity, with gender and age at time of diagnosis distributions similar to other reported populations [36,37]. Genotypes distribution of *TYMS*

polymorphisms was in HWE and was similar to those found for other Caucasian populations [21,38,39]. Nevertheless, and despite the potential of our results, possible study limitations include: 1) relatively reduced population size; 2) presence of other *TYMS* polymorphisms that possibly could alter TS expression or functionality; 3) limited screen of some important genes that codify other enzymes involved in MTX action mechanism.

TYMS genotypes and MTX therapeutic outcome

Non-response. Among our population and regarding 28 bp VNTR polymorphism, 3R allele was associated with risk for non-response to MTX, which increases in the presence of both 3R alleles, in accordance to previous studies [13,40]. Literature describes 3R allele as associated with higher TS levels [19,20] and TS levels as predictive of clinical response to MTX [13,40]. Moreover, 3R allele has been associated with higher MTX doses required [13] and higher RA disease activity [40]. Despite the significant univariate analysis results, multivariate analysis did not confirm them. Additionally, other studies demonstrated associations between 3R homozygotes and response to MTX [39] or showed no association [21,41–43]. It has been suggested by some authors that it is of greater importance to consider the SNP C>G on 3R allele and analyze the TSER polymorphisms instead of studying 28 bp VNTR polymorphism alone. 3RG allele was associated with higher transcriptional activity and translation efficiency due to its increased ability to complex with the USF protein [15,44]. Accordingly, the number of functional E-box in both 2R and 3RC alleles should be the same [6,15], which should reveal that patients with these genotypes would have similar TS expression and, consequently, a resembling clinical response. However, our results seem to demonstrate that 2R and 3RC alleles are different since 3RC3RG genotype was associated with over 3-fold increased risk for non-response to MTX when compared to 2R3RG. In addition, our results showed that 3RC allele was associated with non-response to MTX, when compared to 2R allele, and 3RC3RC genotype has a non-significant trend for non-response to MTX when compared to 2R2R genotype. Nevertheless, no statistically significant differences were observed attending to TSER polymorphisms grouped according to theoretically TS expression levels and to multivariate analysis. Moreover, a previous study demonstrated that non-response to MTX was associated to 3RG3RG patients [21]. Therefore, the putative relationship between TSER polymorphisms and clinical response to MTX outcome needs further clarification.

In relation to 1494del6 polymorphism, our results demonstrated that 6bp− allele was associated with non-response to MTX. Additionally, multivariate analysis showed that 6bp− carriers were associated with about 3-fold increased risk for non-response to MTX. *In vitro* studies have demonstrated that 6bp− allele has decreased mRNA stability and, thereby reduced TS expression [22,23], however, in other previously reported studies in RA Caucasian patients no association was observed [41]. Moreover, one study in Psoriasis, a disease where MTX is used in similar doses than RA, 6bp− allele demonstrated a trend for non-response, however, this study included Caucasian and non-Caucasian patients [45]. Studies in Asiatic patients have reported different results, some of them reported an association between 6bp− allele and response [13,43], while others reported no associations [42,46]. From all of these results it seems that ethnicity could be an important factor to predict the clinical response to MTX.

Toxicity. Regarding the occurrence of MTX-related overall toxicity, our results did not reach significance pertaining to TSER and 1494del6 polymorphisms, in accordance with previously reported studies [21,41,42,46,47]. Nevertheless, other studies reported significant associations of 28 bp VNTR polymorphism with MTX-related toxicity [38,48]. To the best of our knowledge this is the first report evaluating the influence of TSER polymorphisms in MTX-related toxicity in RA.

TYMS haplotypes and MTX therapeutic outcome

Non-response. Haplotypes may have a particular significance in regard to functionality or as genetic markers for unknown functional variants. Therefore, haplotype analysis was performed, to assess of possible consequences on the phenotype in the copresence of several variants of the same gene. As reported by others [6,39,49,50], *TYMS* polymorphisms were in LD, especially 2R6bp+ and 3RG6bp− haplotypes. Univariate haplotype analysis demonstrated that 3R6bp−, 3RC6bp+, 3RC6bp− and 3RG6bp− haplotypes (haplotypes harboring 3R allele for 28 bp VNTR, 3RC allele for TSER and 6bp− allele for 1494del6) were associated with almost 3-fold increased risk for non-response to

MTX. Nevertheless, multivariate analysis showed that haplotypes harboring simultaneously 3R and 6bp− alleles (3R6bp−, 3RC6bp− and 3RG6bp−) were associated with non-response to MTX. This suggests a prominent role of the 3′-UTR polymorphism in predicting the clinical response to MTX and it seems that 6bp− allele can interact differently with 2R and 3R alleles, in agreement with Lurje *et al.* [51]. Additionally, our results suggested that the haplotype revealing more risk for non-response to MTX was 3RC6bp−, which combines the major risk alleles from the 5′UTR (3RC) and from the 3′UTR (6bp−). Only one study in RA has performed haplotype analysis, where an association between 3R6bp− haplotype and response to MTX was demonstrated [39]. Nevertheless, there are some important differences: no reference to SNP C>G; studied population included patients with early RA; and the study evaluated the impact in clinical response to MTX combined therapy with sulfasalazine. Thus, we propose that *TYMS* haplotype analysis should be used in future studies to elucidate the influence of *TYMS* in MTX therapeutic outcome, which could help to interpret these preliminary conflicting data.

Toxicity. Regarding MTX-related toxicity, no differences were observed attending to *TYMS* haplotypes. Despite it was expected that *TYMS* haplotypes follow the same tendency as *TYMS* genotypes, to the best of our knowledge no studies analyzed the *TYMS* haplotypes and the development of toxicity arising from MTX in RA.

The observed discrepancies among different studies could be explained by inter-study variability, ethnicity variability, samples sizes, variety of methods used to measure the MTX therapeutic outcome, different treatment regimens, and different genotyping protocols with limited quality of results. Therefore, functional TS studies in RA should be conducted to better understanding TS expression regulation mechanism and its putative importance in establishing more effective clinical therapeutic strategies when MTX is used in RA patients. To the best of our knowledge, this is the first report regarding the study of the association of *TYMS* polymorphisms with MTX therapeutic outcome in Portuguese RA patients. This study concluded that *TYMS* polymorphisms seem to be important to predict clinical response to MTX in RA patients; *TYMS* genotypes and haplotypes harboring 6bp− allele were associated with non-response to MTX; *TYMS* haplotypes harboring simultaneously 3R and 6bp− alleles seem to be predictors of non-response to MTX; and, to elucidate the role of *TYMS* on MTX therapeutic outcome full haplotypic information should be exploited. Despite the potential of our findings, translation into clinical practice requires larger and multicentric studies in order to clearly endorse the utility of these polymorphisms.

Acknowledgments

The authors wish to acknowledge to *Fundação para a Ciência e Tecnologia* (FCT) for the Doctoral Grant (SFRH/BD/64441/2009) for Aurea Lima and also to the nursing service of Rheumatology Day Hospital of São João Hospital Center and the physicians from the Rheumatology Department of São João Hospital Center, especially to Lucia Costa (MD.) and Francisco Ventura (MD. Ph.D.).

Author Contributions

Conceived and designed the experiments: AL VS MB RM. Performed the experiments: AL RA. Analyzed the data: AL VS. Contributed reagents/materials/analysis tools: AL HS RM. Wrote the paper: AL VS MB RA HS RM.

References

1. O'Dell JR (2004) Therapeutic strategies for rheumatoid arthritis. N Engl J Med 350: 2591–2602.

2. Mikuls TR, O'Dell J (2000) The changing face of rheumatoid arthritis therapy: results of serial surveys. Arthritis Rheum 43: 464–465.

3. Ranganathan P, McLeod HL (2006) Methotrexate pharmacogenetics: the first step toward individualized therapy in rheumatoid arthritis. Arthritis Rheum 54: 1366–1377.

4. Benucci M, Saviola G, Manfredi M, Sarzi-Puttini P, Atzeni F (2011) Cost effectiveness analysis of disease-modifying antirheumatic drugs in rheumatoid arthritis. A systematic review literature. Int J Rheumatol 2011: 845496.

5. Finckh A, Liang MH, van Herckenrode CM, de Pablo P (2006) Long-term impact of early treatment on radiographic progression in rheumatoid arthritis: A meta-analysis. Arthritis Rheum 55: 864–872.

6. Lima A, Azevedo R, Sousa H, Seabra V, Medeiros R (2013) Current approaches for TYMS polymorphisms and their importance in molecular epidemiology and pharmacogenetics. Pharmacogenomics 14: 1337–1351.

7. Lima A, Bernardes M, Sousa H, Azevedo R, Costa L, et al. (2013) SLC19A1 80G allele as a biomarker of methotrexate-related gastrointestinal toxicity in Portuguese rheumatoid arthritis patients. Pharmacogenomics.

8. Chan ES, Cronstein BN (2002) Molecular action of methotrexate in inflammatory diseases. Arthritis Res 4: 266–273.

9. Swierkot J, Szechinski J (2006) Methotrexate in rheumatoid arthritis. Pharmacol Rep 58: 473–492.

10. Kremer JM (2004) Toward a better understanding of methotrexate. Arthritis Rheum 50: 1370–1382.

11. Krajinovic M, Costea I, Primeau M, Dulucq S, Moghrabi A (2005) Combining several polymorphisms of thymidylate synthase gene for pharmacogenetic analysis. Pharmacogenomics J 5: 374–380.

12. Touroutoglou N, Pazdur R (1996) Thymidylate synthase inhibitors. Clin Cancer Res 2: 227–243.

13. Kumagai K, Hiyama K, Oyama T, Maeda H, Kohno N (2003) Polymorphisms in the thymidylate synthase and methylenetetrahydrofolate reductase genes and sensitivity to the low-dose methotrexate therapy in patients with rheumatoid arthritis. Int J Mol Med 11: 593–600.

14. Krajinovic M, Costea I, Chiasson S (2002) Polymorphism of the thymidylate synthase gene and outcome of acute lymphoblastic leukaemia. Lancet 359: 1033–1034.

15. Mandola MV, Stoehlmacher J, Muller-Weeks S, Cesarone G, Yu MC, et al. (2003) A novel single nucleotide polymorphism within the 5′ tandem repeat polymorphism of the thymidylate synthase gene abolishes USF-1 binding and alters transcriptional activity. Cancer Res 63: 2898–2904.

16. Mandola MV, Stoehlmacher J, Zhang W, Groshen S, Yu MC, et al. (2004) A 6 bp polymorphism in the thymidylate synthase gene causes message instability and is associated with decreased intratumoral TS mRNA levels. Pharmacogenetics 14: 319–327.

17. Zhu H, Deng FY, Mo XB, Qiu YH, Lei SF (2014) Pharmacogenetics and pharmacogenomics for rheumatoid arthritis responsiveness to methotrexate treatment: the 2013 update. Pharmacogenomics 15: 551–566.

18. Marsh S, McKay JA, Cassidy J, McLeod HL (2001) Polymorphism in the thymidylate synthase promoter enhancer region in colorectal cancer. Int J Oncol 19: 383–386.

19. Corre S, Galibert MD (2005) Upstream stimulating factors: highly versatile stress-responsive transcription factors. Pigment Cell Res 18: 337–348.

20. Marsh S (2005) Thymidylate synthase pharmacogenetics. Invest New Drugs 23: 533–537.

21. Jekic B, Lukovic L, Bunjevacki V, Milic V, Novakovic I, et al. (2013) Association of the TYMS 3G/3G genotype with poor response and GGH 354GG genotype with the bone marrow toxicity of the methotrexate in RA patients. Eur J Clin Pharmacol 69: 377–383.

22. Pullmann R Jr, Abdelmohsen K, Lal A, Martindale JL, Ladner RD, et al. (2006) Differential stability of thymidylate synthase 3′-untranslated region polymorphic variants regulated by AUF1. J Biol Chem 281: 23456–23463.

23. Zhang Z, Shi Q, Sturgis EM, Spitz MR, Hong WK, et al. (2004) Thymidylate synthase 5′- and 3′-untranslated region polymorphisms associated with risk and progression of squamous cell carcinoma of the head and neck. Clin Cancer Res 10: 7903–7910.

24. Aletaha D, Neogi T, Silman AJ, Funovits J, Felson DT, et al. (2010) 2010 rheumatoid arthritis classification criteria: an American College of Rheumatology/European League Against Rheumatism collaborative initiative. Ann Rheum Dis 69: 1580–1588.

25. Ortiz Z, Shea B, Suarez Almazor M, Moher D, Wells G, et al. (2000) Folic acid and folinic acid for reducing side effects in patients receiving methotrexate for rheumatoid arthritis. Cochrane Database Syst Rev: CD000951.

26. Ortiz Z, Shea B, Suarez-Almazor ME, Moher D, Wells GA, et al. (1998) The efficacy of folic acid and folinic acid in reducing methotrexate gastrointestinal toxicity in rheumatoid arthritis. A metaanalysis of randomized controlled trials. J Rheumatol 25: 36–43.

27. Prevoo ML, van 't Hof MA, Kuper HH, van Leeuwen MA, van de Putte LB, et al. (1995) Modified disease activity scores that include twenty-eight-joint

counts. Development and validation in a prospective longitudinal study of patients with rheumatoid arthritis. Arthritis Rheum 38: 44–48.

28. U.S. department of health and human services (2010) Common Terminology Criteria for Adverse Events (CTCAE) Version 4.03. National Institutes of Health and National Cancer Institute.

29. Lima A, Seabra V, Martins S, Coelho A, Araujo A, et al. (2014) Thymidylate synthase polymorphisms are associated to therapeutic outcome of advanced non-small cell lung cancer patients treated with platinum-based chemotherapy. Mol Biol Rep.

30. Sole X, Guino E, Valls J, Iniesta R, Moreno V (2006) SNPStats: a web tool for the analysis of association studies. Bioinformatics 22: 1928–1929.

31. Excoffier L, Laval G, Schneider S (2005) Arlequin (version 3.0): an integrated software package for population genetics data analysis. Evol Bioinform Online 1: 47–50.

32. Schaid DJ, Rowland CM, Tines DE, Jacobson RM, Poland GA (2002) Score tests for association between traits and haplotypes when linkage phase is ambiguous. Am J Hum Genet 70: 425–434.

33. Sullivan KM, Dean A, Soe MM (2009) OpenEpi: a web-based epidemiologic and statistical calculator for public health. Public Health Rep 124: 471–474.

34. Bland JM, Altman DG (1995) Multiple significance tests: the Bonferroni method. BMJ 310: 170.

35. Schoonjans F, Zalata A, Depuydt CE, Comhaire FH (1995) MedCalc: a new computer program for medical statistics. Comput Methods Programs Biomed 48: 257–262.

36. Gibofsky A (2012) Overview of epidemiology, pathophysiology, and diagnosis of rheumatoid arthritis. Am J Manag Care 18: S295–302.

37. Rindfleisch JA, Muller D (2005) Diagnosis and management of rheumatoid arthritis. Am Fam Physician 72: 1037–1047.

38. Bohanec Grabar P, Logar D, Lestan B, Dolzan V (2008) Genetic determinants of methotrexate toxicity in rheumatoid arthritis patients: a study of polymorphisms affecting methotrexate transport and folate metabolism. Eur J Clin Pharmacol 64: 1057–1068.

39. James HM, Gillis D, Hissaria P, Lester S, Somogyi AA, et al. (2008) Common polymorphisms in the folate pathway predict efficacy of combination regimens containing methotrexate and sulfasalazine in early rheumatoid arthritis. J Rheumatol 35: 562–571.

40. Dervieux T, Furst D, Lein DO, Capps R, Smith K, et al. (2004) Polyglutamation of methotrexate with common polymorphisms in reduced folate carrier, aminoimidazole carboxamide ribonucleotide transformylase, and thymidylate synthase are associated with methotrexate effects in rheumatoid arthritis. Arthritis Rheum 50: 2766–2774.

41. Owen SA, Hider SL, Martin P, Bruce IN, Barton A, et al. (2012) Genetic polymorphisms in key methotrexate pathway genes are associated with response to treatment in rheumatoid arthritis patients. Pharmacogenomics J 13: 227–234.

42. Ghodke Y, Chopra A, Joshi K, Patwardhan B (2008) Are Thymidylate synthase and Methylene tetrahydrofolate reductase genes linked with methotrexate response (efficacy, toxicity) in Indian (Asian) rheumatoid arthritis patients? Clin Rheumatol 27: 787–789.

43. Inoue S, Hashiguchi M, Takagi K, Kawai S, Mochizuki M (2009) Preliminary study to identify the predictive factors for the response to methotrexate therapy in patients with rheumatoid arthritis. Yakugaku Zasshi 129: 843–849.

44. Kawakami K, Watanabe G (2003) Identification and functional analysis of single nucleotide polymorphism in the tandem repeat sequence of thymidylate synthase gene. Cancer Res 63: 6004–6007.

45. Campalani E, Arenas M, Marinaki AM, Lewis CM, Barker JN, et al. (2007) Polymorphisms in folate, pyrimidine, and purine metabolism are associated with efficacy and toxicity of methotrexate in psoriasis. J Invest Dermatol 127: 1860–1867.

46. Takatori R, Takahashi KA, Tokunaga D, Hojo T, Fujioka M, et al. (2006) ABCB1 C3435T polymorphism influences methotrexate sensitivity in rheumatoid arthritis patients. Clin Exp Rheumatol 24: 546–554.

47. Ranganathan P, Culverhouse R, Marsh S, Mody A, Scott-Horton TJ, et al. (2008) Methotrexate (MTX) pathway gene polymorphisms and their effects on MTX toxicity in Caucasian and African American patients with rheumatoid arthritis. J Rheumatol 35: 572–579.

48. Weisman MH, Furst DE, Park GS, Kremer JM, Smith KM, et al. (2006) Risk genotypes in folate-dependent enzymes and their association with methotrexate-related side effects in rheumatoid arthritis. Arthritis Rheum 54: 607–612.

49. Dotor E, Cuatrecases M, Martinez-Iniesta M, Navarro M, Vilardell F, et al. (2006) Tumor thymidylate synthase 1494del6 genotype as a prognostic factor in colorectal cancer patients receiving fluorouracil-based adjuvant treatment. J Clin Oncol 24: 1603–1611.

50. Lima A, Seabra V, Martins S, Coelho A, Araújo A, et al. (2014) Thymidylate synthase polymorphisms are associated to therapeutic outcome of advanced non-small cell lung cancer patients treated with platinum-based chemotherapy. Molecular Biology Reports: 1–9.

51. Lurje G, Zhang W, Yang D, Groshen S, Hendifar AE, et al. (2008) Thymidylate synthase haplotype is associated with tumor recurrence in stage II and stage III colon cancer. Pharmacogenet Genomics 18: 161–168.

3

LD-Aminopterin in the Canine Homologue of Human Atopic Dermatitis

John A. Zebala[1]*, Alan Mundell[2], Linda Messinger[3], Craig E. Griffin[4], Aaron D. Schuler[1], Stuart J. Kahn[1]

1 Syntrix Biosystems, Inc., Auburn, Washington, United States of America, **2** Animal Dermatology Service, Edmonds, Washington, United States of America, **3** Veterinary Referral Center of Colorado, Englewood, Colorado, United States of America, **4** Animal Dermatology Clinic, San Diego, California, United States of America

Abstract

Background: Options are limited for patients with atopic dermatitis (AD) who do not respond to topical treatments. Antifolate therapy with systemic methotrexate improves the disease, but is associated with adverse effects. The investigational antifolate LD-aminopterin may offer improved safety. It is not known how antifolate dose and dosing frequency affect efficacy in AD, but a primary mechanism is thought to involve the antifolate-mediated accumulation of 5-aminoimidazole-4-carboxamide ribonucleotide (AICAR). However, recent *in vitro* studies indicate that AICAR increases then decreases as a function of antifolate concentration. To address this issue and understand how dosing affects antifolate efficacy in AD, we examined the efficacy and safety of different oral doses and schedules of LD-aminopterin in the canine model of AD.

Methods and Findings: This was a multi-center, double-blind trial involving 75 subjects with canine AD randomized to receive up to 12 weeks of placebo, once-weekly (0.007, 0.014, 0.021 mg/kg) or twice-weekly (0.007 mg/kg) LD-aminopterin. The primary efficacy outcome was the Global Score (GS), a composite of validated measures of disease severity and itch. GS improved in all once-weekly cohorts, with 0.014 mg/kg being optimal and significant (43%, $P<0.01$). The majority of improvement was seen by 8 weeks. In contrast, GS in the twice-weekly cohort was similar to placebo and worse than all once-weekly cohorts. Adverse events were similar across all treated cohorts and placebo.

Conclusions: Once-weekly LD-aminopterin was safe and efficacious in canine AD. Twice-weekly dosing negated efficacy despite having the same daily and weekly dose as effective once-weekly regimens. Optimal dosing in this homologue of human AD correlated with the concentration-selective accumulation of AICAR *in vitro*, consistent with AICAR mediating LD-aminopterin efficacy in AD.

Editor: Douglas Thamm, Colorado State University, United States of America

Funding: This study was supported by the National Institutes of Health (no. AR 056547 to SK and JZ). The funders had no role in study design, data collection and analysis, decision to publish, or preparation of the manuscript.

Competing Interests: SJK and JAZ are employees of Syntrix Biosystems and JAZ holds stock in Syntrix Biosystems. LD-aminopterin is a product in development at Syntrix Biosystems. LM is a paid consultant for Antech Diagnostics, and has received research or speaker support from AB Science, Dechra, Greer Laboratories, Novartis Animal Health, Virbac and Zoetis. CG has been a paid consultant, speaker or received research support from the following companies: Elanco, Merck, Novartis Animal Health, Pinnaclife, Sogeval, TEVA animal health, Veterinary Allergy Reference Laboratory and Zoetis.

* Email: jzebala@syntrixbio.com

Introduction

Atopic dermatitis (AD) affects approximately 3% to 5% of the adult population in the western world, and 30% of the worldwide pediatric population [1]. It is a complex, relapsing disease arising from interactions between genes and the environment and is characterized by pruritus, disruption of the epidermal barrier, and IgE-mediated sensitization to food and environmental allergens [2]. The pathogenesis of AD may involve an aberrant Th2 adaptive immune response to innocuous environmental antigens, skin barrier abnormalities, and an inadequate host response to cutaneous microbes [3].

Patients with AD who fail to respond to topical corticosteroids or topical calcineurin inhibitors may require second-line systemic immunosuppressive therapy [4]. Systemic treatment options include cyclosporine, corticosteroids, azathioprine and methotrexate [5,6]. Cyclosporine and prednisolone are appropriate as short-term treatments [5], the former being nephrotoxic and the latter predisposing to osteoporosis, hypertension and other side-effects [7]. Cyclosporine is also almost entirely metabolized by the liver cytochrome P450 IIIA system, and clinically significant sustained drug-drug interactions can occur during long-term therapy [8]. Caution in the use of azathioprine has been highlighted as well [5], given the heightened risk for hepatosplenic T-cell lymphoma, a rare but frequently lethal form of lymphoma [9]. Despite its well-established record of safety and efficacy, methotrexate is not well tolerated in many patients [10]. The limitations of current systemic treatments have prompted the search for improved treatments that might expand the armamentarium of therapeutic options for patients with AD.

LD-Aminopterin (Syntrix Biosystems, Auburn, WA) is the L- and D-enantiomer of N-[4-[[(2,4-diamino-6-pterdinyl)methyl]amino]benzoyl]-glutamic acid (Figure 1A) [11]. The L-enantiomer is an antifolate congener of methotrexate that is stereoselectively absorbed from LD-aminopterin by the intestinal proton coupled folate transporter [12]. Preclinical and clinical studies indicate it may provide improvements on methotrexate, including better bioavailability [13,14], greater cell uptake and conversion to active polyglutamylated metabolites [13,15], less central nervous system toxicity [16,17,18,19,20], and less liver toxicity [13]. Unlike cyclosporine, LD-aminopterin is not metabolized by human liver microsomes, and thus drug-drug interactions at the cytochrome P450 system are unlikely [12].

Methotrexate, L-aminopterin, and their polyglutamylated metabolites inhibit dihydrofolate reductase and enzymes involved in *de novo* purine and thymidylate synthesis (Figure 1B) [21,22]. Proposed anti-inflammatory mechanisms have centered on inhibition of *de novo* thymidylate synthesis [23,24,25], and inhibition of aminoimidazolecarboxamide ribonucleotide transformylase (AICART), an enzyme involved in *de novo* purine synthesis [26,27,28]. Inhibition of *de novo* thymidylate synthesis prevents cell-cycle progression of activated T-cells and induces their apoptosis by a Fas-independent pathway [23,24,25], an effect reproduced by several groups [29,30,31,32]. Inhibition of AICART causes increased levels of its substrate, 5-aminoimidazole-4-carboxamide-1-β-D-ribofuranosyl 5'-monophosphate (AICAR), which together with its dephosphorylated metabolite 5-aminoimidazole-4-carboxamide-1-β-D-ribofuranoside (AICA), inhibit AMP deaminase and adenosine deaminase [33,34], effects that cause an increase in extracellular adenosine [26]. Extracellular adenosine binds adenosine receptors to affect a reduction in

inflammation [35]. AICA is also cytotoxic to T lymphocytes, potentiates the cytotoxicity of methotrexate added to cultured T lymphocytes [34,36,37] and activates AMP-activated kinase [38,39].

Funk *et al.* recently demonstrated AICAR increased 115-fold following exposure of an erythroblastoid cell line to 10 nM methotrexate, but decreased with increasing methotrexate concentrations, declining to baseline with 1000 nM methotrexate [40]. In contrast, the substrate for thymidylate synthase, 2'-deoxyuridine 5'-monophosphate (dUMP), displayed concentration-dependent accumulation over the same range of methotrexate concentration. It was suggested that if clinical response is dependent on the accumulation of AICAR, that these *in vitro* findings might predict a clinical therapeutic response paradoxically related to dose.

Initial trials of methotrexate in AD simply adopted the dose and regimen commonly used to treat psoriasis and rheumatoid arthritis [41,42]. However, given the different underlying pathologic mechanisms between AD and these other autoimmune diseases, it is not clear that the same dosing strategy would be equally applicable. In fact, no study has examined how dose and regimen affect antifolate efficacy in AD, and thus how to best administer antifolate therapy in AD remains a significant unresolved question.

Although mouse models of AD have many practical benefits in the laboratory, they also have significant limitations in how clinically similar their disease is to human AD. In contrast, dogs naturally and commonly develop a pruritic dermatitis that is clinically and immunologically extremely similar to human AD [43]. Like human AD, canine AD is associated with severe pruritus, skin xerosis and increased transepidermal water loss, face and skin fold involvement, spongiotic dermatitis, skin-infiltrating eosinophils, skin infiltration by IgE(+) and CD1c(+) dendritic cells, Th2-dominated immune responses, positive atopy patch test, and IgE-specific responses. Owing to the remarkable similarity with the

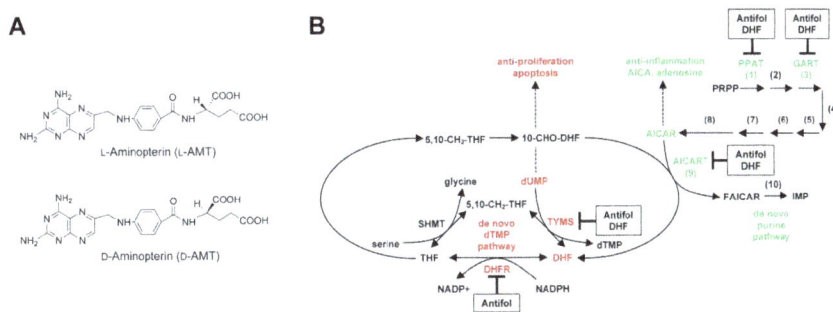

Figure 1. LD-Aminopterin composition and mechanistic model in anti-inflammation. (A) Chemical structure of L-aminopterin (*top*) and D-aminopterin (*bottom*). (B) The anti-inflammatory activity of L-aminopterin and methotrexate have been attributed to inhibition of thymidylate (*red*) and purine (*green*) *de novo* biosynthesis. In the *de novo* pathway of thymidylate (dTMP) synthesis, serine hydroxymethyltransferase (SHMT) catalyzes the conversion of serine and tetrahydrofolate polyglutamates (THF) to 5,10-CH$_2$-THF and glycine. Thymidylate synthase (TYMS) converts 5,10-CH$_2$-THF and deoxyuridine monophosphate (dUMP) to dihydrofolate polyglutamates (DHF) and dTMP. Dihydrofolate reductase (DHFR) completes the cycle by catalyzing the conversion of DHF to THF in an NADPH-dependent reaction. The purine, inosine monophosphate (IMP), is synthesized *de novo* in 10 chemical steps (shown numbered) catalyzed by six enzymes. The six enzymes are phosphoribosylpyrophosphate amidotransferase (PPAT; 1); a trifunctional enzyme composed of glycinamide ribonucleotide synthetase (GARS; 2), GAR formyltransferase (GART; 3) and aminoimidazole ribonucleotide synthetase (AIRS; 5); formylglycinamidine ribonucleotide synthase (FGAMS; 4); a bifunctional enzyme composed of carboxyaminoimidazole ribonucleotide synthase (CAIRS; 6) and succinoaminoimidazolecarboxamide ribonucleotide synthetase (SAICARS; 7); adenylosuccinate lyase (ASL; 8); and a bifunctional enzyme composed of aminoimidazolecarboxamide ribonucleotide transformylase (AICART; 9) and inosine monophosphate cyclohydrolase (IMPCH; 10). Evidence indicates that 10-formyl-7,8-dihydrofolate (10-CHO-DHF) is the predominant *in vivo* substrate for AICART, making AICART and TYMS the only enzymes to produce the DHFR substrate DHF [69]. Inside the cell, L-aminopterin and methotrexate and their polyglutamate metabolites (antifol) bind with high affinity to DHFR, resulting in accumulation of DHF and depletion of the reduced folate pool. Depletion of folates, as well as the direct inhibition by antifol and DHF, have all been implicated in the inhibition of PPAT, GART, AICART and TYMS [22,33,54,70]. In the case of AICART, the accumulation of DHF may cause this reaction to run backwards, since AICAR is normally driven towards the biosynthesis of FAICAR and IMP by the DHFR-catalyzed reduction of DHF to THF, as the equilibrium of this step actually lies in the direction of AICAR formation [60].

human disease, it has been suggested that canine AD can not only help answer mechanistic questions related to disease pathogenesis, but also serve as a model for testing of drugs with clinical potential in humans [43].

Here we report the efficacy and safety results from a 12-week dose-ranging randomized, double-blind, placebo-controlled, multi-center trial that tested the efficacy and safety of orally administered LD-aminopterin given once- or twice-weekly to subjects with canine AD. The objective was to examine how efficacy and safety of antifolate therapy varies as a function of dose and schedule. This study provides insights into how to administer antifolate therapy in canine AD that has implications for treating the human disease with LD-aminopterin based on a mechanism aimed at maximizing AICAR accumulation.

Materials and Methods

Ethics statement

The study was conducted in compliance with the Veterinary International Committee for Harmonization guidance for good clinical practice and was overseen and approved by a local Institutional Animal Care and Use Committee (North Carolina State University) and a centralized Institutional Animal Care and Use Committee (Infectious Disease Research Institute). Owners of subjects provided written consent for subjects to participate in the study and could withdraw from the study at any time.

Study design

Blinded trial. The study was performed as a double-blinded, randomized, placebo-controlled, parallel-group study conducted at four referral-based specialty practices located in the United States (California, Colorado, North Carolina and Washington) (Figure 2).

Subjects were randomized in a 1:1:1:1:1 ratio to receive oral doses of placebo, or LD-aminopterin once-weekly (0.007, 0.014 or 0.021 mg/kg) or twice-weekly (0.007 ×2 mg/kg). Doses are for the free acid of the L-enantiomer. Study drug consisted of either a gelatin capsule containing microcrystalline cellulose (placebo), or a gelatin capsule containing 0.25 mg LD-aminopterin tablets in an appropriate number of whole and/or half tablets to provide the desired dose per subject weight, and backfilled with microcrystalline cellulose. Owners were not required to take any special handling precautions of study drug.

A pre-planned interim efficacy checkpoint at day 56 was instituted based on pilot trial data that indicated responsive subjects achieved the majority of benefit by 4–8 weeks, whereas unresponsive subjects failed to improve with further treatment [44]. Subjects achieving at least 25% GS improvement passed the checkpoint and continued to receive treatment up to day 84. Subjects unable to meet the minimum GS response exited to avoid further futile treatment; their day 56 evaluation became their efficacy endpoint. Efficacy endpoints were therefore from day 56 or 84 per protocol.

Each arm employed a twice-weekly dosing using dummy doses to keep the blind, where the second weekly dose was given 3 days after the first. See below for details on randomization, blinding and dosing compliance. Daily prednisolone (0.5 mg/kg) was offered for the first 14 days without taper to maintain enrollment due to the delayed onset of LD-aminopterin action [44]. No folic

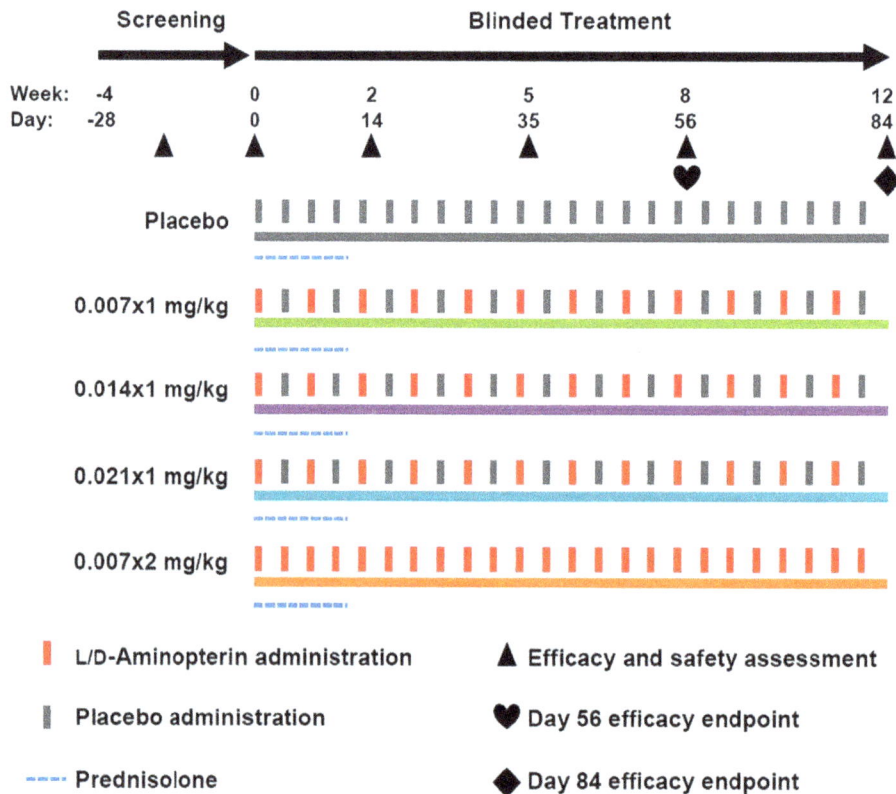

Figure 2. Study flow chart. Randomized subjects with AD were orally administered placebo, or LD-aminopterin once-weekly (0.007 ×1 mg/kg, 0.014 ×1 mg/kg, 0.021 ×1 mg/kg) or twice-weekly (0.007 ×2 mg/kg).

Table 1. Subject demographics and baseline AD characteristics.

| | | LD-Aminopterin | | | | |
| | Placebo | 0.007×1 mg/kg | 0.014×1 mg/kg | 0.021×1 mg/kg | 0.007×2 mg/kg | |
Variable	N=15	N=15	N=15	N=15	N=15	P-value[a]
Age, y	6.7±3.5	4.9±2.5	6.8±3.7	6.0±2.6	5.9±3.0	0.48
Male, N (%)	9 (60.0)	10 (66.6)	10 (66.6)	10 (66.6)	8 (53.3)	0.91
Body weight, kg	28.7±10.1	23.3±12.2	22.7±12.0	26.1±14.3	17.1±10.9	0.11
GS	11.3±8.5	11.4±4.7	12.3±9.0	10.0±5.9	11.9±7.6	0.93
CADESI	160±105	170±64	159±94	130±56	173±99	0.66
PVAS	6.5±1.5	6.7±1.2	7.4±1.5	7.5±1.5	6.6±1.4	0.19
Nonseasonal, N (%)	15 (100.0)	15 (100.0)	14 (93.3)	15 (100.0)	14 (93.3)	0.54

Abbreviations: GS, Global Score; CADESI, Canine Atopic Dermatitis Extent and Severity Index 03; PVAS, Pruritus Visual Analogue Scale.
Data are mean ± SD for continuous variables.
[a]P-values were calculated by chi-square test for categorical data and one-way ANOVA for continuous data.

acid supplementation was specified. Disease activity was assessed at days 0, 14, 35, 56 and 84.

Open-label extension. Subjects from the blinded trial were optionally able to continue on LD-aminopterin in an open-label extension lasting up to 104 weeks. Subjects received other treatments within the standard of care at the discretion of the clinician. Dosing was 0.007–0.021 mg/kg once-weekly at the clinician's discretion.

Study population

Inclusion criteria were (i) a diagnosis of canine AD [45,46]; (ii) moderate-to-severe disease defined by a CADESI score ≥60 and <500 [47]; (iii) age >6 months; (iv) weight 7 to 50 kg; (v) testing to rule out food allergy, flea bite hypersensitivity and external parasites; (vi) absence of fleas and use of a long acting flea adulticide; and (vii) intradermal skin testing or allergen-specific IgE determination confirming the presence of immediate or late-phase hypersensitivity reactions, or reagin immunoglobulins to environmental allergens such as house dust or storage mites, pollens or molds.

Subjects were excluded for (i) pregnancy or lactation; (ii) malignant neoplasia; (iii) diet augmented with fatty acid supplements if the diet was not continued throughout trial; (iv) treatment with long-acting corticosteroids within 6 weeks, oral corticosteroids or cyclosporine within 3 weeks, or oral anti-histamines within 1 week of enrollment; (v) use of anti-allergenic or antipruritic shampoos or conditioners, topical corticosteroids, tacrolimus or cyclosporine within 1 week of enrollment; and (vi) allergen-specific

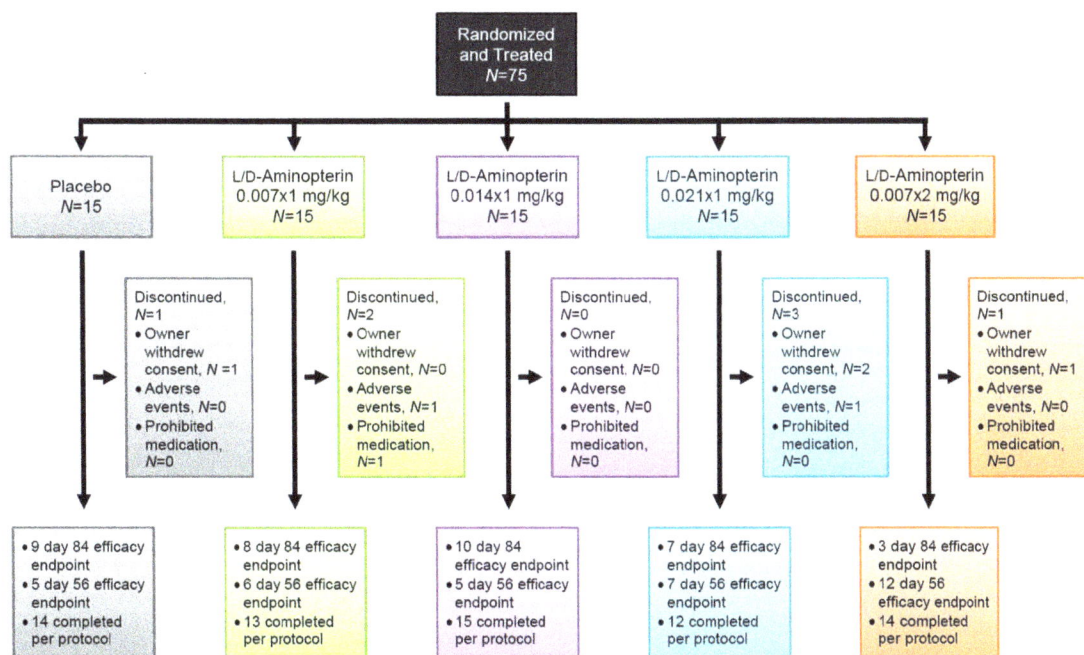

Figure 3. Disposition of subjects. A total of 68 subjects (91%) completed the study per protocol. Discontinuations (9%) were for withdrawal of owner consent (N = 2), owner perceived AE (N = 2), and prohibited medication (N = 1).

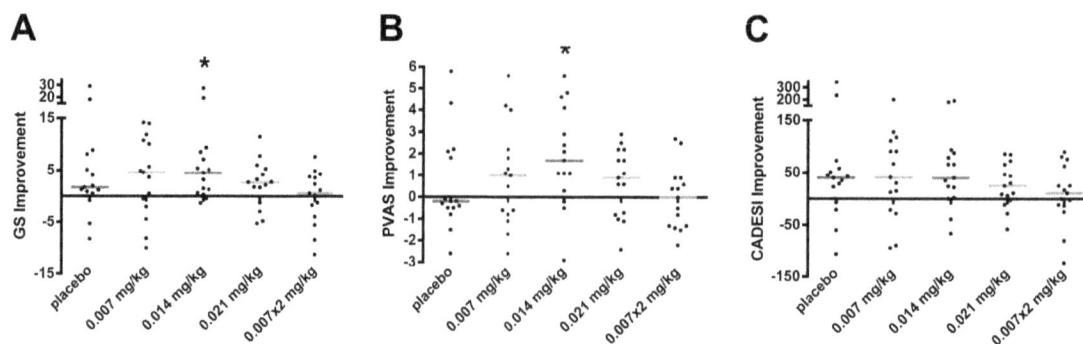

Figure 4. Effect of placebo and LD-aminopterin on canine AD disease measures. Subjects ($N = 75$) with AD were randomized equally to receive placebo, or LD-aminopterin once-weekly (0.007, 0.014 or 0.021 mg/kg) or twice-weekly (0.007 ×2 mg/kg). Improvement in baseline disease measures were determined for (A) GS, (B) PVAS and (C) CADESI (see Materials and Methods). GS and PVAS improved significantly in the 0.014 mg/kg cohort. *$P < 0.05$. Horizontal bars are medians. Abbreviations: GS, Global Score; PVAS, Pruritus Visual Analogue Scale; CADESI, Canine Atopic Dermatitis Extent and Severity Index 03.

immunotherapy started or changed within 6 months of enrollment, or if the allergen-specific immunotherapy was changed during the study. Antibiotics were permitted per protocol to treat skin infections at the discretion of investigators.

Assessments

Blinded trial. Disease activity was assessed using validated disease measures: PVAS to measure itch [48] and the CADESI to measure disease severity [47]. PVAS yields a possible score from 0 to 10, and CADESI yields a possible score from 0 to 1,240. CADESI and PVAS were assessed at study days 0 (baseline), 14, 35, 56 and 84 (i.e. end of weeks 2, 5, 8 and 12). GS is a composite score that is the product of CADESI and PVAS and thus captures the proportional change in CADESI and PVAS, where GS = (CADESI×PVAS)/100.

Safety assessments were performed at study days 0, 14, 35, 56 and 84 and consisted of recording all AEs and serious AEs and noting their severity and relationship to study drug. They included the regular monitoring of hematology, blood chemistry, and urine and physical examination. A central laboratory (Antech Diagnostic GLP, Morrisville, NC) was used for analysis of all specimens

collected and listed below. Hemoglobin, hematocrit, red blood cell (RBC) count, white blood cell (WBC) count with differential (neutrophils including bands, lymphocytes, monocytes, eosinophils, and basophils), and platelet count were measured at all scheduled study visits within the visit window. Serum chemistries including blood urea nitrogen (BUN), creatinine,, alanine transaminase/serum glutamic pyruvate transaminase (ALT/SGPT), alkaline phosphatase, lactate dehydrogenase (LDH), total protein, and albumin, were measured at all scheduled study visits within the visit window. Urinalysis for specific gravity, protein, glucose, blood, ketones, bilirubin and urobilinogen were performed at scheduled visits on day 0 and 84, or day 56 for subjects who exited the study at the interim efficacy checkpoint.

Open-label extension. Safety assessments were every 3 months in the first year and every 6 months in the second year using the same assessments as in the blinded trial.

Study endpoints

Per protocol, the primary efficacy endpoint was the change in baseline GS at study day 56 or 84. The primary study outcome was to assess the efficacy of four LD-aminopterin dosages in

Table 2. Concomitant medications.

		LD-Aminopterin				
	Placebo	0.007 ×1 mg/kg	0.014 ×1 mg/kg	0.021 ×1 mg/kg	0.007 ×2 mg/kg	
Medication	$N = 15$	$N = 15$	$N = 15$	$N = 15$	$N = 15$	*P*-value[a]
Prednisolone, *N* (%)[b]						
Yes	13 (86.6)	13 (86.6)	13 (86.6)	12 (80.0)	11 (73.3)	
No	2 (13.3)	2 (13.3)	2 (13.3)	3 (20.0)	4 (26.6)	0.83
Antibiotics, *N* (%)						
Weeks 0–4	11 (73.3)	12 (80.0)	3 (20.0)	12 (80.0)	12 (80.0)	0.001
Weeks 5–8	7 (46.7)	7 (46.7)	3 (20.0)	6 (40.0)	8 (53.3)	0.397
Weeks 9–12	6 (40.0)	6 (40.0)	3 (20.0)	5 (33.3)	5 (33.3)	0.772
Prohibited, *N* (%)						
Yes	0 (0.0)	1 (6.7)	0 (0.0)	0 (0.0)	0 (0.0)	
No	15 (100.0)	14 (93.3)	15 (100.0)	15 (100.0)	15 (100.0)	0.40

[a]*P*-values calculated by chi-square test.
[b]During first 14 days.

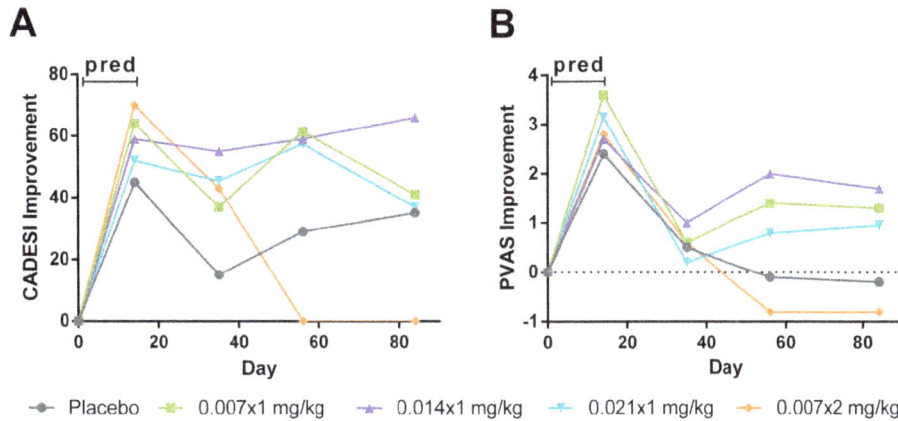

Figure 5. Change in CADESI and PVAS as a function of time in subjects treated with prednisolone and either placebo or LD-aminopterin. Subjects treated with prednisolone (pred) in the first 14 days ($N = 62$) were treated with either placebo ($N = 13$), or LD-aminopterin once-weekly (0.007×1 mg/kg, $N = 13$; 0.014×1 mg/kg, $N = 13$; 0.021×1 mg/kg, $N = 12$) or twice-weekly (0.007×2 mg/kg, $N = 11$). Median improvement in baseline (A) CADESI and (B) PVAS was determined at days 14, 35, 56 and 84. Abbreviations: CADESI, Canine Atopic Dermatitis Extent and Severity Index 03; PVAS, Pruritus Visual Analogue Scale; pred, prednisolone.

subjects with moderate-to-severe canine AD with respect to the primary efficacy endpoint, and determine the most (or least) effective dosage.

Secondary efficacy endpoints evaluated at study day 56 or 84 were the change in baseline CADESI and PVAS. Secondary outcomes included assessing the efficacy of four LD-aminopterin dosages in subjects with moderate-to-severe canine atopic dermatitis with respect to secondary efficacy endpoints, and determine the most (or least) effective dosage; the effect of LD-aminopterin on each secondary efficacy endpoint over time; the safety of LD-aminopterin by clinical and laboratory AEs as a function of dosage and time.

Randomization, blinding and dosing compliance

Randomization was performed centrally by Syntrix Biosystems Drug Supply Management. Subjects were randomized 1:1:1:1:1 into five treatment arms in blocks of five. Randomized blocks were generated using GraphPad QuickCalcs online software (www.graphpad.com/quickcalcs, GraphPad Software, Inc., La Jolla, CA). At randomization, each subject was assigned an identification number that was linked to a treatment arm and a sequentially numbered bottle of blinded study drug. Subject owners did not have contact with one another. All weekly study drug doses were provided in a single similar appearing capsule filled with microcrystalline cellulose. Dosing instructions specified only clear liquids for two hours before taking capsules, except for a small quantity of food to assist in administration. Weight-band-dosing tables were stratified by 1.0 kg increments. To preserve the blind, each arm maintained a schedule of twice-weekly dosing using a dummy dose in the once-weekly treatment schedules, and two dummy doses in the placebo cohort. Dosing compliance was determined by site monitoring and drug accountability (assigned capsules returned). Subject owners, investigator staff, and persons performing the assessments, were blinded to the identity of the treatment.

Statistical analyses

The sample size calculation was based on assessing four dosages of LD-aminopterin and placebo to determine the most or least effective dosage with respect to the primary efficacy endpoint using Hsu's multiple comparisons with the best (Hsu's MCB) test [49].

Assuming a minimum clinically meaningful change in GS of 1.5 (Dr. Thierry Olivry of North Carolina State University), and mean baseline GS of 5.5 and standard deviation of 0.8, both obtained from pilot trial data in subjects (n = 6) with moderate disease [44], a sample size of 15 subjects per cohort was required to achieve a power of 0.9.

The full analysis set consisted of all subjects who were randomized, using the initial randomized dosage, whether the subject ultimately dropped out of the trial or had their dose reduced per protocol. Subjects with missing day 56 or day 84 data were analyzed by the last observation carried forward. Balance in baseline characteristics between cohorts was analyzed by chi-square test for categorical data and one-way ANOVA for continuous data.

The primary outcome, change in baseline GS (absolute and percent change), was analyzed in each cohort by repeated-measures ANOVA. The two-sided type I error was adjusted for multiple cohort comparisons using the Bonferroni correction. The most effective dosage was analyzed using Hsu's MCB [49]. Hsu's MCB compares each cohort mean and the "best" of all the other cohort means to identify the best dosage, or reject a dosage as the best dosage. Hsu's MCB provides joint simultaneous confidence intervals for the differences between the mean baseline change of a dosage cohort minus the maximum of the mean baseline change in each of the other cohorts. If a cohort mean is significantly separated above all other cohort means, it is regarded as 'the best' (i.e., lower confidence limit >0). If a cohort mean has at least one cohort mean significantly separated above it, it is rejected as the best dosage (i.e., upper confidence limit <0). Secondary outcomes for CADESI and PVAS were analyzed as above.

Post hoc testing was by t-test and Mann-Whitney tests, and categorical data on concomitant medications were analyzed by chi-square test with significance claimed at $\alpha = 0.05$. Analyses and sample size calculations were performed with commercial software (PASS and NCSS, NCSS, LLC, Kaysville, UT; and GraphPad Prism version 6.00 for Windows, GraphPad Software, La Jolla, CA).

The safety set included all subjects who took at least one dose of study drug and had at least one post-baseline assessment. AEs were summarized by absolute and relative frequencies stratified by cohort and duration treated.

Table 3. Summary of clinical AEs by cohort[a].

		LD-Aminopterin			
Preferred Term	Placebo	0.007×1 mg/kg	0.014×1 mg/kg	0.021×1 mg/kg	0.007×2 mg/kg
	$N=15$	$N=15$	$N=15$	$N=15$	$N=15$
Subjects with any AE(s)	**10 (66.6)**	**8 (53.3)**	**7 (46.7)**	**10 (66.6)**	**5 (33.3)**
Death	0	0	0	0	0
Serious AEs	0	0	0	0	0
AE led to discontinuation	0	1 (6.7)[b]	0	1 (6.7)[b]	0
All AEs in any cohort	**11 (73.3)**	**12 (80.0)**	**13 (86.7)**	**17 (113.3)**	**8 (53.3)**
Fatigue	1 (6.7)	1 (6.7)	1 (6.7)	2 (13.3)	0
Weight loss	0	0	0	1 (6.7)	0
Diarrhea	2 (13.3)	4 (26.7)	3 (20.0)	5 (33.3)	2 (13.3)
Anorexia	0	1 (6.7)	1 (6.7)	1 (6.7)	1 (6.7)
Vomiting	0	0	1 (6.7)	1 (6.7)	0
Constipation	0	0	1 (6.7)	0	0
Stool increased	1 (6.7)	0	0	0	0
Stool dark color	1 (6.7)	0	0	0	0
Thirst increased	0	2 (13.3)	0	0	0
Halitosis	0	0	1 (6.7)	0	0
Keratoconjunctivitis sicca	0	0	0	1 (6.7)	0
Eye discharge	0	0	0	0	1 (6.7)
Demodicosis[c]	0	1 (6.7)	0	0	1 (6.7)
Pyotraumatic dermatitis	0	0	0	0	1 (6.7)
Skin infection	1 (6.7)	2 (13.3)	2 (13.3)	2 (13.3)	1 (6.7)
Otitis externa	1 (6.7)	0	0	1 (6.7)	1 (6.7)
Urinary incontinence	1 (6.7)	0	1 (6.7)	1 (6.7)	0
Aural hematoma	0	1 (6.7)	0	0	0
Epistaxis	0	0	1 (6.7)	0	0
Anxiety	0	0	0	0	1 (6.7)
Irritability	1 (6.7)	0	0	0	0
Stomach pain	0	0	0	1 (6.7)	0
Dermatitis	1 (6.7)	0	0	0	0
Urticaria	1 (6.7)	0	0	0	0
Tail dysfunction	0	0	0	1 (6.7)	0

Abbreviations: AE, adverse event.
[a]Expressed as *n* and percent of total subjects in each cohort.
[b]AE led to discontinuation by subject owner, not by investigator.
[c]0.007×1 mg/kg cohort: *Demodex canis* at day 44 post 0.5 mg/kg prednisolone on days 0 to 14; and 0.007×2 mg/kg cohort: *Demodex injai* at day 56 post 1.0 mg/kg prednisolone on days 0 to 14. Demodicosis cleared after one dose of milbemycin oxime, and each subject treated with LD-aminopterin for 24 (0.007×1 mg/kg cohort) and 9 (0.007×2 mg/kg cohort) months in the open-label segment without recurrence.

Results

Subject baseline characteristics and disposition in the study

Treatment cohorts were balanced with respect to demographic features and baseline disease characteristics (Table 1). The average disease activity in each cohort was severe, defined by a CADESI≥ 120 [50]. The total population was balanced between moderate ($N = 36$) and severe ($N = 39$) disease. A total of 75 subjects were randomly assigned to receive oral LD-aminopterin or placebo (Figure 3). Four study sites enrolled 5 to 44 subjects each. A total of 68 subjects (90.7%) completed the study per protocol, with 37 subjects treated for 12 weeks and 31 subjects treated up to the interim 8 week efficacy checkpoint. Seven subjects (9.3%) discontinued the study. Drug accountability indicated that 95% ($N = 71$) of all subjects had taken 90% or more of the assigned doses, and this percentage was similar across cohorts.

Administration of weekly oral LD-aminopterin is efficacious in canine AD

The Global Score (GS) improved significantly in the 0.014×1 mg/kg cohort (Figure 4A). The GS improved by a mean (±SD) of 6.1±7.6 points (95% CI, 1.9–10.3), decreasing from 12.3±9.0 at baseline to 6.2±4.8 after treatment ($P<0.05$). The mean (±SD) percent reduction in baseline GS in the

Table 4. Summary of laboratory AEs by cohort[a].

| | Placebo | LD-Aminopterin | | | |
| | | 0.007×1 mg/kg | 0.014×1 mg/kg | 0.021×1 mg/kg | 0.007×2 mg/kg |
Laboratory Abnormality	N= 15	N= 15	N= 15	N= 15	N= 15
Hematocrit Decreased	0	1 (1.3)	1 (1.3)	2 (2.7)	1 (1.3)
RBC Count Decreased	0	0	2 (2.7)	1 (1.3)	0
Thrombocytopenia	1 (1.3)	0	0	0	1 (1.3)
Thrombocytosis	6 (8.0)	7 (9.3)	7 (9.3)	9 (12.0)	4 (5.3)
Leukopenia	1 (1.3)	0	0	0	0
Lymphopenia	2 (2.7)	0	0	1 (1.3)	0
Neutropenia	0	0	1 (1.3)	0	0
Eosinophilia	0	1 (1.3)	0	0	0
BUN Increased	4 (5.3)	4 (5.3)	3 (4.0)	2 (2.7)	2 (2.7)
Creatinine Increased	0	3 (4.0)	1 (1.3)	0	1 (1.3)
Alkaline Phosphatase Increased	6 (8.0)	5 (6.7)	7 (9.3)	6 (8.0)	7 (9.3)
ALT Increased	5 (6.7)	2 (2.7)	3 (4.0)	1 (1.3)	1 (1.3)
Serum Protein Decreased	0	1 (1.3)	0	1 (1.3)	0
Serum Albumin Decreased	2 (2.7)	3 (4.0)	2 (2.7)	1 (1.3)	1 (1.3)
Total	**27 (36.0)**	**27 (36.0)**	**27 (36.0)**	**24 (32.0)**	**18 (24.0)**

Abbreviations: RBC, red blood cell; BUN, blood urea nitrogen; ALT, alanine transaminase.
[a]Expressed as N and percent of 75 total subjects.

0.014×1×mg/kg cohort was 43.2±38.0% (95% CI, 22–64%; $P<$ 0.01).

Treatment with LD-aminopterin also resulted in a significant reduction ($P<0.05$) in itch in the 0.014×1 mg/kg cohort (Figure 4B). The Pruritus Visual Analogue Scale (PVAS) improved by a mean (±SD) of 1.9±2.3 points (95% CI, 0.6–3.2), decreasing from 7.4±1.5 at baseline to 5.5±2.5 after treatment. The mean percent reduction in PVAS in the 0.014×1 mg/kg cohort was 26% (95% CI, 7–43%). Pruritus in 4 of 15 subjects (27%) in the cohort responded with a robust reduction in baseline PVAS≥4 (mean [percent] reduction = 4.8 [65%]).

The change in baseline Canine Atopic Dermatitis Extent and Severity Index 03 (CADESI) was not significant in any cohort, although the 0.014×1 and 0.021×1 mg/kg cohorts had mean (±SD) changes (53±71 and 26±42, respectively) that were significant before adjusting the type I error for multiple comparisons (Figure 4C). There was improvement in mean (±SD) CADESI in the placebo cohort (52±109), but it was not significant even prior to adjusting the type I error for multiple comparisons.

Antibiotics were permitted per protocol to treat skin infections at the discretion of investigators. The mean (±SE) duration of antibiotic treatment was 6.2+3.7 weeks. Antibiotic use was not a confounding factor in the significant efficacy responses to LD-aminopterin in the 0.014×1 mg/kg cohort because antibiotic use was similar across all treatment cohorts and placebo in each consecutive four week treatment period, except in the 0.014×1 mg/kg cohort, where it was lower (Table 2).

Dosing frequency determines optimal efficacy in canine AD

In addition to examining how varying LD-aminopterin dose impacted efficacy in canine AD, this study also examined how the schedule or frequency of administration affected efficacy. Inter-estingly, all endpoints for twice-weekly LD-aminopterin were no better than placebo, and worse than all once-weekly schedules (Figure 4). CADESI in the twice-weekly regimen was notable for being clearly worse than placebo, though not significantly. The 0.007×2 mg/kg cohort was statistically rejected as the best dosage based on GS and PVAS; each endpoint mean was smaller than, and significantly separated from the corresponding endpoint mean in the 0.014×1 mg/kg cohort ($P<0.05$, Hsu's MCB).

A *post hoc* comparison with two weeks of daily prednisolone suggests LD-aminopterin may be highly effective in a subpopulation of canine AD

Per protocol, subjects were optionally treated with prednisolone in the first 14 days (see Materials and Methods). Subjects treated with prednisolone constituted 83% ($N = 62$), and were distributed similarly across cohorts (Table 2). Two independent time-response profiles were clearly evident for CADESI and PVAS in this sub-population, consistent with prednisolone and LD-aminopterin having distinctly different onsets of action (Figure 5). Whereas the action of LD-aminopterin on PVAS required 56 to 84 days to come to full prominence, prednisolone caused a rapid improve-ment in PVAS by day 14 that was lost by the time of the primary efficacy endpoint for LD-aminopterin.

The median (mean±SD) improvement in PVAS at day 14 in the prednisolone-treated population ($N = 62$) was 2.8 (2.9±2.4) points, a treatment effect that was notably consistent among all cohorts (Figure 5). In contrast, the median (mean±SD) improve-ment in PVAS at day 14 in the population not treated with prednisolone ($N = 13$) was 0.0 (0.05±0.6) points. The improve-ment in PVAS at day 14 in the populations treated and not treated with prednisolone were significantly different ($P<0.0001$ for median and mean). Prednisolone treatment thus served not only to maintain enrollment during the onset of LD-aminopterin efficacy, it also provided an internal positive efficacy control that

Table 5. Summary of clinical AEs as a function of 4-week intervals[a].

Preferred Term	0 to 4 Weeks	5 to 8 Weeks	9 to 12 Weeks
All Categories	33 (54.1)	21 (34.4)	7 (11.5)
Constitutional	3 (4.9)	2 (3.3)	1 (1.6)
Fatigue	3 (4.9)	2 (3.3)	0
Weight loss	0	0	1 (1.6)
Gastrointestinal	17 (27.9)	7 (11.5)	4 (6.6)
Diarrhea	10 (16.4)	3 (4.9)	3 (4.9)
Anorexia	2 (3.3)	1 (1.6)	1 (1.6)
Vomiting	1 (1.6)	1 (1.6)	0
ConstipationN	1 (1.6)	0	0
Stool increased	1 (1.6)	0	0
Stool dark color	0	1 (1.6)	0
Thirst increased	2 (3.3)	0	0
Halitosis	0	1 (1.6)	0
Ocular	1 (1.6)	1 (1.6)	0
Keratoconjunctivitis sicca	0	1 (1.6)	0
Eye discharge	1 (1.6)	0	0
Infection	3 (4.9)	9 (14.8)	2 (3.3)
Demodicosis	0	2 (3.3)	0
Pyotraumatic dermatitis	1 (1.6)	0	0
Skin infection	2 (3.3)	5 (8.2)	1 (1.6)
Otitis externa	0	2 (3.3)	1 (1.6)
Renal/Genitourinary	3 (4.9)	0	0
Urinary incontinence	3 (4.9)	0	0
Hemorrhage	2 (3.3)	0	0
Aural hematoma	1 (1.6)	0	0
Epistaxis	1 (1.6)	0	0
Neurology	2 (3.3)	0	0
Anxiety	1 (1.6)	0	0
Irritability	1 (1.6)	0	0
Pain	1 (1.6)	0	0
Stomach pain	1 (1.6)	0	0
Allergy	0	1 (1.6)	0
Dermatitis	0	1 (1.6)	0
Dermatology	0	1 (1.6)	0
Urticaria	0	1 (1.6)	0
Musculoskeletal	1 (1.6)	0	0
Tail dysfunction	1 (1.6)	0	0

[a]Expressed as N and percent of 61 total AEs.

confirmed the reliability and reproducibility of blinded owner-assessed itch using PVAS.

The median PVAS improvement in the 0.014×1 mg/kg cohort ($N = 15$) due to LD-aminopterin was 61% of the median day 14 PVAS improvement due to prednisolone in the total prednisolone-treated population ($N = 62$). However, this difference was not significant ($P = 0.22$). Of the 62 prednisolone-treated subjects, 21 (33.9%) had robust improvement in PVAS≥4 points at day 14. Among the 0.007×1 mg/kg and 0.014×1 mg/kg cohorts, 7 of 30 subjects (23.3%, all with nonseasonal disease) responded at day 84 with improvement in PVAS≥4 points. The fraction of subjects with improvement in PVAS≥4 after LD-aminopterin was not significantly different than after prednisolone ($P = 0.34$). Of the 7 subjects with improvement in PVAS≥4 after LD-aminopterin, 6 were treated with prednisolone, and had a mean improvement due to prednisolone substantially the same as that seen for the larger ($N = 62$) prednisolone-treated population (2.7 ± 2.1 versus 2.9 ± 2.4, respectively). In these 6 subjects, the mean ($\pm$SD) improvement in PVAS due to LD-aminopterin was significantly (77%) greater than from prednisolone (4.8 ± 0.7 versus 2.7 ± 2.1, $P < 0.05$).

LD-Aminopterin is safe and well-tolerated in canine AD

Blinded trial. There was no relationship between clinical (Table 3) or laboratory (Table 4) adverse events (AEs), and either dose or schedule. The incidence of AEs in LD-aminopterin treated cohorts was similar to placebo. The most frequently reported AEs (≥5% of 61 total) across all cohorts were gastrointestinal in nature (45.9% [N = 28]): diarrhea (26.2% [N = 16]) and anorexia (6.6% [N = 4]). All were mild in intensity and self-limiting. Abnormalities in liver function as measured by elevations in serum alanine transaminase were most common in placebo, and in all cases were mild and transient (Table 4). The incidence of AEs decreased as a function of time (Table 5). There were no serious AEs, or AEs that led investigators to discontinue study drug, reduce dose, or deviate from protocol.

Open-label extension. Of the 75 subjects enrolled in the blinded trial, 62 (83%) enrolled in the open-label extension. The doses used in the open-label extension were 0.007 mg/kg (19%), 0.014 mg/kg (57%) and 0.021 mg/kg (24%). Including the 12 weeks of treatment in the blinded trial, 40 (65%) and 23 (37%) subjects were treated for more than 57 and 84 weeks, respectively. The drug was well-tolerated during chronic therapy. There were no clinical serious adverse events or deaths. No clinically significant laboratory adverse events occurred, and there was no dose-dependent trend in the incidence of adverse events for any laboratory test (Table S1). There was no laboratory adverse events that required discontinuation of study drug.

Discussion

This placebo-controlled study examined how dose and schedule of the investigational antifolate LD-aminopterin affected efficacy and safety in canine AD. Oral LD-aminopterin 0.014 mg/kg given once weekly resulted in efficacy in moderate-to-severe canine AD after 8–12 weeks of treatment, causing a significant reduction in GS and PVAS. An exploratory analysis identified ~25% of subjects who were highly responsive to the anti-pruritic effect of LD-aminopterin, and enjoyed a significantly larger mean reduction in itch (65%) than from two weeks of daily prednisolone (4.8 versus 2.7 point reduction, or 77% greater). CADESI was also significantly reduced, but only before correcting for multiple comparisons. CADESI was reduced in placebo but not significantly, an effect likely due to permitted antimicrobials [51], and/ or carry-over effects of prednisolone used in the first 14 days per protocol [52].

Surprisingly, all efficacy endpoints for twice-weekly 0.007 mg/ kg LD-aminopterin were no better than placebo, and worse than all once-weekly schedules. This held whether the once-weekly schedule provided the same daily (0.007 mg/kg) or total weekly (0.014 mg/kg) dose. Based on CADESI, twice-weekly dosing was even worse than placebo, though not significantly. These findings were unexpected and suggest that the schedule of antifolate administration is critical, with a minimum interval between dosings required for efficacy in canine AD.

Like methotrexate, the L-enantiomer of LD-aminopterin potently inhibits dihydrofolate reductase (Figure 1B) [17,53], which results in the rapid accumulation of dihydrofolate poly-glutamates that may reach 20% (~2 μM) of total intracellular folates from an initial undetectable level [21]. Dihydrofolate polyglutamates at these concentrations are capable of inhibiting the first committed step of purine biosynthesis catalyzed by PPAT and the two transformylase reactions catalyzed by GART and AICART [22]. In addition to dihydrofolate polyglutamates, methotrexate polyglutamates have also been implicated as effectors of inhibition of these three steps of de novo purine synthesis [22,33,34,54]. Although AICART inhibition and the accumulation of AICAR and its metabolite AICA have been proposed to mediate anti-inflammatory effects

[26,27,28,34,36,37,39], in vitro studies with leukemia cells and primary human T lymphocytes indicate that PPAT is the primary site of inhibition of purine biosynthesis by methotrexate [22,55]. In particular, levels of 5-phosphoribosyl-1-pyrophosphate, the natural PPAT substrate, increase 5-10-fold from 3 to 12 hours in cells exposed in culture to methotrexate at a concentration (0.1 μM) obtained in the plasma of patients undergoing therapy for inflammation, before decreasing to control levels after 24 hours [56,57]. Thus, methotrexate inhibits PPAT, GART and AI-CART, but empirically induces AICAR accumulation in patients [58,59]. Accumulated AICAR may therefore be derived from either selective inhibition of AICART at low antifolate concentrations [40], or from the pools of intermediates that exist between GART and AICART if both enzymes are inhibited non-selectively [22]. In the latter case, the abundance of intermediates may vary from patient to patient, potentially accounting in part for the variability in antifolate clinical efficacy. Another possibility is that AICAR is derived from FAICAR if the accumulation of dihydrofolate polyglutamates causes the AICART reaction to run backward, as suggested by the fact that the equilibrium of this reaction actually lies in the direction of AICAR formation [60].

Persistent inhibition of PPAT, GART and AICART in subjects would be expected to abrogate the downstream accumulation of AICAR and its AICA metabolite, since each would be eliminated from the body without precursors available for the synthesis of additional AICAR [61]. In patients given a single standard anti-inflammatory dose of methotrexate, Smolenska et al. demonstrated rapid inhibition of de novo purine biosynthesis that was sustained for at least 24–48 hours but that fully reversed by one week after dosing, kinetics that suggest twice-weekly dosing may lead to persistent inhibition of de novo purine synthesis [62]. If the anti-inflammatory effect of LD-aminopterin in AD is due to AICAR, an optimal schedule of therapy would require sufficient time between drug pulses to allow enzymes to cycle between states of complete and incomplete inhibition in order to regenerate intermediates in de novo purine synthesis and maintain optimally elevated and efficacious levels of AICAR. This mechanism could explain why twice-weekly dosing in this study negated efficacy in AD, despite having the same daily and weekly dose as effective once-weekly regimens.

Support for this model comes from recent in vitro studies carried out by Funk et al., who demonstrated a 115-fold increase in AICAR following exposure of an erythroblastoid cell line to 10 nM MTX, but subsequently decreased with increasing MTX concentrations, declining to baseline levels with 1000 nM MTX [40]. In contrast, dUMP displayed concentration-dependent accumulation. These observations led these investigators to predict clinical anti-inflammatory responses due to AICAR might be paradoxically related to antifolate dose, whereas a dose-proportional response would be seen if due to inhibition of thymidylate synthase. Toxicity is observed in all subjects administered a sufficiently high dose of LD-aminopterin or methotrexate [12,14], consistent with the proposal that antifolate toxicity is mediated by thymidylate synthase inhibition [40]. In contrast, the dose-response data for efficacy in this study mirrors the in vitro concentration-response findings for AICAR described by Funk et al. [40], suggesting that LD-aminopterin efficacy in AD is mediated by AICAR accumulation.

Clinical evidence supportive of this model in humans comes from Radmanesh and colleagues, who observed greater efficacy in psoriatics treated with weekly methotrexate given on a single day in three doses (3×5 mg) than when the same weekly dose was administered equally over six days (6×2.5 mg) [63]. Likewise, stepwise increases in methotrexate dose in patients with juvenile idiopathic arthritis who were nonresponders to standard low-dose methotrexate did not result in improved clinical outcomes [64].

The safety of LD-aminopterin in canine AD was also examined. In contrast to efficacy, there was no relationship between safety and either dose or schedule of administration. As discussed above, the discordance between efficacy and toxicity in relation to dose supports distinct mechanisms for each, as previously suggested by *in vitro* studies [40]. The incidence of AEs in cohorts treated with LD-aminopterin were similar to one another and to the placebo-treated group. In a previous dose-ranging toxicology study in the canine [12], we determined that 0.2 mg/kg L-aminopterin given once-weekly was the lowest dose that caused the first signs of mild toxicity. Thus, the optimal therapeutic dose identified in this trial establishes a therapeutic index with a 14-fold margin of safety.

Subjects from the blinded trial were also optionally able to continue on LD-aminopterin in an open-label extension lasting up to 104 weeks. Of the 75 subjects enrolled in the blinded trial, 62 (83%) enrolled in the extension. The doses used in the extension were 0.007 mg/kg (19%), 0.014 mg/kg (57%) and 0.021 mg/kg (24%). Including the 12 weeks of treatment in the blinded trial, 40 (65%) and 23 (37%) subjects were treated for more than 57 and 84 weeks, respectively. The drug was well-tolerated during chronic therapy and no adverse event required discontinuation of study drug.

The safety profile of weekly methotrexate in the canine at anti-inflammatory doses is not well defined. Weekly treatment of five dogs with CAD with an oral anti-inflammatory dose of methotrexate (0.2 mg/kg) for four weeks resulted in severe vomiting in one subject and fatal hepatic necrosis in two subjects (personal communication by Dr. Thierry Olivry, North Carolina State University). Pond and Morrow reported a similar case of fatal hepatic necrosis in a dog with osteosarcoma treated with methotrexate at an oral dose of 5 mg/m^2 (0.25 mg/kg) on the first four days of each week [65]. A four-week toxicology study of LD-aminopterin, L-aminopterin and D-aminopterin in beagle dogs ($N = 6$ per cohort, once-weekly oral gavage of 0.5 mg/kg of each enantiomer or 35-fold the anti-inflammatory dose) found no liver histopathology in any cohort (unpublished data). Although data from controlled studies are needed, these observations suggest methotrexate and LD-aminopterin may have different therapeutic indices in the canine.

Options for systemic treatment of human AD include azathioprine, cyclosporine, and methotrexate [5]. A systematic review and meta-analysis of 15 studies and 602 patients determined that cyclosporine consistently decreased the severity of AD [66]. The pooled mean decrease in disease severity was 22% (95% CI, 8–36%) under low-dose cyclosporine (3 mg/kg), and 40% (95%-CI 29–51%) at dosages ≥4 mg/kg. Although effective, a proportion of patients discontinue cyclosporine because of ineffectiveness or side effects, and long-term use raises concerns of nephrotoxicity [67].

Methotrexate has fewer safety concerns than cyclosporine in humans, and was shown in open-label and randomized controlled trials to be an effective treatment of AD [42,68]. An open-label study evaluated the efficacy and safety of low-dose methotrexate (7.5 mg/week) and cyclosporine (2.5 mg/kg/day) in the treatment of severe AD, and determined there was no statistically significant difference in disease reduction between treatments [41].

Cyclosporine is FDA approved in the United States and elsewhere in the world for the control of CAD. In the pivotal efficacy field trial, four weeks of daily cyclosporine (5 mg/kg) gave a mean (baseline:endpoint) reduction in CADESI (0–360 scale) and PVAS (0–5 scale) in the intent-to-treat population ($N = 262$) of 31.5 (79.0:47.5) and 1.36 (3.75:2.39), respectively [51]. The data from this study show that once-weekly LD-aminopterin (0.014 mg/kg, $N = 15$) resulted in a mean (baseline:endpoint) reduction in CADESI of 53 (159:107), and a reduction in PVAS of 1.9 (7.4:5.5). Qualitatively, cyclosporine and LD-aminopterin appear to have a similar effect on CAD disease activity. Any formal comparison would require a well-controlled and properly powered head-to-head study.

LD-aminopterin may thus provide an additional therapeutic option to treat AD, but with a better safety profile than either methotrexate [16,17,18,19,20], or cyclosporine. The efficacy and safety data for LD-aminopterin from this study go toward supporting the rationale for a human trial and provide insights for optimal antifolate dosing in human AD.

Supporting Information

Table S1 Clinical laboratory adverse events in the open-label trial segment.

Dataset S1 GS scores.

Dataset S2 Percent change in baseline GS scores.

Dataset S3 PVAS scores.

Dataset S4 CADESI scores.

Acknowledgments

We thank Dr. Thierry Olivry and staff at North Carolina State University for enrolling subjects and Dr. Hong Yee and David Coblentz at DF/Net Research, Inc. for biostatistical advice.

Author Contributions

Conceived and designed the experiments: JAZ SJK. Performed the experiments: AM LM CEG. Analyzed the data: JAZ SJK ADS. Contributed reagents/materials/analysis tools: ADS SJK. Wrote the paper: JAZ SJK. Contributed to critical revisions: AM LM CEG ADS.

References

1. Williams H, Robertson C, Stewart A, Ait-Khaled N, Anabwani G, et al. (1999) Worldwide variations in the prevalence of symptoms of atopic eczema in the International Study of Asthma and Allergies in Childhood. J Allergy Clin Immunol 103: 125–138.
2. Sohn A, Frankel A, Patel RV, Goldenberg G (2011) Eczema. Mt Sinai J Med 78: 730–739.
3. Leung DY, Boguniewicz M, Howell MD, Nomura I, Hamid QA (2004) New insights into atopic dermatitis. J Clin Invest 113: 651–657.
4. Brown S, Reynolds NJ (2006) Atopic and non-atopic eczema. BMJ 332: 584–588.
5. Denby KS, Beck LA (2012) Update on systemic therapies for atopic dermatitis. Curr Opin Allergy Clin Immunol 12: 421–426.
6. Proudfoot LE, Powell AM, Ayis S, Barbarot S, Baselgatorres E, et al. (2013) The European treatment of severe atopic eczema in children taskforce (TREAT) survey. Br J Dermatol advance online publication, 16 Jul 2013 doi:10.1111/bjd.12505.
7. Chakravarty K, McDonald H, Pullar T, Taggart A, Chalmers R, et al. (2008) BSR/BHPR guideline for disease-modifying anti-rheumatic drug (DMARD) therapy in consultation with the British Association of Dermatologists. Rheumatology (Oxford) 47: 924–925.

8. Ryan C, Amor KT, Menter A (2010) The use of cyclosporine in dermatology: part II. J Am Acad Dermatol 63: 949–972; quiz 973–944.

9. Parakkal D, Sifuentes H, Semer R, Ehrenpreis ED (2011) Hepatosplenic T-cell lymphoma in patients receiving TNF-alpha inhibitor therapy: expanding the groups at risk. Eur J Gastroenterol Hepatol 23: 1150–1156.

10. Barker J, Horn EJ, Lebwohl M, Warren RB, Nast A, et al. (2011) Assessment and management of methotrexate hepatotoxicity in psoriasis patients: report from a consensus conference to evaluate current practice and identify key questions toward optimizing methotrexate use in the clinic. J Eur Acad Dermatol Venereol 25: 758–764.

11. Zebala J, Maeda DY, Morgan JR, Kahn SJ (2013) Pharmaceutical composition comprising racemic aminopterin. U.S. Patent No. 8,349,837.

12. Menter A, Thrash B, Cherian C, Matherly LH, Wang L, et al. (2012) Intestinal transport of aminopterin enantiomers in dogs and humans with psoriasis is stereoselective: evidence for a mechanism involving the proton-coupled folate transporter. J Pharmacol Exp Ther 342: 696–708.

13. Cole PD, Drachtman RA, Smith AK, Cate S, Larson RA, et al. (2005) Phase II trial of oral aminopterin for adults and children with refractory acute leukemia. Clin Cancer Res 11: 8089–8096.

14. Ratliff AF, Wilson J, Hum M, Marling-Cason M, Rose K, et al. (1998) Phase I and pharmacokinetic trial of aminopterin in patients with refractory malignancies. J Clin Oncol 16: 1458–1464.

15. Smith A, Hum M, Winick NJ, Kamen BA (1996) A case for the use of aminopterin in treatment of patients with leukemia based on metabolic studies of blasts in vitro. Clin Cancer Res 2: 69–73.

16. Cole PD, Beckwith KA, Vijayanathan V, Roychowdhury S, Smith AK, et al. (2009) Folate homeostasis in cerebrospinal fluid during therapy for acute lymphoblastic leukemia. Pediatr Neurol 40: 34–41.

17. Cole PD, Zebala JA, Alcaraz MJ, Smith AK, Tan J, et al. (2006) Pharmacodynamic properties of methotrexate and Aminotrexate during weekly therapy. Cancer Chemother Pharmacol 57: 826–834.

18. Li Y, Vijayanathan V, Gulinello M, Cole PD (2010) Intrathecal methotrexate induces focal cognitive deficits and increases cerebrospinal fluid homocysteine. Pharmacol Biochem Behav 95: 428–433.

19. Li Y, Vijayanathan V, Gulinello ME, Cole PD (2010) Systemic methotrexate induces spatial memory deficits and depletes cerebrospinal fluid folate in rats. Pharmacol Biochem Behav 94: 454–463.

20. Vijayanathan V, Gulinello M, Ali N, Cole PD (2011) Persistent cognitive deficits, induced by intrathecal methotrexate, are associated with elevated CSF concentrations of excitotoxic glutamate analogs and can be reversed by an NMDA antagonist. Behav Brain Res 225: 491–497.

21. Allegra CJ, Fine RL, Drake JC, Chabner BA (1986) The effect of methotrexate on intracellular folate pools in human MCF-7 breast cancer cells. J Biol Chem 261: 6478–6485.

22. Sant ME, Lyons SD, Phillips L, Christopherson RI (1992) Antifolates induce inhibition of amido phosphoribosyltransferase in leukemia cells. J Biol Chem 267: 11038–11045.

23. Genestier L, Paillot R, Fournel S, Ferraro C, Miossec P, et al. (1998) Immunosuppressive properties of methotrexate: apoptosis and clonal deletion of activated peripheral T cells. J Clin Invest 102: 322–328.

24. Paillot R, Genestier L, Fournel S, Ferraro C, Miossec P, et al. (1998) Activation-dependent lymphocyte apoptosis induced by methotrexate. Transplant Proc 30: 2348–2350.

25. Quemeneur L, Gerland LM, Flacher M, Ffrench M, Revillard JP, et al. (2003) Differential control of cell cycle, proliferation, and survival of primary T lymphocytes by purine and pyrimidine nucleotides. J Immunol 170: 4986–4995.

26. Cronstein BN, Eberle MA, Gruber HE, Levin RI (1991) Methotrexate inhibits neutrophil function by stimulating adenosine release from connective tissue cells. Proc Natl Acad Sci USA 88: 2441–2445.

27. Cronstein BN, Naime D, Ostad E (1993) The antiinflammatory mechanism of methotrexate. Increased adenosine release at inflamed sites diminishes leukocyte accumulation in an in vivo model of inflammation. J Clin Invest 92: 2675–2682.

28. Cutolo M, Sulli A, Pizzorni C, Seriolo B, Straub RH (2001) Anti-inflammatory mechanisms of methotrexate in rheumatoid arthritis. Ann Rheum Dis 60: 729–735.

29. Heijden JV, Assaraf Y, Gerards A, Oerlemans R, Lems W, et al. (2013) Methotrexate analogues display enhanced inhibition of TNF-alpha production in whole blood from RA patients. Scand J Rheumatol.

30. Herman S, Zurgil N, Langevitz P, Ehrenfeld M, Deutsch M (2003) The induction of apoptosis by methotrexate in activated lymphocytes as indicated by fluorescence hyperpolarization: a possible model for predicting methotrexate therapy for rheumatoid arthritis patients. Cell Struct Funct 28: 113–122.

31. Spurlock CF 3rd, Aune ZT, Tossberg JT, Collins PL, Aune JP, et al. (2011) Increased sensitivity to apoptosis induced by methotrexate is mediated by JNK. Arthritis Rheum 63: 2606–2616.

32. Swierkot J, Miedzybrodzki R, Szymaniec S, Szechinski J (2004) Activation dependent apoptosis of peripheral blood mononuclear cells from patients with rheumatoid arthritis treated with methotrexate. Ann Rheum Dis 63: 599–600.

33. Allegra CJ, Drake JC, Jolivet J, Chabner BA (1985) Inhibition of phosphoribosylaminoimidazolecarboxamide transformylase by methotrexate and dihydrofolic acid polyglutamates. Proc Natl Acad Sci USA 82: 4881–4885.

34. Baggott JE, Vaughn WH, Hudson BB (1986) Inhibition of 5-aminoimidazole-4-carboxamide ribotide transformylase, adenosine deaminase and 5-adenylate

35. Chan ES, Cronstein BN (2010) Methotrexate–how does it really work? Nat Rev Rheumatol 6: 175–178.

36. Baggott JE, Morgan SL, Ha TS, Alarcon GS, Koopman WJ, et al. (1993) Antifolates in rheumatoid arthritis: a hypothetical mechanism of action. Clin Exp Rheumatol 11 Suppl 8: S101–105.

37. Ha T, Baggott JE (1994) 5-aminoimidazole-4-carboxamide ribotide (AICAR) and its metabolites: metabolic and cytotoxic effects and accumulation during methotrexate treatment. J Nutr Biochem 5: 522.

38. Guigas B, Taleux N, Foretz M, Detaille D, Andreelli F, et al. (2007) AMP-activated protein kinase-independent inhibition of hepatic mitochondrial oxidative phosphorylation by AICA riboside. Biochem J 404: 499–507.

39. Katerelos M, Mudge SJ, Stapleton D, Auwardt RB, Fraser SA, et al. (2010) 5-aminoimidazole-4-carboxamide ribonucleoside and AMP-activated protein kinase inhibit signalling through NF-kappaB. Immunol Cell Biol 88: 754–760.

40. Funk RS, van Haandel L, Becker ML, Leeder JS (2013) Low-dose methotrexate results in the selective accumulation of aminoimidazole carboxamide ribotide in an erythroblastoid cell line. J Pharmacol Exp Ther 347: 154–163.

41. El-Khalawany MA, Hassan H, Shaaban D, Ghonaim N, Eassa B (2013) Methotrexate versus cyclosporine in the treatment of severe atopic dermatitis in children: a multicenter experience from Egypt. Eur J Pediatr 172: 351–356.

42. Schram ME, Roekevisch E, Leeflang MM, Bos JD, Schmitt J, et al. (2011) A randomized trial of methotrexate versus azathioprine for severe atopic eczema. J Allergy Clin Immunol 128: 353–359.

43. Marsella R, Girolomoni G (2009) Canine models of atopic dermatitis: a useful tool with untapped potential. J Invest Dermatol 129: 2351–2357.

44. Olivry T, Paps JS, Bizikova P, Murphy KM, Jackson HA, et al. (2007) A pilot open trial evaluating the efficacy of low-dose aminopterin in the canine homologue of human atopic dermatitis. Br J Dermatol 157: 1040–1042.

45. DeBoer DJ, Hillier A (2001) The ACVD task force on canine atopic dermatitis (XV): fundamental concepts in clinical diagnosis. Vet Immunol Immunopathol 81: 271–276.

46. Willemse T (1986) Atopic skin disease: a review and a reconsideration of diagnostic criteria. J Small Anim Pract 27: 771–778.

47. Olivry T, Marsella R, Iwasaki T, Mueller R, International Task Force On Canine Atopic Dermatitis (2007) Validation of CADESI-03, a severity scale for clinical trials enrolling dogs with atopic dermatitis. Vet Dermatol 18: 78–86.

48. Rybnicek J, Lau-Gillard PJ, Harvey R, Hill PB (2009) Further validation of a pruritus severity scale for use in dogs. Vet Dermatol 20: 115–122.

49. Hsu JC (1996) Multiple Comparisons: Theory and Methods. London: Chapman & Hall. 277 p.

50. Olivry T, Mueller R, Nuttall T, Favrot C, Prelaud P, et al. (2008) Determination of CADESI-03 thresholds for increasing severity levels of canine atopic dermatitis. Vet Dermatol 19: 115–119.

51. Steffan J, Parks C, Seewald W, North American Veterinary Dermatology Cyclosporine Study Group (2005) Clinical trial evaluating the efficacy and safety of cyclosporine in dogs with atopic dermatitis. J Am Vet Med Assoc 226: 1855–1863.

52. Steffan J, Horn J, Gruet P, Strehlau G, Fondati A, et al. (2004) Remission of the clinical signs of atopic dermatitis in dogs after cessation of treatment with cyclosporin A or methylprednisolone. Vet Rec 154: 681–684.

53. Skipper HE, Mitchell JH, Bennett LL (1950) Inhibition of nucleic acid synthesis by folic acid antagonists. Cancer Res 10: 510–512.

54. Lyons SD, Christopherson RI (1991) Antifolates induce primary inhibition of the de novo purine pathway prior to 5-aminoimidazole-4-carboxamide ribotide transformylase in leukemia cells. Biochem Int 24: 187–197.

55. Fairbanks LD, Ruckemann K, Qiu Y, Hawrylowicz CM, Richards DF, et al. (1999) Methotrexate inhibits the first committed step of purine biosynthesis in mitogen-stimulated human T-lymphocytes: a metabolic basis for efficacy in rheumatoid arthritis? Biochem J 342 (Pt 1): 143–152.

56. Buesa-Perez JM, Leyva A, Pinedo HM (1980) Effect of methotrexate on 5-phosphoribosyl 1-pyrophosphate levels in L1210 leukemia cells in vitro. Cancer Res 40: 139–144.

57. Kamal MA, Christopherson RI (2004) Accumulation of 5-phosphoribosyl-1-pyrophosphate in human CCRF-CEM leukaemia cells treated with antifolates. Int J Biochem Cell Biol 36: 545–551.

58. Baggott JE, Morgan SL, Sams WM, Linden J (1999) Urinary adenosine and aminoimidazolecarboxamide excretion in methotrexate-treated patients with psoriasis. Arch Dermatol 135: 813–817.

59. Morgan SL, Oster RA, Lee JY, Alarcon GS, Baggott JE (2004) The effect of folic acid and folinic acid supplements on purine metabolism in methotrexate-treated rheumatoid arthritis. Arthritis Rheum 50: 3104–3111.

60. Wall M, Shim JH, Benkovic SJ (2000) Human AICAR transformylase: role of the 4-carboxamide of AICAR in binding and catalysis. Biochemistry 39: 11303–11311.

61. Dixon R, Fujitaki J, Sandoval T, Kisicki J (1993) Acadesine (AICA-riboside): disposition and metabolism of an adenosine-regulating agent. J Clin Pharmacol 33: 955–958.

62. Smolenska Z, Kaznowska Z, Zarowny D, Simmonds HA, Smolenski RT (1999) Effect of methotrexate on blood purine and pyrimidine levels in patients with rheumatoid arthritis. Rheumatology (Oxford) 38: 997–1002.

63. Radmanesh M, Rafiei B, Moosavi ZB, Sina N (2011) Weekly vs. daily administration of oral methotrexate (MTX) for generalized plaque psoriasis: a randomized controlled clinical trial. Int J Dermatol 50: 1291–1293.

64. Ruperto N, Murray KJ, Gerloni V, Wulffraat N, de Oliveira SK, et al. (2004) A randomized trial of parenteral methotrexate comparing an intermediate dose with a higher dose in children with juvenile idiopathic arthritis who failed to respond to standard doses of methotrexate. Arthritis Rheum 50: 2191–2201.

65. Pond EC, Morrow D (1982) Hepatotoxicity associated with methotrexate therapy in a dog. J small Anim Pract 23: 659–666.

66. Schmitt J, Schmitt N, Meurer M (2007) Cyclosporin in the treatment of patients with atopic eczema - a systematic review and meta-analysis. J Eur Acad Dermatol Venereol 21: 606–619.

67. Behnam SM, Behnam SE, Koo JY (2005) Review of cyclosporine immunosuppressive safety data in dermatology patients after two decades of use. J Drugs Dermatol 4: 189–194.

68. Weatherhead SC, Wahie S, Reynolds NJ, Meggitt SJ (2007) An open-label, dose-ranging study of methotrexate for moderate-to-severe adult atopic eczema. Br J Dermatol 156: 346–351.

69. Baggott JE, Tamura T (2010) Evidence for the hypothesis that 10-formyldihydrofolate is the in vivo substrate for aminoimidazolecarboxamide ribotide transformylase. Exp Biol Med (Maywood) 235: 271–277.

70. Seither RL, Trent DF, Mikulecky DC, Rape TJ, Goldman ID (1989) Folate-pool interconversions and inhibition of biosynthetic processes after exposure of L1210 leukemia cells to antifolates. Experimental and network thermodynamic analyses of the role of dihydrofolate polyglutamylates in antifolate action in cells. J Biol Chem 264: 17016–17023.

4

Role of Somatostatin Receptor-2 in Gentamicin-Induced Auditory Hair Cell Loss in the Mammalian Inner Ear

Yves Brand[᾽], Vesna Radojevic[᾽], Michael Sung, Eric Wei, Cristian Setz, Andrea Glutz, Katharina Leitmeyer, Daniel Bodmer*

Department of Biomedicine, University Hospital Basel, Basel, Switzerland and Clinic for Otolaryngology, Head and Neck Surgery, University Hospital Basel, Basel, Switzerland

Abstract

Hair cells and spiral ganglion neurons of the mammalian auditory system do not regenerate, and their loss leads to irreversible hearing loss. Aminoglycosides induce auditory hair cell death *in vitro*, and evidence suggests that phosphatidylinositol-3-kinase/Akt signaling opposes gentamicin toxicity via its downstream target, the protein kinase Akt. We previously demonstrated that somatostatin—a peptide with hormone/neurotransmitter properties—can protect hair cells from gentamicin-induced hair cell death *in vitro*, and that somatostatin receptors are expressed in the mammalian inner ear. However, it remains unknown how this protective effect is mediated. In the present study, we show a highly significant protective effect of octreotide (a drug that mimics and is more potent than somatostatin) on gentamicin-induced hair cell death, and increased Akt phosphorylation in octreotide-treated organ of Corti explants *in vitro*. Moreover, we demonstrate that somatostatin receptor-1 knockout mice overexpress somatostatin receptor-2 in the organ of Corti, and are less susceptible to gentamicin-induced hair cell loss than wild-type or somatostatin-1/somatostatin-2 double-knockout mice. Finally, we show that octreotide affects auditory hair cells, enhances spiral ganglion neurite number, and decreases spiral ganglion neurite length.

Editor: Prasun K. Datta, Temple University, United States of America

Funding: This study was supported by The Forschungsfonds der Universität Basel (YB) and the Schwerhörigenverein Nordwestschweiz (DB). The funders had no role in study design, data collection and analysis, decision to publish, or preparation of the manuscript.

Competing Interests: The authors have declared that no competing interests exist.

* Email: Daniel.bodmer@usb.ch

᾽ These authors contributed equally to this work.

Introduction

Sensorineural hearing loss is linked to degeneration and death of auditory hair cells (HCs) and their associated spiral ganglion neurons (SGNs), which is irreversible in mammals. Therefore, developing therapeutic strategies for hearing loss prevention requires a better understanding of the survival pathways and molecular events involved in auditory epithelium protection. Until recently, auditory HC and SGN damage has been considered an inevitable consequence of age, genetic conditions, exposure to ototoxic drugs, or certain environmental stimuli. However, several recent studies using *in vitro* aminoglycoside-induced HC death as a model have discovered some of the critical intracellular events that mediate HC damage [1,2,3,4]. Aminoglycoside exposure sets into motion a series of cellular and biochemical alterations in the HCs. Reactive oxygen species have been detected *in vitro* shortly after exposure to aminoglycosides [5]. It has also been demonstrated that small GTPases (e.g., Ras and Rho/Rac/Cdc42) and the c-jun-N-terminal kinase (JNK) signaling pathway are activated in aminoglycoside-exposed cells [6,7,8,9]. Finally, caspases are activated and HCs undergo apoptotic cell death after prolonged aminoglycoside exposure [10,11]. Interestingly, phoshatidylinositol-3-kinase (PI3K) signaling reportedly mediates HC survival and opposes gentamicin toxicity via its downstream target, the protein

kinase Akt [12]. Despite the progress made towards understanding the processes involved in auditory HC death and survival, there is still no available cure for individuals with sensorineural hearing loss; only auditory prosthesis (e.g., hearing aids or cochlear implant) can offer some help to individuals with hearing loss.

The regulatory peptide somatostatin (SST) acts on a wide array of target tissues to modulate neurotransmission, cell secretion, and cell proliferation [13,14]. SST actions are mediated by five subtypes of G protein-coupled receptors (SST receptors 1–5) that are encoded by separate genes [14,15]. These SST receptors modulate several intracellular signaling transduction pathways, including the Mek/Erk, PI3K-Akt, and p38 pathways [16]. Octreotide acetate—a long-acting octapeptide with SST-mimicking pharmacologic actions—inhibits growth hormone, glucagon, and insulin even more potently than SST does. Studies in mice show that SST and its receptors appear to play an important role in cell death. In a retina ischemia model, SST receptor-2 activation protected retinal neurons from damage [17].

Our group has previously studied the somatostatinergic system in the mammalian inner ear [18,19,20]. We demonstrated that SST receptor-1 and -2 are specifically expressed in the outer and inner HCs of the organ of Corti (OC), and in defined supporting cells. Interestingly, SST itself was not expressed in the mammalian cochlea. Most importantly, in *in vitro* studies, we found improved auditory HC survival in OC explants treated with gentamicin and SST compared to in explants treated only with gentamicin. However, the intracellular events mediating these effects remain unknown, and the effects on SGN were not evaluated.

In the current *in vitro* study, we examined whether octreotide could protect mammalian auditory HCs from gentamicin-induced HC death. We also investigated whether this drug increased Akt phosphorylation in the OC. Furthermore, we examined SST receptor-1 knockout mice and confirmed their overexpression of SST receptor-2 in the OC, and evaluated their susceptibility to gentamicin-induced HC loss compared to that of wild-type and SST-1 receptor/SST receptor-2 double-knockout mice. Finally, we evaluated the effects of octreotide on SGN survival and neurite outgrowth.

Material and Methods

Animal procedures

All animal procedures were performed in Basel, Switzerland, following an animal research protocol approved by the Committee on the Ethics of Animal Experiments of Basel (Kantonales Veterinäramt Basel, Permit Number: 2263), in accordance with the European Communities Council Directive of 24 November 1986 (86/609/EEC). Animals were sacrificed prior to all tissue extractions. The procedures used to generate homozygous SST receptor 1 knockout (SST receptor-$1^{-/-}$)/C57BL6J mice and homozygous SST receptor-2 knockout (SST receptor-$2^{-/-}$)/C57BL6J mice have been previously described [21,22]. SST receptor-$1^{-/-}$ mice were crossed with SST receptor-$2^{-/-}$ mice to produce double-knockout mice. Age-matched wild-type mice were produced from the C57BL6J mice used to stabilize the genetic backgrounds of the knockout mice [20]. All experiments used either the above-described mice or Wistar rats (Harlan, Indianapolis, IN, USA).

OC tissue culture

Five-day-old Wistar rat pups (Harlan, Indianapolis, IN, USA) and seven-day-old wild-type, SST receptor-1 knockout, and SST receptor-1/SST receptor-2 double-knockout mice were decapitated, and then cochlear microdissections were performed under a light microscope to isolate the OC and the spiral ganglion (SG) as described by Sobkowicz *et al.* [23]. OCs were initially incubated in cell culture media containing Dulbecco's modified Eagle's medium (DMEM) supplemented with 10% fetal calf serum (FCS), 25 mM HEPES, and 30 U/mL penicillin (Invitrogen, Carlsbad, CA, USA) at 37°C with 5% CO_2, followed by recovery for 24 hours under these conditions.

Next, the OCs were transferred into fresh cell culture media and incubated at 37°C with 5% CO_2 for 20 hours (mouse tissue culture) or for 48 hours with the solution exchanged once after 24 hours (rat tissue culture). HC damage was induced by incubating OCs with gentamicin (Sigma-Aldrich, St. Louis, MO, USA) in the cell culture medium: 48 hours with 50 µM gentamicin for rat tissue cultures, or 20 hours with 0.5 mM gentamicin for mouse tissue culture. Then the rat tissue OCs were pretreated for 24 hours with increasing amounts of octreotide (Novartis Pharma, Switzerland), with final concentrations of 1 µM or 5 µM in the cell culture medium. After this pretreatment, rat OCs were exposed either to 50 µM gentamicin and 1 µM octreotide, 50 µM gentamicin and 5 µM octreotide, or only 5 µM octreotide for 48 hours. Other Wistar rat OCs were not pretreated and were incubated for 48 hours either in culture medium alone (controls) or in culture medium with 50 µM gentamicin. Mouse tissue cultures were prepared similarly but without addition of octreotide to the cell culture media.

Finally, additional experiments using the Akt-inhibitor SH-6 (AG Scientific, San Diego, CA, USA) were performed. Rat OCs were transferred after recovery as described above into fresh culture media and incubated at 37°C with 5% CO_2 for 48 hours. For the gentamicin-alone condition, explants were exposed to culture medium containing 200 µM gentamicin. A new gentamicin batch was used and gentamicin dose was adjusted to provide the same degree of HC damage as in the rat tissue experiments above. It is known that gentamicin toxicity various between different batches of gentamicin [12]. The following conditions were applied: OCs were pretreated for 24 hours with 5 µM octreotide and changed to cell culture media containing 200 µM gentamicin and 5 µM octreotide for 48 hours. OCs were pretreated for 24 hours with 5 µM octreotide and 10 µM SH-6 and changed to cell culture media containing 200 µM gentamicin, 5 µM octreotide and 10 µM SH-6 for 48 hours.

OCs were pretreated for 24 hours with 10 µM SH-6 and changed to cell culture media containing 10 µM SH-6 for 48 hours. In all conditions, the solution was exchanged once after 24 hours.

HC count and statistical analysis

OCs were fixed in 4% paraformaldehyde, and permeabilized with 5% Triton X-100 in phosphate-buffered saline (PBS) containing 10% FCS. The OCs were then incubated with a 1:100 dilution of Texas Red X-phalloidin (Molecular Probes, Eugene, OR, USA) for 45 minutes at room temperature. After fixation, the OCs were visualized and photographed using a fluorescence microscope (Olympus IC71, Center Valley, PA, USA).

Quantitative analysis was performed by evaluating 60 OHCs associated with 20 IHCs in a given microscopic field. Explants were randomly analyzed for the middle and basal turn, with three random microscope fields counted and averaged for each explant. These values were then averaged across the six replications of each experiment in rat OC cultures, and across the 15 replicates of each experiment in mouse OC cultures. Inner HCs and outer HCs were counted and used to assess HC survival. HC counting results were analyzed by analysis of variance (ANOVA), followed by the least-significant difference (LSD) post-hoc test (Stat View 5.0). Differences associated with p values of <0.05 were considered to be statistically significant. All data are presented as mean ± SD.

Assessment of Akt Activation

To assess activation of the PIK3/Akt signaling pathway for each condition, we harvested 10 intact OCs from 5-day-old Wistar rat pups (Harlan) and incubated them overnight in cell culture media, as described above. Subsequently, OC explants were placed in cell culture media with or without 5 µg/mL octreotide (Novartis Pharma) for one hour. Explants were separately collected from the media and lysed with 150 µL T-Per Tissue Protein Extraction Reagent (Thermo Scientific, Rockford, IL, USA) containing 1× phosphatase and protease inhibitors (Roche, Indianapolis, IN, USA). The samples were centrifuged at 14,000 rpm for 10 minutes to separate the cytosolic portion from the membranous components. Supernatants containing the proteins were sonicated for 5 seconds to shear chromosomal DNA. The protein level was then determined using the bicinchoninic acid (BCA) method (Thermo Scientific, Rockford, IL, USA). The samples were diluted 1:4 in 4× NuPAGE LDS sample buffer and 1:10 in 10× NuPAGE reducing agent (Invitrogen Life Technologies, Green Island, New York, USA), and then heated to 95°C for 5 min, and placed on ice to cool. Next, 10 µg of each sample was separated by SDS-PAGE in a 9% acrylamide gel, and electrotransferred to a nitrocellulose membrane. The membrane was blocked with 3% TopBlock (Lubio Science, Lucerne, Switzerland) in TBS-Tween (50 mM

Tris-HCL, pH 7.4; 150 mM NaCl; 0.05% Tween 20) for 1 hour at room temperature. Blots were incubated with primary antibodies in 3% TopBlock/TBS-Tween overnight at 4°C, and then incubated 1 hour with α-DyLight800-coupled α-rabbit and α-mouse antibodies (Pierce). Bands were visualized using an infrared-based laser scanner (LiCor), and blots were evaluated with rabbit anti-p-Akt and rabbit anti-total Akt (Cell Signaling Technology, Beverly, MA, USA), as well as with mouse anti-β-actin (Abcam, Cambridge, UK) as a loading control. The intensity of the bands corresponding to p-Akt were quantified. Band intensity for p-Akt was corrected for intensity of total Akt and then expressed as the percentage increase, compared with non-treated tissue. Western blotting was replicated three times with independent biological replicate. With each biological replicate, western blotting was performed twice. 10 OCs were used per individual blot. Ratio data were analyzed using the Mann–Whitney nonparametric statistical test.

RNA extraction

The cochleas of 14-day-old and 21-day-old wild-type and SST receptor-1 knockout mouse pups were microdissected and then the OC was isolated. Then the OC was placed separately in RNAlater (Qiagen, Hombrechtikon, Switzerland). RNA was isolated using the RNAeasy Minikit (Qiagen) and an Ultra-Turrax T8 tissue homogenizer (IKA-Werke, Staufen, Germany) following the manufacturer's instructions, including DNase treatment. Isolated RNA quantity and quality were determined with a NanoDrop ND 1000 (NanoDrop Technologies, Delaware, USA). All samples had a 260/280-nm ratio of between 1.8 and 2.1.

Real-time PCR

Total RNA (500 ng) was reverse transcribed into cDNA using the First Strand cDNA synthesis kit (Roche Applied Biosciences) following the manufacturer's instructions. The reaction was performed in an ABI Prism 7900 HT Sequence Detection System (Applied Biosystems) using Fast Start Universal SYBR Green Master (Rox; Roche Applied Biosciences Foster City, USA) and 300 nM primer per reaction. The primer sequences were as follows: SSTR2-fwd, 5'-TCTTTGCTTGGTCAAGGTGA-3'; and SSTR2-rev, 5'-TCCTGCTTACTGTCGCTCCT-3' (Microsynth, St. Gallen, Switzerland). The reaction conditions were as follows: 95°C for 10 min, followed by 40 cycles of 95°C for 15 s and 60°C for 60 s.

Statistical analysis of real-time PCR

The relative quantities of specifically amplified cDNA were calculated using the comparative threshold cycle method, with GAPDH used as an endogenous reference (Microsynth). Template-free and reverse-transcription-free controls were analyzed to exclude non-specific amplification and DNA contamination. One-way analyses of variance (ANOVA) followed by Student's t-test with post-hoc Bonferroni's correction was performed. Means were considered significant when $p < 0.05$. The Origin computer program (Microcal Software, Inc., Northampton, MA) was used for statistical analysis and to generate graphs.

Preparation of tissue culture plates for rat SGN experiments

To prepare uniformly coated 24-well cell culture plates (Costar, Corning Inc., Acton, MA, USA), wells were filled with 300 μL of 5 μg/mL poly-L-lysine (PLL) (Sigma-Aldrich) in DMEM (Gibco by Invitrogen, Carlsbad, USA) and incubated at 37°C for 1 hour. The wells were then washed twice with PBS. Next, these prepared wells were filled with 170 μL primary attachment medium, containing DMEM (Gibco), 10% fetal bovine serum (Sigma-Aldrich), 25 mM HEPES buffer (Gibco), and 300 U/mL penicillin (Sigma-Aldrich).

SGN cell culture

The spiral lamina containing the SG was carefully separated from the modiolus, and immediately transferred into primary cell culture medium. It was then cut into equal 300- to 500-μm portions and transferred to the prepared culture plates. The explants were first incubated for 24 h at 37°C in primary attachment medium, after which the culture medium was changed to serum-free maintenance media, comprised of DMEM (Gibco), 25 mM Hepes-Buffer (Gibco), 6 mg/mL glucose (Gibco), 300 U/mL penicillin (Sigma-Aldrich), and 30 μg/mL N_2 supplement (Gibco). The maintenance medium was supplemented with 10 ng/mL recombinant BDNF for trophic support of SG neuron survival and optimization of neurite outgrowth (R&D Systems, Minneapolis, MN, USA). Cultures were kept in a humidified incubator at 5% CO_2 and 37°C for 72 h. In experimental cultures, 1, 10, or 20 μM octreotide (Novartis Pharma) was added to the maintenance media. Maintenance media without octreotide served as a control.

SGN Immunohistochemistry

Explants were fixed with 4% paraformaldehyde for 20 min at room temperature and washed twice with PBS (Gibco). Then they were permeabilized with 5% triton X-100 (Sigma-Aldrich) for 10 min and washed twice with PBS, and non-specific antibody binding was blocked with 5% donkey serum (Sigma-Aldrich). Neurites were labeled for neurofilament using a mouse polyclonal 200-kDa anti-neurofilament primary antibody (1:400; Sigma-Aldrich). After overnight incubation at 4°C with primary antibody, and two washes with PBS, the neurites were visualized by 2.5 h of incubation with fluorescein isothiocyanate (FITC)-conjugated secondary antibodies (1:100; Jackson Immunoresearch, West Grove, PA, USA) against the species of the primary antibody. Staining specificity was confirmed by a series of negative control stainings without primary antibodies.

SGN data analysis

Digital images were obtained using a fluorescence microscope (Olympus IX71, Center Valley, PA, USA) equipped with appropriate excitation and emission filters for FITC. For publication in this manuscript, the images were optimized to achieve uniform brightness and contrast using Adobe Photoshop (Adobe Systems Inc., San Jose, CA, USA). Neurite outgrowth from the SG was evaluated by measuring the number and lengths of the processes. Images of the immunostained cultures were analyzed using NIH ImageJ software (NIH, Bethesda, MD, USA). Each neurite was traced, and the neurite numbers and average lengths per explant were analyzed using a one-way analysis of variance (ANOVA) followed by a Tukey LSD post-hoc test. Data presented in the text and figures are means and standard deviations. Results were considered significant when the likelihood for a type 1 error was less than 5% ($p < 0.05$). Twenty SG explants were analyzed per experimental condition.

Results

Octreotide has no toxic effect on HCs, and highly protects HCs from gentamicin-induced HC damage *in vitro*

The number of surviving HCs in rat OCs cultured with the highest octreotide dosage used in this study (5 μM) for 72 hours did not differ compared to in those cultured without octreotide (Fig. 1), thus excluding a toxic effect of octreotide. Untreated control OCs and those treated with octreotide each showed three orderly rows of outer HCs and a single row of inner HCs (Fig. 2). As expected, gentamicin treatment led to HC loss (Fig. 2). Treatment with both gentamicin (50 μM) and octreotide (1 μM and 5 μM) significantly increased HC survival (Fig. 1; ANOVA, $p < 0.01$ for all conditions)

Octreotide increases Akt phosphorylation *in vitro*

Western blotting revealed specific activation of Akt in OCs treated with octreotide *in vitro*. Blots using anti-p-Akt revealed strongly increased Akt activation after a 1-hour exposure to 5 μM octreotide, with both total Akt and β-actin used as references. Using total Akt as internal control, normalized p-Akt was expressed as % of control. In three replicates, the relative intensity of p-Akt was significantly increased in octreotide treated OCs compared to OCs in culture media only (Fig. 3; Mann-Whitney test, p<0.05).

The Akt inhibitor SH-6 reverses the protective effect of octreotide on gentamicin-induced HC damgage *in vitro*

Treatment with SH-6, an inhibitor of Akt, alone (10 μM) did not result in HC damage. OC treated with SH-6 showed three orderly rows of outer HCs and a single row of inner HCs (Fig. 4). However, the protective effect of octretide on gentamicin-induced HC damage was reversed when SH-6 (10 μM) was added in addition to octreotide to the cell culture media (Fig. 5; ANOVA, p<0.01).

Higher SST receptor-2 expression in SST recptor-1 knockout mice

At postnatal days 14 and 21, the mRNA gene expression of SST receptor-2 in OCs was two-fold higher in SST receptor-1 knockout mice compared to in their age-matched wild-type littermates (Fig. 6; unpaired Student's t-test, $p < 0.01$).

SST receptor-1 knockout is protective against gentamicin-induced HC damage in mouse OC explants

HC survival was increased in explants from SST receptor-1 knockout mice compared to in explants from wild-type and SST receptor-1/SST receptor-2 double-knockout mice after gentamicin (0.5 mM) exposure for 20 h. (Fig. 7; ANOVA, $p < 0.0001$).

Octreotide treatment results in increased SG neurite number

Treatment of neonatal SG explants with the two highest concentrations of octreotide (10 or 20 μM) increased the number of neurites per SG explant compared to in controls (Fig. 8; ANOVA, $p < 0.05$). Fig. 9 shows a representative image of SG explants treated with the different concentrations of octreotide compared to the control.

Octreotide results in decreased length of SG neurites

Treatment of neonatal SG explants with all concentrations of octreotide (1, 10, or 20 μM) significantly decreased the length of SG neurites compared to in controls (Fig. 10; ANOVA, $p < 0.05$ for all conditions).

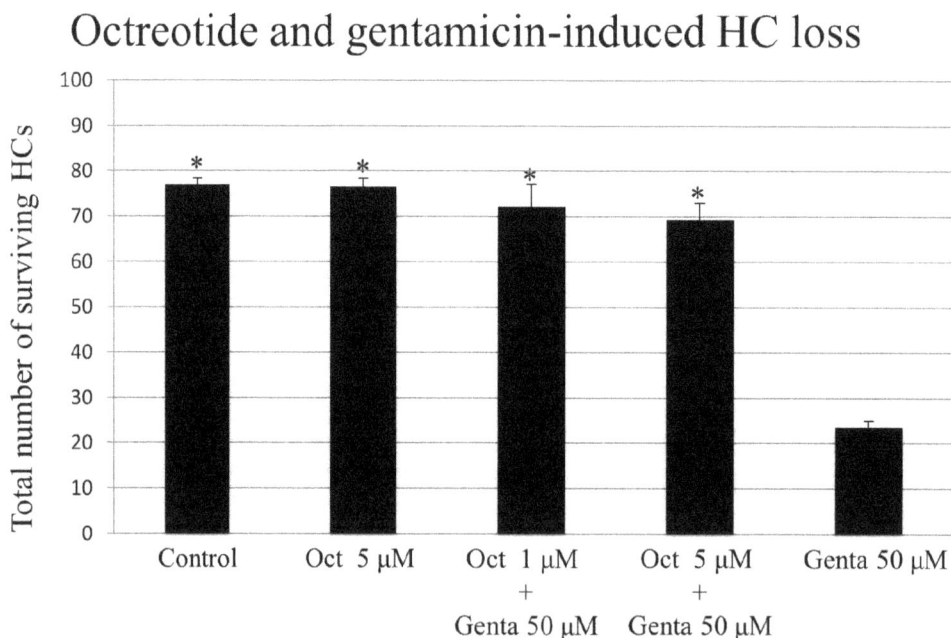

Octreotide and gentamicin-induced HC loss

Figure 1. Quantitative analysis of surviving HCs under different experimental conditions in the basal and middle cochlear turns. HC survival was significantly higher in groups treated with octreotide and gentamicin compared to with gentamicin treatment alone ($p < 0.01$ octreotide 1 μM, $p < 0.01$ octreotide 5 μM). No toxic effects were observed in OCs incubated with octreotide without gentamicin or with culture medium only (control). Asterisks indicate significant difference from treatment with gentamicin only ($p < 0.01$ for all conditions). Data are expressed as the mean number of surviving HCs per 20 inner HCs. Vertical lines represent one standard deviation. $n = 6$ for each experimental condition.

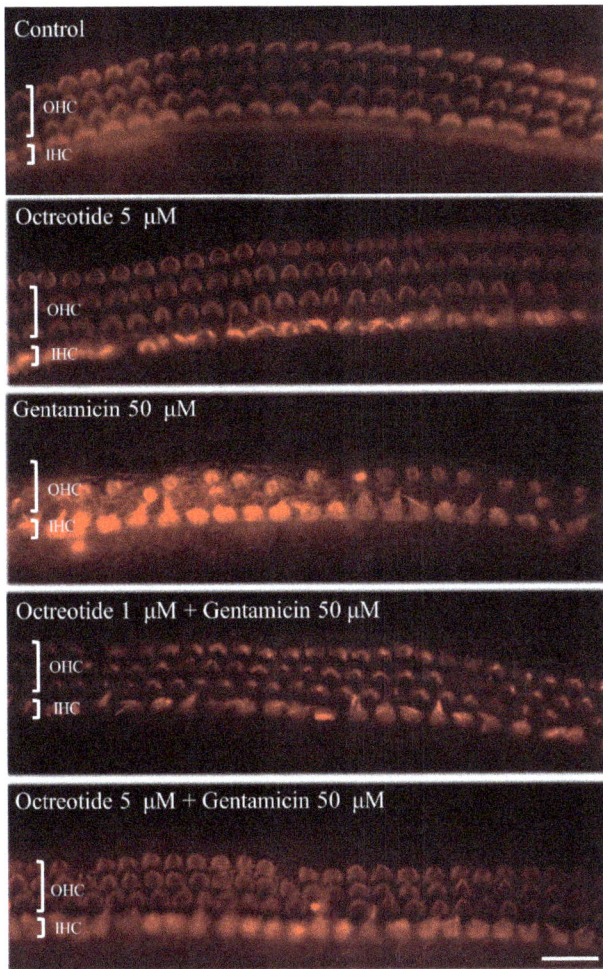

Figure 2. Effects of octreotide on gentamicin-induced HC damage. Three orderly rows of outer HCs (OHC) and a single row of inner HCs (IHC) were observed in control OCs and in OCs exposed to 5 μM octreotide without gentamicin. Comparatively, OCs cultured with gentamicin showed significant loss of HCs. Addition of increasing concentrations of octreotide to gentamicin-treated OCs resulted in significantly decreased HC loss compared to in those treated only with gentamicin. Scale bar = 20 μm.

Discussion

Octreotide protects HCs from gentamicin-induced toxicity and activates Akt signaling *in vitro*

Interestingly, we found that octreotide significantly reduced HC loss in OC samples treated with gentamicin, as compared to in samples treated with gentamicin alone. Moreover, western blotting revealed that octreotide treatment increased Akt phosphorylation *in vitro* and inhibition of Akt by using the Akt inhibitor SH-6 reversed the protective effect of octreotide on gentamicin-induced HC loss *in vitro*. It should be noted that the cochlear explants for our tissue culture experiments were harvested from newborn animals since only newborn animals can be used for extended culture of inner ear HCs; adult HCs do not survive in culture [23]. A large number of studies have used aminoglycosides as HC death inducers, and the immature cochlea is an established *in vitro* model. However, since younger animals are more sensitive to ototoxins [1,4,6], some caution should be exercised in generalizing our present results to adults.

How can the protective effect of octreotide in gentamicin-induced HC loss be explained? To date, there are no previous reports of octreotide- or SST-induced intracellular events in the inner ear. However, there exist some data regarding SST-induced signaling in the visual system. The retina and inner ear share common characteristics: both arise from neuroepithelium, harbor sensitive sensory cells together with supporting cells, and display a complex and highly organized microarchitecture. Therefore, the molecular events involved in HC and SGN damage and death might share features similar to those involved in retinal cell damage and death [24,25,26]. In the retina, SST has pleiotropic effects and activates a variety of signaling mechanisms by its receptors, including PI3K, mitogen-activated kinases, and calcium channels [27].

Gentamicin exposure also activates pathways that promote HC survival, which supports the current opinion that cells exist in a finely tuned balance between survival and cell death [12,28,29]. Several survival pathways that operate in HCs have been defined, including the H-Ras/Raf/MEK/Erk pathway and the PI3K-Akt pathway [9]. Interestingly, Chung *et al*. demonstrated that PI3K/Akt mediates HC survival and opposes gentamicin toxicity in neonatal rat OC explants [12]. Moreover, we previously demonstrated that simvastatin protects HCs from gentamicin-induced toxicity and activates Akt signaling *in vitro* [30]. Therefore, we examined whether the observed protection from gentamicin by octreotide might be due to octreotide's influence on the PI3K/Akt pathway—and, in fact, our present results confirmed activation of Akt by octreotide. However, it is likely that other intracellular processes are involved, since somatostatin receptors modulate several intracellular signaling transduction pathways [16,27]. It is also possible that somatostatin protects HCs from aminoglycoside toxicity-induced cell death through its ability to limit glutamate release or to inhibit glutamate excitotoxicity, as has been previously observed in the retina [31]. We must also consider the possibility that octreotide might interact physically with gentamicin. Although we have no direct evidence excluding physical interaction, this seems unlikely since the octreotide concentration was 10 to 50 times lower than the gentamicin concentration.

Overexpression of SST receptor-2 protects HCs from gentamicin-induced toxicity *in vitro*

SST receptor-1 or SST receptor-2 deletion substantially alters the SST content in the retina [24,25,32]. In retinas from SST receptor KO mice, SST receptor-1 and SST receptor-2 expressions have been found to compensate for each other, such that SST receptor-1 loss results in increased expression of SST receptor-2 [32]. Notably, SST receptors also have a neuroprotective function in the retina. SST and its five receptors are expressed in the retina, predominantly in amacrine cells and bipolar cells [33]. Activation of SST receptor-2 by SST or its analogues reportedly protects retinal neurons against ischemia-induced damage [17,26]. Additionally, studies in mice with genetic alterations of the somatostatinergic system have revealed that increased functional SST receptor-2 expression protects against retinal ischemia [34]. Therefore, SST receptor-2 analogs may have therapeutic benefits in retinal diseases, such as glaucoma or diabetic retinopathy.

Accordingly, our present results showed that SST receptor-1 KO mice exhibited up-regulation of SST receptor-2 in the OC compared to wild-type mice. We previously demonstrated that SST can dose-dependently protect HCs from aminoglycoside toxicity *in vitro* [18]. Interestingly, here we showed highly significant HC protection from aminoglycoside toxicity in the

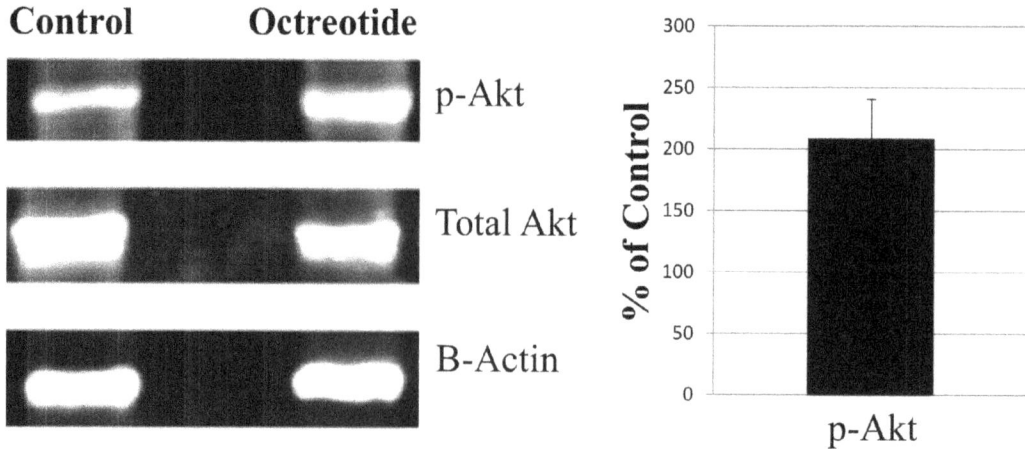

Figure 3. Representative western blots of p- Akt, total Akt, and β-Actin. OCs were exposed for 1 hour to either control media (Control) or media containing 5 μM octreotide. P-Akt levels were normalized against total Akt. Octreotide treated levels are expressed as % of control values. P-Akt levels were significantly increased by octreotide treatment (p<0.05). Bars show the mean ± one standard deviation of 3 independent experiments. In each experiment 10 OCs from 5 animals were used.

cochlea of SST receptor-1 knockout mice compared to in wild-type mice and SST receptor-1/SST receptor-2 double-knockout mice. The elevated level of SST receptor-2 might be responsible for this observed protection. However, up-regulation and interactions of other SST receptors must still be considered.

What mechanism might be responsible for the neuroprotective role of SST receptor-2 in the cochlea? Studies in mouse retinal explants have demonstrated that SST receptor-2 inhibits potassi-um-induced glutamate release [35]. By limiting the amount of glutamate available to glutamate receptors, SST and its analogs may exert neuroprotection against glutamate neurotoxicity, which characterizes many retinal diseases. Glutamate excitotoxicity appears to be mediated by caspase-3 activation, as shown in cerebrocortical neurons [36]. Glutamate excitotoxicity is also involved in HC damage and death in the cochlea [37]. Therefore, it is possible that SST protects HCs from aminoglycoside toxicity,

SH-6, octreotide and gentamicin-induced HC loss

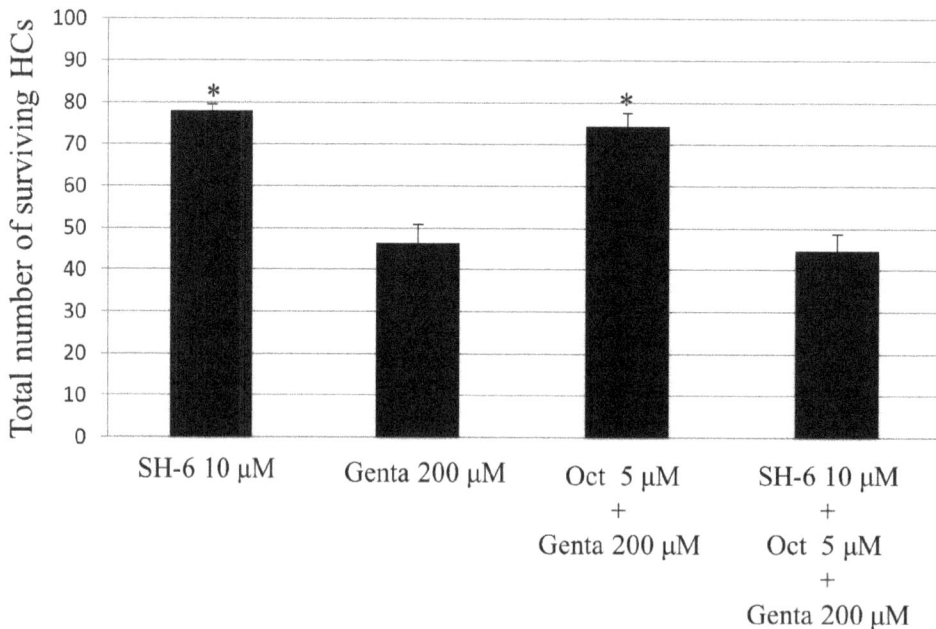

Figure 4. Quantitative analysis of surviving HCs under different experimental conditions in the basal and middle cochlear turns. No toxic effects were observed in OCs incubated with the AKT inhibitor SH-6 at 10 μM. HC survival was significantly higher in groups treated with octreotide and gentamicin compared to gentamicin treatment alone or treatment with SH-6, octreotide and gentamicin. Asterisks indicate significant difference from treatment with gentamicin only (p<0.01 for all conditions). Data are expressed as the mean number of surviving HCs per 20 inner HCs. Vertical lines represent one standard deviation. n=6 for each experimental condition.

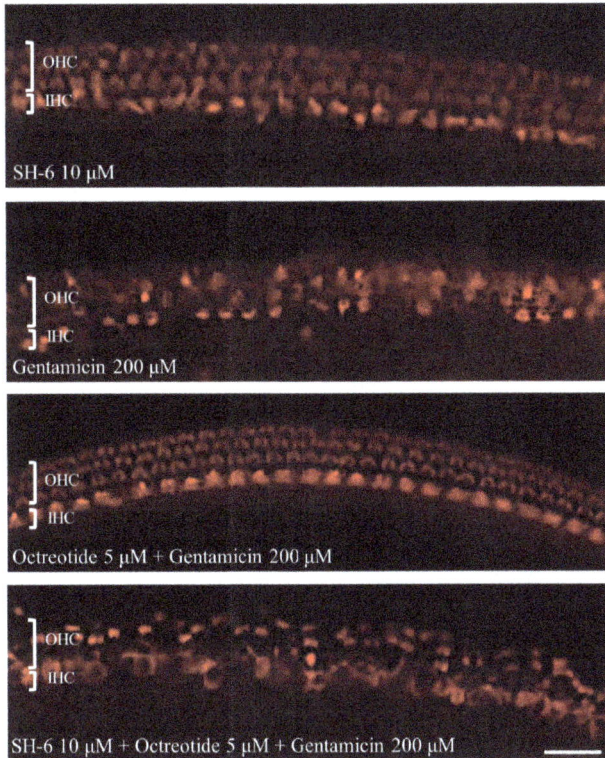

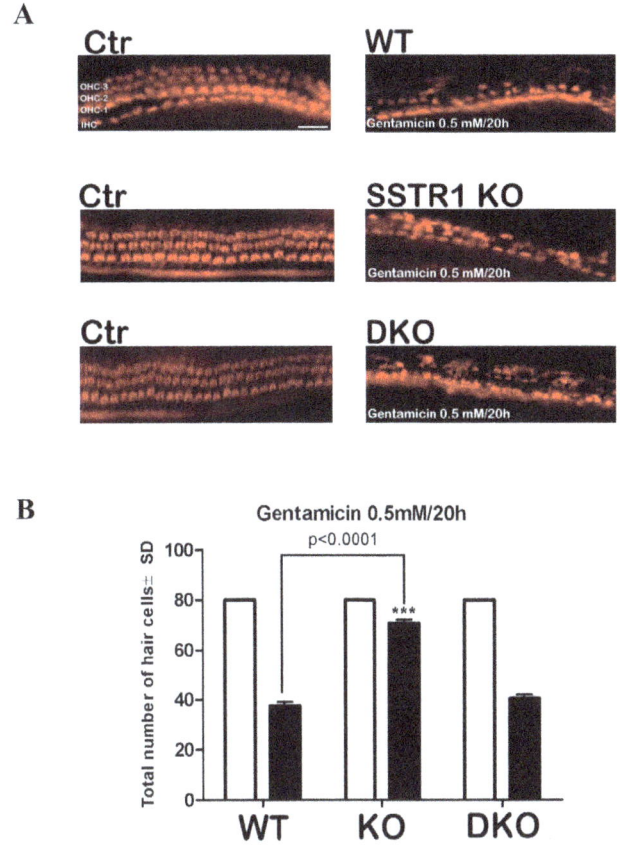

Figure 5. Effects of octreotide and the Akt inhibitor SH-6 on gentamicin-induced HC damage. Three orderly rows of outer HCs (OHC) and a single row of inner HCs (IHC) were observed in OCs exposed to 10 µM SH-6. Comparatively, OCs cultured with gentamicin showed significant loss of HCs. Addition of octreotide to gentamicin-treated OCs resulted in significantly decreased HC loss compared to in those treated with gentamicin only. However, addition of SH-6 and octreotide did not result in decreased hair cell loss. Scale bar = 20 µm.

either by limiting glutamate release or by mitigating the toxic action of excess glutamate on HCs. It is interesting to note that octreotide predominantly activates SST receptor-2 [38,39]. The findings that octreotide offers a high degree of protection from

Figure 7. Quantitative analysis of surviving HCs. (a) Gentamicin-induced HC damage in the OCs of wild-type, SST receptor-1 knockout and SST receptor-1/SST receptor-2 double knockout mice. Photograph of phalloidin-labeled OC. The three outer HC (OHC) rows and a single inner HC (IHC) row can be seen in controls. OCs exposed to 0.5 mM gentamicin demonstrate HC loss. (b) Histograms show a significant difference in the number of surviving HCs in the OCs of SST receptor-1 knockout mice exposed to gentamicin compared to in the gentamicin-treated OCs of wild-type mice ($p > 0.0001$). Histogram and bars represent mean ± one standard deviation. $n = 15$ for each experimental condition.

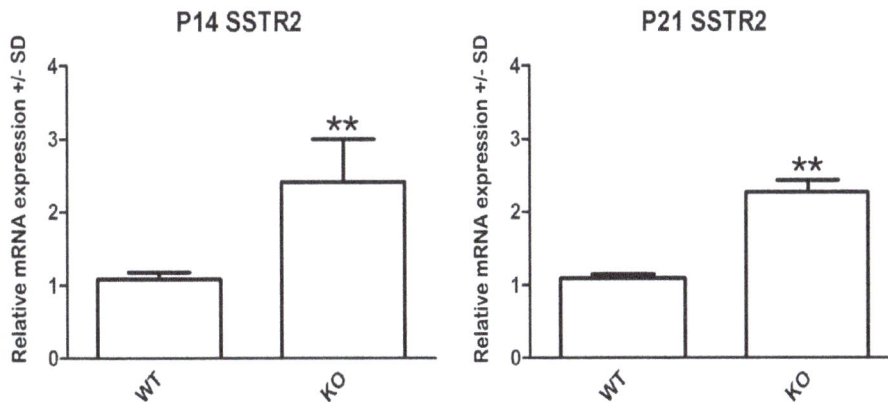

Figure 6. Cochlear gene expression of SST receptor-2 in the OC of postnatal day (P)14 and P21 wild-type (WT) and SST receptor-1 knockout (KO) mice. The relative distribution of SST receptor-2 mRNA expression in OC tissue from wild-type (WT) and SST receptor-1 KO of different postnatal ages was quantified by real-time PCR. GAPDH was used as an endogenous control. Gene expression levels are expressed as the mean (± one standard deviation) fold increase compared to the values obtained in OC explants from P14 and P21 WT and SST receptor-1 knockout mice. Data were obtained from five independent experiments, per condition in each experiment 5 OCs were used. **$p < 0.001$ using Student's t-test.

Octreotide - number of neurites

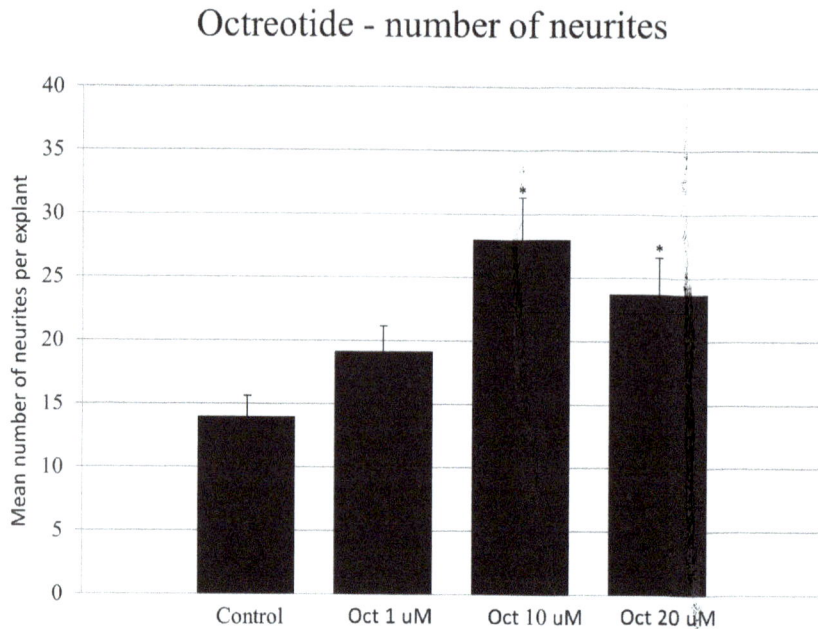

Figure 8. Average number of SG neurites observed from SG explants under the different experimental conditions. The number of neurites observed on control are compared to those seen with three different levels of octreotide (Oct). Lines represent one standard deviation. Asterisks denote statistical difference compared to control ($p<0.05$ for all conditions). $n = 20$ for each experimental condition.

gentamicin toxicity and that HC loss is reduced in SST receptor-1 mice that overexpress SST receptor-2 suggest an important function of SST receptor-2.

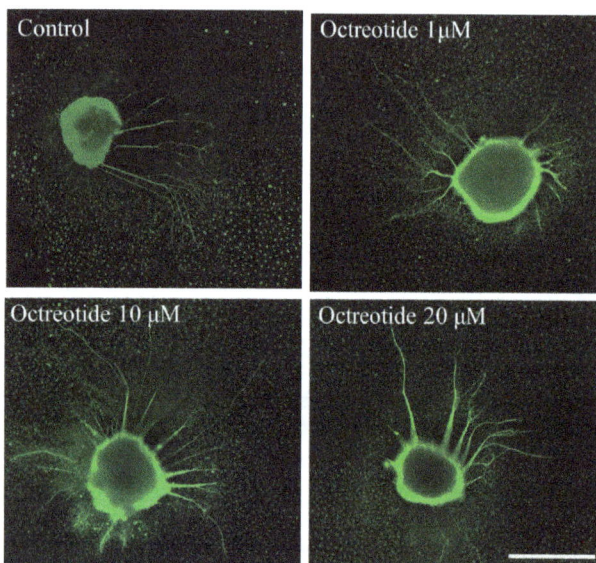

Figure 9. Effects of octreotide on SGNs *in vitro*. SGNs were treated with increasing concentrations of octreotide (1 µM, 10 µM and 20 µM, respectively) and compared to controls without octreotide treatment. Average number of SG neurites and average lenth of SG neurites observed from SG explants were analyzed. Representative SG explants stained with anti-200-kDa neurofilament antibody for each experimental condition are shown. Scale bar = 300 µm.

Octreotide treatment results in increased SG neurite number and a slight decrease in SGN length *in vitro*

We found that octreotide protected HCs from gentamicin-induced HC damage, and we further demonstrated that octreotide induced neurite formation in neonatal rat SG explants. SST receptor-2 is the main receptor–pharmacological target mediating the effects of octreotide [38]. SST receptor-2 modulates several intracellular signaling transduction pathways in other systems, such as the PI3K-Akt and p38 pathways [16,40]. SST receptor-2 reportedly mediates opposing proliferative effects through the activated p38 and activated Akt pathways in CHO-K1 cells *in vitro* [41]. While activation of p38 signaling generally promotes apoptosis [42], there are also several documented examples of survival enhancement by these pathways [43,44]. We previously found that p38 and PI3K/Akt mediate BDNF-induced neurite formation in neonatal cochlear SG explants [45]. However, we did not assess the intracellular signaling pathways involved in octreotide-induced neurite formation in the present study. Therefore, we can only speculate that the p38 and PI3K/Akt intracellular signaling transduction pathways are involved in this process. Further studies are needed to elucidate the intracellular signaling transduction pathways involved in octreotide-induced SG neurite formation.

Interestingly, our data demonstrated a statistically significant decrease in average neurite length observed in octreotide-treated SG explants, indicating that octreotide mediates a negative effect on SG neuronal outgrowth. Tentler *et al.* showed that SST receptor-2 inhibits adenylate cyclase and, consequently, cAMP production in pituitary tumor GH4C1 cells, which ultimately results in decreased protein kinase A (PKA) activity [46]. Inhibition of PKA activity reportedly results in decreased in average neurite length and has no effect on neurite formation on neonatal rat SG explants. However, here we observed that inhibition of PI3K/Akt and Mek/Erk signaling resulted in increased neurite length, indicating that these pathways are

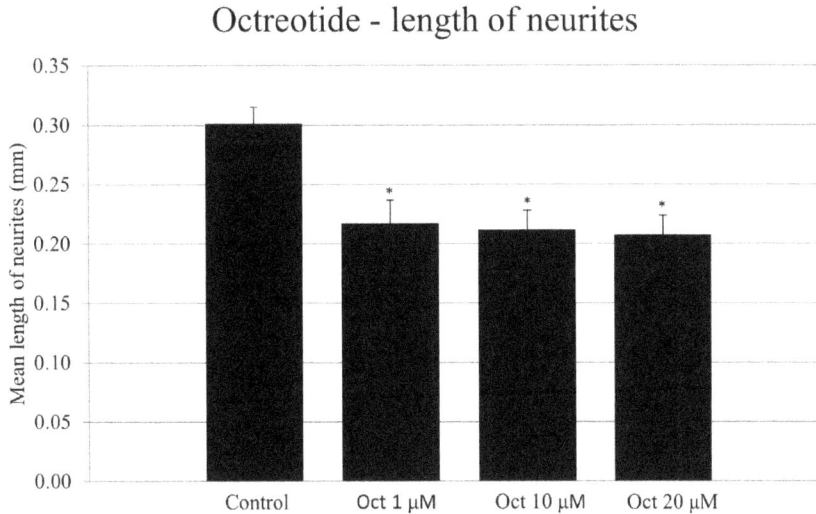

Figure 10. Average length of SG neurites observed on SG explants in the different experimental conditions. The lengths of neurites observed in the controls (media only) are compared to those seen with three different levels of octreotide (Oct). Lines represent one standard deviation. Asterisks denote statistical difference compared to control ($p < 0.05$ for all experimental conditions). $n = 20$ for each experimental condition.

involved in neurite elongation. Octreotide exposure may result in activation of intracellular signaling transduction pathways that have both positive and negative influences on neurite length, resulting in a total negative effect. Although this hypothesis may be too complex to be attractive without additional supporting data, it is at least consistent with our observations.

It should be noted that we could not distinguish between the dentrites and axons of SGNs, since we have not found markers that distinguish between the two in explants. Similarly, we could not distinguish between type I and type II SGN neurites, since peripherin labeling does not distinguish these two neuron classes in rat tissue cultures, due to up-regulation of peripherin type I neurons *in vitro* [47]. However, since 95% of SG neurons are type I cells, it seems likely that this class of neuron dominates our results.

Conclusions

Our findings indicate that the SST analog octreotide highly protects HCs from gentamicin-induced toxicity and activates Akt signaling *in vitro*. Octreotide treatment also resulted in increased

SG neurite number and a decrease in SGN length. Moreover, we discovered that SST receptor-1 knockout mice exhibited protection from aminoglycoside-induced HC loss compared to SST receptor-1/SST receptor-2 double-knockout and wild-type mice. SST receptor-2 was overexpressed in SST receptor-1 knockout mice, and might be responsible for the observed high degree of HC protection from gentamicin-induced HC loss. Our results suggest the somatostatinergic system in the inner ear to be a new and promising target in efforts to protect the mammalian inner ear from cell death.

Acknowledgments

The authors thank Markus Saxer for technical help.

Author Contributions

Conceived and designed the experiments: YB VR DB. Performed the experiments: YB VR MS EW CS AG KL. Analyzed the data: YB VR MS EW CS DB AG KL. Wrote the paper: YB VR DB.

References

1. Lautermann J, Dehne N, Schacht J, Jahnke K (2004) Aminoglycoside- and cisplatin-ototoxicity: from basic science to clinics. Laryngorhinootologie 83: 317–323.
2. Matsui JI, Cotanche DA (2004) Sensory hair cell death and regeneration: two halves of the same equation. Curr Opin Otolaryngol Head Neck Surg 12: 418–425.
3. Cheng AG, Cunningham LL, Rubel EW (2005) Mechanisms of hair cell death and protection. Curr Opin Otolaryngol Head Neck Surg 13: 343–348.
4. Rybak LP, Whitworth CA (2005) Ototoxicity: therapeutic opportunities. Drug Discov Today 10: 1313–1321.
5. Priuska EM, Schacht J (1995) Formation of free radicals by gentamicin and iron and evidence for iron/gentamicin complex. Biochem Pharmacol 50: 1749–1752.
6. Pirvola U, Xing-Qun L, Virkkala J, Saarma M, Murakata C, et al. (2000) Rescue of hearing, auditory hair cells, and neurons by CEP-1347/KT7515, an inhibitor of c-Jun N-terminal kinase activation. J Neurosci 20: 43–50.
7. Bodmer D, Brors D, Bodmer M, Ryan AF (2002) Rescue of auditory hair cells from ototoxicity by CEP-11004, an inhibitor of the JNK signalling pathway. Laryngorhinootologie 81: 853–856.
8. Bodmer D, Brors D, Pak K, Gloddek B, Ryan AF (2002) Rescue of auditory hair cells from aminoglycoside toxicity by Clostridium difficile toxin B, an inhibitor of the small GTPases Rho/Rac/Cdc42. Hear Res 172: 81–86.
9. Battaglia A, Pak K, Brors D, Bodmer D, Frangos J, et al. (2003) Involvement of ras activation in toxic hair cell damage of the mammalian cochlea. Neuroscience 122: 1025–1035.
10. Huang T, Chen AG, Stupak H, Liu W, Kim A, et al. (2000) Oxidative stress-induced apoptosis of cochlear sensory cells: otoprotective strategies. Int J Dev Neurosci 18: 259–270.
11. Okuda T, Sugahara K, Takemoto T, Shimogori H, Yamashita H (2005) Inhibition of caspases alleviates gentamicin-induced cochlear damage in guinea pigs. Auris Nasus Larynx 32: 33–37.
12. Chung WH, Pak K, Lin B, Webster N, Ryan AF (2006) A PI3K pathway mediates hair cell survival an opposes gentamicin toxicity in neonatal rat organ of Corti. J Assoc Res Otolaryngol 7: 372–382.
13. Epelbaum J, Dournaud P, Fodor M, Viollet C (1994) The neurobiology of somatostatin. Crit Rev Neurobiol 8: 25–44.
14. Patel YC (1999) Somatostatin and its receptor family. Front. Neuroendocrinol 20: 157–198.
15. Reisine T (1995) Somatostatin. Am J Physiol 269: 813–820.
16. War SA, Kumar U (2012) Coexpression of human somatostatin receptor-2 (SSTR2) and SSTR3 modulates antiproliferative singaling and apoptosis. J Mol Signal 7: 5.

17. Celiker U, Ilhan N (2002) Nitric oxide and octreotide in retinal ischemia/reperfusion injury. Doc Ophthalmol 105: 327–338.
18. Caelers A, Monge A, Brand Y, Bodmer D (2009) Somatostatin and gentamicin-induced auditory hair cell loss. Laryngoscope 119: 933–937.
19. Radojevic V, Hanusek C, Setz C, Brand Y, Kapfhammer JP, et al. (2011) The somatostatinergic system in the mammalian cochlea. BMC Neurosci 12: 89.
20. Bodmer D, Brand Y, Radojevic V (2012) Somatostatin receptor types 1 and 2 in the developing mammalian cochlea. Dev Neurosci 34: 342–353.
21. Kreienkamp HJ, Akgun E, Baumeister H, Meyerhof W, Richter D (1999) Somatostatin receptor subtype 1 modulates basal inhibition of growth hormone release in somatotrophs. FEBS Lett 462: 464–466.
22. Zheng H, Bailey A, Jiang MH, Honda K, Chen HY, et al. (1997) Somatostatin receptor subtype 2 knockout mice are refractory to growth hormone-negative feedback on arcuate neurons. Mol Endocrinol 11: 1709–1717.
23. Sobkowicz HM, Loftus JM, Slapnick SM (1993) Tissue culture of the organ of Corti. Acta Otolaryngol Suppl 502: 3–36.
24. Casini G (2005) Neuropeptides and retinal development. Arch Ital Biol 143: 191–198.
25. Casini G, Catalani E, Dal Monte M, Bagnoli P (2005) Functional aspects of the somatostatinergic system in the retina and the potential therapeutic role of somatostatin in retinal disease. Histol Histopathol 20: 615–632.
26. Mastrodimou N, Lambrou GN, Thermos K (2005) Effect of somatostatin analogues on chemically induced ischaemia in the rat retina. Arch Pharmacol 371, 44–53.
27. Cerviva D, Casini G, Bagnoli P (2008) Physiology and pathology of somatostatin in the mammalian retina: A current view. Mol Cell Endocrinol 286: 112–122.
28. Boatright KM, Salvesen GS (2003) Caspase activation. Biochem. Soc Symp 70: 233–242.
29. Boatright KM, Salvesen GS (2003) Mechanisms of caspase activation. Curr Opin Cell Biol 15: 725–731.
30. Brand Y, Setz C, Levano S, Listyo A, Chavez E, et al. (2011) Simvastatin protects auditory hair cells from gentamicin-induced toxicity and activates Akt signalling in vitro. BMC Neurosci 12: 114.
31. Pilar G, Gray DB, Meriney SD (1996) Membrane delimited and intracellular soluble pathways in the somatostatin modulation of ACh release. Life Sci 58: 1979–1986.
32. Dal Monte M, Petrucci C, Vasilaki A, Cervia D, Grouselle D, et al. (2003) Genetic deletion of somatostatin receptor 1 alters somatostatinergic transmission in the mouse retina. Neuropharmacology 45: 1080–1092.
33. Bagnoli P, Dal Monte M, Casini (2003) Expression of neuropeptides and their receptors in the developing retina of mammals. Histol Histopathol 18: 1219–1242.
34. Catalani E, Cervia D, Martini D, Bagnoli P, Simonetti E, et al. (2007) Changes in neuronal response to ischemia in retinas with genetic alterations of somatostatin receptor expression. Eur J Neurosci 25: 1447–1459.
35. Dal Monte M, Cammalleri M, Martini D, Casini G, Bagnoli P (2007) Antiangiogenic role of somatostatin receptor 2 in a model of hypoxia-induced neovascularization in the retina: results from transgenic mice. Invest Ophthalmol Vis Sci 48: 3480–3489.
36. Tenneti L, Lipton SA (2000) Involvement of activated caspase-3-like proteases in N-methyl-D-aspartate-induced apoptosis in cerebrocortical neurons. J Neurochem 74: 134–142.
37. Kopke RD, Coleman JK, Liu J, Campbell KC, Riffenburgh RH (2002) Candidate's thesis: enhancing intrinsic cochlear stress defenses to reduce noise-induced hearing loss. Laryngoscope 112: 1515–1532.
38. Weckbecker G, Lewis I, Albert R, Schmid HA, Hoyer D, et al. (2003) Opportunities in somatostatin research: biological, chemical and therapeutic aspects. Nat. Rev Drug Discov 2: 999–1017.
39. Lesche S, Lehmann D, Nagel F, Schmid HA, Schulz S (2009) Differential effects of octreotide and pasireotide on somatostatin receptor internalization and trafficking in vitro. J Clin Endocrinol Metab 94: 654–661.
40. Theodoropoulou M, Stalla GK (2013) Somatostatin receptors: From signalling to clinical practice. Front Neuroendocrinol 34: 228–252.
41. Sellers LA, Alderton F, Carruthers AM, Schindler M, Humphrey PP (2000) Receptor isoforms mediate opposing proliferative effects through gbetagamma-activated p38 or Akt pathways. Mol Cell Biol 20: 5974–5985.
42. Mielke K, Herdegen T (2000) JNK and p38 stress kinases degenerative effectors of signal transduction-cascades in the nervous system. Prog Neurobiol 61: 45–60.
43. Nishina H, Fischer KD, Radvanyi L, Shahinian A, Hakem R, et al. (1997) Stress-signalling kinase Sekl protects thymocytes from apoptosis mediated by CD95 and CD3. Nature 385: 350–353.
44. Du L, Lyle CS, Obey TB, Gaarde WA, Muir JA, et al. (2004) Inhibition of cell proliferation and cell cycle progression by specific inhibition of basal JNK activity: evidence that mitotic Bcl-2 phosphorylation is JNK-independent. J Biol Chem 279: 11957–11966.
45. Mullen LM, Pak KK, Chavez E, Kondo K, Brand Y, et al. (2012) Ras/p38 and PI3K/Akt but not Mek/Erk signalling mediate BDNF-induced neurite formation on neonatal cochlear spiral ganglion explants. Brain Res 1430: 25–34.
46. Tentler JJ, Hadcock JR, Gutierrez-Hartmann A (1997) Somatostatin acts by inhibiting the cyclic 3′, 5′-adenosin monophosphate (cAMT)/protein kinase A pathway, cAMP response element-binding protein (CREB) phosphorylation, and CREB transcription potency. Mol Endocrinol 11: 859–866.
47. Lallemend F, Vandenbosch R, Hadjab S, Bodson M, Breuskin I, et al. (2007) New insights into peripherin expression in cochlear. Neurons Neurosci 150: 212–222.

5

Cyclosporine A Induces Apoptotic and Autophagic Cell Death in Rat Pituitary GH3 Cells

Han Sung Kim[1], Seung-Il Choi[2], Eui-Bae Jeung[3]*, Yeong-Min Yoo[1]*

1 Department of Biomedical Engineering, College of Health Science, Yonsei University, Wonju, Gangwon-do, Republic of Korea, **2** Cornea Dystrophy Research Institute and Department of Ophthalmology, Yonsei University College of Medicine, Seoul, Republic of Korea, **3** Laboratory of Veterinary Biochemistry and Molecular Biology, College of Veterinary Medicine, Chungbuk National University, Cheongju, Republic of Korea

Abstract

Cyclosporine A (CsA) is a powerful immunosuppressive drug with side effects including the development of chronic nephrotoxicity. In this study, we investigated CsA treatment induced apoptotic and autophagic cell death in pituitary GH3 cells. CsA treatment (0.1 to 10 µM) decreased survival of GH3 cells in a dose-dependent manner. Cell viability decreased significantly with increasing CsA concentrations largely due to an increase in apoptosis, while cell death rates due to autophagy altered only slightly. Several molecular and morphological features correlated with cell death through these distinct pathways. At concentrations ranging from 1.0 to 10 µM, CsA induced a dose-dependent increase in expression of the autophagy markers LC3-I and LC3-II. Immunofluorescence staining revealed markedly increased levels of both LC3 and lysosomal-associated membrane protein 2 (Lamp2), indicating increases in autophagosomes. At the same CsA doses, apoptotic cell death was apparent as indicated by nuclear and DNA fragmentation and increased p53 expression. In apoptotic or autophagic cells, p-ERK levels were highest at 1.0 µM CsA compared to control or other doses. In contrast, Bax levels in both types of cell death were increased in a dose-dependent manner, while Bcl-2 levels showed dose-dependent augmentation in autophagy and were decreased in apoptosis. Manganese superoxide dismutase (Mn-SOD) showed a similar dose-dependent reduction in cells undergoing apoptosis, while levels of the intracellular calcium ion exchange maker calbindin-D9k were decreased in apoptosis (1.0 to 5 µM CsA), but unchanged in autophagy. In conclusion, these results suggest that CsA induction of apoptotic or autophagic cell death in rat pituitary GH3 cells depends on the relative expression of factors and correlates with Bcl-2 and Mn-SOD levels.

Editor: Ilya Ulasov, Swedish Medical Center, United States of America

Funding: This research was supported by the Leading Foreign Research Institute Recruitment Program through the National Research Foundation of Korea (NRF) funded by the Ministry of Science, ICT & Future Planning (2010-00757) and by a grant from the Next-Generation BioGreen 21 Program, Rural Development Administration, Republic of Korea (No. PJ00956301). The funders had no role in study design, data collection and analysis, decision to publish, or preparation of the manuscript.

Competing Interests: The authors have declared that no competing interest exist.

* Email: yyeongm@hanmail.net (YMY); ebjeung@cbu.ac.kr (EBJ)

Introduction

Programmed cell death (PCD) can be classified as either type I or type II PCD, based on distinctive morphological and biochemical characteristics. Type I PCD, or apoptosis, is characterized by blebbing, changes to the cell membrane such as loss of membrane asymmetry and attachment, cell shrinkage, nuclear fragmentation, chromatin condensation, and the formation of apoptotic bodies (apoptosomes) [1]. In contrast, type II PCD, or autophagy, is marked by extensive autophagic degradation of intracellular organelles, resulting in lysosome-associated cytoplasmic vacuolation/autophagosome formation [2]. Microtubule-associated protein 1 light chain 3 (LC3) is a marker for the autophagic process during which it is converted from the cytosolic form, LC3-I, to LC3-II, a modified form that is localized to autophagosomal membranes [3].

Currently, there is no known overlap in the pathways that modulate autophagic and apoptotic cell death [4,5]. However, apoptotic and autophagic processes may functionally coordinate in three ways: both apoptosis and autophagy can cooperate to induce cell death; autophagy can act as an antagonist to block apoptotic cell death; and autophagy can act as a precursor or even initiator of apoptosis. Several specific examples of coordination between autophagy and apoptosis have been documented. The proapoptotic molecule TRAIL mediates autophagy [6]. In addition, growth factor deprivation induces an autophagic cell death, which can be inhibited by the anti-apoptotic factor Bcl-2 [7]. These data suggest that autophagic cell death may be induced by HSpin1, a transmembrane protein that interacts with Bcl-2/Bcl-xL [8].

Cyclosporine A (CsA) was first approved by the United States Food and Drug Administration in the early 1980s, and has been used significantly in prophylactic anti-rejection therapy for patients receiving allogeneic transplants (kidney, liver, and heart) for over two decades [8,9]. However, several side effects of CsA have been reported in both transplant and non-transplant (i.e., individuals with autoimmune disorders) patients, including nephrotoxicity, hepatotoxicity, neurotoxicity, hypertension, dyslipidemia, gingival hyperplasia, hypertrichosis, malignancies, and an increased risk of cardiovascular events [10,11]. Recent reports demonstrate that CsA induces autophagy *in vitro* in human

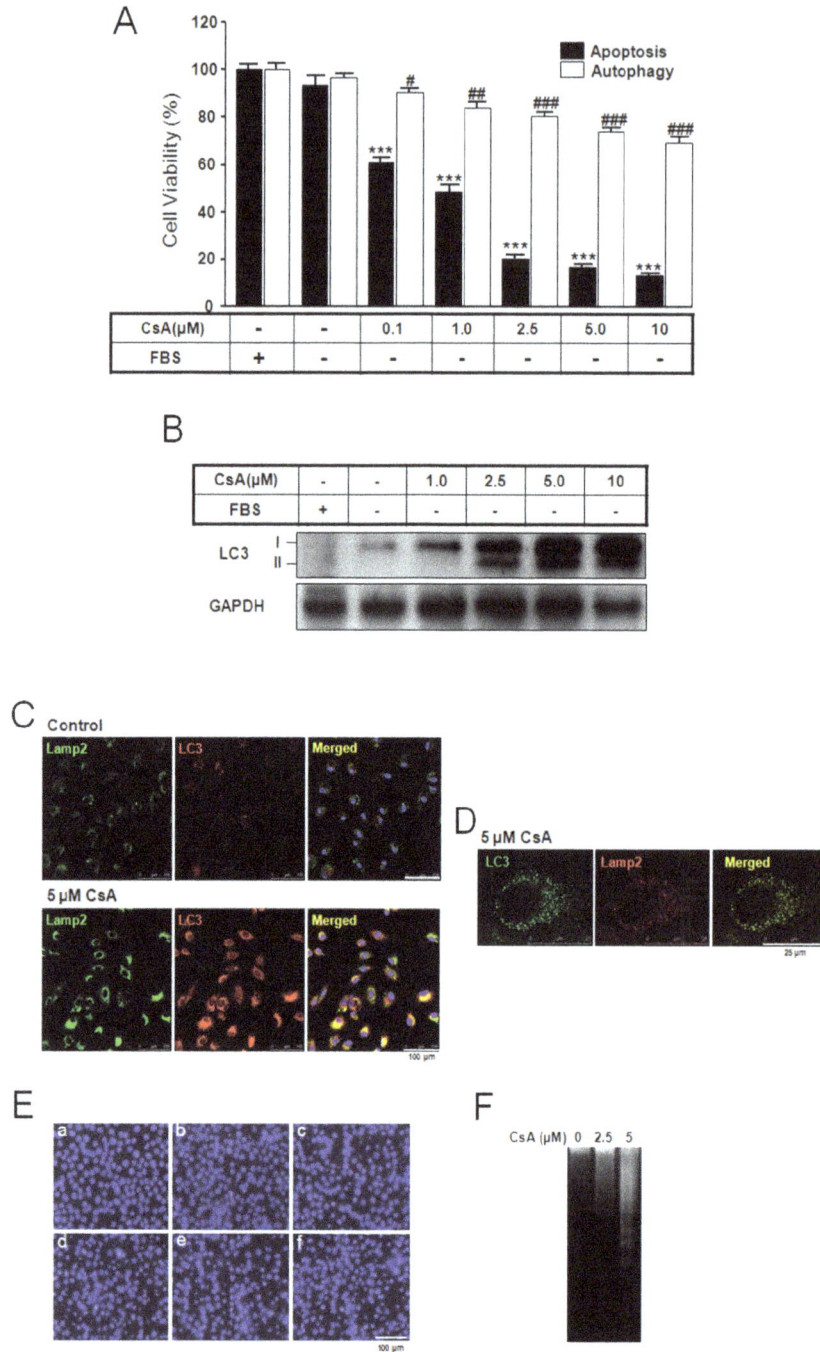

Figure 1. Cell viability and LC3 expression following CsA treatment. GH3 cells were incubated in DMEM with and without 10% fetal bovine serum in the presence or absence of CsA (0 to 10 μM) for 10 h. Cell survival was determined using Cell Counting Kit-8 (A) and LC3 expression was determined by Western blotting (B) as described in Materials and Methods. Immunofluorescence staining of LC3 and Lamp2 (C, D), DAPI staining (E), and DNA fragmentation (F) were captured as described in Materials and Methods. E: a, with FBS; b, without FBS; c, 1.0 μM CsA; d, 2.5 μM CsA; e, 5.0 μM CsA; f, 10 μM CsA. Scale bars: C, 100 μm; D, 25 μm; E, 100 μm. ***$p<0.001$ vs. serum treatment. #$p<0.05$, ##$p<0.01$, ###$p<0.001$ vs. serum treatment.

tubular cells and *in vivo* in rat kidneys, and it has been suggested that autophagy serves as a protective mechanism against cyclosporine toxicity [12–14]. Importantly, it has been shown that cyclosporine-induced autophagy is triggered by endoplasmic reticulum (ER) stress [15]. Other reports show that CsA can induce apoptosis in human proximal tubular cells *in vitro* [16] and

induces apoptosis via prolonged ER stress in an experimental model of chronic nephropathy [17].

In this study, we investigated whether CsA induces apoptotic and/or autophagic cell death in rat pituitary GH3 cells and correlated the levels of several characteristic molecular markers of these two pathways with cell death outcomes.

A

B

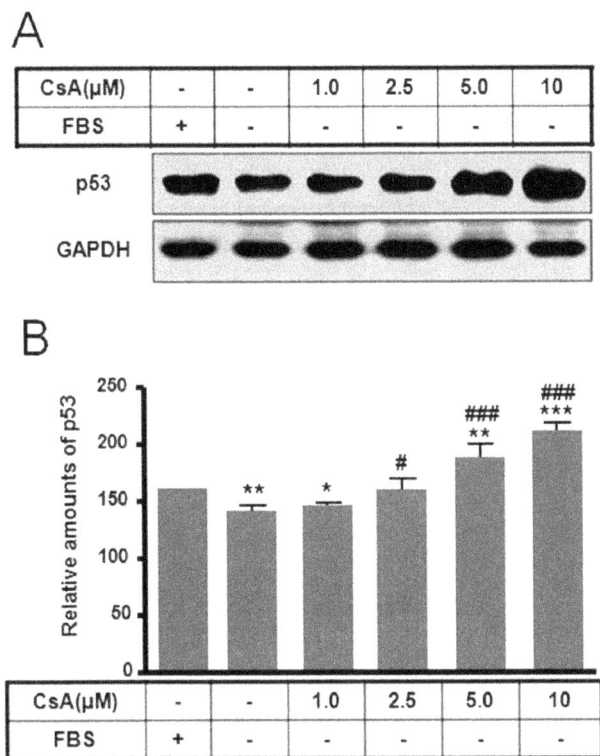

Figure 2. Effect of CsA-mediated apoptotic cell death on p53 expression levels. GH3 cells were incubated in DMEM with or without 10% fetal bovine serum in the presence or absence of CsA (0 to 10 μM) for 10 h. p53 expression (A) was determined by Western blotting and the relative amount (B) was calculated as described in the Materials and Methods. *$p<0.05$, **$p<0.01$ vs. serum treatment. #$p<0.05$, ###$p<0.001$ vs. no serum treatment.

Results

Cell viability and LC3 expression following CsA treatment

Many *in vivo* and *in vitro* studies have shown that CsA induces either autophagy or apoptosis [12–14,16,17]. To further these studies and examine how CsA concentration alters cell death pathway fate, we investigated CsA treatment for 10 h at different CsA doses. For autophagy study, polyethyleneimine was coated on culture dish. CsA treatment induced apoptotic and autophagic cell death with distinct toxicities in rat pituitary GH3 cells. CsA treatment (0.1 to 10 μM) decreased survival of rat pituitary GH3 cells in a dose-dependent manner. Apoptosis resulted in an 87% decrease in cell viability following treatment with 10 μM CsA, whereas autophagy resulted in a 30% reduction in cell viability at the same dose (Fig. 1A). At concentrations ranging from 1.0 to 10 μM, CsA induced a dose-dependent increase in the expression of LC3-I and LC3-II (Fig. 1B). At 5 μM CsA, immunofluorescence staining were performed to detect the co-localization of LC3 and lysosomal-associated membrane protein 2 (Lamp2). The increased LC3-positive granules or puncta were co-localized with the increased Lamp2, indicating increases in autophagosomes (Fig. 1C, D). Together, these data indicate that CsA induces autophagy in rat pituitary GH3 cells. In parallel, some CsA concentrations also induced apoptotic cell death in a dose-dependent manner as assayed by nuclear fragmentation with DAPI staining (Fig. 1E). In particular, treatment with 2.5 to 10 μM CsA induced clear nuclear and DNA fragmentation

(Fig. 1E, F) and an increase in p53 expression (Fig. 2) compared to treatment with or without serum. These data indicate that 1.0 to 10 μM CsA can induce dose-dependent autophagy and apoptosis in GH3 cells.

The differences of molecular levels between apoptosis and autophagy

Levels of several molecules implicated in PCD pathways were examined in cells undergoing CsA induced apoptosis or autophagy. In both apoptosis and autophagy, p-ERK levels were highest following treatment with 1.0 μM CsA and decreased following 2.5 to 10 μM CsA treatment (Fig. 3). In contrast, Bax levels were altered in a dose-dependent fashion that varied with the cell death pathway, showing an increase in autophagy (Fig. 4A, B) and apoptosis (Fig. 4E, F). These changes in p-ERK and Bax levels demonstrate that CsA toxicity can influence survival of GH3 cells depending on the CsA dose. Bcl-2 levels increased during autophagy following treatment with 1.0 and 10 μM CsA as measured by western blot (Fig. 4A, C) and immunofluorescence (Fig. 4D), while Bcl-2 levels decreased during apoptosis (Fig. 4E, G). These data suggest that Bcl-2 protein expression may result in difference between apoptosis and autophagy by CsA.

The differences of Cu/Zn- and Mn-SOD between apoptosis and autophagy

CsA-induced nephrotoxicity may result from oxidative stress, and correspondingly, antioxidant enzymes, including SOD, catalase, and glutathione peroxidase, were found to be reduced in CsA related toxicity [18]. To examine these alterations in relationship to CsA-induced cell death, we assayed the levels of Cu/Zn- and Mn-SOD. CsA-mediated autophagy resulted in slightly lower levels of Cu/Zn-SOD expression, while Mn-SOD expression was relatively unchanged (Fig. 5A–C). In apoptotic cells, Cu/Zn-SOD expression was increased following 2.5 μM CsA treatment and higher doses decreased levels to those observed in the serum free condition (Fig. 5D, E). Mn-SOD expression showed a dose-dependent reduction (Fig. 5D, F). These results suggest that a decline in Mn-SOD levels may induce CsA-mediated apoptotic cell death in GH3 cells.

The difference of calbindin-D9k between apoptosis and autophagy

In a recent report, Pallet et al. [13,14] found that CsA induces ER stress in renal tubular cells. Indeed, the ER appears to be an initiator or a regulator of apoptosis [19]. Stimulation of inositol 1,4,5-triphosphate (IP3) causes Ca^{2+} release from the ER [20], which is involved in apoptotic signal transduction and is required for Ca^{2+}-dependent DNA fragmentation [21]. To examine this pathway, we assayed the levels of the intracellular Ca^{2+} modulator calbindin-D9k. Calbindin-D9k levels were increased in autophagy compared to the serum-free control (Fig. 6A, B). However, calbindin-D9k levels were significantly decreased in apoptosis induced by 1.0 to 5 μM CsA (Fig. 6C, D). This result suggests that calbindin-D9k may be an important molecular determinant of apoptotic or autophagic cell death by CsA in rat pituitary GH3 cells.

Discussion

CsA has been shown to induce autophagic and apoptotic cell death in both *in vivo* and *in vitro* experiments [12–14,16,17]. These researches suggest that the relative levels of ER stress responses may be a determinant of apoptotic cell death by

A

CsA(μM)	-	-	1.0	2.5	5.0	10
FBS	+	-	-	-	-	-

p-ERK

ERK

B

C

CsA(μM)	-	-	1.0	2.5	5.0	10
FBS	+	-	-	-	-	-

p-ERK

ERK

D

Relative amounts of p-ERK (B)

CsA(μM)	-	-	1.0	2.5	5.0	10
FBS	+	-	-	-	-	-

Relative amounts of p-ERK (D)

CsA(μM)	-	-	1.0	2.5	5.0	10
FBS	+	-	-	-	-	-

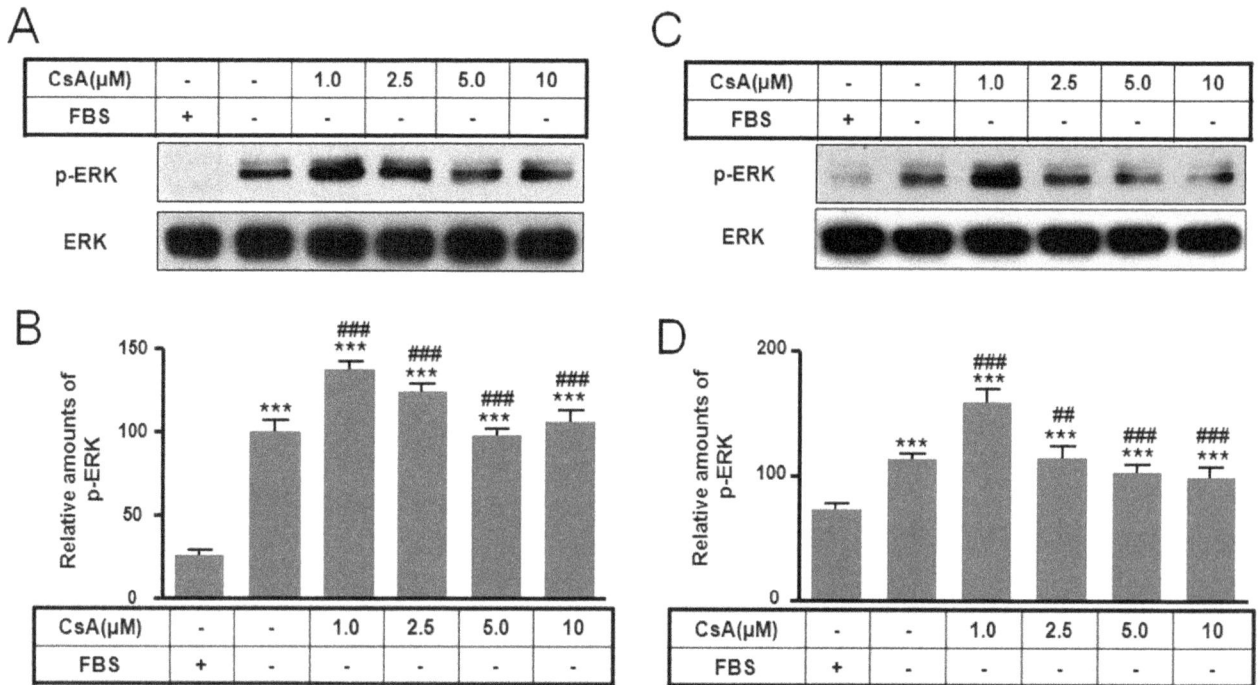

Figure 3. Effect of CsA-mediated autophagic and apoptotic cell death on p-ERK levels. GH3 cells were incubated in DMEM with or without 10% fetal bovine serum in the presence or absence of CsA (0 to 10 μM) for 10 h. Levels of p-ERK for autophagy (A) and apoptosis (C) were determined by Western blotting and the relative amount of p-ERK for autophagy (B) and apoptosis (D) was calculated as described in the Materials and Methods. ***$p<0.01$ vs. serum treatment. ##$p<0.01$, ###$p<0.001$ vs. no serum treatment.

depleting proapoptotic proteins. In other experiments, CsA induced apoptosis by up-regulating the proapoptotic factors p53 and Bax, cleaving PARP, and down-regulating the antiapoptotic factor Bcl-2 in cultured rat mesangial cells and in a rat chronic nephrotoxicity model [22,23].

The mechanisms underlying chronic CsA nephropathy are not completely understood. Activation of the intrarenal renin-angiotensin system, increased release of endothelin-1, inappropriate apoptosis, stimulation of inflammatory mediators, and ER stress have all been implicated in the pathogenesis of chronic CsA nephropathy [10]. Pellet et al. [13] demonstrated that CsA induces autophagy in primary cultured human renal tubular cells through LC3-II expression and autophagosome visualization by electron microscopy. CsA-induced autophagy may occur downstream of ER stress. Various ER stresses activate autophagy, and salubrinal, an inhibitor of eIF2alpha dephosphorylation, both protects cells against ER stress and inhibits LC3-II expression.

The cross-talk between autophagic and apoptotic cell death pathways is complex [4,5]. Three different types of interaction between these pathways have been postulated: (1) an apoptosis/autophagy partnership; (2) autophagy may antagonize apoptosis; or (3) autophagy may enable apoptosis. We demonstrated that CsA treatment increased nuclear fragmentation and Bax-2 levels, and reduced Bcl-2, resulting in apoptotic cell death. Translocation of Bcl-2 family members from the cytoplasm to the mitochondria is a central step in propagation of apoptotic signals to the cytoplasm [24–26]. In contrast, Bcl-2 levels increased during autophagic cell death. This result suggests that autophagy may act to interfere or to attenuate apoptosis and participate in cell protection. Autophagy was shown to be essential for survival during nutrient starvation *in vivo* and *in vitro* [27–30]. Autophagy also protected epithelial cells from apoptotic cell death [31] during

metabolic stress [32], drug treatment, and radiation damage for genotoxic ROS [33].

Our data demonstrate that CsA activates p53 and triggers apoptotic cell death in GH3 cells. These data are consistent with other reports, which showed CsA treatment activates p53, induces nuclear localization, and activates the expression of the p53 target genes Bax, mdm2, and p21^waf1 [24]. Importantly, inhibition of endogenous p53 or lack of functional p53 significantly reduces the extent of CsA-induced apoptosis, indicating that p53 is critical for CsA-mediated cell death. Our data show that an immunosuppressive drug is capable of activating p53 in GH3 cells.

Although a direct link between CsA-mediated ROS generation and adverse renal toxicity in rat renal tubular cells has not been demonstrated [34], it has been suggested that increased ROS levels may contribute to adverse CsA side effects [35,36]. CsA-induced ROS generation and susceptibility to ROS are highly tissue-specific [37]. Autophagy is also regulated by ROS, including superoxide and hydrogen peroxide, though superoxide may be the central ROS that regulates autophagy [38,39]. To examine the role of ROS in CsA mediated cell death, we examined ERK and SOD protein levels, components of the pathways that are activated in response to oxidative stress [40]. In addition, both phosphoinositide 3-kinase/protein kinase B (PI3K/PKB) and ERK pathways have been implicated in survival and death responses of mouse kidney cells [41]. CsA treatment of GH3 cells reduced Cu/Zn-SOD expression in autophagy (Fig. 5A, B) and Mn-SOD expression in apoptosis in a dose dependent fashion (Fig. 5D, F). These results suggest that the changes in Cu/Zn- and Mn-SOD expression are associated with apoptotic and autophagic cell death by CsA treatment in GH3 cells.

The ER appears to function as a key initiator or a regulator of apoptosis with Ca^{2+}-mediated signaling [19,42], An increase in

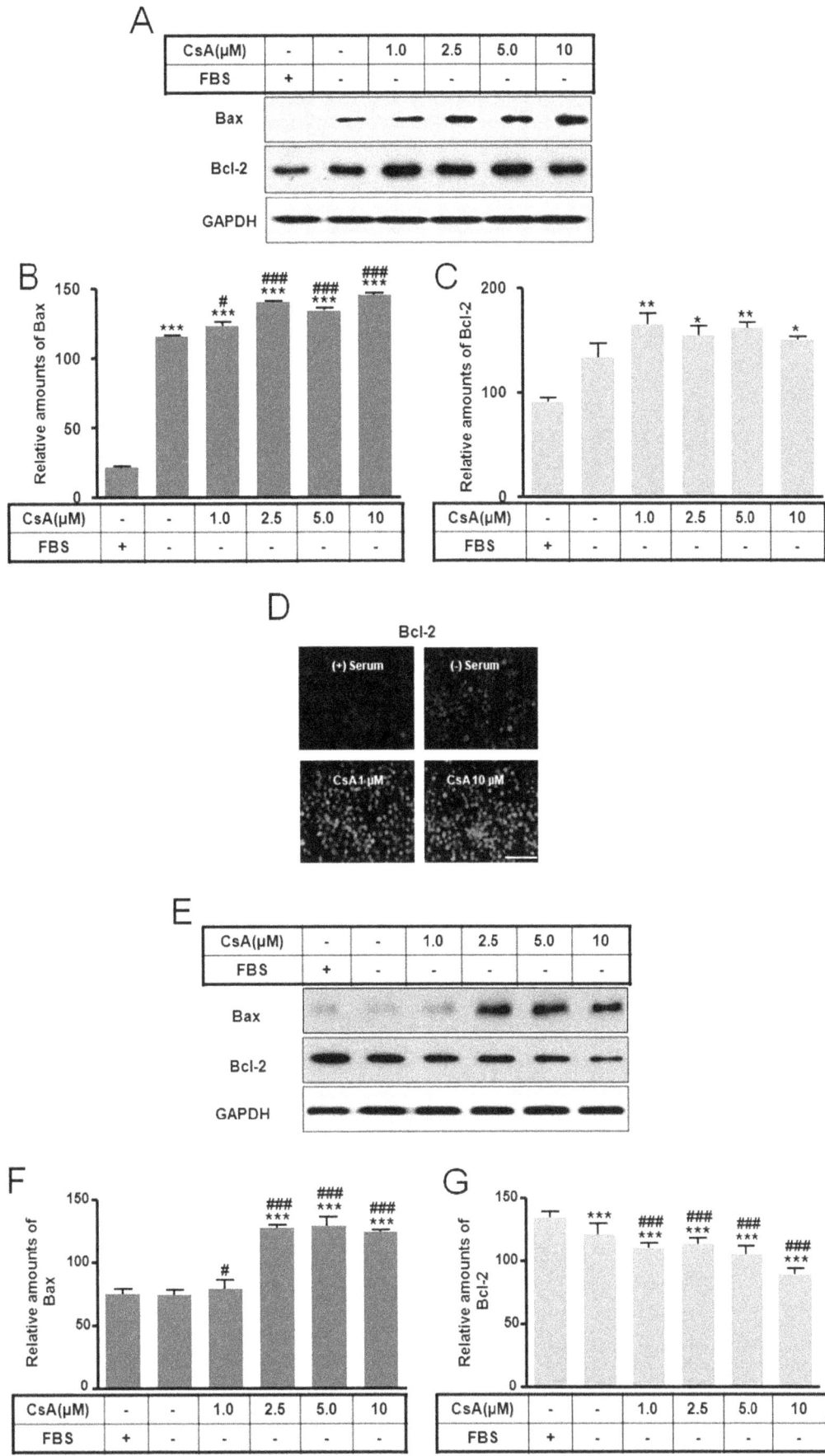

Figure 4. Effect of CsA-mediated autophagic and apoptotic cell death on Bax and Bcl-2 levels. GH3 cells were incubated in DMEM with or without 10% fetal bovine serum in the presence or absence of CsA (0 to 10 μM) for 10 h. Levels of Bax and Bcl-2 for autophagy (A) and apoptosis (E) were determined by Western blotting and the relative amount of Bax and Bcl-2 for autophagy (B, C) and apoptosis (F, G) was calculated as described in the Materials and Methods. Bcl-2 was imaged on an OLYMPUS DP controller and manager using an inverted microscope (D). *$p<0.05$, **$p<0.01$, ***$p<0.001$ vs. serum treatment. #$p<0.05$, ###$p<0.001$ vs. no serum treatment. Scale bar is 100 μm.

mitochondrial matrix Ca^{2+} regulates metabolism, and Ca^{2+} also modulates mitochondrial permeability transition, which is controlled by permeability transition pores [43]. Permeability transition pores have been implicated in apoptotic, autophagic, and necrotic cell death pathways. We examined calbindin-D9k levels indirectly as a marker for intracellular Ca^{2+} modulation. Calbindin-D9k is expressed in the mammalian intestine (duodenum), kidney, pituitary gland, growth cartilage, bone, and female reproductive tissues [44–47]. Uterine calbindin-D9k has been shown to be involved in the regulation of myometrial activity by intracellular calcium [48]. Our results suggest that calbindin-D9k may act as an important intracellular calcium ion exchanger that regulates apoptotic or autophagic cell death by CsA in rat pituitary GH3 cells.

In summary, we found that CsA induces apoptotic and autophagic cell death in rat pituitary GH3 cells. Apoptotic and autophagic cell death could be distinguished both morphologically and molecularly as they displaced distinct levels of Bcl-2 and Mn-SOD expression.

Materials and Methods

Cell culture

Rat pituitary GH3 cells, which originated from the growth hormone-producing tumor of the rat anterior pituitary and somatomammotroph phenotype, were purchased from ATCC (CCL-82.1) and were cultured in Dulbecco's modified Eagle's medium (DMEM, GibcoBRL, Gaithersburg, MD, USA) supplemented with or without 10% heat-inactivated fetal bovine serum (GibcoBRL) at $37°C$ in 5% CO_2 in a humidified atmosphere. Cells were treated with CsA (0 to 100 μM) in serum-free medium. For autophagy studies, culture dishes were coated with polyethyleneimine (20 μg/mL) (Sigma, St. Louis, MO, USA), washed with distilled water, and dried.

Cell viability assays

Cell survival was quantified using Cell Counting Kit-8 (Dojindo Laboratories, Tokyo, Japan). In brief, GH3 cells were cultured in 96-well plates (Corning Inc., Corning, NY, USA) at a density of $5×10^3$ cells per well. The cells were cultured in the presence or absence of melatonin. After 15 h, cells were washed, treated with Cell Counting Kit-8 reagents, incubated in the dark for 4 h, and then absorbance (450 nm) was measured using a plate reader (Molecular Device, Sunnyvale, CA). Percent viability was calculated as the absorbance of the melatonin-treated sample/control absorbance ×100.

Western blot analysis

Cells were harvested, washed two times with ice-cold PBS, and then resuspended in 20 mM Tris-HCl buffer (pH 7.4) containing a protease inhibitor mixture (0.1 mM phenylmethylsulfonyl fluoride, 5 μg/mL aprotinin, 5 μg/mL pepstatin A, 1 μg/mL chymostatin) and phosphatase inhibitors (5 mM Na_3VO_4, 5 mM NaF). Whole cell lysates were prepared with a Dounce homogenizer (20 strokes), followed by centrifugation at 13,000×g for 20 min at $4°C$. Protein concentration was determined using a BCA assay (Sigma). Proteins (50 μg) were separated by 12% SDS-PAGE and then transferred

onto polyvinylidene difluoride (PVDF) membranes. Membranes were hybridized with antibodies specific for LC3 (Santa Cruz Biotechnology, Santa Cruz, CA, USA), p53 (Santa Cruz Biotechnology), Cu/Zn- and Mn-SOD (Cell Signaling Technology, Beverly, MA, USA), p-ERK and ERK (Transduction Laboratories, Lexington, KY, USA), Bax and Bcl-2 (Santa Cruz Biotechnology), calbindin-D9k (Swant, Bellinzona, Switzerland), and GAPDH (Assay Designs, Ann Arbor, MI, USA). Immunoreactive proteins were visualized by exposure of the membrane to X-ray film. The films were scanned and the band intensities and optical densities were corrected by background subtraction, quantified using ImageJ analysis software (version 1.52, Wayne Rasband, NIH, Bethesda, MD, USA), and then normalized.

Immunofluorescence staining

GH3 cells grown on culture slides (BD Falcon Labware, REF 354108) were permeabilized and fixed in methanol at $-20°C$ for 3 min. Cells were washed with phosphate-buffered saline (PBS), blocked with 10% bovine serum albumin (Sigma) with PBS for 10 min, and incubated with primary antibody in blocking buffer for 1 h at room temperature (RT). Cells were hybridized with secondary antibodies for 1 h at RT. The coverslips were mounted on glass slides using Vectashield mounting medium (Vector Labs Inc., Burlingame, CA, USA). Cells were viewed under a Leica TCS SP5 confocal microscope (Leica, Microsystems CMS GmbH, Germany). The following primary antibodies were used: Lamp2 (Cell Signaling Technology) and LC3 (Cell Signaling Technology). The following secondary antibodies were used: Alexa 594 (red)-conjugated anti-rabbit IgG (Vector Laboratories Inc.) and fluorescein isothiocyanate (green)-labeled anti-mouse IgG (Jackson ImmunoResearch Laboratories, West Grove, PA, USA).

Nuclear staining and immunofluorescence microscopy

Cells were fixed in 4% paraformaldehyde buffered with 0.1 M phosphate (pH 7.3) for 30 min and then washed with phosphate-buffered saline (PBS). Cells were permeabilized with 0.3% Triton X-100 for 20 min, washed with PBS, and then stained with 4,6-diamidino-2-phenylindole (DAPI, Santa Cruz, USA) for 10 min. For immunofluorescence microscopy, GH3 cells were grown on culture slides (Nunclon, Gibco cat. no. 176740, NV Invitrogen SA, Merelbeke, Belgium) and then fixed in cold methanol for 10 min at $20°C$. Cells were washed in PBS, blocked with 5% bovine serum albumin in PBS for 30 min, and incubated with primary antibody (Bcl-2, Santa Cruz Biotechnology) in 2% bovine serum albumin (BSA) for 1 h at room temperature. Cells were washed with PBS and subsequently incubated with secondary antibody (fluorescein isothiocyanate (green)-labeled anti-mouse IgG (Jackson ImmunoResearch Laboratories) in 2% BSA for 1 h at room temperature. After washing with PBS, images were acquired at a peak excitation wavelength of 340 nm (OlympusIX71) using an OLYMPUS DP controller and manager (200× magnification).

DNA fragmentation analysis

Cell pellets were resuspended in 750 μL of lysis buffer (20 mm Tris-HCl, 10 mm EDTA, and 0.5% Triton X-100, pH 8.0) and left on ice for 45 min with occasional shaking. DNA was extracted

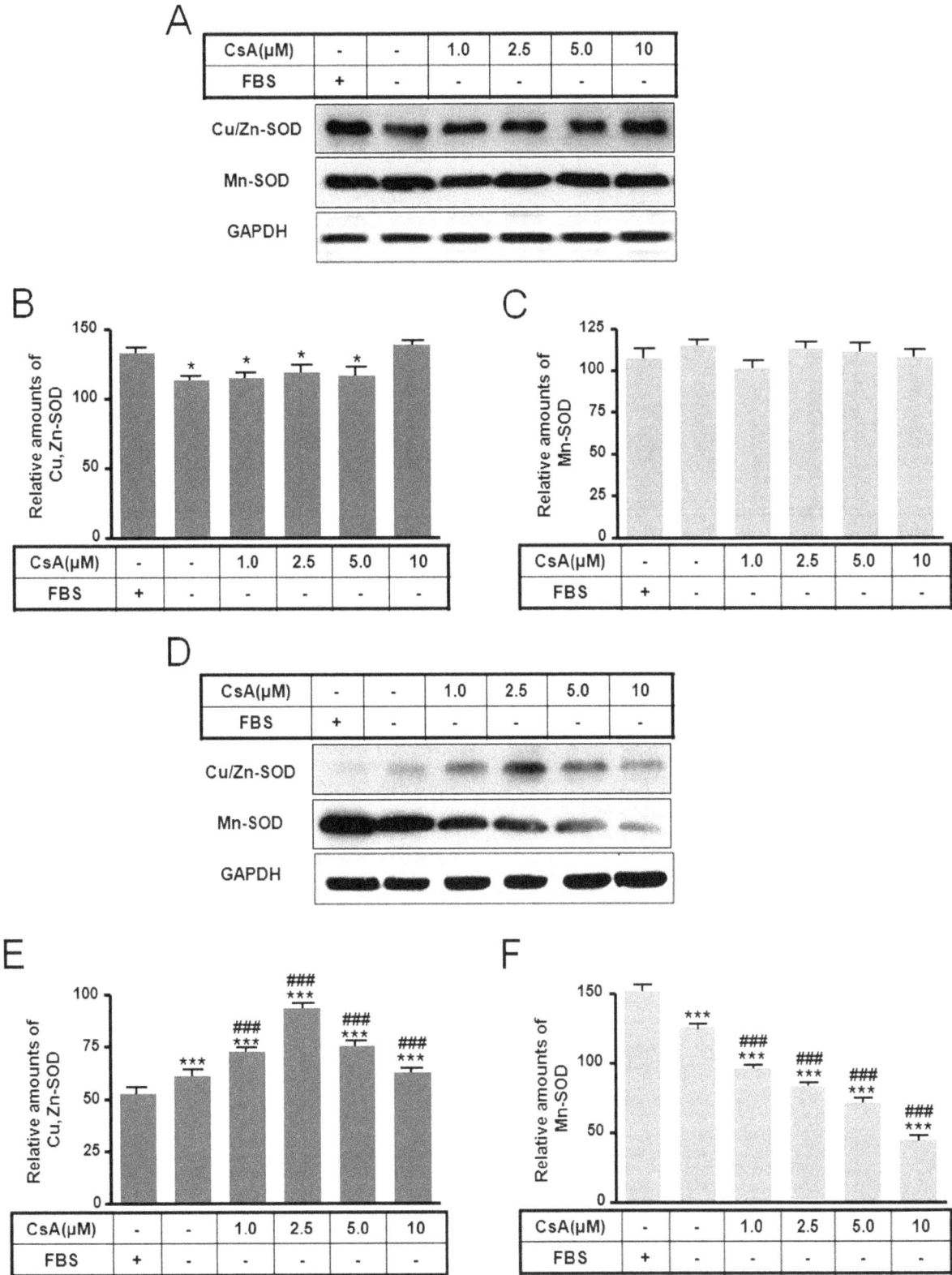

Figure 5. Effect of CsA-mediated autophagic and apoptotic death on Cu/Zn- and Mn-SOD levels. GH3 cells were incubated in DMEM with or without 10% fetal bovine serum in the presence or absence of CsA (0 to 10 μM) for 10 h. Cu/Zn- and Mn-SOD levels for autophagy (A) and apoptosis (D) were determined by Western blotting and the relative amount for autophagy (B, C) and apoptosis (E, F) was calculated as described in the Materials and Methods. *$p<0.05$, ***$p<0.001$ vs. serum treatment. ###$p<0.001$ vs. no serum treatment.

A

C

B

D

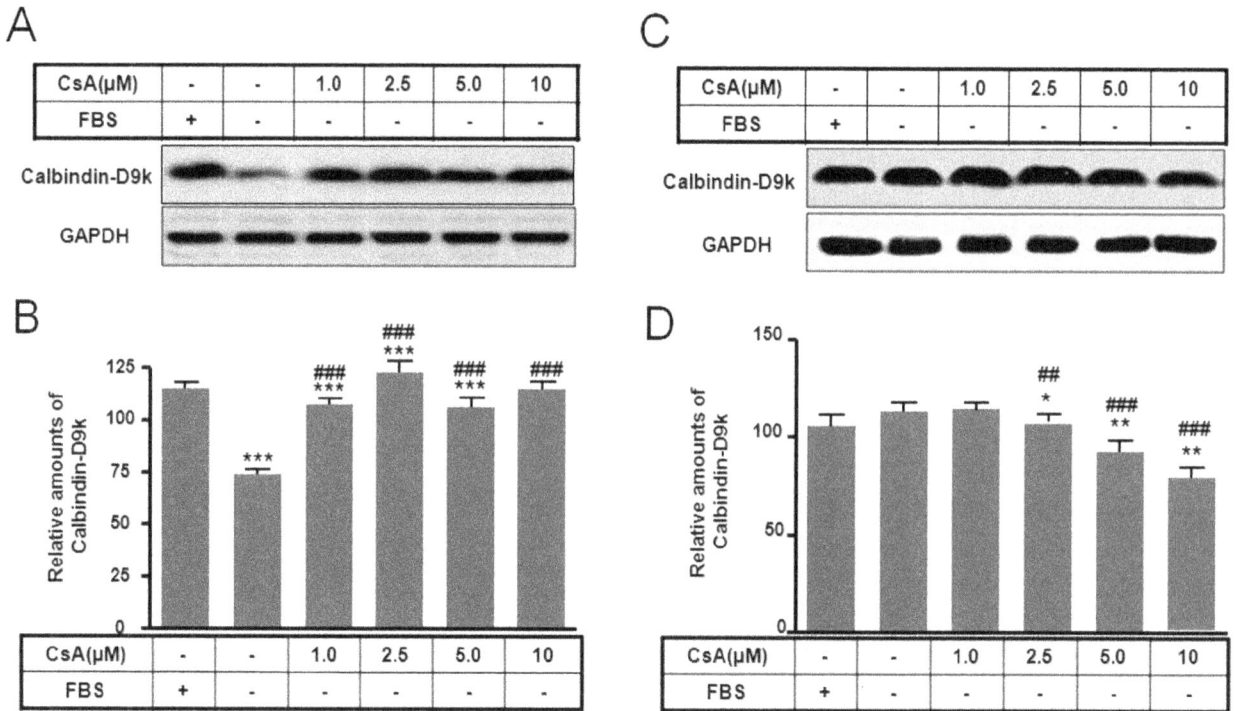

Figure 6. Effect of CsA-mediated autophagic and apoptotic death on calbindin-D9k levels. GH3 cells were incubated in DMEM with or without 10% fetal bovine serum in the presence or absence of CsA (0 to 10 μM) for 10 h. Calbindin-D9k levels for autophagy (A) and apoptosis (C) were determined by Western blotting and the relative amount of calbindin-D9k for autophagy (B) and apoptosis (D) was calculated as described in the Materials and Methods. $*p<0.05$, $**p<0.01$, $***p<0.001$ vs. serum treatment. $\#\#p<0.01$, $\#\#\#p<0.001$ vs. no serum treatment.

with phenol/chloroform and precipitated with alcohol. The precipitate was dried and resuspended in 100 μL of 20 mM Tris-HCl, pH 8.0. After degradation of RNA with RNase (0.1 mg/mL) at 37°C for 1 h, samples (15 μL) were electrophoresed on a 1.2% agarose gel in 450 mM Tris borate-EDTA buffer (TBE, pH 8.0) and photographed under UV light.

Statistical analysis

Significant differences were detected by ANOVA, followed by Tukey's test for multiple comparisons. Analysis was performed using the Prism Graph Pad v4.0 (Graph Pad Software Inc., San Diego, CA, USA). Values are expressed as means ± SD of at least three separated experiments, in which case a representative experiment is depicted in the figures. P values <0.05 were considered statistically significant.

Author Contributions

Conceived and designed the experiments: HSK YMY. Performed the experiments: HSK SIC EBJ YMY. Analyzed the data: HSK EBJ YMY. Contributed reagents/materials/analysis tools: EBJ YMY. Wrote the paper: EBJ YMY.

References

1. Kerr JF (1965) A histochemical study of hypertrophy and ischaemic injury of rat liver with special reference to changes in lysosomes. J Pathol Bacteriol 90: 419–435.
2. Stromhaug PE, Klionsky DJ (2001) Approaching the molecular mechanism of autophagy. Traffic 2: 524–531.
3. Kabeya Y, Mizushima N, Ueno T, Yamamoto A, Kirisako T, et al. (2000) LC3, a mammalian homologue of yeast Apg8p, is localized in autophagosome membranes after processing. EMBO J 19: 5720–5728.
4. Lockshin RA, Zakeri Z (2004) Apoptosis, autophagy, and more. Int J Biochem Cell Biol 36: 2405–2419.
5. Eisenberg-Lerner A, Bialik S, Simon HU, Kimchi A (2009) Life and death partners: apoptosis, autophagy and the cross-talk between them. Cell Death Differ 16: 966–975.
6. Mills KR, Reginato M, Debnath J, Queenan B, Brugge JS (2004) Tumor necrosis factor-related apoptosis-inducing ligand (TRAIL) is required for induction of autophagy during lumen formation in vitro. Proc Natl Acad Sci USA 101: 3438–3443.
7. Cárdenas-Aguayo Mdel C, Santa-Olalla J, Baizabal JM, Salgado LM, Covarrubias L (2003) Growth factor deprivation induces an alternative non-apoptotic death mechanism that is inhibited by Bcl2 in cells derived from neural precursor cells. J Hematother Stem Cell Res 12: 735–748.

8. Yanagisawa H, Miyashita T, Nakano Y, Yamamoto D (2003) HSpin1, a transmembrane protein interacting with Bcl-2/Bcl-xL, induces a caspase-independent autophagic cell death. Cell Death Differ 10: 798–807.
9. Cecka JM, Terasaki PI (1991) The UNOS Scientific Renal Transplant Registry. Clin Transpl 1991; 1–11.
10. Yoon HE, Yang CW (2009) Established and newly proposed mechanisms of chronic cyclosporine nephropathy. Korean J Intern Med 24: 81–92.
11. Olyaei AJ, de Mattos AM, Bennet WM (2001) Nephrotoxicity of immunosuppressive drugs: new insight and preventive strategies. Curr Opin Crit Care 7: 384–389.
12. Pallet N, Anglicheau D (2009) Autophagy: a protective mechanism against nephrotoxicant-induced renal injury. Kidney Int 75: 118–119.
13. Pallet N, Bouvier N, Bendjallabah A, Rabant M, Flinois JP, et al. (2008) Cyclosporine-induced endoplasmic reticulum stress triggers tubular phenotypic changes and death. Am J Transplant 8: 2283–2296.
14. Pallet N, Bouvier N, Legendre C, Gilleron J, Codogno P, et al. (2008) Autophagy protects renal tubular cells against cyclosporine toxicity. Autophagy 4: 783–791.
15. Ciechomska IA, Gabrusiewicz K, Szczepankiewicz AA, Kaminska B (2013) Endoplasmic reticulum stress triggers autophagy in malignant glioma cells undergoing cyclosporine a-induced cell death. Oncogene 32: 1518–1529.
16. Daly PJ, Docherty NG, Healy DA, McGuire BB, Fitzpatrick JM, et al. (2009) The single insult of hypoxic preconditioning induces an antiapoptotic response

in human proximal tubular cells, in vitro, across cold storage. BJU Int 103: 254–259.

17. Han SW, Li C, Ahn KO, Lim SW, Song HG, et al. (2008) Prolonged endoplasmic reticulum stress induces apoptotic cell death in an experimental model of chronic cyclosporine nephropathy. Am J Nephrol 28: 707–714.

18. Longoni B, Migliori M, Ferretti A, Origlia N, Panichi V, et al. (2002) Melatonin prevents cyclosporine-induced nephrotoxicity in isolated and perfused rat kidney. Free Radic Res 36: 357–363.

19. Rutkowski DT, Kaufman RJ (2004) A trip to the ER: coping with stress. Trends Cell Biol 14: 20–28.

20. Oakes SA, Scorrano L, Opferman JT, Bassik MC, Nishino M, et al. (2005) Proapoptotic BAX and BAK regulate the type 1 inositol trisphosphate receptor and calcium leak from the endoplasmic reticulum. Proc Natl Acad Sci USA 102: 105–110.

21. Boulares AH, Zoltoski AJ, Contreras FJ, Yakovlev AG, Yoshihara K, et al. (2002) Regulation of DNAS1L3 endonuclease activity by poly(ADP-ribosyl)ation during etoposide-induced apoptosis. Role of poly(ADP-ribose) polymerase-1 cleavage in endonuclease activation. J Biol Chem 277: 372–378.

22. Han SY, Chang EJ, Choi HJ, Kwak CS, Park SB, et al. (2006) Apoptosis by cyclosporine in mesangial cells. Transplant Proc 38: 2244–2246.

23. Shihab FS, Andoh TF, Tanner AM, Yi H, Bennett WM (1999) Expression of apoptosis regulatory genes in chronic cyclosporine nephrotoxicity favors apoptosis-Adult male Sprague-Dawley rats. Kidney Int 56: 2147–2159.

24. Pyrzynska B, Serrano M, Martínez-A C, Kaminska B (2002) Tumor suppressor p53 mediates apoptotic cell death triggered by cyclosporin A. J Biol Chem 277: 14102–14108.

25. Goping IS, Gross A, Lavoie JN, Nguyen M, Jemmerson R, et al. (1998) Regulated targeting of BAX to mitochondria. J Cell Biol 143: 207–215.

26. Wolter KG, Hsu YT, Smith CL, Nechushtan A, Xi XG, et al. (1997) Movement of Bax from the cytosol to mitochondria during apoptosis. J Cell Biol 139: 1281–1292.

27. Komatsu M, Waguri S, Ueno T, Iwata J, Murata S, et al. (2005) Impairment of starvationinduced and constitutive autophagy in Atg7-deficient mice. J Cell Biol 169: 425–434.

28. Kuma A, Hatano M, Matsui M, Yamamoto A, Nakaya H, et al. (2004) The role of autophagy during the early neonatal starvation period. Nature 432: 1032–1036.

29. Boya P, González-Polo RA, Casares N, Perfettini JL, Dessen P, et al. (2000) Inhibition of macroautophagy triggers apoptosis. Mol Cell Biol 25: 1025–1040.

30. Lum JJ, Bauer DE, Kong M, Harris MH, Li C, et al. (2005) Growth factor regulation of autophagy and cell survival in the absence of apoptosis. Cell 120: 237–248.

31. Fung C, Lock R, Gao S, Salas E, Debnath J (2008) Induction of autophagy during extracellular matrix detachment promotes cell survival. Mol Biol Cell 19: 797–806.

32. Karantza-Wadsworth V, Pate S, Kravchuk O, Chen G, Mathew R, et al. (2007) Autophagy mitigates metabolic stress and genome damage in mammary tumorigenesis. Genes Dev 21: 1621–1635.

33. Ito H, Daido S, Kanzawa T, Kondo S, Kondo Y (2005) Radiation-induced autophagy is associated with LC3 and its inhibition sensitizes malignant glioma cells. Int J Oncol 26: 1401–1410.

34. Galletti P, Di Gennaro CI, Migliardi V, Indaco S, Della Ragione F, et al. (2005) Diverse effects of natural antioxidants on cyclosporin cytotoxicity in rat renal tubular cells. Nephrol Dial Transplant 20: 1551–1558.

35. Pérez de Lema G, Arribas I, Prieto A, Parra T, de Arriba G, et al. (1998) Cyclosporin A-induced hydrogen peroxide synthesis by cultured human mesangial cells is blocked by exogenous antioxidants. Life Sci 62: 1745–1753.

36. Wolf A, Trendelenburg CF, Diez-Fernandez C, Prieto P, Houy S, et al. (1997) Cyclosporine A-induced oxidative stress in rat hepatocytes. J Pharmacol Exp Ther 280: 1328–1334.

37. Buetler TM, Cottet-Maire F, Krauskopf A, Ruegg UT (2000) Does cyclosporin A generate free radicals? Trends Pharmacol Sci 21: 288–290.

38. Chen Y, Azad MB, Gibson SB (2009) Superoxide is the major reactive oxygen species regulating autophagy. Cell Death Differ 16: 1040–1052.

39. Lim SW, Hyoung BJ, Piao SG, Doh KC, Chung BH, et al. (2012) Chronic cyclosporine nephropathy is characterized by excessive autophagosome formation and decreased autophagic clearance. Transplantation 94: 218–225.

40. Blanc A, Pandey NR, Srivastava AK (2003) Synchronous activation of ERK 1/2, p38mapk and PKB/Akt signaling by H_2O_2 in vascular smooth muscle cells: potential involvement in vascular disease. Int J Mol Med 1: 229–234.

41. Sarró E, Tornavaca O, Plana M, Meseguer A, Itarte E (2008) Phosphoinositide 3-kinase inhibitors protect mouse kidney cells from cyclosporine-induced cell death. Kidney Int 73: 77–85.

42. Oakes SA, Opferman JT, Pozzan T, Korsmeyer SJ, Scorrano L (2003) Regulation of endoplasmic reticulum Ca^{2+} dynamics by proapoptotic BCL-2 family members. Biochem Pharmacol 66: 1335–1340.

43. Smaili SS, Hsu YT, Carvalho AC, Rosenstock TR, Sharpe JC, et al. (2003) Mitochondria, calcium and pro-apoptotic proteins as mediators in cell death signaling. Braz J Med Biol Res 36: 183–190.

44. Choi KC, Leung PC, Jeung EB (2005) Biology and physiology of Calbindin-D9k in female reproductive tissues: involvement of steroids and endocrine disruptors. Reprod Biol Endocrinol 3: 66.

45. Lee GS, Choi KC, Park SM, An BS, Cho MC, et al. (2003) Expression of human Calbindin-D(9k) correlated with age, vitamin D receptor and blood calcium level in the gastrointestinal tissues. Clin Biochem 36: 255–261.

46. Nguyen TH, Lee GS, Ji YK, Choi KC, Lee CK, et al. (2005) A calcium binding protein, calbindin-D9k, is mainly regulated by estrogen in the pituitary gland of rats during estrous cycle. Brain Res Mol Brain Res 141: 166–173.

47. Tinnanooru P, Dang VH, Nguyen TH, Lee GS, Choi KC, et al. (2008) Estrogen regulates the localization and expression of calbindin-D9k in the pituitary gland of immature male rats via the ERalpha-pathway. Mol Cell Endocrinol 285: 26–33.

48. Choi KC, Jeung EB (2008) Molecular mechanism of regulation of the calcium-binding protein calbindin-D9k, and its physiological role(s) in mammals: a review of current research. J Cell Mol Med 12: 409–420.

6

Elevation of Proteasomal Substrate Levels Sensitizes Cells to Apoptosis Induced by Inhibition of Proteasomal Deubiquitinases

Chao Sun[1], Peristera Roboti[2], Marjo-Riitta Puumalainen[2¤], Mårten Fryknäs[3], Xin Wang[1], Padraig D'Arcy[1], Malin Hult[4], Stephen High[2], Stig Linder[1], Eileithyia Swanton[2]*

1 Cancer Center Karolinska, Department of Oncology and Pathology, Karolinska Institute, Stockholm, Sweden, 2 Faculty of Life Sciences, University of Manchester, Manchester, United Kingdom, 3 Department of Medical Sciences, Division of Clinical Pharmacology, Uppsala University, Uppsala, Sweden, 4 Center for Inherited Metabolic Diseases, Karolinska University Hospital, Stockholm, Sweden

Abstract

Inhibitors of the catalytic activity of the 20S proteasome are cytotoxic to tumor cells and are currently in clinical use for treatment of multiple myeloma, whilst the deubiquitinase activity associated with the 19S regulatory subunit of the proteasome is also a valid target for anti-cancer drugs. The mechanisms underlying the therapeutic efficacy of these drugs and their selective toxicity towards cancer cells are not known. Here, we show that increasing the cellular levels of proteasome substrates using an inhibitor of Sec61-mediated protein translocation significantly increases the extent of apoptosis that is induced by inhibition of proteasomal deubiquitinase activity in both cancer derived and non-transformed cell lines. Our results suggest that increased generation of misfolded proteasome substrates may contribute to the mechanism(s) underlying the increased sensitivity of tumor cells to inhibitors of the ubiquitin-proteasome system.

Editor: Ruby John Anto, Rajiv Gandhi Centre for Biotechnology, India

Funding: Funding was provided by Cancerfonden www.cancerfonden.se, Cancerföreningen in Stockholm www.rahfo.se, Vetenskapsrådet www.vr.se, EU FP6 (LSHC-CT-2007-037665) http://ec.europa.eu/programmes/horizon2020/, King Gustav V Jubilee Foundation and the Wellcome Trust (081671/B/06/Z) www.wellcome.ac.uk The funders had no role in study design, data collection and analysis, decision to publish, or preparation of the manuscript.

Competing Interests: Translocation inhibitor cpd A was kindly provided by Novartis.

* Email: lisa.swanton@manchester.ac.uk

¤ Current address: Institute of Pharmacology and Toxicology, University of Zürich-Vetsuisse, Zürich, Switzerland

Introduction

It has been estimated that as much as one-third of all proteins are destroyed within minutes of synthesis at the ribosomes [1–3]. These highly labile polypeptides include defective ribosomal translation products, as well as proteins that fold incorrectly during or shortly after synthesis. Misfolded proteins containing non-native structures are inherently cytotoxic [4], and quality control systems operate to identify and rapidly eliminate such aberrant proteins in order to maintain cellular homeostasis. Malignant transformation and tumor growth are associated with disregulated protein translation [5], which together with adverse intracellular conditions commonly experienced in the tumor environment, such as acidification [6] and increased levels of reactive oxygen species [7], may well result in increased generation of misfolded proteins. This hypothesis is further supported by the observation that tumor cells frequently exhibit signs of proteotoxic stress, including increased expression of Hsp70 and Hsp90 chaperones [8–10] and activation of the unfolded protein response (UPR). The level of proteotoxic stress in tumor cells may also be further exacerbated by aneuploidy and the resulting imbalance in components of protein complexes [11,12].

The ubiquitin proteasome system (UPS) is the major intracellular protein degradation system responsible for the removal of defective and misfolded polypeptides in eukaryotes [13]. The 26S proteasome complex consists of a 20S core particle, which contains chymotrypsin-like, trypsin-like and peptidylglutamyl peptide hydrolysing activities [14], and two associated 19S regulatory particles, which control access to the proteolytic core. Proteins are targeted to the proteasome for degradation when they become modified with ubiquitin. Ubiquitin is a highly conserved 76 amino acid protein that is covalently attached to target proteins via a series of enzymatic steps, which culminate in the formation of an isopeptide bond between the C-terminus of ubiquitin and a lysine residue in the target protein [15]. Ubiquitin itself contains 7 lysine residues and additional ubiquitin monomers may be attached to any of these lysine residues, thus building up a polyubiquitin chain on the target protein. Chains of 4 or more ubiquitin molecules, typically linked through lysine 48 of ubiquitin, form highly specific signals for proteasomal degradation [16]. Subunits of the 19S particle act as ubiquitin receptors that bind these polyubiquitin chains and present the ubiquitinated proteasomal substrate to the 20S proteolytic core [16]. Ubiquitin is removed from substrate proteins prior to degradation by the action of deubiquinase (DUB) enzymes, which catalyse hydrolysis of the isopeptide bond and regenerate free ubiquitin monomers [15]. In humans, substrate deubiquitination is catalysed by three proteasome-associated DUBs, USP14 and UCHL5 (or UCH37), which are cysteine proteases, and a metalloprotease RPN11 (or POH1). The relationship between these proteasomal DUBs and their precise roles in regulating substrate degradation are complex and not yet fully understood [17].

Interfering with the UPS in cancer cells has been successfully exploited for therapeutic purposes. Bortezomib (Velcade) is a selective inhibitor of the 20S proteasome that shows cytotoxic activity against several malignant cell types and has been approved by the FDA for the treatment of patients with multiple myeloma [18]. A second protesome inhibitor, carfilzomib, was recently approved for relapsed multiple myeloma, and a number of additional agents are being developed. Despite their demonstrated therapeutic value, the mechanisms underlying the cytotoxicity of proteasome inhibitors are not well defined. A common view is that proteasome inhibition results in the stabilization of proteins that inhibit cell survival [18–21]. NF-κB is one such protein, and this transcription factor has received considerable attention with regard to its potential role in apoptosis induced by proteasome inhibitors [18]. Likewise, the involvement of Myc and Noxa in this process has been investigated [22,23]. Another potential scenario is that the accumulation of aberrant proteasomal substrates mediates the cytotoxic effects of proteasome inhibitors, either as a consequence of their inherent toxicity, or via the activation of stress signalling pathways such as the UPR [24]. Yet another mechanism was recently proposed whereby a fatal depletion of amino acids, due to reduced recycling of amino acids through proteasomal protein degradation, underlies proteasome inhibitor-induced cell death [25].

We recently identified compound b-AP15, an inhibitor of the USP14 and UCHL5 cysteine deubiquitinases of the 19S proteasome [26]. Similar to inhibitors of the 20S proteasome, b-AP15 inhibits proteasome function in cells, leading to an accumulation of polyubiquitinated proteasome substrates, and the compound is effective in a number of solid tumor models and in multiple myeloma tumor models [19,27]. One of the hallmarks of the DUB inhibitor b-AP15 is that it is more effective at inducing apoptosis of tumor cells than non-malignant cells [26], and here we have examined potential mechanisms underlying this selective cytotoxicity. We show that b-AP15 exposure induces greater accumulation of high molecular weight polyubiquitinated proteins in HCT116 colon carcinoma cells than in untransformed cells, suggesting that the cellular load of ubiquitinated proteasomal substrates may modulate the sensitivity of cells to b-AP15. To test this hypothesis, we used an inhibitor of protein translocation at the endoplasmic reticulum to increase the production of misfolded proteins, and found that this increased the sensitivity of cells to b-AP15. In contrast, cellular cysteine levels did not appear to be altered following exposure to b-AP15 and amino acid supplementation did not protect against b-AP15-induced cell death. Together, our results suggest that the polyubiquitinated proteasome substrates, which accumulate upon exposure of cells to b-AP15 contribute to the cytoxic effects of this drug.

Results

Elevated accumulation of ubiquitin conjugates in HCT116 tumor cells treated with an inhibitor of proteasomal deubiquitinases

In order to gain insight into the mechanisms through which the USP14/UCHL5 inhibitor b-AP15 kills cells, we compared the effect of this compound on HCT116 colon cancer cells and non-malignant hTERT-RPE1 cells. Colon cancer cells are among the most sensitive to b-AP15 in the NCI_{60} panel (HCT116, HT29 and HCT15 cell lines show similar levels of sensitivity [26]), and hence

were chosen for studies addressing the mechanism of b-AP15-induced apoptosis. Cells were treated with 1 μM b-AP15 for increasing periods of time, and lysates analysed by immunoblotting with antibodies that recognise K48-linked ubiquitin. A dramatic increase in cellular levels of high-molecular weight ubiquitin-conjugated proteins was observed within 1 hour of drug treatment (**Fig. 1A, right**). A similar time-course of accumulation of ubiquitin-conjugated proteins was observed when HCT116 cells were treated with bortezomib, a clinically used inhibitor of the 20S proteasome (**Fig. 1A, left**). The accumulation of these high molecular weight species required ongoing protein synthesis (not shown), suggesting that they include nascent polypeptides that were ubiquitinated during or soon after synthesis and were en route to proteasomal degradation. Both b-AP15 and bortezomib treatment induced the expression of Hsp70B', a stress-induced chaperone [28], and led to an increase in levels of the proteasome substrate p21 (**Fig. 1A; Fig. S1**). In contrast, Hsp90 expression was not obviously increased following treatment with b-AP15 (**Fig. S1**). These observations are consistent with the compounds blocking proteasomal degradation and thus increasing cellular levels of misfolded proteasome substrates. Continued exposure of HCT116 cells to b-AP15 resulted in apoptosis, as demonstrated by cleavage of poly(ADP-ribose) polymerase (PARP) (**Fig. 1B**). Thus, treatment of HCT116 colon cancer cells with b-AP15 leads to a rapid accumulation of poly-ubiquitinated proteins, upregulation of Hsp70B', and ultimately induction of programmed cell death. We previously demonstrated that untransformed cells are less sensitive to b-AP15-induced apoptosis than cancer cells such as HCT116 cells [26]. Consistent with this, cleavage of PARP was not detected in immortalized epithelial hTERT-RPE1 cells following exposure to b-AP15, despite being clearly apparent in HCT116 cells treated under the same conditions (**Fig. 1B**).

Interestingly, the accumulation of polyubiquitinated proteins induced by b-AP15 was also lower in hTERT-RPE1 cells than in HCT116 cells treated in parallel (**Fig. 1B**). Thus, the increased sensitivity of HCT116 cells to b-AP15-induced cell death correlated with greater accumulation of polyubiquitin conjugates (**Fig. 1B**). Cancer cells are characterised by disregulated translational control and increased rates of protein synthesis, and thus the production of polypeptides that require proteasomal degradation is also likely to be greater in these cells [5]. On this basis, we speculated that it might be the load of proteasome substrates generated by HCT116 cells that determines their sensitivity to b-AP15, a hypothesis that is supported by the correlation between the accumulation of polyubiquitin-conjugated proteins and PARP cleavage following exposure to b-AP15 (**Fig. 1B**).

An inhibitor of protein translocation at the endoplasmic reticulum increases the accumulation of ubiquitin conjugates in cells treated with UPS inhibitors

In order to address the role of proteasomal substrate levels more directly, we examined the effect of increasing the production of putative substrates using cpdA, an inhibitor of Sec61-mediated protein translocation at the endoplasmic reticulum (ER) [29]. The effect of cpdA on ER translocation in HCT116 cells was determined by monitoring the fate of an endogenous glycoprotein, prosaposin. In the absence of cpdA, the prosaposin precursor (preprosaposin, prepSAP) is efficiently translocated into the ER lumen, the signal peptide removed and the protein becomes N-glycosylated at five sites, generating prosaposin (**Fig. 2A, DMSO, labelled pSAP-5**). Treatment with Endoglycosidase H or inhibition of N-glycosylation with tunicamycin produced a un-glycosylated form of prosaposin (**Fig. 2A, labelled pSAP-0**). Inhibition of ER translocation by cpdA has previously been shown

A

B

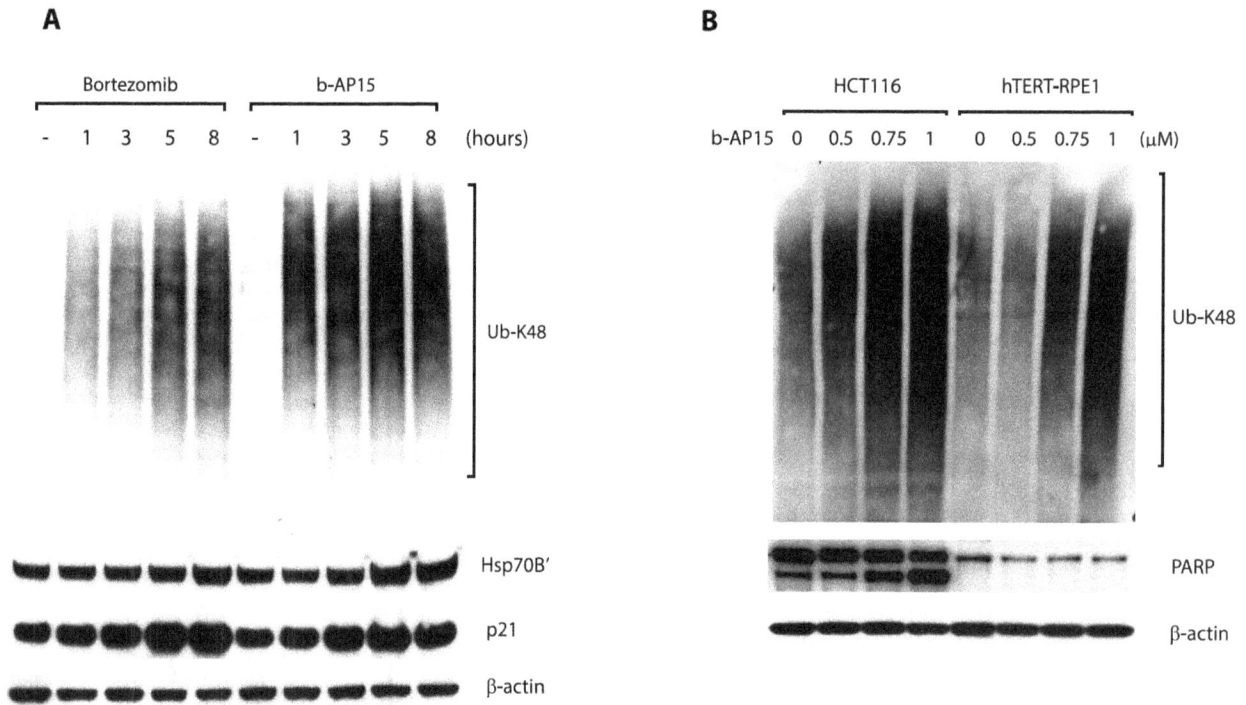

Figure 1. b-AP15 treatment causes accumulation of high molecular weight polyubiquitinated proteins and induces apoptosis. A. HCT116 cells were cultured in the presence of 1 μM b-AP15 or 100 nM bortezomib and harvested at the indicated time points. Lysates were subjected to immunoblotting for K48-linked polyubiquitin, HSP-70B', p21, PARP or β-actin (loading control). **B.** HCT116 cells or hTERT-RPE1 cells were exposed to different concentrations of b-AP15 for 1 hour, followed by washing and incubation for 16 hours in drug-free medium. Lysates were subjected to immunoblotting for K48-linked polyubiquitin, PARP and β-actin (loading control).

to result in the mislocalisation of precursor proteins to the cytosol, where they are targeted for proteasomal degradation [30]. In line with these observations, treatment with cpdA led to the appearance of an additional form of prosaposin in HCT116 cells (**Fig. 2A, labelled prepSAP-0**). The additional form generated in the presence of cpdA was not sensitive to cleavage by Endoglycosidase H (**Fig. 2B**), showing that it lacks nascent N-glycans. Furthermore, this species migrates more slowly than the un-glycosylated prosaposin (**Fig. 2B, compare pSAP-0 and prepSAP-0**), suggesting that it still retains its signal peptide. Thus, we conclude that this additional form represents preprosaposin that has failed to be translocated across the ER membrane, and is therefore unable to undergo post-translational modifications within the ER lumen. These results suggest that cpdA inhibits the translocation of endogenous proteins across the ER membrane of HCT116 cells, leading to the generation of mislocalised polypeptides. The steady state levels of non-translocated pre-prosaposin were not obviously increased by exposure to b-AP15 or bortezomib (**Fig. 2C**), consistent with its rapid polyubiquitination [30] to generate a range of higher molecular weight species that are not readily visible. Indeed, HCT116 cells pre-treated with 10 μM cpdA for 16 h appeared to contain slightly increased levels of polyubiquitin conjugates than untreated cells (**Fig. 2D, 2E, compare lanes 1 and 6**). Taken together, these data suggest that the treatment of HCT116 cells with cpdA leads to increased generation of proteasome substrates.

We next examined the effect of b-AP15 treatment on HCT116 cells in the presence of cpdA to increase the load of endogenous proteasomal substrates. In the presence of cpdA, b-AP15 treatment leads to an even greater increase in the levels of

polyubiquitinated proteins, as compared to cells treated with b-AP15 alone (**Fig. 2D**). A similar effect of cpdA to promote the accumulation of polyubiquitin conjugates was also observed upon treatment with bortezomib (**Fig. 2E**). This increase in the level of polyubiquitinated components upon co-exposure to cpdA was particularly evident 1 h after addition of b-AP15 or bortezomib (**Fig. 2D, 2E, compare lanes 2 and 7**).

In order to examine the effect of cpdA on the accumulation of proteasomal substrates more directly, we utilized Ub^{G76V}-YFP. This model fusion protein is constitutively degraded by the proteasome, and hence steady state levels of Ub^{G76V}-YFP can be used to quantify ubiquitin proteasome-dependent proteolysis in live cells [31]. Treatment of MelJuSo cells stably expressing Ub^{G76V}-YFP with b-AP15 for 6 hours resulted in a marked accumulation of the reporter (**Fig. 3A**), confirming effective inhibition of proteasomal degradation. Although most of the Ub^{G76V}-YFP detected by immunoblotting was not polyubiquiti-nated (**Fig. 3A**), most, but not all, cells that became Ub^{G76V}-YFP positive after exposure to b-AP15 also stained with an antibody to lysine 48-linked polyubiquitin chains (**Fig. 3B**). The effect of b-AP15 on the accumulation of Ub^{G76V}-YFP was determined in real-time using an IncuCyte instrument to quantify the number of YFP-positive cells (**Fig. 3C**). Accumulation of Ub^{G76V}-YFP occurred rapidly following addition of the drug, and YFP-positive cells could be detected in cells treated with 0.25 μM b-AP15 (**Fig. 3C, red trace**). The number of YFP positive cells increased up to 1 μM b-AP15 (**Fig. 3C, blue trace**), and then Ub^{G76V}-YFP accumulation decreased at higher concentrations of b-AP15, potentially due to its toxicity at these levels. Interestingly, when cells were co-treated with cpdA, YFP-positive cells were now

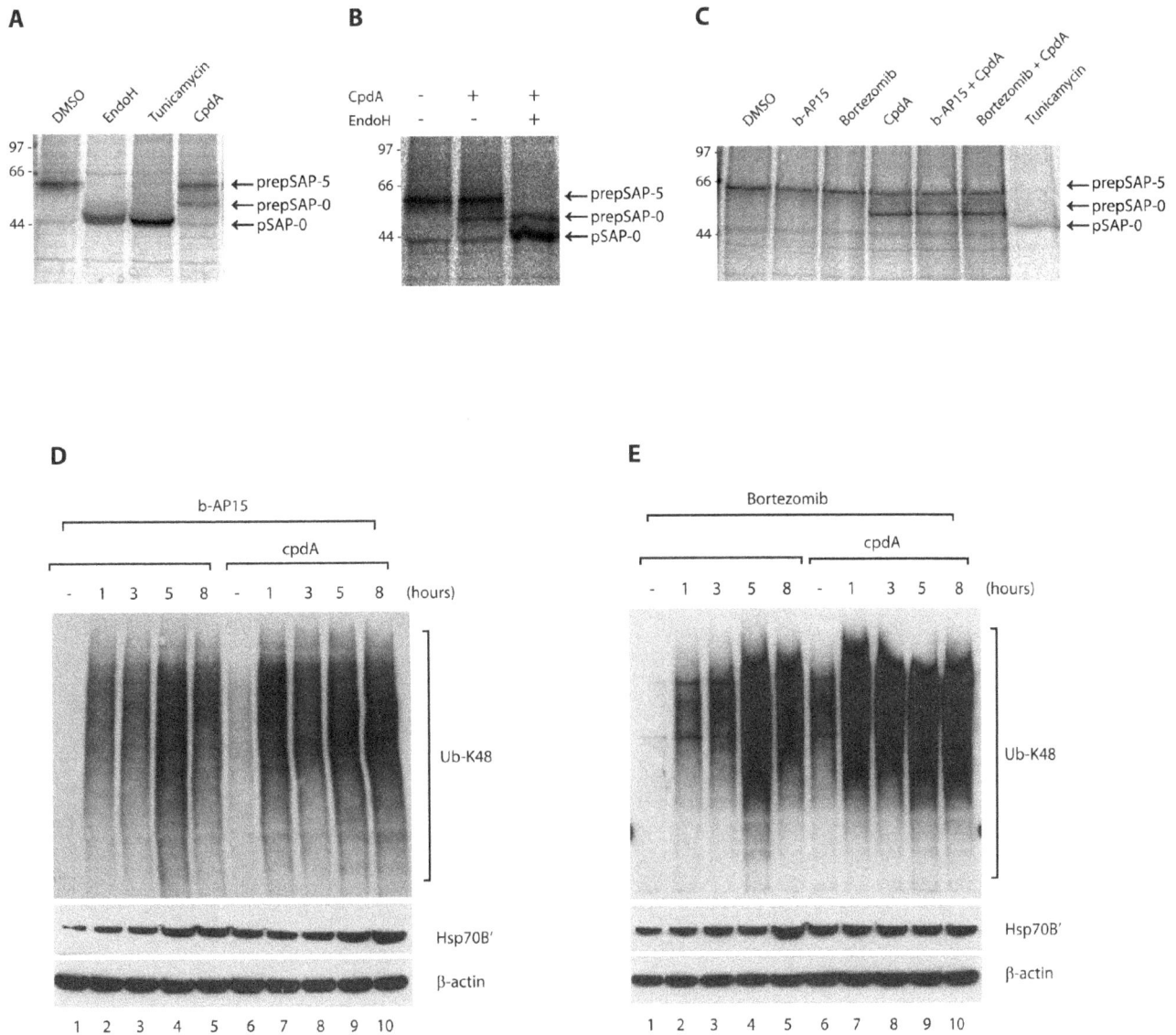

Figure 2. CpdA inhibits co-translational translocation of endogenous prosaposin into the ER. A. and B. HCT116 cells were treated for 1 h with DMSO, tunicamycin (10 μg/ml) or CAM741 (10 μM) before pulse-labelling for 10 min with [^{35}S] Met/Cys. Endogenous prosaposin (pSAP) was recovered by immunoprecipitation and newly synthesised pSAP species were visualised by phosphorimaging. In DMSO-treated cells, pSAP was fully glycosylated (pSAP-5), whereas Endo H digestion or tunicamycin treatment yielded non-glycosylated pSAP (pSAP-0). Inhibition of protein translocation into the ER by CAM741 resulted in the appearance of a pSAP species that migrated more slowly than the non-glycosylated protein and was Endo H-resistant. This species may represent signal sequence-containing preprosaposin (prepSAP-0) that has failed to translocate across the ER membrane. **C.** Distinct forms of endogenous pSAP in HCT116 cells treated with DMSO, b-AP15 (1 μM), bortezomib (20 nM), cpdA (10 μM) or tunicamycin (10 μg/ml) were recovered by immunoprecipitation and visualised by phosphorimaging. Treatment with cpdA specifically inhibits the co-translational translocation of pSAP into the ER as judged by the appearance of prepSAP-0 species.

detected following treatment with only 0.13 μM b-AP15 (**Fig. 3D, orange trace**). Hence, in the presence of cpdA, the threshold concentration of b-AP15 required for detection of Ub^{G76V}-YFP accumulation was decreased by approximately one half (**Fig. 3C compared with Fig. 3D**). In contrast, treatment with 10 μM cpdA alone did not induce accumulation of Ub^{G76V}-YFP (**Fig. 3C, D, compare yellow traces**). We therefore conclude that cpdA increases the accumulation of proteasomal substrates, such as Ub^{G76V}-YFP, induced by b-AP15. Together these results suggest a mechanism whereby cpdA enhances the ability of b-AP15 to inhibit proteasomal degradation by increasing the burden of proteasome substrates, which may potentially

saturate or block the proteasome. Hence, cpdA may promote accumulation of polyubiquitinated proteins both by increasing the production of mislocalised proteasome substrates and also by reducing proteasomal degradation.

To gain further information about the nature of the poly-ubiquitinated proteins that accumulated in the presence of b-AP15 and cpdA, we examined the subcellular distribution of ubiquitin using indirect immunofluorescence microscopy. HeLa cells transiently expressing FLAG-tagged ubiquitin were treated with compounds for 6 h, fixed and stained with anti-FLAG to visualise ubiquitin and anti-calnexin as a subcellular marker for the ER. The distribution of FLAG-ubiquitin in cells treated with 10 μM

A

B

C

D

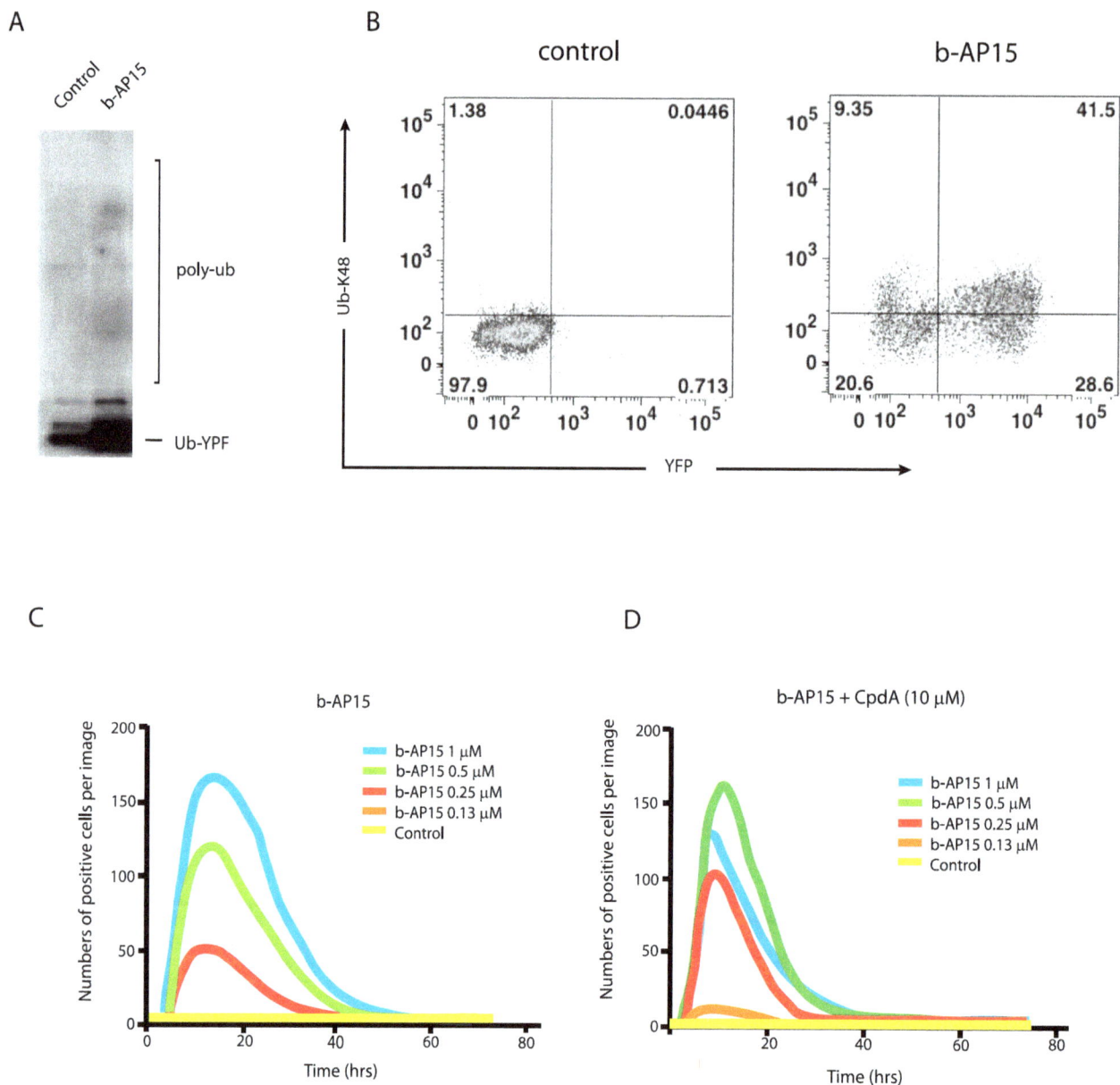

Figure 3. CpdA enhances the ability of b-AP15 to inhibit proteasomal degradation. A. MelJuSo cells stably expressing Ub^{G76V}-YFP were treated with 1 μM b-AP15 for 6 hours or left untreated (control), and cell lysates subjected to immunoblotting using anti-YFP. The migration of Ub^{G76V}-YFP and the expected position of higher molecular weight polyubiquitinated forms (poly-Ub) are indicated. **B.** MelJuSo-Ub^{G76V}-YFP cells were exposed to 1 μM b-AP15 for 8 hours or left untreated (control). Cells were labeled with an anti-K48 polyubiquitin antibody followed by an allophycocyanin conjugated secondary antibody and analyzed by FACS. **C. and D.** MelJuSo-Ub^{G76V}-YFP cells were exposed to different concentrations of b-AP15 in the presence or absence of 10 μM cpdA as indicated. Changes in the number of fluorescence-positive cells/field following addition of the compounds were monitored using an IncuCyte-FLR microscope.

cpdA for 6 hours resembled that of untreated cells (**Fig 4; not shown**). Hence, FLAG-ubiquitin was enriched in the nucleus and was also apparent throughout the cytoplasm (**Fig. 4, top row**). In contrast, treatment with 1 μM b-AP15 resulted in the appearance of numerous ubiquitin-positive foci. These punctate structures were of variable size, and were distributed throughout the cytoplasm and in the perinuclear region (**Fig. 4, middle row**). FLAG-ubiquitin containing foci were also evident in cells co-treated with a combination of b-AP15 and cpdA (**Fig. 4, bottom row**). However, their appearance was qualitatively different, with fewer larger FLAG-ubiquitin containing punctae being observed

in the presence of cpdA. The formation of similar ubiquitin-positive foci has been observed in response to a variety of conditions that disrupt protein folding and/or degradation [32]. In many cases, these structures contain misfolded proteins. Thus, the ubiquitin-positive puncta observed here may represent deposition sites for the polyubiquitinated proteins that accumulate following treatment with b-AP15. The subtle difference in the appearance of the punctae formed in the presence of cpdA may reflect the accumulation of a distinct type of ubiquitinated proteasome substrates, including mislocalised ER targeted proteins such as preprosaposin.

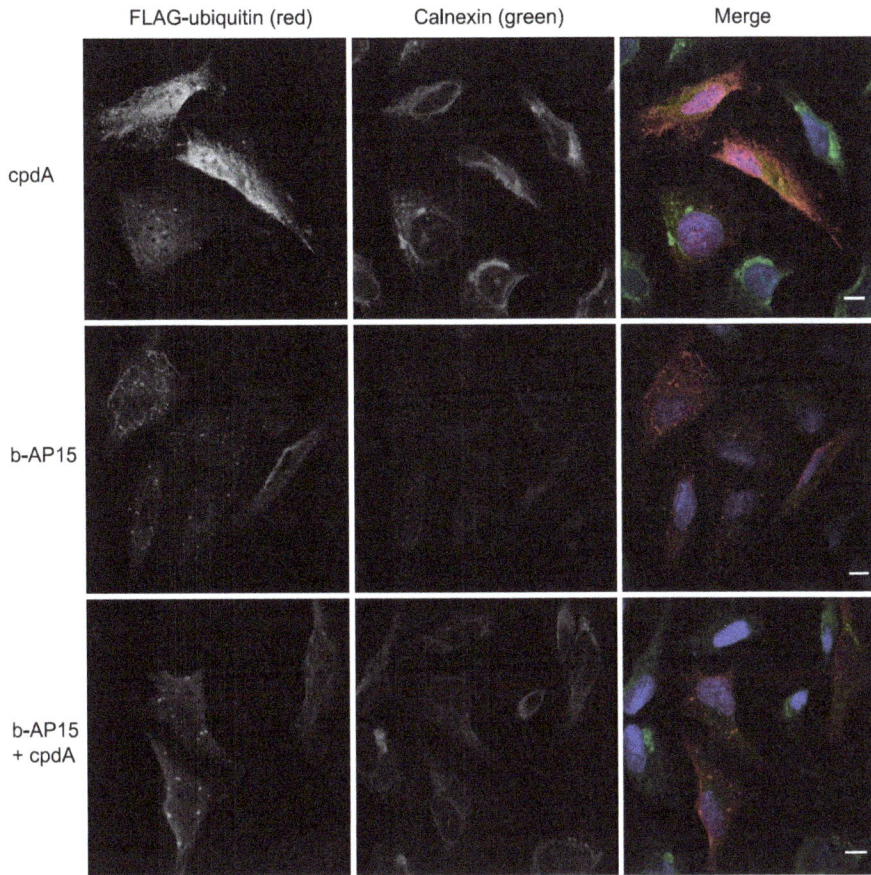

Figure 4. b-AP15 treatment alters the subcellular distribution of ubiquitin and causes the appearance of ubiquitin positive inclusions. HeLa-M cells transiently transfected with FLAG-ubiquitin were treated with 10 µM cpdA (top panels), 1.0 µM b-AP15 (middle panels) or 0.8 µM b-AP15 plus 10 µM cpdA (bottom panels) for 16 h. Cells were fixed and stained with anti-FLAG and anti-calnexin antibodies followed by fluorescently labelled secondary antibodies. Confocal images were collected and images show the combined optical stacks. Scale bar = 10 µM.

Inhibition of protein translocation with cpdA sensitizes cells to b-AP15-induced apoptosis

In order to examine the consequences of increased accumulation of polyubiquitinated conjugates and decreased proteasomal degradation, we next measured the effect of cpdA treatment on b-AP15-induced cell death. In line with previous observations [26], exposure of HCT116 cells to b-AP15 rapidly (within 16 hours) induced cleavage of PARP (**Fig. 5A**) and cell death (**Fig. 5B**) in a dose-dependent manner. Interestingly, pre-treatment of cells with cpdA prior to exposure to b-AP15 significantly increased the extent of PARP cleavage (**Fig. 5A**) and cell death (**Fig. 5B**) induced by b-AP15. Hence, in the absence of cpdA, 0.75 µM b-AP15 was required to effectively stimulate PARP cleavage, whilst 0.5 µM was sufficient in cells pretreated with cpdA (**Fig. 5A**). More strikingly, cpdA pre-treatment greatly sensitized HCT116 cells to b-AP15 induced cell death, with the percentage of dead cells increasing from approximately 20% after exposure to 0.75–1.0 µM b-AP15 to 40–50% in the presence of cpdA (**Fig. 5B**). A direct correlation between the accumulation of polyubiquitinated proteins and the extent of cell death during these treatment regimens was observed, with cpdA enhancing the build-up of lysine 48-linked polyubiquitin conjugates (**Fig. 5A, B**). This correlation also extended to non-transformed hTERT-RPE1 cells, in which lower levels of polyubiquitin conjugates accumulated and b-AP15 failed to induce PARP cleavage or evident cell death in the

absence of cpdA even after 24 hours (**Fig. 5C, D; Fig. S2**). However, when hTERT-RPE1 cells were exposed to cpdA the levels of polyubiquitin conjugates induced by b-AP15 were increased, and both PARP cleavage and cell death were now observed (**Fig. 5C, D**). Primary human fibroblasts were even more resistant than RPE1 cells to b-AP15-induced cell death (**Fig. S2**). However, 36 h after drug treatment, a similar synergistic toxicity of cpdA and b-AP15 was apparent (**Fig. 5E, F**), showing that both untransformed cell types can be sensitized to b-AP15-induced apoptosis by treatment with cpdA. Taken together, these data are consistent with the hypothesis that the amount of proteasome substrates that are generated in cells modulates sensitivity to b-AP15.

b-AP15 treatment does not deplete the intracellular cysteine pool in HCT116 cells

The data presented above support a model whereby the cytotoxicity of b-AP15 is due to an inhibition of proteasome function and the resulting accumulation of polyubiquitinated proteins. These species will include misfolded and short lived proteins en route to proteasomal degradation, and may be directly toxic to cells. Alternatively, reduced proteasomal degradation following exposure to b-AP15 may result in a lethal depletion of amino acids, most notably cysteine, due to the deficient recycling of amino acids from UPS substrates, as has been found to occur

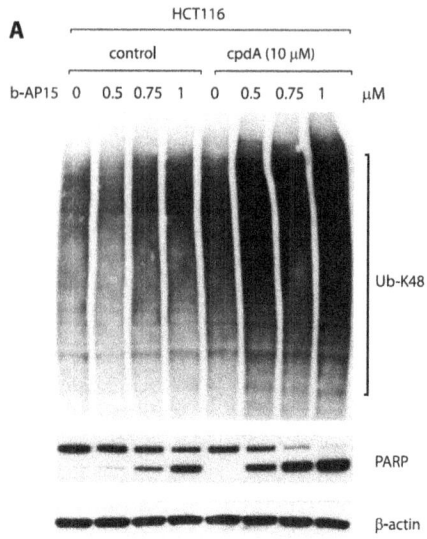

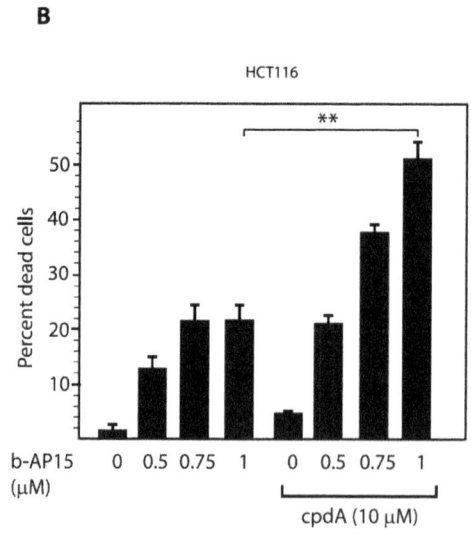

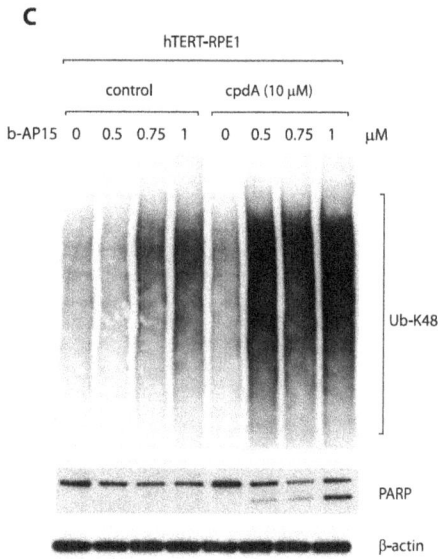

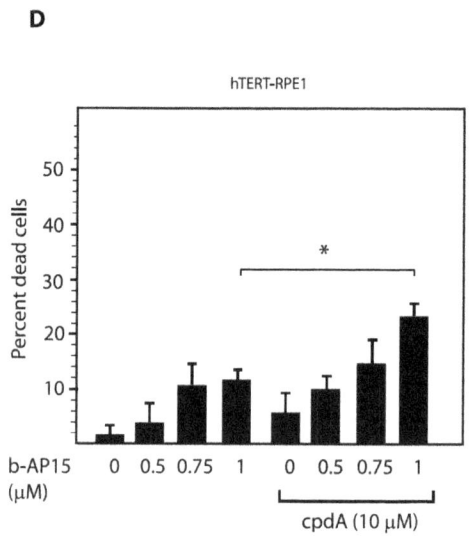

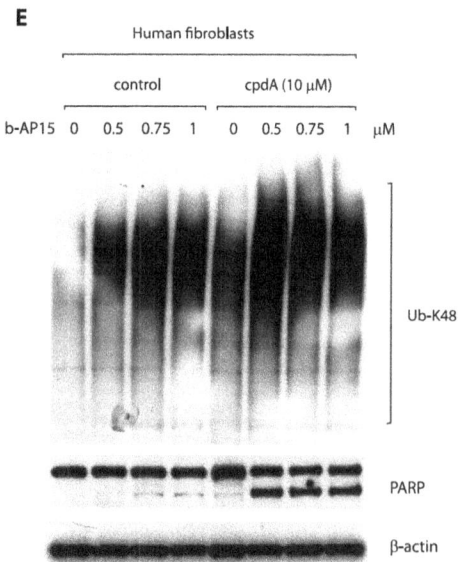

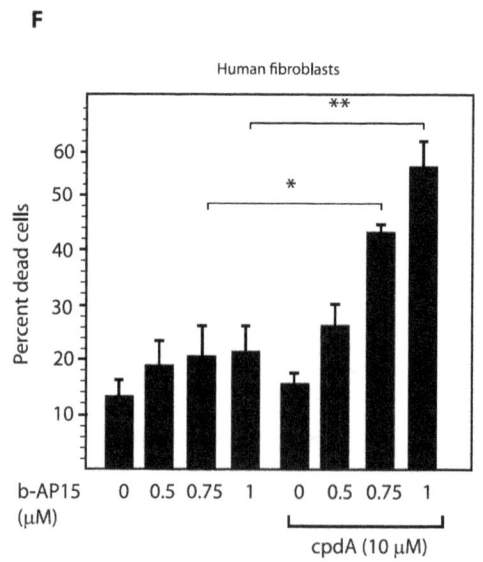

Figure 5. CpdA increases cellular levels of polyubiquitinated proteins and sensitizes cells to b-AP15 induced cell death. A. HCT116 cells were pre-treated with or without (control) 10 μM cpdA for 16 hours, exposed to the indicated concentration of b-AP15 for 1 hour, then incubated for a further 16 hours in drug-free medium. Lysates were subjected to immunoblotting for K48-linked polyubiquitin, PARP and β-actin (loading control). **B.** HCT116 cells were treated as above. Following treatment, the number of dead cells was measured by Trypan-blue staining. Bar chart shows mean +/− SD of three independent experiments. Statistical significance was calculated using the Student's t-test. P values * = 0.05 and ** = 0.01. **C. and D.** hTERT-RPE1 cells were treated and analyzed as above. **E. and F.** Human diploid fibroblasts were treated and analyzed as above, and the number of dead cells was determined 36 hours after drug treatment.

when proteasome activity is inhibited directly [25]. Indeed, we found that b-AP15 treatment resulted in an increase in LC3-II levels, indicative of autophagy, as might be expected under conditions of amino acid shortage (**Fig. 6A**). In order to address this issue more directly, we therefore examined the effects of b-AP15 and cpdA on the cellular pools of cysteine (**Table 1**). However, no decrease in cysteine levels could be detected following exposure of HCT116 cells to b-AP15 for 8 h, either alone or in combination with cpdA. Furthermore, supplementing cultures with 1 mM cysteine and other essential amino acids did not protect HCT116 cells from b-AP15-induced cell death (**Fig. 6B**), or reduce the induction of autophagy observed after b-AP15 treatment (**Fig. 6A**). Similar results were obtained using bortezomib (**Fig. 6A, B**), and we found that cellular cysteine levels were actually higher in HCT116 cells treated with bortezomib (**Table 1**). The reasons for the apparent discrepancy with recent work [25] are not clear, but may well be due to differences in cell type (i.e. human tumor versus mouse NIH3T3 cells), the precise experimental conditions used, or the stage in proteasomal degradation that is inhibited by b-AP15. Nonetheless, these results strongly suggest that cysteine depletion is not the sole cause of apoptosis in HCT116 cells exposed to b-AP15.

Discussion

The UPS is a major pathway for protein degradation in eukaryotic cells. Tumor cells depend heavily on UPS function, and inhibitors of the proteolytic activity of the proteasome have demonstrated efficacy for the treatment of certain types of cancer.

We recently identified b-AP15, a novel type of UPS inhibitor that targets DUB enzymes of the 19S regulatory particle of the proteasome, but does not inhibit the activity of the 20S core particle [26]. b-AP15 displays anti-cancer activity in several models of cancer including solid-tumors, suggesting that inhibition of 19S DUB activity may provide an alternative therapeutic strategy to proteasome inhibitors [26]. Treatment of cancer cells with b-AP15 has been shown to inhibit degradation of UPS substrates and to induce accumulation of very high molecular weight ubiquitin-conjugated proteins [26]. Thus, it was suggested that a build-up of proteasome substrates may underlie the cytotoxicity of b-AP15 [17,26]. Here we have addressed this hypothesis and provide evidence that sensitivity to b-AP15-induced cell death is related to the accumulation of high molecular weight polyubiquitinated proteins induced by this inhibitor of proteasomal DUBs.

Previous work demonstrated that b-AP15 preferentially kills cancer cells [26], and we confirm here that non-transformed hTERT-RPE1 cells and primary human fibroblasts are less sensitive to b-AP15-induced apoptosis than HCT116 colon carcinoma cells. Furthermore, we identify a correlation between the sensitivity of cells to b-AP15-induced cell death and the accumulation of high molecular weight ubiquitin conjugates, consistent with the suggestion that a harmful build-up of ubiquitinated proteasome substrates contributes to the cytotoxic effects of b-AP15. Hence, in contrast to HCT116 cells, non-malignant cells accumulated much lower levels of polyubiquitinated proteins upon exposure to b-AP15 and did not undergo extensive apoptosis following drug treatment.

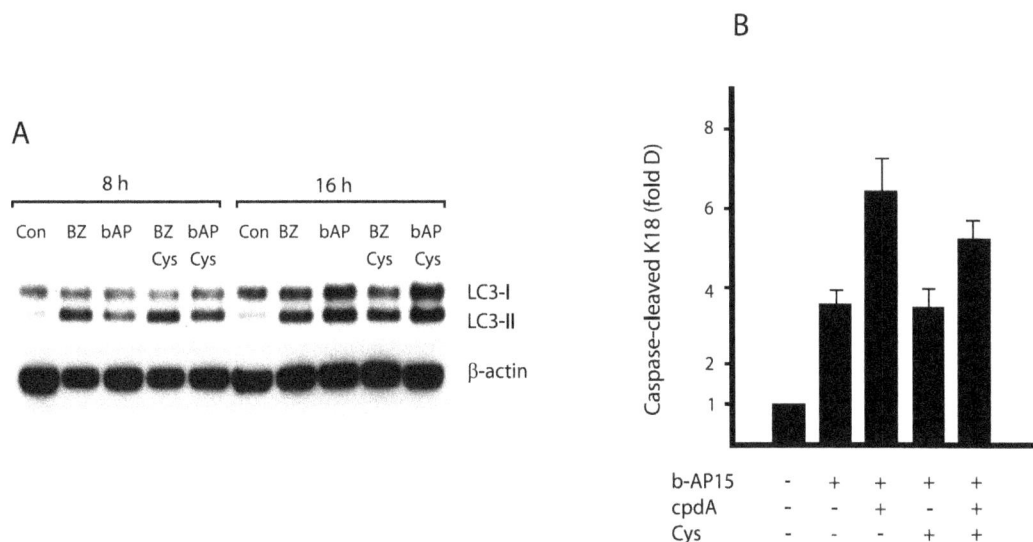

Figure 6. Cysteine supplementation does not protect HCT116 cells from b-AP15 toxicity. A. HCT116 cells were exposed to 1 μM b-AP15 or 100 nM bortezomib (BZ) for 8 hours or 16 hours, in the presence or absence of 1 mM cysteine as indicated. Cell lysates were subjected to immunoblotting for LC3-I/II and β-actin (loading control). **B.** HCT116 cells were exposed to 1 μM b-AP15 for 16 hours in the presence or absence of 1 mM cysteine. Where indicated, cells were pre-treated with 10 μM cpdA for 16 hours prior to b-AP15 treatment. Apoptosis was determined by the appearance of caspase-cleaved K18. Bar chart shows mean values ± SD of of three independent experiments.

Table 1. Treatment of HCT116 cells with b-AP15 and cpdA does not deplete cellular cysteine levels.

Sample	Control	b-AP15	BZ	b-AP15 + cpdA	BZ + cpdA
Cysteine (µmol/g)	0.46±0.11	0.65±0.17	1.35±0.48	0.54±0.18	0.53±0.22

HCT116 cells were treated with 1 µM b-AP15 or 100 nM bortezomib (BZ) for 8 hours Where indicated, cells were pre-treated with 10 µM cpdA for 16 hours prior to b-AP15 or bortezomib treatment. Cells were collected and intracellular cysteine levels measured as described. Data are mean ± SD of three independent experiments.

We provide further evidence to support this hypothesis using an inhibitor of protein translocation at the ER to show that increasing the cellular load of proteasome substrates modulates sensitivity to b-AP15. Up to a third of cellular proteins are translocated into or across the ER membrane during their biosynthesis, and those that fail to translocate correctly are ubiquitinated and targeted for proteasomal degradation [30]. Thus, we reasoned that inhibition of ER translocation would promote the production of UPS substrates and increase the accumulation of polyubiquitinated proteins in cells treated with b-AP15. The small molecule cpdA was originally identified as an inhibitor of vascular cell adhesion molecule 1 (VCAM1) translocation [29,33], and at low micromolar concentrations causes a more general inhibition of ER protein translocation [34,35]. Consistent with an inhibition of cotranslational translocation in HCT116, treatment with cpdA caused production of a mislocalised form of the endogenous glycoprotein saposin. Furthermore, pre-treatment with cpdA exacerbated the build-up of polyubiquitinated proteins induced by bortezomib or by b-AP15, suggesting that the cellular load of proteasome substrates was indeed increased under these conditions. Strikingly, cpdA also significantly sensitized cells to b-AP15 induced cell death, suggesting a scenario whereby the accumulation of polyubiquitinated proteasome substrates contributes to the toxicity of b-AP15.

Although the precise identity of the polyubiquitinated proteins that accumulate in response to b-AP15 treatment is unknown, we speculate that these will include misfolded and other non-native proteins en route to proteasomal degradation. Such species often possess exposed hydrophobic regions and thus have a tendency to form aggregates. Indeed, we observed the appearance of ubiquitin-positive inclusions within cells following b-AP15 treatment. The formation of similar structures has been observed in the presence of 20S inhibitors and under conditions that disrupt protein folding [36], and we suggest that they represent deposition sites for the polyubiquitinated proteins that accumulate in the presence of b-AP15. It is well established that the accumulation of misfolded proteins and protein aggregation is potentially harmful and can impair cell function, a process termed proteotoxicity or proteotoxic stress. Proteotoxicity of misfolded proteins has been best documented in relation to neurodegenerative diseases, and numerous mechanisms have been proposed to underlie cellular dysfunction in these situations [37]. Prominent among these is the propensity of non-native proteins to engage in inappropriate interactions and thus sequester key cellular factors and/or interfere with essential cellular processes [38,39]. In addition, studies have shown that protein aggregates can disrupt membrane integrity, induce mitochondrial dysfunction and increase production of reactive oxygen species [40–45]. Which, if any, of these mechanisms contribute to b-AP15-induced cell death will be the focus of our future studies.

An alternative possibility is that b-AP15 exerts its anti-tumor activity by inhibition of anti-apoptotic proteins such as NF-κB, by stabilization of tumor suppressors such as p53, or by stabilization of Myc and Noxa, as has been suggested for the killing of cancer cells by 20S proteasome inhibitors [18–23]. Whilst it is conceivable that these mechanisms contribute to b-AP15-induced cell death, they do not readily explain how cpdA increases b-AP15 toxicity. Hence, we favour a model whereby proteotoxicity of undegraded proteasome substrates underlies the ability of b-AP15 to kill cancer cells. In tumor cells, increased translational flux and adverse intra- and extracellular conditions may lead to an increased production of misfolded proteins, and in these situations inhibition of protesomal degradation would be predicted to cause greater toxicity. Our interpretation that cpdA potentiates b-AP15 cytotoxicity by increasing cellular levels of misfolded proteins is consistent with previous work showing that inducing protein misfolding, for example by downregulation of molecular chaperones [28,46] or hyperthermia [8], enhances the severity of proteasome inhibitor-induced proteotoxic stress and cell death.

Depletion of essential amino acids due to inefficient recycling has recently been shown to play a key role in proteasome inhibitor mediated cell death in yeast, fruit flies and mouse cells [25]. However, a lack of amino acids does not appear to be the major underlying cause of b-AP15-induced apoptosis in the HCT116 colon cancer cells studied here. In this respect it is noteworthy that in human cancer cells, the outcome of treatment with proteasome inhibitors appears to depend on the balance between the accumulation of misfolded proteins and the protective upregulation of chaperones [28,46]. We observed strong induction of Hsp70B' in HCT116 cells treated with b-AP15. Hsp70B' is highly inducible and considered as the final defence line against proteotoxic stress in human cells, whereas the mouse genome does not encode Hsp70B'. Hence, such cells may respond differently to the accumulation of proteasome substrates as compared to the human cells we have employed here.

b-AP15 kills cancer cells by inducing apoptosis [26,27,47]. Our recent work shows that the intrinsic pathway is engaged, with conformational activation of Bak and mitochondrial depolarisation evident upon b-AP15 treatment [48]. Interestingly, the toxic activity of b-AP15 is distinct from that of 20S core particle inhibitors. In contrast to bortezomib, b-AP15 induces cell death independently of Bcl-2, Bax and Bak, exhibits anti-tumor activity in solid tumor models and shows greatest activity towards colon carcinoma and CNS lineage cells [17,48]. Although the reasons for these differences are not known, one intriguing possibility is that specific properties of the ubiquitin conjugates that accumulate in the presence of b-AP15 contribute to its selective toxicity. Indeed, the polyubiquitinated proteins observed in b-AP15 treated cells are typically of higher molecular weight than those that accumulate as a result of inhibition of the 20S core particle [26]. It is possible that such extensively polyubiquitinated species have distinct effect on cell function compared to lower order ubiquitin conjugates. In addition, cytoplasmic aggregates containing misfolded proteins have been found to compromise cellular function by sequestering essential cellular components including molecular chaperones [38,39,41]. Hence, it will be important to determine the identity of other cellular factors that are present in the ubiquitin positive inclusions formed upon b-AP15 treatment.

Similarly, it is possible that the nature of the proteasome substrates generated in the presence of cpdA contribute to its ability to potentiate b-AP15 cytotoxicity. Mislocalised polypeptides may have an increased risk of aggregation in the cytosol due to the presence of hydrophobic transmembrane domains and/or unprocessed signal peptides, and thus may induce greater proteotoxicity. Indeed, a dedicated quality control system for proteins possessing hydrophobic 'degrons' has recently been identified [30,49], underscoring the requirement for timely and efficient elimination of such species. In this respect, it is noteworthy that we observed a subtle difference in the subcellular localisation of ubiquitin in cells treated with b-AP15 and cpdA, with larger ubiquitin positive inclusions detected than with b-AP15 alone. Furthermore, co-treatment with cpdA slightly enhanced the ability of b-AP15 to inhibit degradation of the UPS reporter substrate Ub^{G76V}-YFP. These observations are interesting as they raise the possibility that the mislocalised proteins generated in the presence of cpdA may interfere with proteasome function and/or sequester other factors required for proteasomal degradation as recently demonstrated for other protein aggregates [39,41].

In principal, enhancement of b-AP15-induced apoptosis by cpdA co-treatment could provide a potential strategy for improved cancer therapy. Supporting such a combinatorial approach, induction of protein misfolding by genetic or pharmacological inhibition of molecular chaperones has been shown to enhance the anti-tumor effect of bortezomib [50]. Increasing the levels of proteasome substrates in cells by inhibiting protein translocation may thus represent a novel strategy to increase proteotoxic stress and sensitize cells to bortezomib-induced apoptosis. However, we found that cpdA also increased bortezomib toxicity in two non-malignant cell types, and it is therefore unclear whether this combination of drugs could be clinically useful in practice. Rigorous testing of this possibilty will require animal experiments and clinical studies.

Experimental Procedures

Antibodies and plasmids

Monoclonal anti-FLAG (M2), anti-HSPA6, anti-β-actin and polyclonal anti-calnexin antibodies were purchased from Sigma Aldrich. Goat anti-saposin antibody was from Konrad Sandhoff (University of Bonn, Germany). The K48-linked polyubiquitin antibody (clone Apu2) was from Millipore. Antibodies for LC3 and GFP were from Cell Signalling Technology. Anti-PARP, anti-Caspase3 and anti-HSP90 were from Becton Dickinson. The anti-HSP70B' was from Lifespan. The plasmid encoding FLAG-tagged ubiquitin was kindly provided by Sylvie Urbe (University of Liverpool).

Cell culture

HCT116 colon carcinoma cells (ATCC) were maintained in McCoy's 5A modified medium/10% fetal calf serum, hTERT-RPE1 cells [26] in DMEM:F12 medium/10% fetal calf serum, human diploid fibroblasts and HeLa cells (European Collection of Cell Cultures) were maintained in DMEM medium/10% fetal calf serum and MelJuSo-Ub^{G76V}-YFP cells [51] in Iscove's DMEM medium/10% fetal calf serum.

Treatment with small molecules

b-AP15 (NSC687852) was obtained from the Developmental Therapeutics Program of the US National Cancer Institute (http://www.dtp.nci.nih.gov/) or from OncoTargeting AB, bortezomib was from the Department of Oncology, Karolinska Hospital, and cpdA was kindly provided by Novartis. Stock solutions of all compounds were made in DMSO and were added directly to culture media at the concentrations indicated.

SDS-PAGE and Western blot analysis

Following treatment, cells were lysed in RIPA buffer (140 mM NaCl, 10 mM Tris-Cl pH 8.0, 1 mM EDTA, 0.5 mM EGTA, 1% Triton X-100, 0.1% sodium deoxycholate, 0.1% SDS) containing protease inhibitor cocktail (Sigma). Lysates were resolved on Tris-Acetate PAGE gels (Invitrogen, Carlsbad, CA) and transferred onto a polyvinylidene difluoride (PVDF) membrane for Western blotting.

Assessment of apoptosis by ELISA

HCT116 cells were seeded in 96-well microtiter plates at 10,000 cells per well and incubated overnight. Drugs were then added and cells incubated further. At the end of the incubation period, NP40 was added to the tissue culture medium to 0.1% and 25 μl of the content of each well was assayed using the M30-Apoptosense ELISA [52]. This ELISA is based on a specific antibody against a neoepitope of cytokeratin 18 that is generated by the action of caspase-3, -7 and -9 activated in response to apoptosis [53].

FACS

MelJuSo-Ub^{G76V}-YFP cells treated with or without b-AP15 were collected and fixed in 2% paraformaldehyde for 15 min at room temperature. After permeabilization with 0.2% Triton X-100/2% BSA in PBS, cells were stained with an anti-K48 antibody (Milipore, 1:700) for 1 hour at room temperature in PBS/2% BSA, washed 3 times with PBS and incubated with an anti-rabbit antibody conjugated to allophycocyanin (Thermo Scientific, 1:200) for 30 min in PBS/2% BSA at room temperature. Flow cytometry was performed using an BD LSR II instrument.

Immunofluorescence microscopy

For immunofluorescence microscopy, HeLa M cells were grown on coverslips and transfected with jetPEI (Polyplus Transfection) according to the manufacturer's instructions. Eight hours post-transfection, cells were treated with compounds for 16 h. Cells were fixed in methanol at −20°C for 4 min, and then incubated for 1h at room temperature with primary antibodies diluted in PBS. Cells were washed three times with PBS and incubated for a further hour at room temperature with secondary antibodies conjugated to Alexa Fluor 488/594 (Invitrogen). The DNA dye DAPI (1 μg/ml) was included in the second incubation. Coverslips were mounted with Mowiol, allowed to dry, and images were acquired on a Nikon C1 confocal on an upright 90i microscope with a Nikon Apo oil 60x/1.40NA objective and the following settings: pinhole 30 μm, scan speed 1.68 s unidirectional, format 512×512. Images for DAPI, FITC and Texas Red were excited with the 405 nm, 488 nm and 543 nm laser lines, respectively. Image analysis was performed using Adobe Photoshop CS4.

Quantification of Ub^{G76V}-YFP fluorescence

Fluorescence of MelJuSo-Ub^{G76V}-YFP cells was recorded as positive cells/field using an IncuCyte-FLR 20X phase contrast/fluorescence microscope (Essen Instruments, Ann Arbor, MI). Average object summed intensity was calculated (triplicate wells, 4 images/well) using the Incucyte software (Essen Instruments).

Metabolic labelling and immunoprecipitation

Cells were starved in methionine- and cysteine-free DMEM (Invitrogen) supplemented with 2 mM L-glutamine for 20 min at 37°C, and then incubated in fresh starvation medium containing

22 µCi/ml [^{35}S]Met/Cys protein labelling mix (PerkinElmer; specific activity >1,000 Ci/mmol) for 10 min. When the effects of different compounds were studied, the cells were pretreated for 1 h prior to starvation, and the compound(s) were included throughout the starvation and pulse. Cells were washed twice with PBS and solubilised with Triton X-100 lysis buffer containing 10 mM Tris-HCl pH 7.6, 140 mM NaCl, 1 mM EDTA, 1% Triton X-100, and a protease inhibitor cocktail. Clarified lysates were denatured with 1% SDS at 37°C for 30 min, then diluted 5-fold with Triton X-100 lysis buffer containing 8 mM cold Met/Cys and 1 mM PMSF, and pre-cleared by incubation with pansorbin (Calbiochem) at 4°C for 1 h. The pre-cleared lysates were incubated overnight with antibodies specific for saposin D and immune complexes were captured using Protein A-Sepharose beads (GenScript, USA). Beads were washed three times with Triton X-100 buffer before eluting proteins with SDS-PAGE sample buffer. The immunoprecipitated material was denatured at 37°C for 1 h, and then analysed directly or digested with 0.5 µl of Endo H and incubated at 37°C overnight. Samples were resolved by SDS-PAGE, and visualised by phosphorimaging (FLA-3000; Fuji). Phosphorimaging exposures were processed using AIDA v3.52.

Measurement of intracellular cystein levels

The method for measurement of non-protein bound total cystine (cysteine and cystine) was developed from that of Araki and Sako [54]. HCT116 cells were washed in PBS and resuspended in 800 µl ice-cold water, then immediately disrupted by ultrasonication on ice/water slush (Soniprep 150 MSE, three times for 3 sec, amplitude 10 microns, rested for 30 sec between bursts). Cell debris was removed by centrifugation for 3 min at 5000 g, 4°C, and protein concentration of the resulting samples determined using the Bradford assay. To precipitate protein, samples were mixed with an equal volume of 10% ice-cold trichloroacetic acid (TCA), and precipitated protein removed by centrifugation for 5 min at 5000 g, 4°C. Samples were stored at −70°C until further analysis. To reduce cystine to cysteine, 250 µl sample was mixed with 25 µl 75 mM Tris-(2-carboxyethyl) hydrochloride (TCEP) and incubated for 30 min at RT. Thiol groups were derivatized by incubating 200 µl of the reduced sample with an equal volume of 1 mg/ml 7-fluoro-4-sulfobenofuran (SBD-F, CAS 84806-27-9),

2.5 M boric acid pH 10, for 60 min at 60°C in a shaking water bath. The samples were then cooled on ice, and the supernatant was collected after centrifugation for 5 min at 3000 g. Derivatized samples were separated on a Spherisorb 5u ODS (2), 250×4.6 mm column (Phenomenex) with 0.55 M acetate buffer pH 4.0 mobile phase at a rate of 1 ml/min, on an Ultimate 3000 HPLC instrument with Chromeleon evaluation software (Dionex). The cystine content was determined by fluorimetry (excitation 385 nm, emission 515 nm) by comparison to a standard curve (0.6–12 µM cystine ($C_6H_{12}N_2O_4S_2$, Sigma C8755) in boric acid buffer [0.1 M H_3BO_3, 2 mM EDTA, pH 9.5]), and expressed as µmol cystine/g protein.

Supporting Information

Figure S1   HCT116 cells were exposed to different concentrations of b-AP15 for 1 hour, followed by washing and incubation in drug-free medium for 16 hours. Cell lysates were subjected to immunoblotting for K48-linked ubiquitin, HSP70, HSP90 and β-actin (loading control).

Figure S2   Cells were treated with or without (control) 10 µM cpdA for 16 hours, exposed to the indicated concentration of b-AP15 for 1 hour, then incubated for a further 16 or 24 hours in drug-free medium. Following treatment, the number of dead cells was measured by Trypan-blue staining.

Acknowledgments

We thank Novartis for generously providing translocation inhibitor cpd A, and Peter March (University of Manchester, UK) for assistance with bioimaging.

Author Contributions

Conceived and designed the experiments: ES SH SL SC MF PD. Performed the experiments: CS PR MRP XW MH. Analyzed the data: ES SH SL SC MF PD. Wrote the paper: ES SH SL CS PR.

References

1. Schubert U, Anton LC, Gibbs J, Norbury CC, Yewdell JW, et al. (2000) Rapid degradation of a large fraction of newly synthesized proteins by proteasomes. Nature 404: 770–774.
2. Yewdell JW, Nicchitta CV (2006) The DRiP hypothesis decennial: support, controversy, refinement and extension. Trends Immunol 27: 368–373.
3. Duttler S, Pechmann S, Frydman J (2013) Principles of cotranslational ubiquitination and quality control at the ribosome. Mol Cell 50: 379–393.
4. Bucciantini M, Giannoni E, Chiti F, Baroni F, Formigli L, et al. (2002) Inherent toxicity of aggregates implies a common mechanism for protein misfolding diseases. Nature 416: 507–511.
5. Silvera D, Formenti SC, Schneider RJ (2010) Translational control in cancer. Nat Rev Cancer 10: 254–266.
6. Swietach P, Vaughan-Jones RD, Harris AL (2007) Regulation of tumor pH and the role of carbonic anhydrase 9. Cancer Metastasis Rev 26: 299–310.
7. Medicherla B, Goldberg AL (2008) Heat shock and oxygen radicals stimulate ubiquitin-dependent degradation mainly of newly synthesized proteins. J Cell Biol 182: 663–673.
8. Neznanov N, Komarov AP, Neznanova L, Stanhope-Baker P, Gudkov AV (2011) Proteotoxic stress targeted therapy (PSTT): induction of protein misfolding enhances the antitumor effect of the proteasome inhibitor bortezomib. Oncotarget 2: 209–221.
9. Solimini NL, Luo J, Elledge SJ (2007) Non-oncogene addiction and the stress phenotype of cancer cells. Cell 130: 986–988.
10. Dai C, Whitesell L, Rogers AB, Lindquist S (2007) Heat shock factor 1 is a powerful multifaceted modifier of carcinogenesis. Cell 130: 1005–1018.
11. Oromendia AB, Amon A (2014) Aneuploidy: implications for protein homeostasis and disease. Dis Model Mech 7: 15–20.
12. Torres EM, Sokolsky T, Tucker CM, Chan LY, Boselli M, et al. (2007) Effects of aneuploidy on cellular physiology and cell division in haploid yeast. Science 317: 916–924.
13. Hershko A, Ciechanover A (1998) The ubiquitin system. Annu Rev Biochem 67: 425–479.
14. Groll M, Heinemeyer W, Jager S, Ullrich T, Bochtler M, et al. (1999) The catalytic sites of 20S proteasomes and their role in subunit maturation: a mutational and crystallographic study. Proc Natl Acad Sci U S A 96: 10976–10983.
15. Komander D, Rape M (2012) The ubiquitin code. Annu Rev Biochem 81: 203–229.
16. Finley D (2009) Recognition and processing of ubiquitin-protein conjugates by the proteasome. Annu Rev Biochem 78: 477–513.
17. D'Arcy P, Linder S (2012) Proteasome deubiquitinases as novel targets for cancer therapy. Int J Biochem Cell Biol 44: 1729–1738.
18. Caravita T, de Fabritiis P, Palumbo A, Amadori S, Boccadoro M (2006) Bortezomib: efficacy comparisons in solid tumors and hematologic malignancies. Nat Clin Pract Oncol 3: 374–387.
19. Williams SA, McConkey DJ (2003) The proteasome inhibitor bortezomib stabilizes a novel active form of p53 in human LNCaP-Pro5 prostate cancer cells. Cancer Res 63: 7338–7344.
20. Adams J (2002) Proteasome inhibitors as new anticancer drugs. Curr Opin Oncol 14: 628–634.
21. Chen S, Blank JL, Peters T, Liu XJ, Rappoli DM, et al. (2010) Genome-wide siRNA screen for modulators of cell death induced by proteasome inhibitor bortezomib. Cancer Res 70: 4318–4326.
22. Orlowski RZ, Kuhn DJ (2008) Proteasome inhibitors in cancer therapy: lessons from the first decade. Clin Cancer Res 14: 1649–1657.

23. Nikiforov MA, Riblett M, Tang WH, Gratchouck V, Zhuang D, et al. (2007) Tumor cell-selective regulation of NOXA by c-MYC in response to proteasome inhibition. Proc Natl Acad Sci U S A 104: 19488–19493.

24. Obeng EA, Carlson LM, Gutman DM, Harrington WJ, Jr., Lee KP, et al. (2006) Proteasome inhibitors induce a terminal unfolded protein response in multiple myeloma cells. Blood 107: 4907–4916.

25. Suraweera A, Munch C, Hanssum A, Bertolotti A (2012) Failure of amino acid homeostasis causes cell death following proteasome inhibition. Mol Cell 48: 242–253.

26. D'Arcy P, Brnjic S, Olofsson MH, Fryknas M, Lindsten K, et al. (2011) Inhibition of proteasome deubiquitinating activity as a new cancer therapy. Nat Med.17 1636–1640.

27. Tian Z, D'Arcy P, Wang X, Ray A, Tai YT, et al. (2014) A novel small molecule inhibitor of deubiquitylating enzyme USP14 and UCHL5 induces apoptosis in multiple myeloma and overcomes bortezomib resistance. Blood 123: 706–716.

28. Noonan EJ, Place RF, Giardina C, Hightower LE (2007) Hsp70B' regulation and function. Cell Stress Chaperones 12: 219–229.

29. Besemer J, Harant H, Wang S, Oberhauser B, Marquardt K, et al. (2005) Selective inhibition of cotranslational translocation of vascular cell adhesion molecule 1. Nature 436: 290–293.

30. Hessa T, Sharma A, Mariappan M, Eshleman HD, Gutierrez E, et al. (2011) Protein targeting and degradation are coupled for elimination of mislocalized proteins. Nature 475: 394–397.

31. Dantuma NP, Lindsten K, Glas R, Jellne M, Masucci MG (2000) Short-lived green fluorescent proteins for quantifying ubiquitin/proteasome-dependent proteolysis in living cells. Nat Biotechnol 18: 538–543.

32. Weisberg SJ, Lyakhovetsky R, Werdiger AC, Gitler AD, Soen Y, et al. (2012) Compartmentalization of superoxide dismutase 1 (SOD1G93A) aggregates determines their toxicity. Proc Natl Acad Sci U S A 109: 15811–15816.

33. Garrison JL, Kunkel EJ, Hegde RS, Taunton J (2005) A substrate-specific inhibitor of protein translocation into the endoplasmic reticulum. Nature 436: 285–289.

34. Harant H, Wolff B, Schreiner EP, Oberhauser B, Hofer L, et al. (2007) Inhibition of vascular endothelial growth factor cotranslational translocation by the cyclopeptolide CAM741. Mol Pharmacol 71: 1657–1665.

35. Kang SW, Rane NS, Kim SJ, Garrison JL, Taunton J, et al. (2006) Substrate-specific translocational attenuation during ER stress defines a pre-emptive quality control pathway. Cell 127: 999–1013.

36. Salomons FA, Menendez-Benito V, Bottcher C, McCray BA, Taylor JP, et al. (2009) Selective accumulation of aggregation-prone proteasome substrates in response to proteotoxic stress. Mol Cell Biol 29: 1774–1785.

37. Douglas PM, Dillin A (2010) Protein homeostasis and aging in neurodegeneration. J Cell Biol 190: 719–729.

38. Park SH, Kukushkin Y, Gupta R, Chen T, Konagai A, et al. (2013) PolyQ Proteins Interfere with Nuclear Degradation of Cytosolic Proteins by Sequestering the Sis1p Chaperone. Cell 154: 134–145.

39. Olzscha H, Schermann SM, Woerner AC, Pinkert S, Hecht MH, et al. (2011) Amyloid-like aggregates sequester numerous metastable proteins with essential cellular functions. Cell 144: 67–78.

40. Bence NF, Sampat RM, Kopito RR (2001) Impairment of the Ubiquitin-Proteasome System by Protein Aggregation. Science 292: 1552–1555.

41. Chakrabarti O, Hegde RS (2009) Functional depletion of mahogunin by cytosolically exposed prion protein contributes to neurodegeneration. Cell 137: 1136–1147.

42. Chakrabarti O, Rane NS, Hegde RS (2011) Cytosolic aggregates perturb the degradation of nontranslocated secretory and membrane proteins. Mol Biol Cell 22: 1625–1637.

43. Campioni S, Mannini B, Zampagni M, Pensalfini A, Parrini C, et al. (2010) A causative link between the structure of aberrant protein oligomers and their toxicity. Nat Chem Biol 6: 140–147.

44. Lashuel HA (2005) Membrane permeabilization: a common mechanism in protein-misfolding diseases. Sci Aging Knowledge Environ 2005: pe28.

45. Bennett EJ, Bence NF, Jayakumar R, Kopito RR (2005) Global impairment of the ubiquitin-proteasome system by nuclear or cytoplasmic protein aggregates precedes inclusion body formation. Mol Cell 17: 351–365.

46. Gabai VL, Budagova KR, Sherman MY (2005) Increased expression of the major heat shock protein Hsp72 in human prostate carcinoma cells is dispensable for their viability but confers resistance to a variety of anticancer agents. Oncogene 24: 3328–3338.

47. Berndtsson M, Beaujouin M, Rickardson L, Havelka AM, Larsson R, et al. (2009) Induction of the lysosomal apoptosis pathway by inhibitors of the ubiquitin-proteasome system. Int J Cancer 124, 1463–1469.

48. Brnjic S, Mazurkiewicz M, Fryknas M, Sun C, Zhang X, et al. (2013) Induction of Tumor Cell Apoptosis by a Proteasome Deubiquitinase Inhibitor Is Associated with Oxidative Stress. Antioxid Redox Signal. DOI: 10.1089/ars.2013.5322.

49. Leznicki P, High S (2012) SGTA antagonizes BAG6-mediated protein triage. Proc Natl Acad Sci U S A 109: 19214–19219.

50. Mitsiades CS, Mitsiades NS, McMullan CJ, Poulaki V, Kung AL, et al. (2006) Antimyeloma activity of heat shock protein-90 inhibition. Blood 107: 1092–1100.

51. Menendez-Benito V, Verhoef LG, Masucci MG, Dantuma NP (2005) Endoplasmic reticulum stress compromises the ubiquitin-proteasome system. Hum Mol Genet 14: 2787–2799.

52. Hagg M, Biven K, Ueno T, Rydlander L, Bjorklund P, et al. (2002) A novel high-through-put assay for screening of pro-apoptotic drugs. Invest New Drugs 20: 253–259.

53. Leers MP, Kolgen W, Bjorklund V, Bergman T, Tribbick G, et al. (1999) Immunocytochemical detection and mapping of a cytokeratin 18 neo-epitope exposed during early apoptosis. J Pathol 187: 567–572.

54. Araki A, Sako Y (1987) Determination of free and total homocysteine in human plasma by high-performance liquid chromatography with fluorescence detection. J Chromatogr 422: 43–52.

7

A Multicenter Phase I/II Study of Obatoclax Mesylate Administered as a 3- or 24-Hour Infusion in Older Patients with Previously Untreated Acute Myeloid Leukemia

Aaron D. Schimmer[1]*, **Azra Raza**[2], **Thomas H. Carter**[3], **David Claxton**[4], **Harry Erba**[5], **Daniel J. DeAngelo**[6], **Martin S. Tallman**[7], **Carolyn Goard**[1], **Gautam Borthakur**[8]

1 Princess Margaret Cancer Centre, Toronto, Ontario, Canada, 2 Columbia University Medical Center, New York, New York, United States of America, 3 The University of Iowa, Iowa City, Iowa, United States of America, 4 Penn State, Hershey, Pennsylvania, United States of America, 5 University of Alabama at Birmingham, Birmingham, Alabama, United States of America, 6 Dana-Farber Cancer Institute, Boston, Massachusetts, United States of America, 7 Leukemia Service, Memorial Sloan-Kettering Cancer Center, Weill Cornell Medical College, New York, New York, United States of America, 8 The University of Texas MD Anderson Cancer Center, Houston, Texas, United States of America

Abstract

Purpose: An open-label phase I/II study of single-agent obatoclax determined a maximum tolerated dose (MTD) and schedule, safety, and efficacy in older patients ($\geq$70 yr) with untreated acute myeloid leukemia (AML).

Experimental Design: Phase I evaluated the safety of obatoclax infused for 3 hours on 3 consecutive days (3 h$\times$3 d) in 2-week cycles. Initial obatoclax dose was 30 mg/day (3 h$\times$3 d; n = 3). Obatoclax was increased to 45 mg/day (3 h$\times$3 d) if $\leq$1 patient had a dose-limiting toxicity (DLT) and decreased to 20 mg/day (3 h$\times$3 d) if DLT occurred in $\geq$2 patients. In the phase II study, 12 patients were randomized to receive obatoclax at the dose identified during phase I (3 h$\times$3 d) or 60 mg/day administered by continuous infusion over 24 hours for 3 days (24 h$\times$3 d) to determine the morphologic complete response rate.

Results: In phase I, two of three patients receiving obatoclax 30 mg/day (3 h$\times$3 d) experienced grade 3 neurologic DLTs (confusion, ataxia, and somnolence). Obatoclax was decreased to 20 mg/day (3 h$\times$3 d). In phase II, no clinically relevant safety differences were observed between the 20 mg/day (3 h$\times$3 d; n = 7) and 60 mg/day (24 h$\times$3 d; n = 5) arms. Neurologic and psychiatric adverse events were most common and were generally transient and reversible. Complete response was not achieved in any patient.

Conclusions: Obatoclax 20 mg/day was the MTD (3 h$\times$3 d) in older patients with AML. In the schedules tested, single-agent obatoclax was not associated with an objective response. Evaluation in additional subgroups or in combination with other chemotherapy modalities may be considered for future study.

Trial Registration: ClinicalTrials.gov NCT00684918

Editor: Maria R. Baer, University of Maryland, United States of America

Funding: The study that is the subject of this manuscript was sponsored by Gemin X Pharmaceuticals, Inc., which was acquired by Cephalon, Inc., a wholly-owned subsidiary of Teva Pharmaceutical Industries, Ltd (TEVA). The development and publication of this manuscript has been financially supported by TEVA. Employees of TEVA were actively involved in the development of the manuscript, including providing data as well as review and comment of manuscript drafts. Powered 4 Significance LLC was contracted by TEVA to provide medical writing assistance in preparing an initial draft of the manuscript and editorial assistance.

Competing Interests: The authors have read the journal's policy and have the following competing interests: ADS, GB, and THC report research support from GeminX, during the conduct of the study; HE reports compensation from Celgene, Novartis, Celator, Sunesis, Seattle Genetics, and Incyte, outside the submitted work; AR, DC, DJD, CG, and MST have no competing interests to declare.

* Email: aaron.schimmer@utoronto.ca

Introduction

Acute myeloid leukemia (AML) is a heterogeneous hematologic malignancy that results from the clonal expansion of primitive myeloid precursor cells [1]. AML is the most common form of acute leukemia in adults, and has a high mortality. In the United States, a total of 14,590 new AML diagnoses were projected to occur in 2013, with an estimated 10,370 deaths [2]. AML is predominantly a disease of older adults with a median age at diagnosis of 66 years [3]. Compared with younger patients (<55 years), AML in older patients ($\geq$55 years) is more frequently

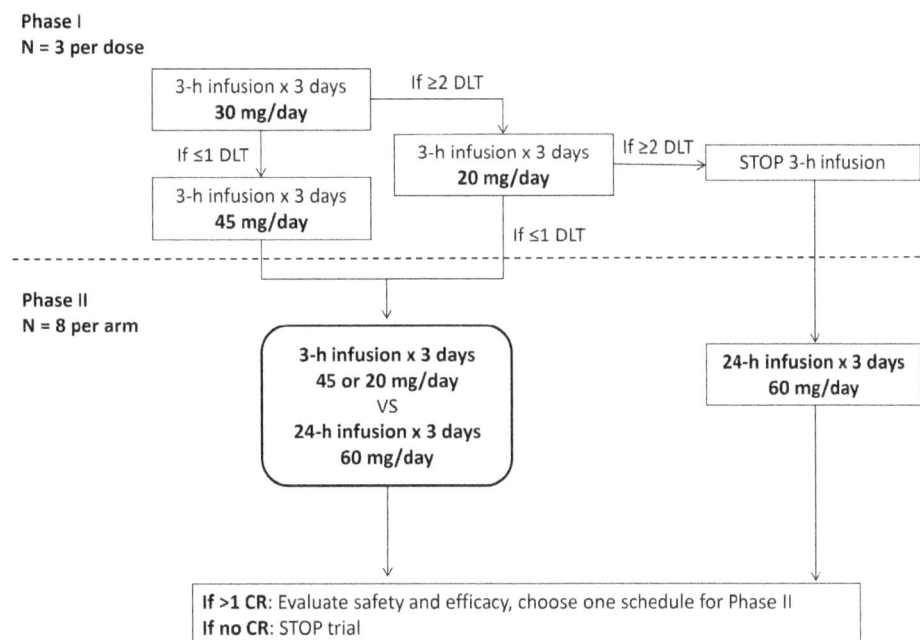

Figure 1. Study design. CR, complete response; DLT, dose-limiting toxicity.

associated with a poor prognosis, in part due to decreased response rates and increased toxicity of standard induction chemotherapy. Consequently, there is an unmet need for therapies that provide efficacy and favorable tolerability in older patients with AML.

Development of small-molecule inhibitors specific for anti-apoptotic proteins is a novel approach to the treatment of hematologic cancers. Anti-apoptotic B cell-chronic lymphocytic leukemia/lymphoma 2 (Bcl-2) family members (Bcl-2, Bcl-XL, Bcl-w, Bcl-b, A1/Bfl-1, and Mcl-1) are overexpressed in many cancers and inhibit apoptosis by sequestering pro-apoptotic members of the family (BH3-only proteins, and Bax and Bak) [4,5]. Importantly, emerging evidence suggests that anti-apoptotic Mcl-1 is critical for sustained survival and expansion of human AML and plays a role in drug resistance in this disease [6]. Since interactions between anti- and pro-apoptotic family members are mediated by the BH3 domain protein interaction motif [4,5], small molecules that bind to the BH3 binding groove may induce apoptosis by inhibiting sequestration of pro-apoptotic factors [7].

Obatoclax mesylate (obatoclax, also known as GX15-070) is a novel anticancer therapeutic for hematologic malignancies and solid tumors. The compound, which acts as a BH3 mimetic, was developed as a pan-inhibitor of anti-apoptotic members of the Bcl-2 family, including Mcl-1, to trigger cell death [8,9]. Preclinical investigations demonstrated that obatoclax induces apoptosis and reduces proliferation in AML cell lines and primary AML cells, in part by inhibition of Mcl-1 sequestration of Bax [10].

In phase I trials of single-agent obatoclax, antitumor activity was observed in several hematologic malignancies, including AML, myelodysplastic syndrome, and Hodgkin's and non-Hodgkin's lymphoma [11–14]. Although there were a limited number of objective responses in these early clinical studies, hematologic improvement was observed in a larger proportion of treated patients. One striking clinical response occurred in a 70-year-old woman with previously untreated AML who achieved a complete response (CR) after receiving 20 mg/m² obatoclax as a 24-hour infusion [14]. Her CR was maintained over 8 months and

suggested that a subset of treatment-naive patients with AML might benefit from obatoclax therapy.

Continuous infusion of obatoclax 60 mg/day for 3 days (24 h×3 d) in 2-week cycles has previously been evaluated in phase II trials in patients with myelofibrosis or Hodgkin's lymphoma [13,15]. An accelerated 3-hour infusion, 3-day (3 h×3 d) regimen has not yet been evaluated in a 2-week cycle in patients with hematologic malignancies, nor have these regimens been formally compared. The objectives of this multicenter phase I/II study were to expand on previous experiences with obatoclax and to evaluate the dose and schedule of single-agent obatoclax for safety and efficacy in older patients with previously untreated AML.

Patients and Methods

Study Design

An open-label, multicenter, phase I/II study was conducted. The protocol for this trial and supporting CONSORT checklist are available as supporting information; see **Checklist S1** and **Protocol S1**. The phase I portion of the study consisted of a nonrandomized safety evaluation, followed by a randomized phase II evaluation of different treatment schedules (**Figure 1**). The phase I safety evaluation assessed obatoclax as a 3-hour infusion over 3 consecutive days (3 h×3 d). Because the optimal schedule for obatoclax treatment was unknown, the phase II evaluation utilized a randomized open-label design to assess 3-hour or 24-hour infusion schedules, over 3 consecutive days (3 h×3 d or 24 h×3 d). The selected dose for the 24-hour infusion was the previously defined maximum tolerated dose (MTD) of 60 mg/day [16].

Ethics Statement

The study was conducted in accordance with the October 2000 version of the Declaration of Helsinki, as well as Good Clinical Practice and International Conference on Harmonisation guidelines. An accredited institutional review board approved this study prior to its initiation and all patients provided informed written consent. This study was registered at clinicaltrials.gov (NCT00684918).

Patient Eligibility

Patients at least 70 years of age with histologically confirmed AML were eligible to participate. In the phase I portion of the study, patients may have received one previous therapy. In the phase II portion of the study, no prior therapy for AML was allowed except for hydroxyurea. Additional eligibility requirements included Eastern Cooperative Oncology Group (ECOG) performance status ≤ 2 and normal hepatic and renal function (total bilirubin ≤ 2 mg/dL unless resulting from hemolysis; aspartate transaminase/alanine transaminase $\leq 2.5 \times$ institutional upper limit of normal; creatinine within normal institutional limits or creatinine clearance ≥ 50 mL/min/1.73 m^2 for patients with creatinine levels above institutional normal).

Patients were excluded if they had a history of allergy to components of the formulated product. Comorbidities requiring patient exclusion included a history of seizure disorder or central nervous system leukemia or other symptomatic neurologic illness; uncontrolled systemic infection considered opportunistic, life-threatening, or clinically significant; symptomatic congestive heart failure; unstable angina pectoris; cardiac arrhythmia; significant pulmonary disease or hypoxia; psychiatric illness/social situation that would limit compliance with study requirements; or infection with human immunodeficiency virus.

Treatments

Obatoclax mesylate (30 mg) was diluted with 5% dextrose, USP and a final concentration of 11.54% polyethylene glycol 300, 0.46% polysorbate 20 for intravenous (IV) infusion. Treatments were evaluated as depicted in **Figure 1**. In the phase I portion of the study, the first 3 patients enrolled received obatoclax 30 mg/day over 3 hours for 3 consecutive days (3 h×3 d). Dose-limiting toxicities (DLTs) were defined as grade ≥ 3 infusion-related neurologic adverse events (AEs) and nonhematologic AEs not responsive to symptom-directed therapy. If (during cycle 1) ≤ 1 of the 3 patients experienced a DLT, subsequent enrolled patients would receive 45 mg/day for 3 consecutive days (3 h×3 d). If, however, ≥ 2 of 3 patients experienced a DLT, an additional group of 3 patients would be enrolled and would receive 20 mg/day over 3 hours for 3 consecutive days (3 h×3 d). If ≤ 1 of 3 patients experienced a DLT at 20 mg/day, this dose would be utilized for the 3 h×3 d phase II study. If ≥ 2 of the 3 patients experienced a DLT at 20 mg/day, the 3 h×3 d schedule would be halted, and the phase II study would use only the 24 h×3 d schedule for obatoclax administration, which has previously been shown to be well tolerated and produced a CR in a patient with AML [14].

In the phase II portion of the study, patients were randomized (1:1) into two arms. One arm received obatoclax at the dose identified from phase I at 3 h×3 d; the other arm received 60 mg/day (24 h×3 d). In both phases, obatoclax was administered in two 2-week cycles as induction therapy. Any patient achieving a CR could receive four additional treatment cycles as consolidation therapy every 2 weeks for a total of six cycles of obatoclax treatment. Patients who did not achieve CR after two cycles of obatoclax were to be removed from study.

Prophylaxis with H-1 and H-2 blockers was recommended prior to each cycle, given the known prevalence of acute hypersensitivity reactions associated with obatoclax exposure [14]. Full supportive care was offered to treat acute nausea, vomiting, or DLTs as appropriate, including anti-emetic prophylaxis, blood products, antibiotics, IV immunoglobulins, and hematopoietic growth factors.

Endpoints and Assessments

Endpoints. Safety endpoints included number of DLTs, treatment-emergent AEs, including serious AEs, and clinical laboratory values. AEs were recorded according to the National Cancer Institute Common Terminology Criteria for Adverse Events, version 3.0 [17].

Clinical response was assessed using standard criteria [18]. The primary efficacy endpoint was rate of morphologic CR, or cytogenetic CR in patients with abnormal cytogenetics at baseline. Morphologic CR was defined as neutrophils $>1000/\mu L$, platelets $>100,000/\mu L$, and bone marrow blasts $<5\%$. Cytogenetic CR was defined as neutrophils $>1000/\mu L$, platelets $>100,000/\mu L$, bone marrow blasts $<5\%$, and normal cytogenetics. Additional efficacy endpoints included molecular CR, partial remission (PR), and morphologic leukemia-free state, and change in bone marrow blasts from baseline to post-induction therapy (day 28).

Schedule of assessments. AEs were recorded from baseline screening to 28 days after the last obatoclax dose. Complete blood counts (absolute neutrophil count, lymphocytes, monocytes, eosinophils, and basophils) were obtained on day 1 of each cycle, every 2–3 days for the first week of cycle 1, and on day 8 of each cycle. Physical and neurologic examinations, including vital signs, body weight, ECOG performance status, and serum chemistries were performed at baseline and on days 1 and 8 of each cycle. Chest radiographs, pulmonary function tests, and urinalysis were conducted at baseline and at the 28-day follow-up visit. An electrocardiogram was obtained at baseline, 30 minutes before the end of infusion on day 3 in cycle 1, and at the 28-day follow-up visit, and repeated as clinically indicated.

Bone marrow aspirates and biopsies were conducted at baseline, on day 28, after obatoclax consolidation therapy, and as clinically indicated. CR was documented by repeat bone marrow examination on day 28 or earlier. Bone marrow cytogenetics were assessed at baseline, repeated on occurrence of CR, and as clinically indicated.

Statistical Analysis

Safety was assessed in all patients who received any amount of study drug per National Cancer Institute Common Terminology Criteria for Adverse Events, version 3. Efficacy was assessed in all patients with at least one post-baseline efficacy assessment. All outcomes were summarized descriptively. For categorical variables, summary tabulations of the number and percentage in each parameter were provided. For continuous variables, the mean, median, standard deviation, minimum, and maximum were presented. To determine the rate of morphologic CR, a two-stage design was used, powered to detect a CR rate of $\geq 15\%$ against a non-interesting rate of 5%, with alpha $= 0.05$ and a power of 90%. Stage 1 was to enroll 37 patients under this design, and if ≥ 3 patients achieved CR at the end of Stage 1, an additional 47 patients would be enrolled.

Results

Demographics and Patient Disposition

A total of 19 patients were enrolled in the study from March 2008 to March 2009; one patient did not receive obatoclax treatment. Demographics and disease characteristics were similar across all regimens (**Table 1**). Overall, the mean age was 81 years (range 72–90 years). The median time from AML diagnosis to study entry was 0.6 months (range 0–37 months). Most (56%)

Table 1. Demographics and baseline characteristics of obatoclax-treated patients (N = 18).

	Phase I		Phase II		All (N = 18)
	30 mg/d (3 h×3 d) (n = 3)	20 mg/d (3 h×3 d) (n = 3)	20 mg/d (3 h×3 d) (n = 7)	60 mg/d (24 h×3 d) (n = 5)	
Median (range) age, years	83 (81–85)	74 (72–90)	82 (72–89)	80 (76–86)	81.5 (72–90)
Male, n (%)	1 (33)	1 (33)	3 (43)	3 (60)	8 (44)
ECOG PS, n (%)					
0	1 (33)	0	3 (43)	1 (20)	5 (28)
1	1 (33)	1 (33)	4 (57)	2 (40)	8 (44)
2	1 (33)	2 (67)	0	2 (40)	5 (28)
Median time since AML diagnosis, months (range)	1 (0.5–7.4)	1.2 (1–16.7)	0.45 (−0.2–0.6)	0.2 (0–37.2)	0.6 (−0.2–37.2)
AML classification, n (%)					
M1	1 (33)	0	3 (43)	0	4 (22)
M2	2 (67)	1 (33)	3 (43)	4 (80)	10 (56)
M3	0	1 (33)*	0	0	1 (6)
M4	0	1 (33)	0	1 (20)	2 (11)
Missing	0	0	1 (14)	0	1 (6)
Cytogenetics, n (%)					
Abnormal	1 (33)	3 (100)	3 (43)	3 (60)	10 (56)
Normal	1 (33)	0	2 (29)	2 (40)	5 (28)
Missing	1 (33)	0	2 (29)	0	3 (17)
Median leukocyte count (range), $10^3/\mu L$	1.3 (0.5–1.4)	29.2 (19.3–64.3)	4.3 (1.1–7.3)	2.8 (1.1–27.2)	4.25 (0.5–64.3)
Median platelet count (range), $10^3/\mu L$	23 (8–154)	126 (69–227)	63 (19–518)	63 (19–259)	66 (8–518)
Median hemoglobin (range), g/L	86 (86–102)	105 (99–110)	101 (86–117)	93 (90–100)	97.5 (86–117)
Median neutrophil count (range), $10^3/\mu L$	0.2 (0.2–0.2)[a]	1.95 (0.6–3.3)[b]	0.39 (0.1–2.6)[c]	6.98 (0.5–13.5)[d]	0.55 (0.1–13.5)[e]

*Classified as acute promyelocytic leukemia.
[a]Evaluated in 1 patient.
[b]Evaluated in 2 patients.
[c]Evaluated in 5 patients.
[d]Evaluated in 2 patients.
[e]Evaluated in 10 patients.
AML, acute myeloid leukemia; ECOG PS, Eastern Cooperative Oncology Group performance status.

patients had a French-American-British classification of M2; the distribution of AML classification was similar for all regimens. Approximately 56% of patients had an abnormal karyotype; further details of the cytogenetic abnormalities were not available, thus precluding classification of cytogenetics into risk groups.

Patient disposition is shown in **Figure 2**. The most common reasons for withdrawal from the study were failure to achieve CR following induction therapy (eight patients, 42%), adverse events (including DLT; four patients, 21%), and disease progression (three patients, 16%). Two patients were granted waivers for laboratory abnormalities (hyperuricemia) at baseline.

The number of treatment cycles administered in the phase I/II studies is shown in **Table 2.** Overall, the median number of cycles administered was two (range 1–11). In both the phase I and II studies, a total of 15 patients (83%) were treated for at least two cycles. Four patients also received additional treatment cycles as consolidation therapy as they experienced disease stabilization or decreased blast counts in the marrow, although these continued cycles were considered protocol deviations. Of these four patients, one patient received a total of eight cycles of therapy, two patients received a total of four cycles, and one patient received a total of 11 cycles.

Safety

In the phase I safety study, two of three patients treated with 30 mg/day (3 h×3 d) experienced grade 3 neurologic events that led to discontinuation of treatment and were classified as DLTs (**Table 3**). The first patient experienced grade 3 somnolence (day 1) and grade 3 confusion. This patient also experienced dizziness, mood alteration, and speech disorder during infusion (all grade < 3; **Table 4**). The second patient experienced grade 3 ataxia and grade 1 confusion, euphoria, and somnolence on day 1 of cycle 1. In both patients, neurologic DLTs were assessed as definitely related to study drug and resolved within 24 hours. The third patient in this group also experienced euphoria and ataxia (grade <3) and discontinued due to leukemic infiltrate. One serious AE (grade 3 neutropenic fever) was reported for this regimen, but was judged to be unrelated to study drug (**Table 3**).

Per protocol, the obatoclax dose in the 3 h×3 d regimen was decreased to 20 mg/day in a subsequent cohort of three patients. No additional DLTs were observed in this group and only one grade 3 serious AE (pneumonia) was reported and was considered unrelated to study drug. In addition, one patient experienced grade 2 cytokine release syndrome, which was also considered a serious AE (**Table 3**). Therefore, the 20 mg/day (3 h×3 d) regimen was chosen for the randomized phase II portion of the

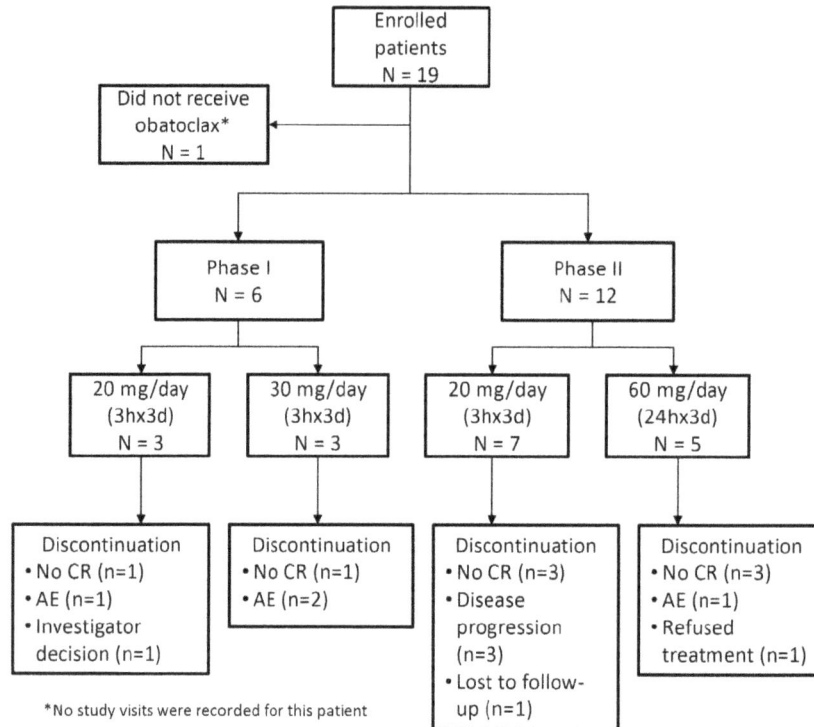

Figure 2. Patient disposition. AE, adverse event; CR, complete response.

study for comparison with obatoclax 60 mg/day (24 h×3 d), as previously defined in a small study of 18 patients [16].

In the phase II study, most patients receiving either obatoclax 20 mg/day (3 h×3 d) or 60 mg/day (24 h×3 d) experienced mild to moderate (grade <3), transient, neurologic AEs such as euphoria, somnolence, ataxia, dizziness, and confusion (**Table 4**). AEs of grade ≥3 that were reported in more than one patient included febrile neutropenia (n = 3), dizziness (n = 2), atrial fibrillation (n = 2), and acute myocardial infarction (n = 2). The events of dizziness, atrial fibrillation, and acute myocardial infarction were considered to be at least possibly related to

obatoclax. Acute myocardial infarction resulted in treatment discontinuation in one patient.

Combining both the phase I and II components of the study, all 18 (100%) patients who received obatoclax experienced at least 1 AE, the most common of which were neurologic (n = 14; 77.8%) or psychiatric (n = 16; 88.9%); most were transient and mild, and resolved without sequelae. Ten patients experienced serious AEs; details are provided in **Table 3**. Evidence for trends between severity of AEs (grade ≥3) and dose or schedule was not observed (**Table 4**). AEs (any grade) with the highest reported incidence included euphoria (67%), somnolence (44%), and ataxia (39%).

Table 2. Study drug exposure in patients treated with obatoclax (N = 18).

	Phase I		Phase II		All (N = 18)
	30 mg/d (n = 3)	**20 mg/d (n = 3)**	**3-h infusion (20 mg/d) (n = 7)**	**24-h infusion (60 mg/d) (n = 5)**	
Median number of cycles (range)	2 (1–2)	4 (4–8)	2 (2–11)	2 (1–2)	2 (1–11)
Total cycles, n (%)					
1	3 (100)	3 (100)	7 (100)	5 (100)	18 (100)
2	2 (67)	3 (100)	7 (100)	3 (60)	15 (83)
3	0	3 (100)	1 (14)	0	4 (22)
4	0	3 (100)	1 (14)	0	4 (22)
5	0	1 (33)	1 (14)	0	2 (11)
6	0	1 (33)	1 (14)	0	2 (11)
7	0	1 (33)	1 (14)	0	2 (11)
8	0	1 (33)	1 (14)	0	2 (11)
>9	0	0	1 (14)	0	1 (6)

Table 3. Summary of dose-limiting toxicities and serious adverse events.

| | Phase I | | Phase II | |
| | | | 3-h infusion (20 mg/d) (n = 7) | 24-h infusion (60 mg/d) (n = 5) |
	30 mg/d (n = 3)	20 mg/d (n = 3)		
Dose-limiting toxicity	Grade 3 somnolence/confusion (1); grade 3 ataxia (1)	0	0	0
Serious TEAE, n (grade, attribution)				
Febrile neutropenia	1 (gr 3, NR)		1 (gr 3, NR)	
Atrial fibrillation			2 (gr 3, NR; gr 3, NR)	
Acute myocardial infarction			1 (gr 3, NR)	1 (gr 4, PS)
Cough			2 (gr 1, NR; gr 2, PS)	
Catheter site infection			1 (gr 3, NR)	1 (gr 3, NR)
Cytokine release syndrome		1 (gr 2, PR)		
Pneumonia		1 (gr 3, NR)	1 (gr 3, NR)	
Acute sinusitis			1 (gr 1, NR)	
Dyspnea			1 (gr 2, NR)	
Fatigue				1 (gr 5, NR)
Dizziness				1 (gr 3, PR)

AE, adverse event; DLT, dose-limiting toxicity (DLTs were defined as grade ≥3 infusion-related neurologic AEs and nonhematologic AEs not responsive to symptom-directed therapy); PR, probably related; PS, possibly related; NR, not related; TEAE, treatment-emergent adverse event.

Somnolence was the most commonly reported grade ≥3 AE (17%). The most common grade ≥3 toxicities based on laboratory data were leukocytosis and thrombocytopenia (each n = 9). No clinically meaningful differences were observed across regimens for laboratory reports of grade ≥3 hematologic findings.

Two patients died during the study for reasons unrelated to obatoclax administration. One patient on the 60 mg/day (24 h×3 d) regimen died on day 23 of the study from progressive disease. The second patient, receiving 20 mg/day (3 h×3 d), died on day 41; the cause of death is unknown. An additional six patients died more than 30 days after the last obatoclax dose. Causes of death in these patients were progressive disease (n = 2), sepsis (n = 1), and unknown (n = 3).

Efficacy

CR was not achieved with obatoclax induction. However, three patients on the 20 mg/day (3 h×3 d) regimen demonstrated a 7% to 17% decrease in bone marrow blast percentage between the baseline assessment and the end of cycle 2 (**Table 5**). It is noteworthy that two of these patients with decreased marrow blasts also demonstrated an increase of 33% to 57% in neutrophil count. Of these three patients, one patient in phase I receiving obatoclax 20 mg/day (3 h×3 d arm) had a decrease in marrow blasts from 27% to 10% with increased neutrophils from $9344×10^3/\mu L$ to $14,700×10^3/\mu L$. This patient received eight cycles of study treatment and withdrew from the study on the advice of the investigator. One additional patient in the phase II portion of the study had increased neutrophil (pre-treatment: $230×10^3/\mu L$; end of cycle 2: $2500×10^3/\mu L$) and platelet counts (pre-treatment: $73×10^3/\mu L$; end of cycle 2: $250×10^3/\mu L$) without a significant change in the marrow blasts. This patient remained stable and received 11 cycles of obatoclax.

Discussion

The results of the current study demonstrate that the safety profile of obatoclax administered for 3 consecutive days (every 2 weeks) by 3-hour or 24-hour infusion was generally mild and similar to previous reports [11–14]. AEs were typically transient, neurologic, or psychiatric findings that resolved without sequelae. Based on the safety profile, 20 mg/day was determined to be the MTD of obatoclax when administered over 3 hours/day for 3 consecutive days in older AML patients. Two patients treated with the 30 mg/day (3 h×3 d) regimen experienced DLTs consisting of grade 3 confusion, somnolence, or ataxia, which led to premature discontinuation and selection of the 20 mg/day dose for further evaluation. In the phase II comparison of obatoclax 20 mg/day (3 h×3 d) to the previously evaluated 60 mg/day (24 h×3 d) regimen, both dosing schedules demonstrated similar safety profiles.

The mechanism(s) underlying the development of neurologic or psychiatric symptoms is uncertain. However, it is plausible that these symptoms represent an on-target effect, since Bcl-2 promotes neuron survival and Bcl-XL plays a role in synaptic plasticity [19,20]. Alternatively, the neurologic and psychiatric effects of obatoclax may reflect binding to targets other than Bcl-2 family members.

Based on laboratory data, grade ≥3 leukocytosis and thrombocytopenia occurred in 50% of patients in this study. These hematologic abnormalities are likely related to underlying disease rather than obatoclax as they were present at baseline. However, it should be noted that inhibition of Bcl-XL by obatoclax may result in thrombocytopenia. Because patients also received platelet transfusions during the study as supportive care, the impact of obatoclax on platelet production may have been obscured.

Six patients experienced cardiac events during this study, four of which were assessed as at least possibly related to the study drug. The causal relationship with ischemic cardiac events, if any, is unclear because patients in this study were older with multiple comorbidities. In other clinical studies evaluating obatoclax, QTc prolongation has been reported by automated electrocardiogram. In a study of patients with relapsed small cell lung cancer, the interval between obatoclax doses was extended to 3 days in a

Table 4. Treatment-emergent adverse events occurring in more than one patient.

n	Phase I 20 mg/d (n = 3)		Phase I 30 mg/d (n = 3)		Phase II 3-h infusion (20 mg/d) (n = 7)		Phase II 24-h infusion (60 mg/d) (n = 5)		All (N = 18)	
	All grade	Grade ≥3	All grade	Grade ≥3	All grade	Grade ≥3	All grade	Grade ≥3	All grade	Grade ≥3
Euphoria	3	0	2	0	6	1	1	0	12	1
Somnolence	0	0	2	2	4	1	2	0	8	3
Ataxia	0	0	3	1	3	0	1	1	7	2
Dizziness	1	0	1	0	2	1	1	1	5	2
Confusion	0	0	2	1	1	0	2	0	5	1
Constipation	1		0		1		3		5	
Fever	0	0	0	0	3		2	0	5	
Diarrhea	1		1		1	1	1		4	1
Peripheral edema	0		0		2		2		4	
Febrile neutropenia	0	0	1	1	2	2	1	1	4	4
Disorientation	2		0		1		0		3	
Cough	0		1		2		0		3	
Unsteady gait	0		1		2		0		3	
Insomnia	1		0		2		0		3	
Dyspnea	0		0		1		2		3	
Hypoxia	0	0	0	0	3	1	0	0	3	1
Dysarthria	1		0		2		0		3	
Fatigue	0	0	0	0	2	0	1	1	3	1
Headache	0		0		2		1		3	
Ecchymosis	0		0		3		0		3	
Hypotension	0		0		1		1		2	
Tachycardia	0	0	0	0	2		0		2	
Cardiac murmur	0		0		2				2	
Atrial fibrillation	0	0	0	0	2	2	0	0	2	2
Pneumonia	1	1	0	0	1	1	0	0	2	2
Loose stool	1		0		1		0		2	
Gingival pain	0		1		1		0		2	
Cytokine release syndrome	1		0		1		0		2	
Abnormal breath sounds	0		0		1		1		2	
Acute MI	0	0	0	0	1	1	1	1	2	2
Dry mouth	0		0		1		1		2	
Crackles (lung)	0		0		1		1		2	
Thrush	0		0		2		0		2	

Table 4. Cont.

n	Phase I 20 mg/d (n = 3)		Phase I 30 mg/d (n = 3)		Phase II 3-h infusion (20 mg/d) (n = 7)		Phase II 24-h infusion (60 mg/d) (n = 5)		All (N = 18)	
	All grade	Grade ≥3	All grade	Grade ≥3	All grade	Grade ≥3	All grade	Grade ≥3	All grade	Grade ≥3
Hypocalcemia	0		1		1		0		2	
Hypokalemia	0		1		1		0		2	
Slurred speech	0		0		2		0		2	
Agitation	0	0	0	0	1	1	1	0	2	1
Muscular weakness	0		0		1		1		2	
Anxiety	0		1		1		0		2	

Table 5. Improvement in marrow blast count, neutrophil count, and platelet count in patients receiving obatoclax.

Patient	Treatment (phase)	AML Classification	Blast count, % (aspirate)		Neutrophils, ×10^3/μL		Platelets, ×10^3/μL	
			Base-line	End Cycle 2	Baseline	End Cycle 2	Baseline	End Cycle 2
04.002*	20 mg/d (I)	M4	27	10	9344	14,700	227	206
04.003	20 mg/d (I)	M3	27	20	3300	4400	126	139
05.002	20 mg/d (II)	M1	33	23	390	360	49	47

*Received eight cycles of obatoclax.
AML, acute myeloid leukemia.

single patient who experienced QTc prolongation during the first cycle [14,21]. In another phase I dose escalation study in advanced hematologic malignancies, grade 3 QTc prolongation was observed in 3 patients, but was confounded by the presence of QTc prolongation at baseline [14]. Notably, an imbalance in the ratio of anti-apoptotic to pro-apoptotic Bcl-2 proteins appears causal in the development of cardiovascular disease, including ischemic heart disease [22], so relationship to study drug cannot be excluded.

In the current study, four patients had a clinical response of stable disease and were treated for up to 11 cycles. We did not observe a CR in this study. This contrasts with our previous report of a CR achieved with single-agent obatoclax (20 mg/m^2 over 24 hours) in one older treatment-naive patient with AML with a mixed-lineage leukemia (MLL) t(9;11) translocation [14]. Although such dramatic single-agent activity in previously untreated AML was not confirmed by our data, it is possible that MLL-associated leukemia may be particularly sensitive to Bcl-2 family inhibitors. Preclinical studies have shown that inhibition of MLL expression using siRNA corresponded with reduced Bcl-XL levels and leukemic proliferation that may be mediated by HoxA9 [23,24]. Thus, Bcl-2 proteins may play an important role in the proliferation of MLL-associated leukemia. In addition, a limitation of our study is that pharmacodynamic activity or pharmacokinetic parameters were not assessed. Integrating these evaluations in future clinical trials of obatoclax may provide further insight into the clinical potential of obatoclax as a single agent.

In addition to obatoclax, several other small-molecule BH3 mimetics are under investigation. ABT-737 and ABT-263 (navitoclax), for example, bind three of six Bcl-2 family members with high affinity [25,26]. Navitoclax has been evaluated in phase I trials in lymphoid malignancies [27,28]. However, these inhibitors do not bind Mcl-1 with as high affinity, and their therapeutic potential is constrained by dose-limiting thrombocytopenia associated with potent Bcl-XL inhibition in platelets [29]. Obatoclax was developed as a promiscuous Bcl-2 family inhibitor and also inhibits Mcl-1, which is essential for development and sustained growth of AML [6]. Follow-on analysis of the correlation between anti-apoptotic Bcl-2 proteins and obatoclax response was not conducted in this study, and will be important to include in future phase II evaluations.

It is conceivable, given their mechanism of action, that Bcl-2 family inhibitors might be most active in combination with other inducers of cell death. For example, obatoclax induced apoptosis in OCI-AML3 leukemic cells when used in combination with ABT-737 and synergistically induced apoptosis in combination with cytosine arabinoside in leukemic cell lines and in primary AML samples [10]. Preclinical investigations also suggest that obatoclax potentiates the effect of established drugs in AML [30–32], and several clinical trials of obatoclax combined with conventional chemotherapeutic agents have been completed or are ongoing in a range of solid tumors and hematologic malignancies (e.g., NCT00612612, NCT00521144 at clinicaltrials.gov).

In conclusion, based on the safety profile described in this study, 20 mg/day is the MTD of obatoclax when administered by 3-hour infusion over 3 consecutive days in an older AML population, and it has similar tolerability to a 60 mg/day (24 h×3 d) regimen. Although the current study does not support the efficacy of obatoclax as a single agent in an unselected group of treatment-naive AML patients, additional studies may reveal activity in select subgroups, particularly in combination with other chemotherapeutics.

Acknowledgments

ADS is a Leukemia and Lymphoma Society Scholar in Clinical Research. The authors thank Powered 4 Significance LLC for assistance with medical writing in preparing an initial draft of the manuscript and editorial assistance.

Author Contributions

Conceived and designed the experiments: ADS GB. Performed the experiments: ADS AR THC DC HE DJD MST CG GB. Analyzed the data: ADS AR THC DC GB. Contributed reagents/materials/analysis tools: ADS AR THC DC HE DJD MST GB. Wrote the paper: ADS AR THC HE DJD MST CG.

References

1. Estey E, Dohner H (2006) Acute myeloid leukaemia. Lancet 368: 1894–1907.
2. Siegel R, Naishadham D, Jemal A (2013) Cancer statistics, 2013. CA Cancer J Clin 63: 11–30.
3. National Cancer Institute. SEER Stat Fact Sheets: Leukemia. Available: http://seer.cancer.gov/statfacts/html/leuks.html. Accessed: 2014 Apr 17.
4. Chipuk JE, Moldoveanu T, Llambi F, Parsons MJ, Green DR (2010) The BCL-2 family reunion. Mol Cell 37: 299–310.
5. Cory S, Adams JM (2002) The Bcl2 family: regulators of the cellular life-or-death switch. Nat Rev Cancer 2: 647–656.
6. Glaser SP, Lee EF, Trounson E, Bouillet P, Wei A, et al. (2012) Anti-apoptotic Mcl-1 is essential for the development and sustained growth of acute myeloid leukemia. Genes Dev 26: 120–125.
7. Willis SN, Fletcher JI, Kaufmann T, van Delft MF, Chen L, et al. (2007) Apoptosis initiated when BH3 ligands engage multiple Bcl-2 homologs, not Bax or Bak. Science 315: 856–859.
8. Nguyen M, Marcellus RC, Roulston A, Watson M, Serfass L, et al. (2007) Small molecule obatoclax (GX15-070) antagonizes MCL-1 and overcomes MCL-1-mediated resistance to apoptosis. Proc Natl Acad Sci U S A 104: 19512–19517.
9. Zhai D, Jin C, Satterthwait AC, Reed JC (2006) Comparison of chemical inhibitors of antiapoptotic Bcl-2-family proteins. Cell Death Differ 13: 1419–1421.
10. Konopleva M, Watt J, Contractor R, Tsao T, Harris D, et al. (2008) Mechanisms of antileukemic activity of the novel Bcl-2 homology domain-3 mimetic GX15-070 (obatoclax). Cancer Res 68: 3413–3420.

11. Hwang JJ, Kuruvilla J, Mendelson D, Pishvaian MJ, Deeken JF, et al. (2010) Phase I dose finding studies of obatoclax (GX15-070), a small molecule pan-BCL-2 family antagonist, in patients with advanced solid tumors or lymphoma. Clin Cancer Res 16: 4038–4045.
12. O'Brien SM, Claxton DF, Crump M, Faderl S, Kipps T, et al. (2009) Phase I study of obatoclax mesylate (GX15-070), a small molecule pan-Bcl-2 family antagonist, in patients with advanced chronic lymphocytic leukemia. Blood 113: 299–305.
13. Oki Y, Copeland A, Hagemeister F, Fayad LE, Fanale M, et al. (2012) Experience with obatoclax mesylate (GX15-070), a small molecule pan-Bcl-2 family antagonist in patients with relapsed or refractory classical Hodgkin lymphoma. Blood 119: 2171–2172.
14. Schimmer AD, O'Brien S, Kantarjian H, Brandwein J, Cheson BD, et al. (2008) A phase I study of the pan bcl-2 family inhibitor obatoclax mesylate in patients with advanced hematologic malignancies. Clin Cancer Res 14: 8295–8301.
15. Parikh SA, Kantarjian H, Schimmer A, Walsh W, Asatiani E, et al. (2010) Phase II study of obatoclax mesylate (GX15-070), a small-molecule BCL-2 family antagonist, for patients with myelofibrosis. Clin Lymphoma Myeloma Leuk 10: 285–289.
16. Raza A, Galili N, Borthakur G, Carter TH, Claxton DF, et al. (2009) A safety and schedule seeking trial of Bcl-2 inhibitor obatoclax in previously untreated older patients with acute myeloid leukemia (AML). J Clin Oncol 27: suppl; abstr 3579.

17. National Cancer Instititute. Common Terminology Criteria for Adverse Events v3.0 (CTCAE). Available: http://ctep.cancer.gov/protocolDevelopment/electronic_applications/docs/ctcaev3.pdf. Accessed: 2014 Apr 17.

18. Cheson BD, Bennett JM, Kopecky KJ, Buchner T, Willman CL, et al. (2003) Revised recommendations of the International Working Group for Diagnosis, Standardization of Response Criteria, Treatment Outcomes, and Reporting Standards for Therapeutic Trials in Acute Myeloid Leukemia. J Clin Oncol 21: 4642–4649.

19. Li H, Chen Y, Jones AF, Sanger RH, Collis LP, et al. (2008) Bcl-xL induces Drp1-dependent synapse formation in cultured hippocampal neurons. Proc Natl Acad Sci U S A 105: 2169–2174.

20. Offen D, Beart PM, Cheung NS, Pascoe CJ, Hochman A, et al. (1998) Transgenic mice expressing human Bcl-2 in their neurons are resistant to 6-hydroxydopamine and 1-methyl-4-phenyl-1,2,3,6- tetrahydropyridine neurotoxicity. Proc Natl Acad Sci U S A 95: 5789–5794.

21. Paik PK, Rudin CM, Pietanza MC, Brown A, Rizvi NA, et al. (2011) A phase II study of obatoclax mesylate, a Bcl-2 antagonist, plus topotecan in relapsed small cell lung cancer. Lung Cancer 74: 481–485.

22. Dewson G, Kluck RM (2010) Bcl-2 family-regulated apoptosis in health and disease. Cell Health Cytoskel 2: 9–22.

23. Thomas M, Gessner A, Vornlocher HP, Hadwiger P, Greil J, et al. (2005) Targeting MLL-AF4 with short interfering RNAs inhibits clonogenicity and engraftment of t(4;11)-positive human leukemic cells. Blood 106: 3559–3566.

24. Izon DJ, Rozenfeld S, Fong ST, Komuves L, Largman C, et al. (1998) Loss of function of the homeobox gene Hoxa-9 perturbs early T-cell development and induces apoptosis in primitive thymocytes. Blood 92: 383–393.

25. Rooswinkel RW, van de Kooij B, Verheij M, Borst J (2012) Bcl-2 is a better ABT-737 target than Bcl-xL or Bcl-w and only Noxa overcomes resistance mediated by Mcl-1, Bfl-1, or Bcl-B. Cell Death Dis 3: e366.

26. Tse C, Shoemaker AR, Adickes J, Anderson MG, Chen J, et al. (2008) ABT-263: a potent and orally bioavailable Bcl-2 family inhibitor. Cancer Res 68: 3421–3428.

27. Roberts AW, Seymour JF, Brown JR, Wierda WG, Kipps TJ, et al. (2012) Substantial susceptibility of chronic lymphocytic leukemia to BCL2 inhibition: results of a phase I study of navitoclax in patients with relapsed or refractory disease. J Clin Oncol 30: 488–496.

28. Wilson WH, O'Connor OA, Czuczman MS, LaCasce AS, Gerecitano JF, et al. (2010) Navitoclax, a targeted high-affinity inhibitor of BCL-2, in lymphoid malignancies: a phase 1 dose-escalation study of safety, pharmacokinetics, pharmacodynamics, and antitumour activity. Lancet Oncol 11: 1149–1159.

29. Vogler M, Hamali HA, Sun XM, Bampton ET, Dinsdale D, et al. (2011) BCL2/BCL-X(L) inhibition induces apoptosis, disrupts cellular calcium homeostasis, and prevents platelet activation. Blood 117: 7145–7154.

30. Brem EA, Thudium K, Khubchandani S, Tsai PC, Olejniczak SH, et al. (2011) Distinct cellular and therapeutic effects of obatoclax in rituximab-sensitive and -resistant lymphomas. Br J Haematol 153: 599–611.

31. Campas C, Cosialls AM, Barragan M, Iglesias-Serret D, Santidrian AF, et al. (2006) Bcl-2 inhibitors induce apoptosis in chronic lymphocytic leukemia cells. Exp Hematol 34: 1663–1669.

32. Rahmani M, Aust MM, Attkisson E, Williams DC, Jr., Ferreira-Gonzalez A, et al. (2012) Inhibition of Bcl-2 antiapoptotic members by obatoclax potently enhances sorafenib-induced apoptosis in human myeloid leukemia cells through a Bim-dependent process. Blood 119: 6089–6098.

8

Variations in Metal Tolerance and Accumulation in Three Hydroponically Cultivated Varieties of *Salix integra* Treated with Lead

Shufeng Wang[1,2]⁹, **Xiang Shi**[2]⁹, **Haijing Sun**[2]*, **Yitai Chen**[2], **Hongwei Pan**[2], **Xiaoe Yang**[1]*, **Tariq Rafiq**[1]

1 MOE Key Lab of Environmental Remediation and Ecosystem Health, College of Environmental and Resource Sciences, Zhejiang University, Zijingang Campus, Hangzhou, P.R. China, **2** Research Institute of Subtropical Forestry, Chinese Academy of Forestry, Fuyang, Hangzhou, P.R. China

Abstract

Willow species have been suggested for use in the remediation of contaminated soils due to their high biomass production, fast growth, and high accumulation of heavy metals. The tolerance and accumulation of metals may vary among willow species and varieties, and the assessment of this variability is vital for selecting willow species/varieties for phytoremediation applications. Here, we examined the variations in lead (Pb) tolerance and accumulation of three cultivated varieties of *Salix integra* (Weishanhu, Yizhibi and Dahongtou), a shrub willow native to northeastern China, using hydroponic culture in a greenhouse. In general, the tolerance and accumulation of Pb varied among the three willow varieties depending on the Pb concentration. All three varieties had a high tolerance index (TI) and EC50 value (the effective concentration of Pb in the nutrient solution that caused a 50% inhibition on biomass production), but a low translocation factor (TF), indicating that Pb sequestration is mainly restricted in the roots of *S. integra*. Among the three varieties, Dahogntou was more sensitive to the increased Pb concentration than the other two varieties, with the lowest EC50 and TI for root and above-ground tissues. In this respect, Weishanhu and Yizhibi were more suitable for phytostabilization of Pb-contaminated soils. However, our findings also indicated the importance of considering the toxicity symptoms when selecting willow varieties for the use of phytoremediation, since we also found that the three varieties revealed various toxicity symptoms of leaf wilting, chlorosis and inhibition of shoot and root growth under the higher Pb concentrations. Such symptoms could be considered as a supplementary index in screening tests.

Editor: Manuel Reigosa, University of Vigo, Spain

Funding: This work was financially supported by Ministry of Science and Technology of China (#2010DFB33960 and #2012AA100602), the National Natural Science Foundation of China (#31400526), and the Fundamental Research Funds from the Central Universities (# 2014FZA6000). The funders had no role in study design, data collection and analysis, decision to publish, or preparation of the manuscript.

Competing Interests: The authors have declared that no competing interests exist.

* Email: sunhaijing@163.com (HS); xyang@zju.edu.cn (XY)

⁹ These authors contributed equally to this work.

Introduction

Heavy metal contamination in water, air, or soil is a major environmental concern worldwide [1–3]. Excessive levels of heavy metals can be introduced into the environment by mining, smelting, electroplating, use of pesticides or fertilizers, industrial discharge, etc. [1,4–7]. In China, twenty million hectares of agricultural land have been polluted with heavy metals, and it has been reported that each year over 12 million tons of grain are contaminated by toxic metals [8]. Heavy metals cannot be degraded but can be stabilized or extracted by plants, in a process known as "phytoremediation" [1,4,9,10]. The phytoremediation of heavy metals is affected by several factors, such as the species of plant, solubility in soil solution, and the translocation of the heavy metals from the soil to the harvestable plant parts, etc. [11,12]. Of these factors, the plant species plays a crucial role, and therefore, selecting appropriate plant species is important for the successful application of plant-based remediation techniques.

Lead (Pb) is one of the most toxic metals, and its concentration in agricultural soil has rapidly increased due to various anthropogenic inputs [13]. Lead is not a bio-essential element, but is easily absorbed by and accumulated in plants [14,15]. Many studies have shown that the patterns of Pb uptake, transport and accumulation in plants are strongly governed by plant factors, and that Pb accumulation and distribution vary largely among different plant organs [3,16]. Most studies have shown that, for the majority of plant species, Pb uptake is restricted to roots, with only a small portion being translocated to the shoots [12,17–19]. Whereas hyperaccumulation is an exception, for instance, *Thlaspi rotundifolium*, a cadmium hyperaccumulator, is capable of accumulating up to 8,200 mg kg^{-1} Pb in the shoots [10]. Despite the abilities of these plants to accumulating high amounts of Pb in above-ground tissues, their application in contaminated soil remediation is often limited because of their slow growth rate and low biomass yield [20]. In contrast, other species e.g., *Brassica juncea* (cv.426308) have been shown to accumulate high Pb concentrations in shoots (34,500 mg kg^{-1}) and in addition have high biomass which allows high metal removals [12]. However, their growth performance and metal-accumulating abilities can also be limited by low pH and high Pb concentration (>1, 500 mg kg^{-1}) in the soil [21].

Plant species are selected for use in the remediation of heavy metal contamination based on factors, such as high biomass production, ability to take up and accumulate metals, deep root systems, high growth rate, and ease of planting and maintenance [1,22–24]. Trees, especially those fast-growing woody species, with their large biomass and deeper, more integrated root systems, have provided a unique means for deep phytoremediation of soil or water during recent decades [11,25,26]. In recent years, willow species, which could be grown intensively for use in energy production, have been suggested for use in the remediation of metal contaminated soils. A number of *Salix* species have been studied for their ability to tolerate and accumulate heavy metals [27–30]. Ali *et al.* [9] demonstrated that *Salix acmophylla* can accumulate considerable amounts of Cu, Ni and Pb in different plant parts and exhibits high tolerance to these metals. Tlustos *et al.* [30] reported that *S. smithiana* Willd. is able to accumulate 456 mg kg^{-1} and 26.6 mg kg^{-1} Pb in the roots and trunk, respectively, but no more than 10 mg kg^{-1} Pb was observed in the twigs and leaves. Most of the studies confirmed that *Salix* species are able to accumulate a high concentration of Pb in the root systems and have the potential for phytostablization of Pb, and that the uptake patterns and Pb accumulation in *Salix* vary among species and varieties [30,31]. However, the critical concentration causing Pb-induced phytotoxicity or growth inhibition in different *Salix* species or varieties is still unknown. Hydroponic methods are effective in the rapid screening for heavy metal tolerance and accumulation in plants, and have been widely used in evaluating the phytoremediation potential of willow species in recent years [18,29].

The aim of our study was to detect the Pb accumulation potential, especially the critical toxicity thresholds based on EC50 estimates in three cultivated varieties of *Salix integra* using hydroponic methods. The results are important for effective selection of willow species for phytoremediation application.

Salix integra, a shrub willow native to northeastern China, Japan, Korea, and Primorsky Krai in the far southeast of Russia, has been identified as a Cd-accumulating plant [32,33]. Cultivation and planting of *S. integra* for shoot and biomass production have a long history in China, and many cultivated varieties are grown in northeastern China. Yizhibi, Dahongtou and Weishanhu are three cultivated varieties of *S. integra* with high biomass production and easy cultivation [34–36]. In previous studies, we have confirmed that *S. integra* was able to tolerate and accumulate high concentrations of Cd and Zn in hydroponic culture, and we measured their differences in Cd uptake and accumulation, as well as the tolerance indices, in the *S. integra* varieties [35–37].

In the present study, we hypothesized that the three varieties of *S. integra* could take up and accumulate Pb in their roots and above-ground parts, and that Pb accumulation would vary among the varieties according to the Pb contamination level. To test this hypothesis, hydroponic culture experiments were conducted to compare the Pb tolerance and accumulation among the three varieties of *S. integra* and to determine the critical level of Pb for their normal growth. The results are expected to improve our understanding of metal accumulation patterns in *Salix* species and provide guidance for future application of *S. integra* in the phytoremediation of Pb contaminated soils and/or water.

Materials and Methods

Experimental site and willow preparation

The experiment was conducted in a greenhouse in Fuyang, Hangzhou, Zhejiang Province, P. R. China (30°03′N, 119°57′E)

in May, 2011. The greenhouse temperature was maintained between 20–25°C, with a natural photoperiod. The *S. integra* varieties Weishanhu, Yizhibi and Dahongtou were obtained from native habitats in Shandong Province. No specific permits were required to extract samples from this site, which is not privately-owned, and this study did not involve protected species. Cuttings (8–10 cm) from 1-year-old stems in nursery beds at the Institute of Subtropical Forestry were selected for uniformity based on the diameter (Φ 0.4–0.5 cm) and the number of buds (4–6 buds per cutting). Cuttings were rooted in tap water for 4 wk in hydroponic pots (50 cm×35 cm×15 cm, length × width × height) and then transferred to 15 L aerated Knop's solution [38] (6.1 mM Ca (NO$_3$)$_2$, 2.5 mM KNO$_3$, 1.6 mM KCl, 1.8 mM KH$_2$PO$_4$, 2.1 mM MgSO$_4$ and 3.8 µM FeCl$_3$) in each pot, maintaining a constant pH of 5.5 using 1 M HCl or 1 M NaOH.

To avoid possible Pb precipitation caused by the presence of phosphorus (P) in the nutrient solution, the P concentration was kept at a maximum of 0.04 µM according to Zhivotovsky *et al.* [18].

Experimental design

After 2 wk of plant growth in the aerated solution, Pb treatments with the final concentration of 47 µM (T1), 123 µM (T2), 178 µM (T3), and 196 µM Pb (T4) were applied as Pb(NO$_3$)$_2$ for 14 days, with normal nutrient solution as the control. Each treatment was replicated three times, with each replicate consisting of one pot containing six plants for each variety. These pots were arranged in a completey randomized block design.

The metal and nutrient solutions were replaced every 2 d to ensure consistency. Metal-related phytotoxicity symptoms were recorded throughout the experiment. At the end of the growth and Pb treatment period, the plant tissues were separated into roots, wood (the original cuttings), new shoots, and leaves, washed with tap water and rinsed with deionized water. The root length, surface area, volume, diameter and number of root tips of each plant were determined using root scan apparatus (Epson V700), equipped with WinRHIZO software (Regent Instruments Co.). The root tissues were then washed in 1 M HCl solution and rinsed again with deionized water. Samples were dried at 70°C for 48 h and the dry weights were recorded.

Estimation of chlorophyll content

As a non-destructive measurement, we used an Opti-Sciences CCM-200 chlorophyll-meter to estimate the chlorophyll content by recording the Chlorophyll Concentration Index (CCI).

Pb analysis in plant tissues

All plant tissue samples were ashed at 500°C in a muffle furnace, and the ash was dissolved in 10 ml of 1 M HCl solution, and diluted in double deionized water to a 50 ml volume prior to analysis. The total Pb concentration in the liquid samples was determined using inductively coupled plasma optical emission spectrometer (Varian 725-ES, Palo Alto, CA, USA). Certified reference materials (Mixed shoots of shrubs from Pb-Zn mine tailings, GBW 07602, China) were used to ensure the quality of analyses. Good agreement was obtained between our method and certified values. The total metal contents (mg plant^{-1}) in the roots and above-ground tissues (wood, new shoots, and leaves) were calculated by multiplying the tissue dry weight by the metal concentration.

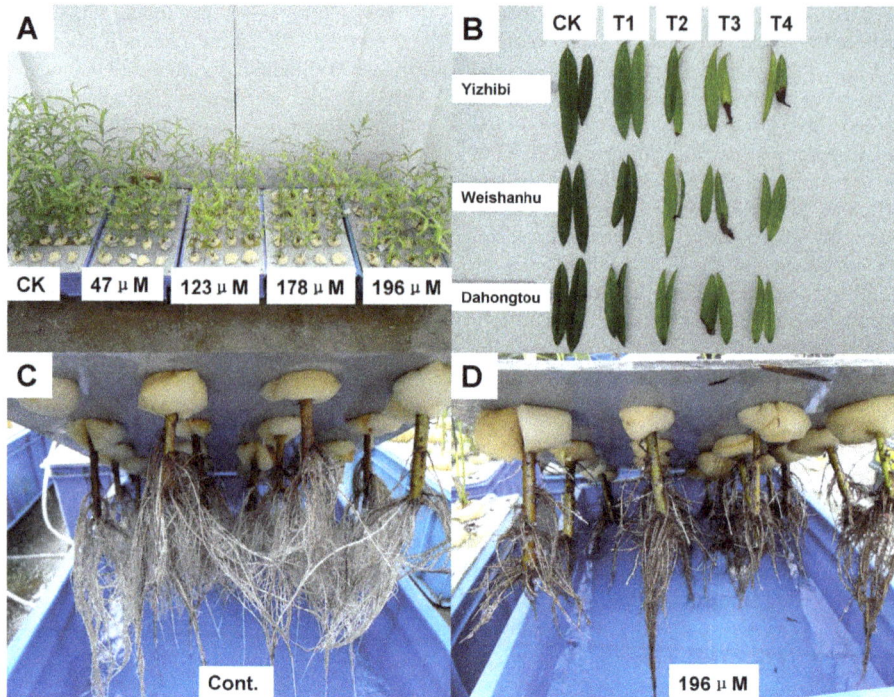

Figure 1. Growth development of *S. integra* exposed to different Pb concentrations for 14 days and metal related symptoms in roots and leaves. (A) Growth of *S. integra* after exposure to 0, 47, 123, 178, and 196 μM of Pb treatments for 14 days. (B) Leaf symptoms of three varieties under different Pb treatments for 14 days. T1, T2, T3 and T4 represent 47, 123, 178, and 196 μM of Pb respectively. (C) Root systems of willows grown in the control. (D) Root systems of willows grown under 196 μM Pb treatment.

Determination of tolerance index

The tolerance index (TI) was determined to assess the ability of the willow varieties to grow in the presence of a given concentration of Pb according to the following equation [39]:

$$TI(\%) = (DW \ of \ treated \ plants)/(DW \ of \ control \ plants) \times 100$$

where DW is the dry weight of the roots or above-ground tissues of the willow.

Determination of translocation factor

The translocation factor (TF) indicates the efficiency of the plant to translocate the accumulated metal from its roots to the aerial parts. It is calculated as follows [40]:

$$TF = \frac{C_{aerial \ parts}}{C_{roots}} \times 100$$

where C_{aerial} parts is the concentration of metal in the above-ground tissues and C_{roots} is the concentration of metal in the roots.

Determination of EC50

The EC50 (effective concentration) is the concentration of the metal that causes a 50% decrease in plant biomass, as compared with the control. Critical toxicity thresholds based on EC50 estimates were determined for the roots and above-ground tissues (combined wood, new shoots, and leaves) for each variety using non-linear regressions to fit curves [41]. A multivariate model

Table 1. Chlorophyll Concentration Index (CCI) of leaves of three varieties of *S. integra* grown in various Pb concentrations for 14 days.

	Varieties		
Pb (μM)	Yizhibi	Weishanhu	Dahongtou
0	13.05±0.93a	13.77±0.16a	11.79±1.14a
47	8.50±0.53b	8.84±0.53b	8.40±0.08b
123	5.58±0.23c	6.39±1.01c	6.20±1.47c
178	6.13±0.78c	6.18±0.68c	6.13±0.72c
196	4.78±0.13c	5.43±0.16c	5.27±0.13c

Values are mean ± S.D. (n=6), and data with different letters in the same column indicate a significant difference at $P<0.05$ according to Fisher's LSD test.

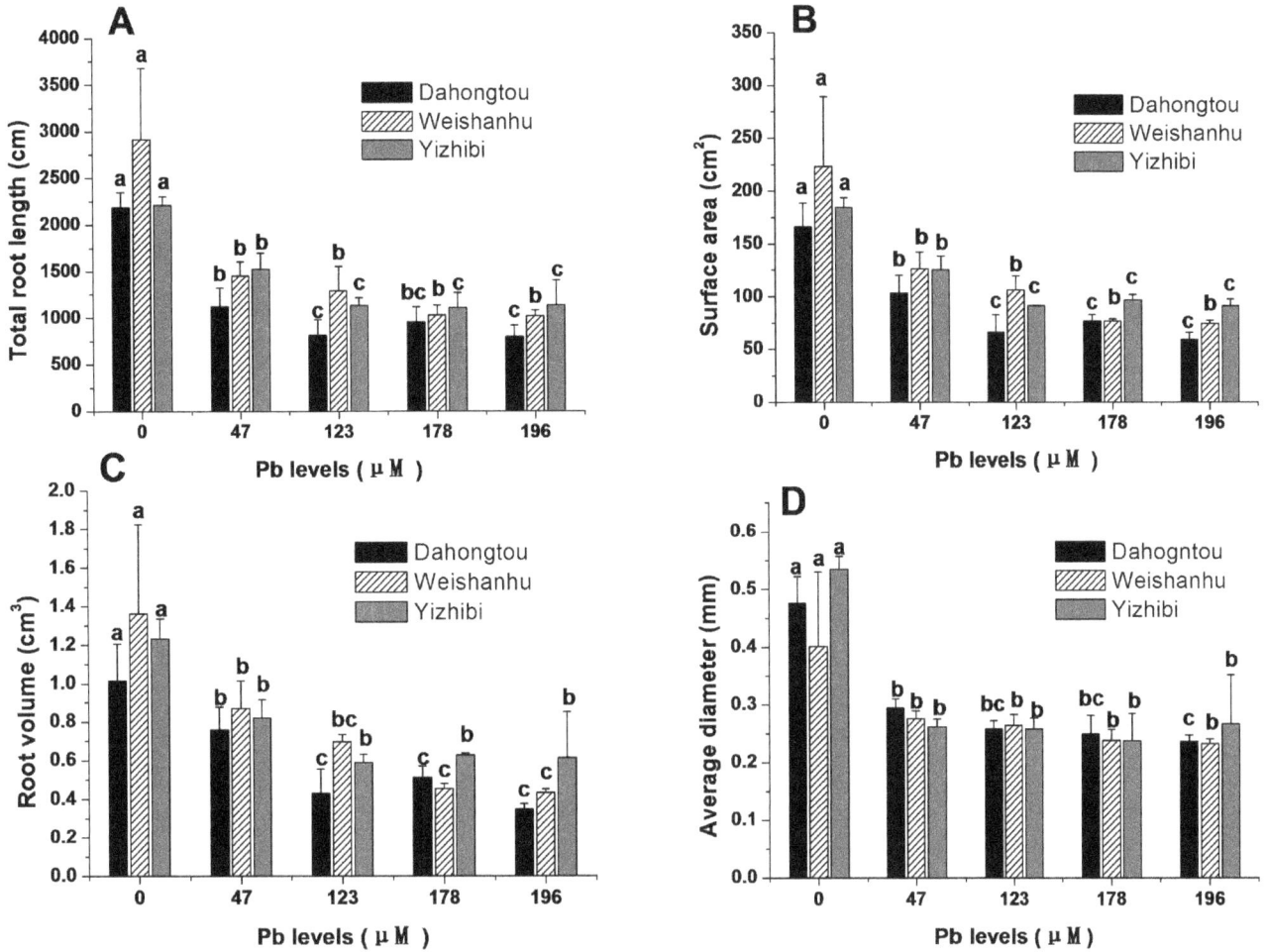

Figure 2. Root characteristics of three *S. integra* varieties exposed to different Pb concentrations for 14 days. (A) Total root length. (B) Surface area. (C) Root volume. (D) Average diameter. Data points and error bars represent means ± S.D. of three replicates (*n* = 3). Different letters indicate significant differences (*P*<0.05) across the treatments according to Fisher's LSD test.

using different variances was used to compare the EC50s among willow varieties [42]:

$$g(x) = g_1 \times e^{-g_2{}^x}$$

where "*g*" is the measured biomass. "g_1" and "g_2" are the parameters to be estimated and "*x*" is the treatment.

Statistical methods

Data were analyzed using the statistical package Data Processing System (DPS 13.01) [43], Origin7.5 and Excel 2003 for Windows. All data were tested for homogeneity of variance and normality. Differences among treatments and cultivars were analyzed by one-way or two-way ANOVA (*P*<0.05) according to Fisher's LSD test.

Results

Visual symptoms

After 5 d of Pb exposure, the foliage of willow plants treated with 123, 178 and 196 μM began to yellow and wilt from the leaf tip. Weishanhu treated with 123 μM Pb exhibited the earliest leaf

wilting. Dahongtou exhibited more pronounced symptoms, including interveinal chlorosis of the basal leaf part, which became more severe on the day of harvest (Fig 1B). The Chlorophyll Concentration Index (CCI) at harvesting was significantly lowered (*P*<0.05) by Pb in all three varieties, and no significant differences were observed among the three varieties (Table 1). In addition, the willow plants had smaller leaves with higher Pb concentrations (178 and 196 μM Pb) (Figure 1B).

After 14 d of Pb treatment, the shoot growth was significantly reduced (Figure 1A) in all varieties for all Pb treatments, compared with the control, but no significant differences were observed among the 47, 123, 178, and 196 *μ*M Pb treatments.

The roots exhibited different levels of blackening and their growth was stunted after exposure to Pb regardless of the Pb concentration, or willow variety (Figure 1C, D). The symptoms became more severe with increasing Pb concentrations. However, no significant differences were observed among the three varieties of *S. integra*. Besides blackening, the total root length and surface area of three varieties were also reduced significantly by the Pb treatments (*P*<0.05), but the reduction was more pronounced in Dahongtou than in Weishanhu or Yizhibi (Figure 2A, B). There was a similar reduction in root volume to that observed in the total

Table 2. Dry weight (DW) (g) of root and aboveground tissues (combined wood, new shoots and leaves) and shoot length (cm) of three willow varieties grown in various Pb concentrations for 14 days.

Pb treatment (µM)	Varieties		
	Yizhibi	Weishanhu	Dahongtou
DW of Aboveground tissue (g)			
0	1.64±0.10ab	1.88±0.27a	1.90±0.07a
47	1.70±0.19a	1.66±0.13ab	1.50±0.15b
123	1.38±0.12bc	1.62±0.08abc	1.46±0.25bc
178	1.43±0.10cd	1.56±0.09bc	1.35±0.09bc
196	1.15±0.14d	1.36±0.09c	1.20±0.09c
DW of Root (g)			
0	0.093±0.007a	0.12±0.01a	0.10±0.01a
47	0.083±0.013a	0.12±0.03a	0.08±0.00a
123	0.052±0.003b	0.08±0.01b	0.06±0.01b
178	0.057±0.003b	0.07±0.01b	0.06±0.00b
196	0.062±0.007b	0.08±0.02b	0.05±0.01b
Shoot length (cm)			
0	40.03±1.15a	45.34±1.77a	39.94±4.73a
47	31.47±0.79b	32.37±3.22b	28.97±0.87b
123	31.31±3.30b	31.67±1.17b	28.72±1.68b
178	28.39±2.36bc	20.56±9.85bc	27.78±3.50b
196	25.81±3.64c	26.28±2.01c	17.17±3.33c

Values are mean ± S.D. (n = 6), and data with different letters in the same column indicate a significant difference at $P<0.05$ according to Fisher's LSD test.

root length and surface area in the three varieties (Figure 2C). The root diameters were significantly reduced by the Pb treatments (Figure 2D), but there were no significant differences among varieties or Pb levels.

Biomass production and tolerance Index

The dry weights of the roots and above-ground tissues (wood, new shoots, and leaves) of *S. integra* varieties are presented in Table 2. The largest root and above-ground tissue biomass for the Pb treatments were recorded for Weishanhu (0.12 and 1.66 g plant^{-1} respectively). Compared with the control, the dry weight of the above-ground tissues of three varieties reduced significantly ($P<0.05$) at higher Pb treatment (178 and 196 µM). While at lower Pb treatment (47 µM), no significant differences were observed in the dry weight of the above-ground tissues of Weishanhu and Yizhibi, whereas there was a significant (P<

0.05) reduction in that of Dahongtou under the same Pb treatments.

For all varieties, there were no significant differences in the dry weight of the roots between the control and the 47 µM Pb treatment, whereas a significant decrease ($P<0.05$) in root dry weight was observed in all three varieties exposed to Pb concentrations of 123,178 and 196 µM, compared with the control. However, the root biomass of three varieties did not change significantly across the Pb concentraitons.

Shoot growth was significantly inhibited ($P<0.05$) by Pb addition in all the varieties especially at the highest Pb concentration. The longest shoot length for all the treatments, including the control, was measured in Weishanhu (45.34 cm).

The TIs observed in each variety for the different Pb concentrations are listed in Table 3. For Yizhibi and Weishanhu, the TIs decreased significantly at 196 µM Pb, whereas at 123 and

Table 3. Tolerance index (TI) in three *S. integra* varieties after 14 days of exposure to increasing concentrations of Pb.

Pb treatment (µM)	Varieties		
	Yizhibi	Weishanhu	Dahongtou
47	95.80±4.50a	89.20±5.90a	79.91±10.04a
123	81.37±9.09ab	85.83±9.38ab	76.69±15.92a
178	80.92±1.76ab	82.72±12.89ab	70.43±2.32a
196	66.87±9.72b	73.15±15.83b	62.78±2.54a

Values are mean ± S.D. (n = 6), and data with different letters in the same column indicate a significant difference at $P<0.05$ according to Fisher's LSD test.

Table 4. Calculated EC50 toxicity thresholds, models, R and P. for roots and the aboveground tissues (combined wood, new shoots and leaves) of three willow varieties exposed to increasing levels of lead.

S. integra varieties	Model	R	P	EC50 (µM)
Roots				
Yizhibi	$y = 0.091727 * e^{-0.002786x}$	0.8925	0.0416	242.6
Weishanhu	$y = 0.123656 * e^{-0.002779x}$	0.9566	0.0108	258.5
Dahogntou	$y = 0.095516 * e^{-0.003416x}$	0.9897	0.0013	204.5
Aboveground tissue				
Yizhibi	$y = 1.7132 * e^{-0.001531x}$	0.8685	0.0561	481.3
Weishanhu	$y = 1.8468 * e^{-0.001252x}$	0.9132	0.0303	537.5
Dahogntou	$y = 1.8086 * e^{-0.001921x}$	0.9298	0.0221	335.9

178 µM Pb, they were only slightly reduced compared to the treatment with 47 µM Pb. For Dahongtou, there were no significant differences in TI among the treatments. The TI varied among the three varieties under the different treatments. Weishanhu had the highest TI at all Pb concentrations, except at 47 µM, when the highest TI was measured for Yizhibi.

Toxicity thresholds

The Pb toxicity thresholds of the three *Salix* varieties were determined for the roots and above-ground tissues (Table 4). All EC_{50} values were >200 µM with the highest EC_{50} value recorded in the roots (258.5 µM) and above-ground tissues (537.5 µM) of Weishanhu. Dahongtou had the lowest EC_{50} values in these tissues. For all three varieties, the EC_{50} values in the roots were lower than those in the above-ground tissues ($P<0.05$).

Metal concentration in plant tissue and translocation factor

The Pb concentrations in the roots, wood, new shoots and leaves varied significantly ($P<0.05$) across treatments, but no significant differences were observed among the varieties, with the exception of wood. For the roots and leaves, there were significant differences ($P<0.05$) in Pb concentration across the varieties × treatment interactions (Table 5).

Lead concentrations in the control were generally below the detection limit. The highest Pb concentrations were found in the roots for all varieties and ranged from 6,799–24,597 mg kg^{-1} (Table 6). All varieties had the lowest Pb concentration in their roots with the 47 µM Pb treatment, ranging from 6,799–9,632 mg kg^{-1}. There was a steady increase in the root Pb concentration in Yizhibi and Dahongtou with increasing Pb treatment concentration, and a significant increase ($P<0.05$) was observed in all varieties at 47 µM Pb treatment. The Pb root concentration in Weishanhu increased significantly with 123 µM Pb, whereas decreased when the Pb treatments ≥123 µM. No significant differences in the root Pb concentration of the three varieties were observed between the 178 µM and 196 µM Pb treatments.

The Pb concentration in the wood tissue for all the varieties and treatments ranged from 216–930 mg kg^{-1}, with the highest Pb concentration displayed by Dahongtou with the 196 µM Pb treatment. With increasing Pb concentration, the wood tissue Pb concentration significantly increased ($P<0.05$) in all three varieties.

The shoot Pb concentration in all three varieties increased significantly ($P<0.05$) under all Pb treatments compared with the control, however, no significant differences were observed among treatments except in Weishanhu, whose shoot Pb concentration decreased significantly with the 123 µM Pb treatment. The highest Pb concentration in new shoots was observed in Dahogntou (49.4 mg kg^{-1}) under 196 µM Pb, while the lowest was observed in Weishanhu (14.6 mg kg^{-1}) under 123 µM Pb.

The highest Pb concentrations in the leaves for all varieties were observed with treatment of 123 µM Pb. The leaf Pb concentrations declined significantly ($P<0.05$) with increasing Pb treatment concentration.

The TF can be used to evaluate the capacity of a plant to translocate heavy metals from the roots to the harvested parts. There were significant ($P<0.05$) differences in the TFs of three varieties among the different tissue types and among different Pb treatments. For all the varieties, the TFs of wood were much higher than those of leaf or new shoots, and increased gradually with increasing Pb treatment concentration, indicating more effective translocation in S. *integra* from the roots to the wood than to the leaves or new shoots (Fig 3). The TFs of leaves and new shoots were extremely low (<0.05) and decreased with increasing Pb concentration in solutions, with the exception of Dahogntou, which displayed a distinctive increase under 123 µM Pb treatment (Fig 3A).

Metal contents in above-ground tissues and roots

The Pb concent of the above-ground tissues (the combined wood, new shoots and leaves) varied significantly across the treatments and varieties × treatment interactions ($P<0.05$). The variety × treatment interaction effects were also significant ($P<0.05$) in the roots (Table 7), whereas no significant differences in the Pb concent were observed in the roots or above-ground tissues among the varieties.

The highest root Pb content in Weishanhu was found in the treatment of 123 µM Pb concentration (Table 8), whereas Yizhibi and Dahongtou had the highest Pb content in their roots with the 178 or 196 µM Pb treatments.

The highest Pb contents in the above-ground tissues for both Yizhibi and Weishanhu were displayed under 178 µM Pb treatment, whereas the highest Pb content for Dahongtou was exhibited under 196 µM Pb treatment. All varieties had the lowest Pb content in their roots and aboveground tissues with the 47 µM Pb treatment.

Table 5. Two-way ANOVA analysis for Pb concentration in root, new shoots, wood, and leaves.

Variable	root		New shoots		wood		leaves	
	F	P	F	P	F	P	F	P
Variety	1.04	0.3960	3.22	0.0942	5.81	0.0277	0.60	0.5697
Treatment	19.64	0.0003	14.52	0.0010	82.81	0.0000	7.87	0.0071
Variety × treatment	5.67	0.0003	1.65	0.1561	1.14	0.3664	9.65	0.0000

Discussion

Excess Pb causes a number of toxicity symptoms in plants, and the non-specific symptoms include stunted growth, chlorosis and inhibition of root growth [44]. In our study, at Pb concentrations ≥ 123 μM, willows exhibited stunted growth, leaf dehydration and chlorosis, and severely reduced root biomass. Similar Pb toxicity symptoms were observed in previous studies [45–48]. According to Sharma and Dubey [44], these toxicity symptoms are some of the physiological responses to metal treatments exhibited by plants, because the presence of Pb in the cell, even in small amounts, can potentially cause a wide range of adverse effects on physiological processes, such as inhibition of enzymatic activities, disturbed mineral nutrition, water imbalance, altered hormonal status, and altered membrane permeability.

After exposure to different Pb concentrations, there were significant variations in the dry weights of the three varieties of *S. integra*. In general, the root biomass was significantly lower than the above-ground biomass among the varieties and treatments. The root dry weights of the three varieties did not change significantly at the lowest Pb concentration of 47 μM, while the above-ground biomass decreased significantly at the same concentration. This suggests that the shoot growth of *S. integra* was more inhibited than the root's at lower Pb concentrations. In contrast, in a previous study [18], the roots of *S. lucida*, *S. serissima*, *S. sachalinensis L.*, *S. miyabeana L.*, and *S. nigra* clones showed a significant reduction in biomass at a similar concentration of Pb (48 μM). This indicates that there is inter-specific variability in sensitivity to environmental stress such as Pb increasing concentrations in soil solution. The roots are in direct contact with Pb, and provide the primary route for metal ion penetration [49]. Plant roots rapidly respond to absorbed Pb, through a reduction in root growth and changes in branching pattern [44]. In our study, the development of root morphology of the three varieties were significantly inhibited by all Pb treatments, which is consistent with the behavior of other species, e.g., *Picea abies* [50] and *Zea mays* [51]. Studies have shown that the inhibition of root growth under Pb stress is a result of Pb-induced inhibition of cell division in root tips [52]. Weirzbicka [53] reported that the roots of onion (*Allium cepa*) exhibited a reduction in root growth, mitotic irregularities and chromosome stickiness when exposed to different concentrations of Pb nitrate. Yang *et al.* [54] also observed a disturbance in alignment of microtubules in *Oryza sativa* in the presence of Pb, and damage to microtubules is now considered one of the key components of Pb-induced damage in plants [44,52].

Lead has also been reported to decrease the chlorophyll content by imparing the uptake of essential elements such as Mg and Fe by plants [55], or by increasing chlorophyllase activity in Pb-treated plants [56]. In our study, we also observed severe chlorosis in the leaves of the three varieties of *S. integra* and a significant decrease in the chlorophyll concentration index, however, further study is required to investigate the reasons for these damages in leaves with Pb treatment.

Assessing metal tolerance is of paramount importance when selecting plants for utilization in phytoremediation [40]. To characterize metal tolerance in plants, one of the most common parameters used is the tolerance index (TI) [57]. Willows have shown significant variations in tolerance across species and clones. In previous studies, significant variations in metal tolerance were found among willow species and clones exposed to cadmium, copper, or arsenic [29,38,58,59]. *S. integra* was also confirmed to have a high capacity for cadmium and zinc uptake [32,33,35], but no information about the Pb tolerance and accumulation was

Table 6. Average Pb concentrations (mg kg^{-1}) in dry plant tissues of *S. integra* exposed to various Pb treatments for 14 days.

Plant tissue	Pb µM	Varieties		
		Yizhibi	Weishanhu	Dahongtou
Leaves	0	ND	ND	ND
	47	29.3±11.6b	27.7±5.4a	20.0±8.4b
	123	44.9±7.5a	36.9±9.4a	89.3±19.6a
	178	15.4±4.7c	14.3±3.6b	16.7±5.8bc
	196	13.7±2.7c	9.9±3.8b	8.6±3.8bc
New shoots	0	ND	ND	ND
	47	29.8±9.3a	33.4±4.7a	36.1±10.4a
	123	17.2±1.7a	14.6±2.9b	27.4±10.5a
	178	19.7±9.8a	43.0±7.2a	34.1±1.1a
	196	25.9±8.5a	34.1±13.1a	49.4±23.5a
Wood	0	ND	ND	ND
	47	244.±82.4c	216.0±37.5c	328.0±118.3c
	123	548.3±147.8b	470.3±49.7b	574.0±114.2b
	178	765.0±147.5a	655.0±117.9a	788.7±135.1ab
	196	865.0±69.1a	611.0±64.6a	930.3±199.5a
Root	0	ND	ND	ND
	47	9,632.3±2,268.6b	6,799.7±1,918.1c	8,318.7±1,827.6c
	123	18,271.0±3,334.5a	19,629.0±1,799.5a	13,999.0±9,64.2b
	178	18,074.7±4,816.4a	15,340.7±3,954.3ab	23,998.0±2,760.0a
	196	21,471.0±1,274.9a	14,391.7±3,738.0b	24,597.0±1,855.1a

Values are mean ± S.D. (n = 6), and data with different letters in the same column indicate a significant difference at $P<0.05$ according to Fisher's LSD test.

reported. There are many cultivated varieties of *S. integra* in China, and the three varieties evaluated in this study were chosen because of their different tolerances to Cd treatment [35–37]. Although TI varies with increasing Pb concentration, all three varieties of *S. integra* tested can be defined as highly tolerant (TI >60) to Pb according to the scheme proposed by Lux *et al.* [60], with Weishanhu being the most tolerant variety in our study.

The EC50 value is another important parameter that describes the tolerance of plant species in multiple concentration tests [57]. In our study, we observed that the EC50 values of the roots were lower than those of the above-ground parts of the plants, suggesting that the roots of *S. integra* were more sensitive to Pb treatment. A similar behavior was observed in *S. lucida, S. serissima, S. sachalinensis* L., *S. miyabeana* L., and *S. nigra* clones [18]. However, more interesting are the extremely high EC50 values for the roots and aboveground tissues of the three varieties which we observed in *S. integra*. Previously, EC50 values for willows in the presence of Pb in hydroponic culture were recorded as 6–32 µM [18], while we measured values >200 µM. High tolerance to metals is a key factor for a plant to successfully colonize metallic soil. Based on the observed EC50 values, Weishanhu was found to be the most tolerant of the three varieties, and could be a suitable candidate for Pb phytostabilization in contaminated areas.

A number of studies have shown that Pb accumulates preferentially in roots [19,61], and our results are consistent with this observation. Sharma and Dubey [44] reported that Pb is confined in roots probably because it binds to ion exchangeable sites on the cell wall, with further extracellular precipitation as Pb carbonates. It has also been reported that Pb has a strong ability to bind to the carboxyl groups of galacturonic and glucuronic acids in the cell wall, which limits the apoplastic transport of this metal [62]. In addition, the Casparian strip of the endodermis plays an important role in restricting Pb transport across the endodermis into other tissues [63]. In addition to the physical barrier that causes poor translocation of Pb from the roots to new shoots, some researchers reported that a short-term exposure to a heavy metal could also result in poor translocation of Pb [64]. Zhivotovsky [18] conducted a short-term hydroponical test to show that *Salix* may need more time to adapt to a specific environment to improve the transport of Pb to above-ground plant tissues. In our study, we also used a short-term screening test, and found Pb accumulation patterns were similar to those of previous studies. However, it is interesting that we observed higher concentrations of Pb in the wood cuttings than in the new shoots and leaves, which showed that Pb absorbed by the roots was transported to the woody parts more readily than to the new shoots and leaves. We also observed that Dahogntou exposed to 123 µM Pb was most efficient in translocating Pb into the leaves. In addition, the highest Pb root concentration we recorded (24,597 mg kg^{-1}) was higher than those that have been reported in the literature for *Salix* (4,164–23,023 mg kg^{-1}) grown in either soil or solution [18,65]. Our results indicate that the roots of *S. integra* can both tolerate and accumulate high Pb concentrations; however, the tolerance mechanism in the root system is still unknown. Further research is required to evaluate whether this tolerance to Pb in roots is acquired by blocking the entrance of Pb into the root cells, or by some metabolic mechanisms that detoxify Pb toxicity.

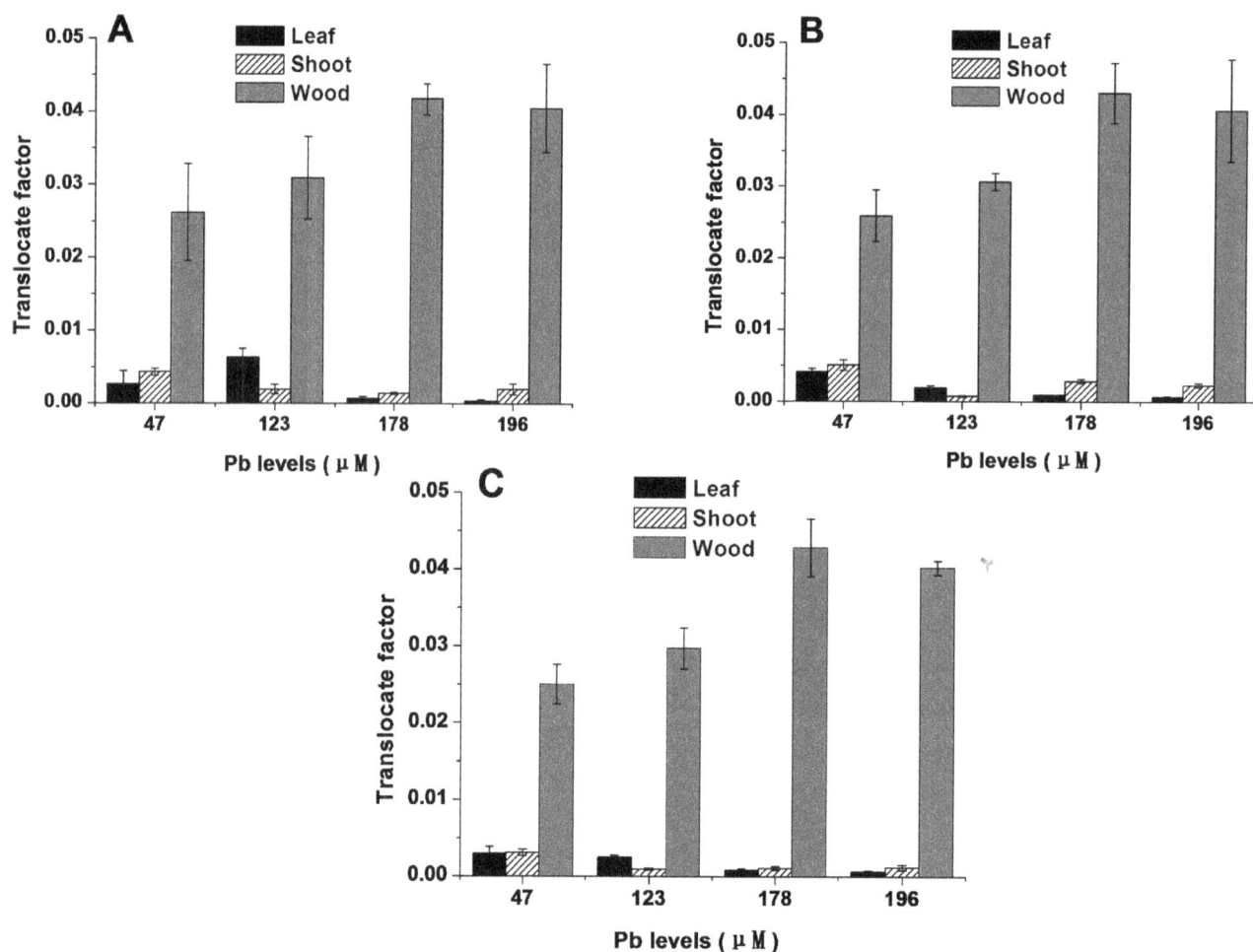

Figure 3. Translocation factor (TF) of leaf, new shoots and wood in three varieties of *S. integra* for the different Pb treatment. (A) Dahongtou. (B) Weishanhu. (C) Yizhibi.

Conclusions

Willows have potential for phytoremediation due to their high biomass productivity, high transpiration rate, and species-specific heavy metal uptake [28,66,67]. *S. integra* is a fast-growing shrub willow, and in China, it is generally cultivated by short-rotation plantation, with the new shoots harvested for use in weaving. Yizhibi, Dahogntou and Weishanhu are three cultivated varieties of *S. integra*, that display excellent in biomass production and environmental adaptation [34,35]. We demonstrated that *S. integra* can tolerate, transport and accumulate various levels of Pb when exposed to different Pb concentrations, and there is no significant difference in the tolerance to Pb among the varieties. For effective phytoextration, we are interested in plants that exhibit not only high tolerance to heavy metals, but also high accumulation of metals in tissues with high biomass production. We observed that Dahongtou had higher Pb contents in the above-ground tissues and the highest TF in leaves under 123 μM Pb. These results bring new insight into the selection of candidates for Pb phytoextraction. However, the results are based on a short-term study, the potential effectiveness of *S. integra* Dahongtou in phytoextraction should be tested in the long-term and open-field studies. In summary, all the varieties, especially Weishanhu and Yizhibi, have high TI and EC50, demonstrating their high

Table 7. Two-way ANOVA analysis for Pb concent in root and the aboveground tissues.

Variable	root		aboveground	
	F	P	F	P
Variety	0.33	0.7290	5.00	0.0527
Treatment	1.89	0.2327	28.47	0.0006
Variety × treatment	4.44	0.0044	2.92	0.0298

Table 8. Lead content (mg plant^{-1}) in root and aboveground tissue (wood, new shoots, and leaves) of three *S. integra* varieties exposed to different Pb concentrations for 14 days.

plant tissue	Pb μM	Varieties		
		Yizhibi	Weishanhu	Dahongtou
Root	47	0.85±0.26c	0.77±0.16b	0.69±0.15c
	123	0.95±0.21bc	1.56±0.29a	0.83±0.20bc
	178	1.01±0.21b	1.11±0.22b	1.33±0.13a
	196	1.32±0.17a	1.11±0.38b	1.17±0.17ab
Aboveground tissue	47	0.22±0c	0.23±0.01c	0.33±0.06c
	123	0.50±0.09b	0.52±0.03b	0.63±0.06b
	178	0.70±0.12a	0.68±0.07a	0.74±0.07ab
	196	0.65±0.05ab	0.55±0.04b	0.88±0.13a

Values are mean ± S.D. (n = 6), and data with different letters within the same column indicate a significant difference at $P<0.05$ according to Fisher's LSD test.

tolerance and lower sensitivity to Pb and, thus, their suitability for phytostabilization of Pb.

Acknowledgments

We sincerely thank the Academic Editor and anonymous reviewers for their valuable comments on improving the paper. We acknowledge Dr. Zhenli He from University of Florida, and Xiaojiao Han from Research Institute of Subtropical Forestry for their suggestions and correction of English writing.

Author Contributions

Conceived and designed the experiments: SW XS XY. Performed the experiments: SW XS. Analyzed the data: SW XS HS. Contributed reagents/materials/analysis tools: YC HP. Wrote the paper: SW XS TR. Designed the software used in analysis: SW XS HS. Plant materials cultivation before treatment: XS HP. Plant growth monitoring: SW XS.

References

1. Ali H, Khan E, Sajad MA (2013) Phytoremediation of heavy metals—Concepts and applications. Chemosphere 91: 869–881.

2. Kachenko AG, Singh B, Bhatia NP (2007) Heavy metal tolerance in common fern species. Australian journal of botany 55: 63–73.

3. Yoon J, Cao X, Zhou Q, Ma LQ (2006) Accumulation of Pb, Cu, and Zn in native plants growing on a contaminated Florida site. Science of the Total Environment 368: 456–464.

4. Chehregani A, Malayeri B (2007) Removal of heavy metals by native accumulator plants. International Journal of Agriculture and Biology 9: 462–465.

5. Fulekar M, Singh A, Bhaduri AM (2009) Genetic engineering strategies for enhancing phytoremediation of heavy metals. African Journal of Biotechnology 8: 529–535.

6. Mehmood T, Chaudhry M, Tufail M, Irfan N (2009) Heavy metal pollution from phosphate rock used for the production of fertilizer in Pakistan. Microchemical Journal 91: 94–99.

7. Okieimen FE (2011) Heavy metals in contaminated soils: a review of sources, chemistry, risks and best available strategies for remediation. ISRN Ecology 2011: 1–21.

8. Cheng S (2003) Heavy metal pollution in China: origin, pattern and control. Environmental Science and Pollution Research 10: 192–198.

9. Ali M, Vajpayee P, Tripathi R, Rai U, Singh S, et al. (2003) Phytoremediation of lead, nickel, and copper by Salix acmophylla Boiss: role of antioxidant enzymes and antioxidant substances. Bulletin of environmental contamination and toxicology 70: 0462–0469.

10. Baker A, Reeves R, McGrath S (1991) In situ decontamination of heavy metal polluted soils using crops of metal-accumulating plants-a feasibility study. In situ bioreclamation Boston, Butterworth-Heinemann: 600–605.

11. Komives T, Gullner G (2006) Dendroremediation: the use of trees in cleaning up polluted soils. Phytoremediation Rhizoremediation: Springer. pp. 23–31.

12. Kumar PN, Dushenkov V, Motto H, Raskin I (1995) Phytoextraction: the use of plants to remove heavy metals from soils. Environmental Science & Technology 29: 1232–1238.

13. Li X, Wai OW, Li Y, Coles BJ, Ramsey MH, et al. (2000) Heavy metal distribution in sediment profiles of the Pearl River estuary, South China. Applied Geochemistry 15: 567–581.

14. Freitas EV, Nascimento CW, Souza A, Silva FB (2013) Citric acid-assisted phytoextraction of lead: A field experiment. Chemosphere: 213–217.

15. Islam E, Yang X, Li T, Liu D, Jin X, et al. (2007) Effect of Pb toxicity on root morphology, physiology and ultrastructure in the two ecotypes of *Elsholtzia argyi*. Journal of hazardous materials 147: 806–816.

16. Yang Q, Shu W, Qiu J, Wang H, Lan C (2004) Lead in paddy soils and rice plants and its potential health risk around Lechang Lead/Zinc Mine, Guangdong, China. Environment International 30: 883–889.

17. Grejtovský A, Markušová K, Nováková L (2008) Lead uptake by Matricaria chamomilla. L Plant, Soil and Environment 54: 47–54.

18. Zhivotovsky OP, Kuzovkina JA, Schulthess CP, Morris T, Pettinelli D, et al. (2010) hydroponic screening of willows (Salix L.) for lead tolerance and accumulation. International journal of phytoremediation 13: 75–94.

19. Wierzbicka M (1999) Comparison of lead tolerance in *Allium cepa* with other plant species. Environmental Pollution 104: 41–52.

20. Ebbs SD, Kochian LV (1997) Toxicity of zinc and copper to Brassica species: implications for phytoremediation. Journal of Environmental Quality 26: 776–781.

21. Kuzovkina YA, Zhivotovsky OP, Schulthess CP, Pettinelli D (2010) Plant selection for a pilot phytoremediation study at a former skeet range. Remediation Journal 20: 93–105.

22. Ghosh M, Singh S (2005) A review on phytoremediation of heavy metals and utilization of it's by products. Asian J Energy Environ 6: 1–18.

23. Marques AP, Rangel AO, Castro PM (2009) Remediation of heavy metal contaminated soils: phytoremediation as a potentially promising clean-up technology. Critical Reviews in Environmental Science and Technology 39: 622–654.

24. Suresh B, Ravishankar G (2004) Phytoremediation-A novel and promising approach for environmental clean-up. Critical reviews in biotechnology 24: 97–124.

25. Kuzovkina YA, Volk TA (2009) The characterization of willow (Salix L.) varieties for use in ecological engineering applications: Co-ordination of structure, function and autecology. Ecological Engineering 35: 1178–1189.

26. Rockwood D, Naidu C, Carter D, Rahmani M, Spriggs T, et al. (2004) Short-rotation woody crops and phytoremediation: Opportunities for agroforestry? New Vistas in Agroforestry: Springer. pp. 51–63.

27. Mleczek M, Łukaszewski M, Kaczmarek Z, Rissmann I, Golinski P (2009) Efficiency of selected heavy metals accumulation by *Salix viminalis* roots. Environmental and Experimental Botany 65: 48–53.

28. Pulford I, Watson C (2003) Phytoremediation of heavy metal-contaminated land by trees—a review. Environment international 29: 529–540.

29. Purdy JJ, Smart LB (2008) Hydroponic screening of shrub willow (Salix spp.) for arsenic tolerance and uptake. International Journal of Phytoremediation 10: 515–528.

30. Tlustoš P, Száková Ji, Vysloužilová M, Pavlíková D, Weger J, et al. (2007) Variation in the uptake of arsenic, cadmium, lead, and zinc by different species

of willows Salix spp. grown in contaminated soils. Central European Journal of Biology 2: 254–275.

31. Zimmer D, Kruse J, Baum C, Borca C, Laue M, et al. (2011) Spatial distribution of arsenic and heavy metals in willow roots from a contaminated floodplain soil measured by X-ray fluorescence spectroscopy. Science of the Total Environment 409: 4094–4100.

32. Harada E, Hokura A, Takada S, Baba K, Terada Y, et al. (2010) Characterization of cadmium accumulation in willow as a woody metal accumulator using synchrotron radiation-based X-ray microanalyses. Plant Cell Physiol 51: 848–853.

33. Liu Y, Chen G-C, Zhang J, Shi X, Wang R (2011) Uptake of cadmium from hydroponic solutions by willows (Salix spp.) seedlings. African Journal of Biotechnology 10: 16209–16218.

34. Tian Y LWW, Fang S Z. (2012) Effects of harvest rotation on the growth, wicker production and quality of willow shrub.pdf. Chinese Journal of Nanjing Forestry University (Natural Science Edition) 36: 86–90.

35. Yang WD, Chen YT (2008) Differences in Uptake and Tolerance to Cadmium in Varieties of Salix integra [J]. Forest Research 6: 857–861.

36. Yang WD, Chen YT (2009) Tolerance of different varieties of Salix integra to high zinc stress. Zhongguo Shengtai Nongye Xuebao/Chinese Journal of Eco-Agriculture 17: 1182–1186.

37. Wang SF, Shi X, Sun HJ, Chen YT, Yang XE (2013) Metal uptake and root morphological changes for two varieties of Salix integra under cadmium stress. Acta Ecologica Sinica 33: 6065–6073.

38. Magdziak Z, Kozlowska M, Kaczmarek Z, Mleczek M, Chadzinikolau T, et al. (2011) Influence of Ca/Mg ratio on phytoextraction properties of Salix viminalis. II. Secretion of low molecular weight organic acids to the rhizosphere. Ecotoxicology and environmental safety 74: 33–40.

39. Wilkins D (1978) The measurement of tolerance to edaphic factors by means of root growth. New Phytologist 80: 623–633.

40. Zacchini M, Pietrini F, Mugnozza GS, Iori V, Pietrosanti L, et al. (2009) Metal tolerance, accumulation and translocation in poplar and willow clones treated with cadmium in hydroponics. Water, Air, and Soil Pollution 197: 23–34.

41. Neter J, Wasserman W, Kutner MH (1996) Applied linear statistical models: Irwin Chicago.

42. Schabenberger O, Tharp BE, Kells JJ, Penner D (1999) Statistical tests for hormesis and effective dosages in herbicide dose response. Agronomy Journal 91: 713–721.

43. Su S, Cai F, Si A, Zhang S, Tautz J, et al. (2008) East learns from West: Asiatic honeybees can understand dance language of European honeybees. PLoS One 3: e2365.

44. Sharma P, Dubey RS (2005) Lead toxicity in plants. Brazilian Journal of Plant Physiology 17: 35–52.

45. HO W, Ang LH, LEE DK (2008) Assessment of Pb uptake, translocation and immobilization in kenaf Hibiscus cannabinus L.) for phytoremediation of sand tailings. Journal of Environmental Sciences 20: 1341–1347.

46. Liu W, Zhou Q, Zhang Y, Wei S (2010) Lead accumulation in different Chinese cabbage cultivars and screening for pollution-safe cultivars. Journal of environmental management 91: 781–788.

47. Rezvani M, Zaefarian F (2011) Bioaccumulation and translocation factors of cadmium and lead in Aeluropus littoralis. Australian Journal of Agricultural Engineering 2: 114–119.

48. Verma S, Dubey R (2003) Lead toxicity induces lipid peroxidation and alters the activities of antioxidant enzymes in growing rice plants. Plant Science 164: 645–655.

49. Piechalak A, Tomaszewska B, Baralkiewicz D, Malecka A (2002) Accumulation and detoxification of lead ions in legumes. Phytochemistry 60: 153–162.

50. Godbold D, Kettner C (1991) Lead Influences Root Growth and Mineral Nutrition of Picea abies Seedlings. Journal of plant physiology 139: 95–99.

51. Obroucheva N, Bystrova E, Ivanov V, Antipova O, Seregin I (1998) Root growth responses to lead in young maize seedlings. Plant and soil 200: 55–61.

52. Eun SO, Shik Youn H, Lee Y (2000) Lead disturbs microtubule organization in the root meristem of Zea mays. Physiologia Plantarum 110: 357–365.

53. Wierzbicka M (1994) Resumption of mitotic activity in Allium cepa L. root tips during treatment with lead salts. Environmental and experimental botany 34: 173–180.

54. Yang Y-Y, Jung J-Y, Song W-Y, Suh H-S, Lee Y (2000) Identification of rice varieties with high tolerance or sensitivity to lead and characterization of the mechanism of tolerance. Plant Physiology 124: 1019–1026.

55. Burzynski M (1987) Influence of lead and cadmium on the absorption and distribution of potassium, calcium, magnesium and iron in cucumber seedlings. Acta Physiologiae Plantarum 9: 229–238.

56. Drazkiewicz M (1994) Chlorophyllase: occurrence, functions, mechanism of action, effects of external and internal factors (Review). Photosynthetica 30: 321–331.

57. Köhl K, Lösch R (1999) Experimental characterization of heavy metal tolerance in plants. Heavy metal stress in plants: Springer. pp. 371–389.

58. Kuzovkina YA, Knee M, Quigley MF (2004) Cadmium and copper uptake and translocation in five willow (Salix L.) species. International Journal of Phytoremediation 6: 269–287.

59. Punshon T, Dickinson N (1999) Heavy metal resistance and accumulation characteristics in willows. International Journal of Phytoremediation 1: 361–385.

60. Lux A, Šottníková A, Opatrná J, Greger M (2004) Differences in structure of adventitious roots in Salix clones with contrasting characteristics of cadmium accumulation and sensitivity. Physiologia Plantarum 120: 537–545.

61. Geebelen W, Vangronsveld J, Adriano DC, Van Poucke LC, Clijsters H (2002) Effects of Pb-EDTA and EDTA on oxidative stress reactions and mineral uptake in Phaseolus vulgaris. Physiologia Plantarum 115: 377–384.

62. Rudakova E, Karakis K, Sidorshina K (1988) Role of plant cell membranes in uptake and accumulation of metal ions. Fiziologiia i biokhimiia kul'turnykh rastenii = Physiology and biochemistry of cultivated plants 20: 3–12.

63. Seregin I, Ivanov V (1997) Histochemical investigation of cadmium and lead distribution in plants. Russian Journal of Plant Physiology 44: 791–796.

64. Dos Santos Utmazian MN, Wieshammer G, Vega R, Wenzel WW (2007) Hydroponic screening for metal resistance and accumulation of cadmium and zinc in twenty clones of willows and poplars. Environmental Pollution 148: 155–165.

65. Borišev M, Pajević S, Nikolić N, Pilipović A, Krstić B, et al. (2009) Phytoextraction of Cd, Ni, and Pb using four willow clones (Salix spp.). Polish Journal of Environmental Studies 18: 553–561.

66. Evangelou MW, Deram A, Gogos A, Studer B, Schulin R (2012) Assessment of suitability of tree species for the production of biomass on trace element contaminated soils. Journal of hazardous materials 209: 233–239.

67. Jensen JK, Holm PE, Nejrup J, Larsen MB, Borggaard OK (2009) The potential of willow for remediation of heavy metal polluted calcareous urban soils. Environmental Pollution 157: 931–937.

9

Antinematode Activity of Violacein and the Role of the Insulin/IGF-1 Pathway in Controlling Violacein Sensitivity in *Caenorhabditis elegans*

Francesco Ballestriero[1], Malak Daim[1], Anahit Penesyan[2], Jadranka Nappi[1], David Schleheck[3], Paolo Bazzicalupo[4], Elia Di Schiavi[4¤], Suhelen Egan[1]*

1 School of Biotechnology and Biomolecular Sciences and Centre for Marine Bio-Innovation, University of New South Wales, Sydney, New South Wales, Australia, 2 Department of Chemistry and Biomolecular Sciences, Macquarie University, Sydney, New South Wales, Australia, 3 Biology Department, University of Konstanz, Konstanz, Germany, 4 Institute of Genetics and Biophysics "Adriano Buzzati Traverso", National Research Council, Naples, Italy

Abstract

The purple pigment violacein is well known for its numerous biological activities including antibacterial, antiviral, antiprotozoan, and antitumor effects. In the current study we identify violacein as the antinematode agent produced by the marine bacterium *Microbulbifer* sp. D250, thereby extending the target range of this small molecule. Heterologous expression of the violacein biosynthetic pathway in *E. coli* and experiments using pure violacein demonstrated that this secondary metabolite facilitates bacterial accumulation in the nematode intestine, which is accompanied by tissue damage and apoptosis. Nematodes such as *Caenorhabditis elegans* utilise a well-defined innate immune system to defend against pathogens. Using *C. elegans* as a model we demonstrate the DAF-2/DAF-16 insulin/IGF-1 signalling (IIS) component of the innate immune pathway modulates sensitivity to violacein-mediated killing. Further analysis shows that resistance to violacein can occur due to a loss of DAF-2 function and/or an increased function of DAF-16 controlled genes involved in antimicrobial production (*spp-1*) and detoxification (*sod-3*). These data suggest that violacein is a novel candidate antinematode agent and that the IIS pathway is also involved in the defence against metabolites from non-pathogenic bacteria.

Editor: Raffi V. Aroian, UMASS Medical School, United States of America

Funding: These authors have no support or funding to report.

Competing Interests: The authors have declared that no competing interests exist.

* Email: s.egan@unsw.edu.au

¤ Current address: Institute of Bioscience and BioResources, National Research Council, Naples, Italy

Introduction

Parasitic nematodes are an important group of human, animal and plant pathogens representing a major threat not only to public health, but also to livestock and agricultural industries around the globe [1]. Yet heavy reliance on the few available chemotherapeutic agents has resulted in the development of nematode resistance and little progress has been made in the search for new treatments [2,3]. Thus there is an urgent requirement for the discovery of new antinematode compounds that can be developed into chemotherapeutic drugs.

The nematode *C. elegans* is a powerful model organism used broadly across the fields of cellular biology, developmental biology and neurobiology, and more recently also as a model organism for the study of host-microbial interactions with a focus on pathogenesis and drug discovery [4,5]. In a recent functional screen of genomic libraries of marine bacteria, a number of fosmid clones expressing high toxicity towards *C. elegans* were identified [6]. One of the highly toxic clones (designated 20G8) with a sequence-insert originating from the marine bacterium *Microbulbifer* sp. D250, expressed a violet pigment. Genetic analysis of the insert

revealed that the clone 20G8 contained genes encoding for the synthesis of the indole-antibiotic violacein (*vioA-E*) [6], suggesting that this metabolite is responsible for its toxic phenotype.

Violacein is produced by several bacterial species thriving in a range of habitats such as terrestrial, marine, fresh water and glacier environments. Some of the known violacein producing organisms include *Chromobacterium violaceum* [7], *Collimonas* sp. [8], *Duganella* sp. [9], *Janthinobacterium lividum* [10,11] and *Pseudoalteromonas* spp. [12,13]. Violacein exhibits several biological activities with ecological relevance. Firstly, violacein has been suggested to be involved in oxidative stress resistance in *C. violaceum* [14]. Secondly, violacein producing bacteria (namely *J. lividum*) have been implicated in the natural defence of amphibians to fungal disease [15]. Finally, Matz and colleagues [12] have demonstrated that in the marine bacterium *Pseudoalteromonas tunicata*, violacein production can act as an antipredator defence mechanism against protozoan grazers. In addition to its ecological significance, violacein has also gained increasing importance for its potential medical and industrial applications. The biological activities of this compound include antioxidant, leishmanicidal, trypanocidal, antifungal, antiviral, antibacterial and antiprotozoal effects, as well as antitumoral and apoptosis-inducing activities in mammalian cancer cells (reviewed in [16]).

Although violacein has a broad range of activities, the direct molecular and cellular targets remain unknown. In addition there is limited understanding of its activity in multicellular eukaryotes and whether metazoans have the capacity to mount a defence to neutralize its activity or not. Whilst lacking adaptive immune strategies, nematodes, such as *C. elegans*, are suitable models to investigate metazoan defence as they posses a sophisticated innate immune system that protects against toxic microorganisms [17]. Key features of the nematodes defence include the conserved immune regulatory pathways, *i.e.*, the p38 mitogen activated protein kinase (MAPK) and the insulin/IGF-1 signalling (IIS) pathways [18]. In particular the IIS pathway with the gene regulators DAF-2 and DAF-16, is increasingly recognised for its important role in stress response, aging and immune homeostasis across nematodes, insects and mammals [18,19]. Furthermore, the IIS pathway is known to play a key role in the innate immune response against different pathogen-induced stresses, including colonization [17,20,21] and bacterial virulence factors [22].

In *C. elegans* the binding of insulin to DAF-2 triggers a phosphorylation cascade that results in activation of PDK-1 (3-phosphoinositide-dependent kinase 1) and eventual retention of the DAF-16 transcriptional activator in the cytoplasm [23]. De-activation or loss of DAF-2 function allows DAF-16 to move to the nucleus where it enhances the expression of genes including among others *sod-3* (superoxide dismutase), *spp-1* (SaPosin-like Protein) and *lys-7*, which are involved in detoxification, antimicrobial peptide expression and antimicrobial lysozyme production, respectively [24,25,26]. Moreover recent data indicates that the canonical IIS signalling diverges at PDK-1 into a second arm of the pathway mediated by the protein WWP-1 (WW domain protein 1) [22]. In the present study, we hypothesised that *C. elegans* makes use of this immune response pathway not only in the situation of infection by pathogenic organisms, but also to neutralize the effect of toxic bacterial secondary metabolites, such as violacein that originate from non-pathogens. Furthermore studying the mechanisms in which *C. elegans* mediates resistance to bacterial metabolites may shed further light into their molecular/cellular targets. To address this hypothesis, we first confirm that violacein is responsible for the toxic activity against *C. elegans* in clone 20G8 and its parental strain *Microbulbifer* sp. D250. We further show that the expression of enzymes that synthesize violacein in *E. coli* facilitates bacterial accumulation in the host intestine and induces apoptosis in the nematode. Finally we demonstrate that the IIS immune pathway modulates *C. elegans* sensitivity to violacein toxicity, most likely via the control of genes involved in detoxification and antimicrobial production.

Materials and Methods

Strains and culture conditions

All bacterial strains and vectors used in this study are listed in Table 1. Bacteria were grown in Luria broth (LB10), nematode growth medium (NGM) [27] or marine broth (Difco Laboratories, Maryland) [28] as indicated, and stored in 30% (v/v) glycerol at −80°C. Solid medium was prepared by the addition of 19 g of agar (Oxoid, Australia) per litre of culture fluid. All strains were grown at 25°C. Where required (see Table 1), chloramphenicol (12.5 μg/ml), kanamycin (100 μg/ml), and L-arabinose (0.02%, w/v) were added to the media. *C. elegans* strains (listed in Table 2) were maintained at 20°C on NGM agar plates spread with *E. coli* OP50 as a food source [29,30]. *C. elegans* strains were stored in glycerol (70:30 vol/vol) at −80°C [30].

Fosmid analysis and transposon mutant library screening

A transposon mutant library of the antinematode fosmid clone 20G8 was generated using an *in vitro* transposon mutagenesis kit (EZ-Tn5 insertion kit; Epicentre) following the manufacturers' instructions. The DNA fosmid sequence for clone 20G8 is available from the National Center for Biotechnology Information (NCBI) public database (GenBank) via accession number JX523957. The subsequent library of 96 *E. coli* transposon mutants was replicated on LB10 Omnitray plates (Nunc, Denmark), and screened for loss of toxic activity towards *C. elegans* as previously described [31]. Clones that were partially or totally grazed by the nematodes were chosen for further characterization in the nematode killing assay (below). The disrupted genes were identified by outward sequencing from the transposon using the KAN-2 forward and reverse primers (Epicentre) (KAN-2 Forward Primer 5' ACCTACAACAAAGCTCTCATCAACC 3', KAN-2 Reverse Primer 5' GCAATGTAACATCAGAGATTTTGAG 3') and sequences were subjected to BLAST analysis [32].

Purification and identification of violacein as the antinematode agent produced by *Microbulbifer* sp. D250

Violacein was purified from an overnight culture of *Microbulbifer* sp. D250, cells were collected by centrifugation and the cell pellets repeatedly extracted with 100% methanol. The resulting (pooled) crude extract was applied to a C18 solid phase extraction column (pre-packed C18 columns, 10 g, Alltech); after several washing steps with methanol:water (20 to 60% methanol), the violacein was eluted with 100% methanol. The violacein fraction from the solid phase extraction was further purified by preparative high performance liquid chromatography (HPLC), when using a Prodigy ODS3 column (Phenomenex, 150×4.6 mm, 5 μm particle size) and a methanol gradient from 0–100% methanol.

Liquid chromatography electrospray ionisation ion-trap mass spectrometry (LC-MS/MS) was employed to confirm that the purified purple pigment (see above) is violacein. Briefly, the LC-MS/MS was performed on a LCQ Deca SP Iontrap-MS/MS system (Thermo Finnigan). Up to 20 μl of the extract were loaded on a Nucleosil C18 column (125×3 mm, 5 μm particle size, Macherey-Nagel, Germany). The mobile phases used were (A) water acidified with 0.1% formic acid and (B) acetonitrile, at a flow rate of 0.2 ml/min. The gradient program was started at 20% B, and after 3 minutes, increased to 100% B over 13 min, and maintained at 100% B. For the MS conditions, the electrospray voltage was -5kV with a current of 12 μA; the sheath gas flow rate was 34l/min; the capillary was maintained at 275°C. For the MS/MS fragmentation, the mass width for isolation of precursor ions was 1.0Da, and the relative collision energy set at 40%. The MS chromatograms were recorded in the positive ion mode. The purified purple pigment eluted at 13.9 min, with an absorption scan corresponding to violacein as observed by HPLC-diode array detection (maxima at 260, 378 and 570 nm), and this peak in the MS exhibited a protonated molecule ([M+H]$^+$) corresponding to violacein (MW 343Da; mass of the observed [M+H]$^+$ ion, 344 Da); the MS/MS fragmentation pattern of this [M+H]$^+$ ion also corresponded to violacein, with ions observed at (% basepeak) 344 (64), 326 (45), 316 (100), 301 (49), 299 (24), 273 (5), 251 (20), 211 (4), 183 (2), 158 (1) and 132 (1) Da (e.g, due to loss of water, elimination of CO and nitrogen species, and cleavages at the rings).

Table 1. Bacterial strains and vectors used in this study.

Strain/Vector	Relevant characteristic or genotype	Source or reference
E. coli EPI300-T1^R	F-mcrA Δ(mrrhsdRMSmcrBC) φ80dlacZΔM15ΔlacX 74 recA1 endA1 araD139 Δ(ara, leu) 7697galU galK λ- rpsL nupG trfA tonA dhfr	Epicentre
E. coli 20G8	Fosmid 20G8 cloned in EPI300-T1^R; Cm^r	This study
E. coli 20G8vioA$^-$	Fosmid 20G8 mutated in vioA gene and cloned in EPI300-T1^R; Cm^r, Kan^r	This study
E. coli 20G8vioB$^-$	Fosmid 20G8 mutated in vioB gene and cloned in EPI300-T1^R; Cm^r, Kan^r	This study
E. coli 20G8vioC$^-$	Fosmid 20G8 mutated in vioC gene and cloned in EPI300-T1^R; Cm^r, Kan^r	This study
E. coli 20G8vioD$^-$	Fosmid 20G8 mutated in vioD gene and cloned in EPI300-T1^R; Cm^r, Kan^r	This study
E. coli OP50	Uracil auxotroph	[29]
Microbulbifer sp. D250	Wild type strain	[62]
Microbulbifer sp. D250 dv2	D250 strain mutated in vioB gene; Kan^r	This study
pLof/Tn10 KM	Mini-Tn10 (Kan^r); Amp^r	[34]
pCC1FOS^a	Fosmid backbone for genomic library; Cm^r	Epicentre

^{a}Copy number inducible by arabinose.

Transposon mutagenesis of *Microbulbifer* sp. D250

To generate mutants of strain D250 that are unable to produce violacein, a random transposon mutagenesis was performed using the Tn-10-Km^R mini-transposon systems as described previously [33]. Briefly, a spontaneous streptomycin (Sm) resistant mutant of *Microbulbifer* sp. D250 was generated (D250-Sm^R) and used as the recipient. *E. coli* containing the Tn10 based delivery plasmid pLOF/Km, encoding a kanamycin (Km) resistance gene marker [34], was used as a transposon donor. Donor cells were conjugated with Sm resistant recipient cells of isolate D250 (D250-Sm^R) on filter discs in 1:1 ratio and incubated for 12 hours at 30°C. The conjugation mix was resuspended in marine broth and serial dilutions of this mixture were spread on, marine agar medium supplemented with 200 µg/ml Sm and 100 µg/ml Km. Transposon mutants were allowed to grow at room temperature for 48 to 72 hours and visually screened for the loss of purple pigmentation; the selected transposon mutants and wild type cells were extracted with methanol (as described above) and loss of pigmentation in the mutants confirmed by measuring the absence of absorbance at 575 nm. The DNA flanking the transposon insertion in the relevant mutants was sequenced using a "pan-handle" method as described previously [33].

Table 2. *C. elegans* strains used in this study.

Strain name	Genotype/allele designation	Relevant characteristics	Source or reference
N2 Bristol	C. elegans wild isolate	Wild type isolate	CGC^a
CU1546	smIs34	ced-1p::ced-1::GFP + rol-6(su1006)	CGC^a
CB1370	daf-2(e1370) III	Mutated in the insulin-like receptor DAF-2. Temperature sensitive dauer constitutive	CGC^a
IU10	daf-16(mgDf47) I; rrf-3(pk1426) II	Mutated in the FOXO-family transcription factor DAF-16	CGC^a
TJ356	zIs356 IV	Integrated DAF-16::GFP roller strain. Daf-c, Rol, fluorescent DAF-16::GFP. Overexpression of DAF-16	CGC^a
JT9609	pdk-1 (Sa680) x	Mutation in the gene encoding for 3-phosphoinositide-dependent protein kinase	CGC^a
RB1178	wwp-1(ok1102) I.	Mutation in the gene encoding for the WW domain protein 1	CGC^{ab}
TM127	daf-2(e1370) III; sod-3(sj134) X	Double mutant in the insulin-like receptor DAF-2 and in the superoxide dismutase SOD-3	CGC^a
MQ876	daf-2(e1370) III; lys-7(ok1384) V	Double mutant in the insulin-like receptor DAF-2 and in the putative antimicrobial lysozyme LYS-7	CGC^a
MQ513	daf-2(e1370) III; spp-1(ok2703) III	Double mutant in the insulin-like receptor DAF-2 and in the antimicrobial peptide caenopore SPP-1	CGC^a
GA186	sod-3(tm760) X	Mutated in the iron/manganese superoxide dismutase SOD-3	CGC^a
RB1286	lys-7(ok1384) V	Mutated in the putative antimicrobial lysozyme LYS-7	CGC^a
RB2045	spp-1(ok2703) III	Mutated in the antimicrobial peptide caenopore SPP-1	CGC^a

^{a}*Caenorhabditis* Genetics Center, the University of Minnesota.
^{b}*C. elegans* Gene Knockout Project http://www.celeganskoconsortium.omrf.org.

Nematode killing assay

E. coli clones were pre-grown overnight at 37°C in LB10, and 10 μl of the cultures were spread onto 3.5 cm diameter LB10 agar plates supplemented with selective antibiotic and L-arabinose as required, followed by incubation at 25°C for four days. L4-stage nematodes were added to the bacterial lawns (30 to 40 per plate), incubated at 20°C, and scored for live and dead nematodes every 24 hours for 20 days. In order to avoid multiple generations of nematodes on the same plate, which may have lead to errors when scoring, the nematodes where transferred to a fresh plate each day. A random non-toxic *E. coli* clone was used as a negative control under the same conditions. A nematode was considered dead when it failed to respond to touch. Since the *C. elegans daf-2* and *pdk-1* mutants are temperature sensitive (at 20°C 15% of the population enter in the resistant dauer stage), all the *daf-2* and *pdk-1* mutant assays were carried out at 15°C. Control *C. elegans* strains were also tested at 15°C in order to avoid temperature bias in nematode's life span. Each assay was carried out in independent triplicate plates.

Since marine agar does not support *C. elegans* growth due to its high osmolarity, the standard nematode killing assay had to be modified in order to assess the toxicity of the marine isolates (*i.e.*, strain D250 and violacein deficient mutant dV2). The marine bacteria were pre-grown overnight at 20°C in liquid marine broth, and 10 μL of the cultures were spread onto 2 cm paper filters (0.22 μm, Millipore, Maryland) and the filters placed on 3.5 cm marine agar plates supplemented with selective antibiotics as required, followed by incubation at 20°C for four days. On day four, the paper filters were removed from marine agar plates and placed onto 3.5 cm LB10 agar plates, to which the L4-stage nematodes were added (30 to 40 per plate). *p* values were calculated on the pooled data of all of the experiments done in each set by using the log-rank (Mantel–Cox) method [35,36] with the Prism software version 6.0c (GraphPad Software, La Jolla, CA, USA). A *p* < 0.05 was considered significant.

Violacein dose response assay

In order to estimate the dose response of *C. elegans* towards violacein, a nematode killing assay was carried out using 96 well agar plates as previously described [31]. *E. coli* clone 20G8 mutated in *vioA* gene (hereafter referred to as 20G8*vioA*⁻ mutant, see Table 1) was pre-grown overnight in LB10 at 37°C and inoculated (2 μL) to each LB10 agar well supplemented with selective antibiotic and L-arabinose as required. Plates were thereafter incubated at 25°C for four days to grow the bacterial lawns, and then each 5 μL of the purified violacein preparation (in methanol, see above) was added at various concentrations to the side of the bacterial lawn and the methanol was allowed to evaporate before the violacein was gently mixed with the lawn by the addition of 10 μl of sterile water. *E. coli* mutant clone 20G8*vioA*⁻ alone, and pure methanol added to a 20G8*vioA*⁻ lawn, were used as negative controls. Methanol was allowed to evaporate before L4-stage nematodes were added (5–15 per well). The lethal concentration at which 50% of the nematodes were killed (LC₅₀) was calculated by inference with the Prism software version 6.0c (GraphPad Software, La Jolla, CA, USA). The assays were carried out in independent triplicate plates. In order to determine if there was a significant difference in the survival of nematodes, a Students *t*-test was performed on the number of surviving nematodes at day seven. A *p* < 0.05 was considered significant.

Heat killing of bacterial strains

Bacterial lawns pre-grown on agar plates were heat killed at 65°C for one hour, and the plates cooled to 20°C before adding nematodes. In order to confirm that heat killed bacteria were dead, samples of the bacterial lawns were streaked on LB10 agar plates before the nematode killing assay, and 20 days after the start of the incubation with nematodes. These plates were each incubated at 37°C for 24 to 48 hours and subsequently checked for an absence of bacterial colonies.

Microscopy

Overnight liquid cultures of each bacterial strain were spread on 3.5 cm LB10 agar plates and grown for four days. Nematode strains were exposed to bacteria for 24 hours and up to four days, depending on the assay, and placed on a microscope slide that had been immersed in 0.1 M sodium azide solution. Nematodes were examined under an Olympus DP70 digital camera system (Japan) with differential interference contrast (DIC) microscope optics. In order to detect the GFP signal indicative of an apoptosis induction, the CED-1::GFP reporter nematodes (strain CU546, Table 2) were exposed to a bacterial lawn of the 20G8 clone and 20G8*vioA*⁻ and 20G8*vioC*⁻ mutants for 24 hours. Nematodes were visualized under epifluorescence microscopy, and the percentage of nematodes with a GFP signal calculated for each treatment.

Results

Violacein is a toxic metabolite that mediates antagonistic interactions between the bacterium *Microbulbifer* sp. D250 and *C. elegans*

Screening of a random transposon mutant library of the fosmid clone 20G8 for the loss of toxic activity towards *C. elegans* resulted in four mutant clones (Tables 1 and 3 and Figure 1). The nematode killing assay demonstrated that the killing phenotype of the four mutant clones was significantly reduced (*p* < 0.0001) when compared to the wild type clone 20G8 (Figure 1). The activity of mutant 20G8*VioA*⁻ was similar to the negative control (*p* = 0.803) and was therefore considered non-toxic for *C. elegans*. Sequencing of the four non-active mutants revealed that in all cases transposons had inserted in open reading frames with high sequence identity to the *vioABCDE* gene cluster of *C. violaceum*, which has previously been shown to be involved in the synthesis of violacein [37] (Table 3). Specifically, two mutants that had lost the violet pigmentation, had insertions in the *vioA* and *vioB* genes, and the two clones that expressed a grey pigmentation typical of violacein precursors, were mutated in the *vioC* and *vioD* genes (Table 3 and Figure 1) [37].

To further support the involvement of violacein in nematode toxicity, *Microbulbifer* sp. D250, the parent organism for the fosmid clone 20G8, was shown to rapidly kill *C. elegans* (*p* < 0.0001) when compared to negative control OP50 (Figure 2A). In contrast, the nematode's life span was significantly improved when exposed to a violacein deficient mutant dV2 (*p* < 0.0001, Figure 2A). Sequencing of the transposon insertion site in strain dV2 supported the loss of violacein production, by demonstrating that the homologue to the violacein biosynthesis gene *vioB* (GenBank AFT64168) had been disrupted in the dV2 mutant. Notably, survival of the nematodes fed with the dV2 mutant strain was significantly reduced (*p* < 0.0001) compared to the negative control using *E. coli* OP50 (see Figure 2A), which might indicate that other antinematode activities are also present in this bacterium. However, the involvement of violacein in toxicity towards *C. elegans* was further confirmed by the direct chemical

A

B

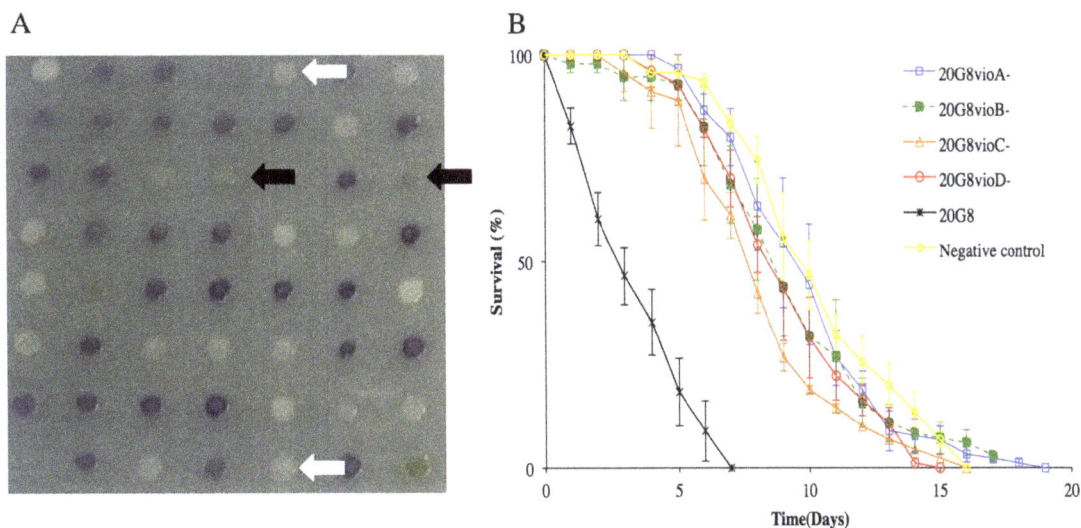

Figure 1. Characterization of the transposon mutant library of violacein producing clone 20G8. (A) Several mutants exhibit changes in violet pigmentation typical of the presence of violacein: 20G8$vioA^-$ and 20G8$vioB^-$ mutants have lost pigmentation (white arrows). 20G8$vioC^-$ and 20G8$vioD^-$ mutants express green-grey pigmentation (black arrows). (B) Nematode killing assay (N2 animals vs bacterial strains 20G8 clone and 20G8$vioABCD^-$ mutants). Survival kinetics of nematodes fed with either the 20G8 clone or the 20G8$vioABCD^-$ mutants deficient in violacein production. The killing phenotype of the four mutant clones is significantly reduced ($p < 0.0001$) when compared to the wild type clone 20G8. A randomly chosen clone from the library with no activity is used as a negative control. Each data point represents means ± the standard error of three replicate plates. p values were calculated on the pooled data of all of the plates in each experiment by using the log-rank (Mantel–Cox) method.

identification of the purified violet pigment produced by *Micro-bulbifer* sp. D250 wild type strain. The pigment was extracted from cultures, purified and analysed by liquid chromatography-mass spectrometry (LC-MS/MS). The purified purple pigment eluted as one peak in the liquid chromatography, and this peak represented violacein, as was identified by the matching mass of the protonated molecule ([M+H]$^+$) and its characteristic MS/MS fragmentation pattern (see Material and Methods).

Finally, we used pure violacein directly in the toxicity assays (see Material and Methods). The 50% survival (LC$_{50}$) of nematodes exposed to violacein falls between the range of 7.5 µM and 75 µM of pure violacein (LC$_{50}$ = 31.13 µM calculated from Figure 2B), when added to a viable bacterial lawn of the violacein non-producing mutant 20G8$vioA^-$ (Figure 2B). Interestingly, the survival of nematodes improved when violacein preparations were added to lawns of heat killed 20G8$vioA^-$ bacteria compared to violacein added to lawns of viable 20G8$vioA^-$ cells (*i.e.* $p = 0.001$

Table 3. Summary of the effects of violacein-producing clone 20G8 and its violacein deficient mutants on *C. elegans*.

Compound predicted to be expressed by the mutant[a]	L-tryptophan	prodeoxyviolacein	proviolacein	violacein
Proposed gene function involved in the biosynthesis[a,b]	*vioAB*: tryptophan 2-monooxygenase	*vioD*: hydroxylase	*vioC*: monooxygenases	N/A
Mutant/clone name and NCBI accession number	20G8$vioA^-$ (AFT64169) 20G8$vioB^-$ (AFT64168)	20G8$vioD^-$ (AFT64166)	20G8$vioC^-$ (AFT64167)	20G8 (JX523957)
BLASTp analysis[c]	VioA tryptophan 2-monooxygenase; VioB polyketide synthase	VioD hydroxylase	VioC monooxygenase	N/A
Nematode survival (p values)[d]	$p < 0.0001$	$p < 0.0001$	$p < 0.0001$	N/A
Colonization[e]	0%	N/A	0%	76% ± 1.6
Apoptosis[f]	0%	N/A	0%	70.3% ± 5.7
Clone pigmentation	white	grey	grey	violet

[a]According to [37,45].
[b]Genes were ordered based on enzymatic activity in violacein biosynthesis and not on the locus position within the cluster [37].
[c]Only hits with 100% query coverage, 100% identity and E value of 0 were considered.
[d]p values of the survival of nematodes exposed to violacein mutants compared to the wild type clone 20G8 are reported.
[e]Percentage of alive nematodes colonized after four days.
[f]Percentage of CU1546 *C. elegans* strain displaying GFP signal indicative of apoptosis induction. All studies were carried out in *C. elegans* strain N2 except for apoptosis studies where CU1546 strain was employed. Each data represents means ± the standard error of three replicates. N/A = not applicable.

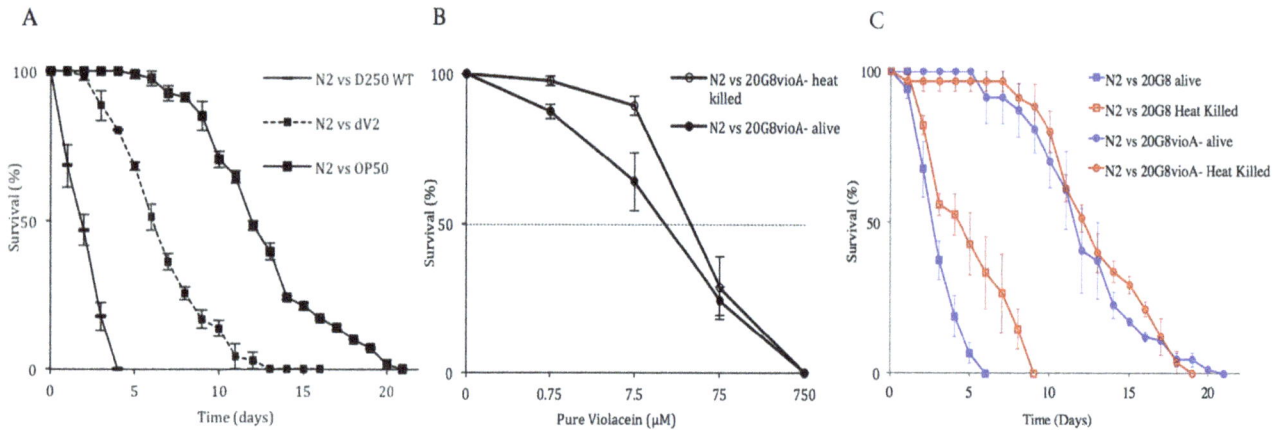

Figure 2. Nematode killing assay and dose response assay (N2 animal vs the alive and heat killed 20G8 clone and D250 strains). (A) Killing kinetics of *Microbulbifer* sp. D250 and dV2 mutant deficient in violacein production. Negative control OP50 is a non-pathogenic strain of *E. coli*. (B) Dose response of *C. elegans* to pure violacein added to 20G8*vioA*⁻ mutant bacteria alive and heat killed. Each point on the graph represents the average survival of worms after seven days exposure to violacein (C) Kinetics of nematode killing when nematodes are fed either the live or heat killed 20G8 clone or 20G8*vioA*⁻ mutant. Each data point represents means ± the standard error of three replicate plates. *p* values were calculated on the pooled data of all of the plates in each experiment by using the log-rank (Mantel–Cox) method.

and $p = 0.043$ for nematodes exposed to 0.75 μM and 7.5 μM of pure violacein, respectively).

Bacterial accumulation in the nematode intestine is involved in the killing activity of the violacein-producing *E. coli* clone

Given that the presence of live bacteria contributed to the decrease in the survival of nematodes in the presence of violacein (see above), we looked for evidence of bacterial accumulation in the nematodes intestine. The gut of 76% (n = 47) of the nematodes exposed to violacein-producing *E. coli* clone 20G8 had an accumulation of bacterial cells expressing a violet pigmentation in their intestine (Figure 3A). In contrast, nematodes exposed for four days to the violacein deficient mutants 20G8*vioA*⁻ (n = 42) and 20G8*vioC*⁻ (n = 57) showed no accumulation of bacteria in the intestinal lumen (Figure 3B and C). Microscopic analysis further showed tissue damage with enlargement of the intestinal lumen and enlargement of extracellular regions in nematodes fed with the 20G8 clone (arrowheads in Figure 3A). These data suggest that accumulation of bacteria in the intestinal lumen is one factor in the killing activity of clone 20G8. In order to further assess this, a nematode killing assay was performed using live and heat killed 20G8 cells. Heat inactivation of bacterial cells showed that nematode survival significantly increases ($p<0.0001$) in the presence of heat killed 20G8 cells compared to viable 20G8 bacteria (Figure 2C). In contrast, there was no significant effect ($p>0.05$) of heat killing on the survival of nematodes exposed to the violacein deficent mutant 20G8*vioA*⁻ (Figure 2C). In order to confirm that bacterial cells were successfully inactivated by heat, they were streaked on LB10 agar plates and monitored for growth. Bacterial growth was not detected confirming that the heat treatment was sufficient to kill the bacteria. Together these data suggest that the toxic effect of violacein is significantly increased by the presence of live violacein-producing bacteria that accumulate in the intestine.

Violacein-producing bacteria induce apoptosis in *C. elegans*

To further elucidate the process of killing, transgenic nematodes with a GFP marker for apoptosis (CU1546 strain, Table 2) were

exposed to a bacterial lawn of the violacein producing clone 20G8 and the violacein deficient mutant clones 20G8*vioA*⁻ and 20G8*vioC*⁻ for 24 hours, and thereafter visualised using differential interference contrast (DIC) and epifluorescence microscopy. The GFP signal was detected in somatic cells only in animals incubated with the 20G8 clone (70.3% of animals, n = 20), but not with 20G8*vioA*⁻ and 20G8*vioC*⁻ mutants (0%, n = 13 and n = 15 respectively) (Figure 4) showing that only the violacein-producing bacteria are capable of inducing apoptosis.

Mutations in the IIS pathway influence the sensitivity of *C. elegans* to violacein-producing bacteria and pure violacein

The IIS pathway with gene regulators DAF-2 and DAF-16 plays an important role in *C. elegans* pathogen defence [18]. We therefore questioned if *C. elegans* uses this pathway to also protect itself against violacein-mediated killing. To address this question, *C. elegans* loss of function mutants in the genes *daf-2* (CB1370), *pdk-1* (JT9609), *daf-16* (IU10) and *wwp-1* (RB1178), and a DAF-16 over-expressing strain (TJ356) (Table 2) were fed with the violacein-producing clone 20G8 in the nematode killing assay. The life span of *C. elegans* strains carrying loss of function mutations in *daf-2* and *pdk-1* and *wwp-1* were significantly increased compared to both wild type animals and *daf-16* mutant ($p<0.0001$) (Figure 5A, 5B and 5C). In contrast the *C. elegans* IU10 strain with loss of function mutation in the FOXO-family transcription factor DAF-16 displayed significantly reduced ($p<0.0005$) survival compared to wild type N2 animals (Figure 5A). Whereas viability of transgenic DAF-16::GFP nematodes that overexpress DAF-16 was significantly improved compared to both wild type and *daf-16* mutant animals ($p<0.0001$) (Figure 5A).

Similar results were observed for the *C. elegans* strains fed with the violacein deficient mutant (20G8*vioA*⁻) supplemented with pure violacein in a dose response assay (Figure 5D). The toxic effect of violacein was dose dependent in wild type N2 nematodes and in *daf-2* and *daf-16* mutant animals (significant difference when nematodes of the same strain were exposed to 7.5, 75 or 750 μM of pure violacein, $p<0.05$). In these assays the lethal dose of violacein required to kill 50% of the population was higher for *daf-2* mutants (75 μM<LC₅₀<750 μM) than for the wild type

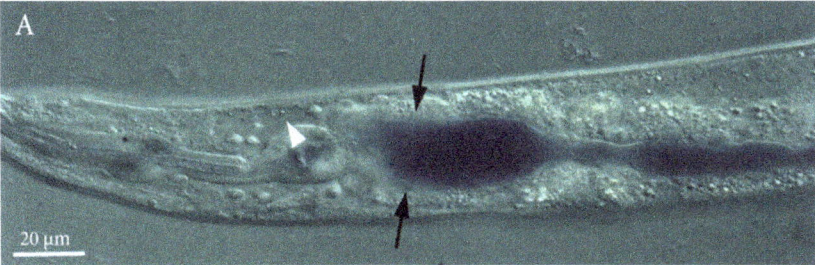

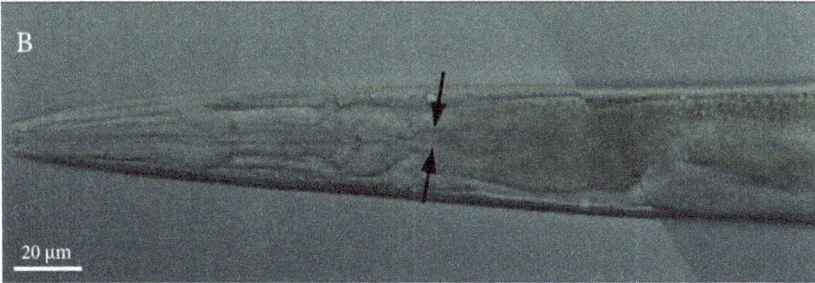

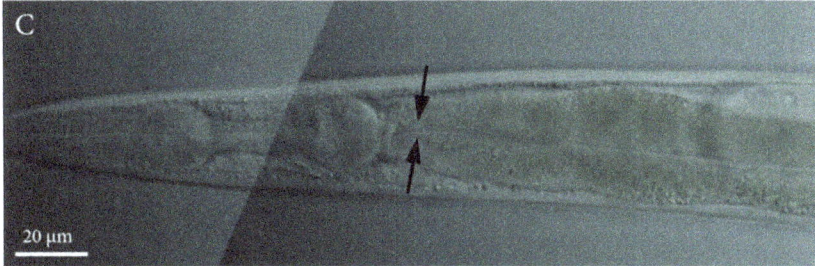

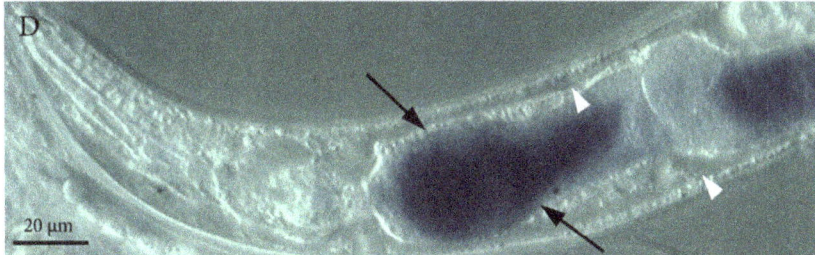

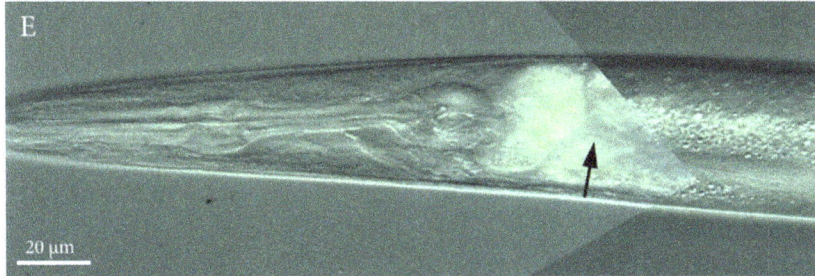

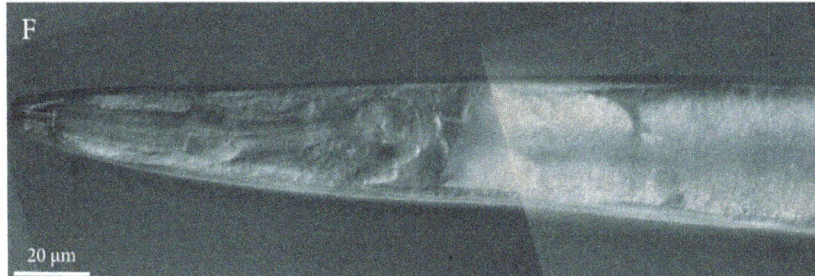

Figure 3. Visualization of bacterial accumulation in the nematode intestine by the *E. coli* clones (wild type animals, the *daf-2* and *daf-16* null mutant animals and the DAF-16 overexpressing nematodes vs the *E. coli* clone 20G8 and 20G8*vioA*⁻ and 20G8*vioC*⁻ mutant clones). All images present *C. elegans* anterior to the left and show the pharynx and first part of the intestinal lumen by differential interference contrast (DIC) microscopy. Accumulation by 20G8 cells in the nematodes intestine (arrows in panels A and D) and extensive enlargement of extracellular regions (white arrowheads panels A and D) in N2 wild type (panel A) and *daf-16* mutant (panel D) animals. No change in phenotype was observed in N2 nematodes fed with violacein deficient mutant clones 20G8*vioA*⁻ (negative control-panel B) and 20G8*vioC*⁻ (panel C). Similarly, no bacterial cells or enlargement of extracellular regions were detected in *daf-2* and DAF-16 over-expressing mutant nematodes exposed to the violacein producing clone 20G8 (arrows in panel E and F respectively). Each panel was assembled from multiple photomicrographs taken with the same magnification and same acquisition settings.

(7.5 µM<LC_{50}<75 µM) or the *daf-16* mutant animals (7.5 µM<LC_{50}<75 µM) (Figures 2B and 5D), further indicating that sensitivity or resistance to violacein is at least partially mediated by the IIS pathway.

C. elegans daf-2 mutant and DAF-16 overexpressing strains are resistant to intestinal accumulation by the violacein-producing clone 20G8

Given that bacterial intestinal accumulation is in part responsible for the toxic phenotype of violacein producing cells and that the IIS pathway has previously been shown to influence bacterial accumulation in the nematode intestine [21], we aimed to determine if elements of the IIS pathway mediate resistance to violacein by preventing bacterial accumulation.

Firstly, we compared the sensitivity of various *C. elegans* strains to violacein when fed viable or heat killed bacteria. The presence of viable bacterial cells significantly increased the sensitivity of *daf-16* mutant animals to violacein compared to when the mutant was exposed to heat killed bacteria (significant difference when pure violacein was added to viable or heat killed bacteria $p<0.05$, Figure 5D). In contrast, heat killed bacteria had no or little impact on the survival of *daf-2* mutant ($p>0.05$, Figure 5D).

Secondly, we assessed the ability of *E. coli* clone 20G8 to accumulate in the intestine of the various *C. elegans* strains. Ninety one percent of *daf-16* mutant nematodes (n = 35) showed evidence of bacterial accumulation when exposed to 20G8 cells, in contrast no or little bacterial accumulation was present in the *daf-2*-loss of function and DAF-16 over-expressing mutants under the same treatment (0% n = 45 and 1.1% n = 31, respectively). Enlargement of extracellular regions was present in wild type and *daf-16* mutant nematodes (90% n = 20 and 92% n = 25, respectively) fed with clone 20G8 (arrowheads in Figure 3 panel A and panel D respectively) while *daf-2*-loss of function and DAF-16 over-expressing mutant animals showed normal anatomical structures and no tissue injury was detected (Figure 3 panel E 0% n = 19 and Figure 3 panel F 0% n = 16, respectively). Together these data indicate that the resistance to violacein by *daf-2* null mutants and DAF-16 over-expressing *C. elegans* strains is, in part, due to their improved ability to prevent accumulation of bacteria in the presence of violacein.

Loss of function in IIS controlled genes results in increased sensitivity to violacein

In *C. elegans* DAF2/DAF16 controls the expression of various effector genes including those relevant for detoxification and antimicrobial activity such as the superoxidase dismutase gene *sod-3* and antimicrobial genes *spp-1* and *lys-7* [23,38]. Thus given that the precise molecular target/s for violacein in *C. elegans* are unknown we sought to determine which, if any, of these relevant downstream genes are required for the increased resistance to violacein observed in *daf-2* null and DAF-16 over-expressing strains. Specifically we chose to test violacein sensitivity in *C. elegans* mutants defective in *sod-3*, *spp-1* and *lys-7* (Table 2)

because of the previous reported involvement of these genes in immunity to bacterial accumulation [39,40,41]. We found that *daf-2;spp-1* and *daf-2;sod-3* double mutants displayed significantly reduced survival compared to the single mutant *daf-2* ($p<0.0001$, Figure 6A) when exposed to the 20G8 clone in a nematode killing assay. No reduction in viability was detected in the *daf-2;lys-7* double mutant when compared to the single mutant *daf-2* ($p = 0.937$, Figure 6A). Interestingly a single mutation in gene *spp-1* significantly reduced the nematode's life span when compared to wild type animals ($p<0.0001$), while the viability of the nematode was not affected by mutations in the *lys-7* and *sod-3* genes ($p>0.05$, Figure 6B). These data indicate that resistance to violacein in *daf-2* mutants is at least in part driven by SPP-1 and SOD-3, with the antimicrobial LYS-7 having little or no involvement.

Discussion

Violacein has antinematode activity

In this study, we identified violacein as the metabolite responsible for the antinematode activity of *Microbulbifer* sp. D250. Violacein is arguably best known for its antibacterial properties and its activity as a potentially novel therapeutic against a range of tumors [16]. However, to the best of our knowledge, and with the exception of studies related to cancer therapy, this is the first report of violacein toxicity towards a multicellular eukaryote, thus adding to the list of biological functions for this natural metabolite.

Genetic analysis identified an operon of five conserved biosynthesis genes *vioA-E* that have been identified across all violacein-producing strains studied to date [7,12,42,43]. Interestingly, mutations in *vioC* or *vioD* result in a grey colony pigmentation, reminiscent of the accumulation of the violacein precursor pro-violacein [7,43,44]. Pro-violacein differs from violacein by the absence of one oxo-group in the C15 position (indolyl instead of indolone, see Table 3 and reference [7]). This minor chemical difference, however, has a substantial impact on the toxicity, as the grey-pigmented *vioC* mutant did not kill nematodes in our study. Similar observation have also been recently made for *E. coli* K12 strains producing either violacein or pro-violacein, which were and were not resistant, respectively, to protozoan predation [45].

C. elegans is intrinsically more resistant to the effects of violacein ($LC_{50}>30$ µM) than other bacterial grazers such as flagellates and amoebae, which were found to be effective at concentrations of 10 µM [46] and 1 µM [12] respectively. Nevertheless, violacein appears to be more potent towards *C. elegans* than toxins derived from known bacterial pathogens. For example, small phenazine molecules, including phenazine-1-carboxylic acid recently identified from *Pseudomonas aeruginosa*, are toxic to *C. elegans* only at concentrations greater than 70 µM. The effective concentration of violacein against *C. elegans* observed here is also similar to that of recent studies investigating novel antihelminthic therapies. For example, in a screen of existing drug leads Taylor *et al* [47]

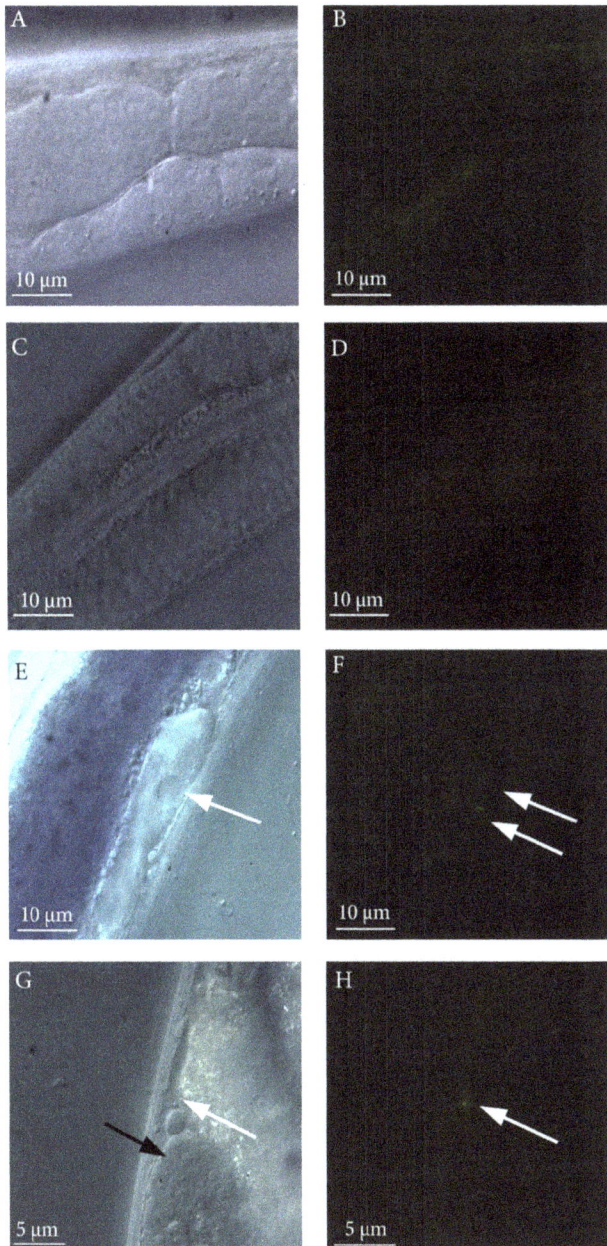

Figure 4. Apoptosis analysis in CED-1::GFP transgenic nematode strain CU1546. Visualization of GFP apoptosis marker inside the nematodes exposed to the violacein deficient mutant clones (20G8vioA⁻ panels A and B; 20G8vioC⁻ panels C and D) or violacein producing clone (20G8, panels E and F and G and H). Images show a section of the intestine and the gonads of C. elegans by DIC (panels A, C, E, G) and by epifluorescence microscopy (panels B, D, F, H). A ring-shaped GFP signal was detected in somatic cells of CU1546 nematodes exposed to 20G8 clone (two white arrows in panel F and white arrow in panel H) but not to 20G8vioA⁻ and 20G8vioC⁻ mutants (panel B and D). The ring-shaped GFP signal is associated with the expression of CED-1 receptor on the cell wall of engulfing cells only when phagocytosis of apoptotic cells is taking place. The ring-shaped GFP signal is apoptosis-specific and is visible in somatic cells of the nematodes (white arrows in E and G), the germ line (gonad) of the nematode is shown in panel G (black arrow). A non-specific fluorescence signal is visible in panels B and D.

identified 18 candidate compounds having detectable phenotypes against C. elegans with EC_{50} ranging between 0.7 μM and > 192 μM, including those already approved as cancer therapeutics such as Dasatinib (LC_{50} 22.3 μM) and Flavopiridol (LC_{50} 48.3 μM). Despite its many biological activities, the toxicity of violacein on (non-tumoral) mammalian cells is quite low. Recently, it has been reported that intraperitoneal doses of violacein of up to 1 mg kg^{-1} are not toxic to mouse blood, kidneys, or liver, thus enabling a potential in vivo use of violacein and its derivatives as a therapeutic compound with few side effects [48]. Given the relatively low effective concentration of violacein towards C. elegans determined here, we propose that the investigation of violacein as an antiparasitic compound is a reasonable prospect and that further studies regarding the activity of violacein against model parasitic nematodes, could reinforce the therapeutic potential of this drug.

Bacterial accumulation in the intestine of C. elegans is a key factor involved in the toxicity of violacein

We observed that in the presence of violacein the otherwise non-pathogenic E. coli has the ability to accumulate in the intestine and eventually kill C. elegans. The exact mechanism by which violacein treatment leads to bacterial accumulation and reduced nematode viability is yet to be determined, however recent reports have demonstrated a link between nematode longevity and intestinal colonization [21]. Specifically, Portal-Celhay et al [21] showed that the capacity to control bacterial accumulation in the gut was dependent on the immunological status and age of the individual animal. Heavy bacterial accumulation has also been shown to reduce the lifespan of the nematodes depending on the bacterial strain used [21,49]. Thus it is possible that exposure to violacein compromises the nematode's defence resulting in a reduced capacity to control bacteria in the gut and, thus, increasing the mortality rate. This is supported by similar observations recently made in various Bacillus species, in which treatment with the Bacillus pore-forming crystal protein (Cry PFP) seemingly sensitizes C. elegans to bacterial infection [50]. An alternative explanation is that the presence of violacein allows bacteria to penetrate the intestinal tissue resulting in a lethal infection. Whilst we did not observe bacteria within the tissue of nematodes exposed to violacein, previous reports have suggested internal infection as a possible cause of death in older nematodes [49,51] and so this alternative possibility should not be dismissed.

Violacein induces apoptosis in C. elegans

In addition to increased susceptibility to bacterial accumulation, violacein is capable of inducing apoptosis in C. elegans (Figure 4). Previous studies have revealed that apoptosis is also involved in violacein-mediated cell death in mammalian cell lines [16] and amoebae [12]. Thus although the exact molecular target of violacein in the eukaryotic cell is yet to be elucidated, the induction of an apoptosis-like cell death mechanism in multiple, distantly related eukaryotic systems (mammalian cells, amoeba and nematodes) suggest that an ancient, common eukaryotic cell process may be an additional target of violacein-driven toxicity.

The insulin/IGF-1 signalling pathway contributes to the native defence against violacein

The innate immune response, including the IIS pathway with regulators DAF-2 and DAF-16, is a key component of the nematodes first line of defence particularly against pathogens [18,19,52]. Here we add to that knowledge by demonstrating that the IIS pathway also contributes to the nematodes native defence

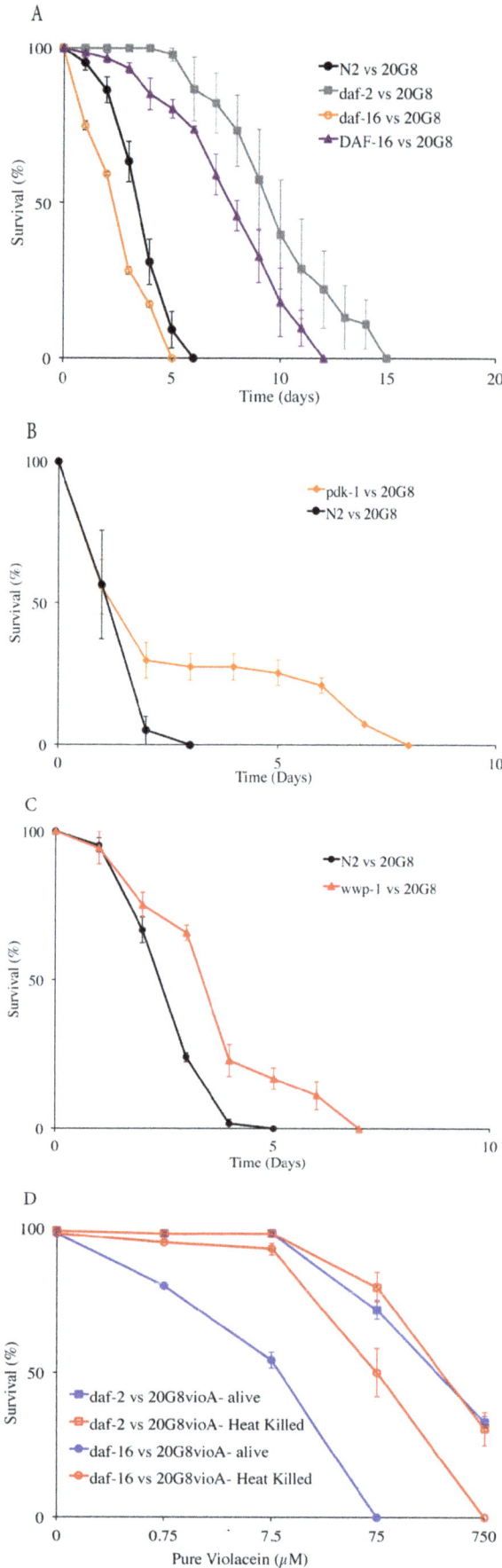

Figure 5. Nematode killing assay and dose response assay (wild type animals and *daf-2*, *daf-16*, *pdk-1*, *wwp-1* and DAF-16 mutant nematodes vs 20G8 clone). The survival of nematodes fed with 20G8 clone was measured for (A) *daf-2*, *daf-16* and DAF-16 mutant animals. (B) *pdk-1* mutant animals (C) *wwp-1* mutant animals (D) Dose response of *C. elegans daf-2* and *daf-16* strains to pure violacein added to 20G8*vioA⁻* mutant bacteria alive and heat killed. Each data point represents means ± the standard error of three replicate plates *p* values were calculated on the pooled data of all of the plates in each experiment by using a Students *t*-test on the number of surviving nematodes at day seven.

against secondary metabolites derived from non-pathogenic bacteria. The *C. elegans daf-2* and *pdk-1* deficient animals were significantly more resistant to the toxicity of violacein and the associated bacterial accumulation as compared to the wild type strain. These finding are consistent with the "long-lived" phenotype of the *C. elegans daf-2* mutant [53], which is also known to mediate the immune defence to bacterial infections [52,54].

Since the molecular target of violacein-mediated toxicity in *C. elegans* remains to be elucidated the mechanisms involved in the nematode immune response towards violacein is unknown. However recent studies using *E. coli* expressing the *Pseudomonas aeruginosa* translational inhibitor exotoxin A (ToxA), have demonstrated that *C. elegans* induces an immune response towards ToxA, which the nematode detects indirectly via the toxin-mediated damage [55]. Others have also demonstrated activation of immunity and detoxification genes in response to damage to a variety of cellular functions [56], many of which could result from exposure to bacterial toxins. Thus such an effector-triggered immunity is likely to be widespread in animals and may function to enable bacteriovorus organisms such as *C. elegans* to discriminate between commensal and pathogenic bacteria [57]. Therefore once the molecular target of violacein is established it will be of interest to determine if *C. elegans* responds directly to the presence of violacein or rather to the associated inhibition of, or damage to, specific cellular functions.

Identifying genes under DAF-2/DAF-16 control that are involved in the increased resistance to violacein may provide further insight into the molecular/cellular target of this compound. Indeed assessment of violacein sensitivity of selected *C. elegans daf-2* double mutant strains in the current study indicates that while the antimicrobial lysozyme (LYS-7) is not involved in violacein resistance, both the superoxide dismutase SOD-3 and the antimicrobial peptide SPP-1 seem to play a role in host defence against violacein. The potential role of SOD-3 is consistent with previous findings that violacein can cause oxidative stress in human cancer cell lines [58]. However other studies have indicated that violacein acts as an antioxidant for the producing bacteria [59]. This apparent contradiction may result from the same molecule producing different responses depending on the target cells. *Spp-1* is expressed in the nematode intestinal cells and has previously been shown to be involved in the immune response against pathogens *Salmonella typhimurium* [41] and *P. aeruginosa* [60]. Results from this study support the idea that SPP-1 could also control the violacein induced intestinal accumulation of *E. coli* in *C. elegans* and further suggest a wider target spectrum for this peptide.

Despite the large body of knowledge surrounding the IIS pathway of the innate immune response in *C. elegans* and its role in stress resistance and pathogenesis, there is a paucity of information regarding its role in response to bacterial toxins. In one of the few studies to date, Chen *et al.* [22] demonstrated that

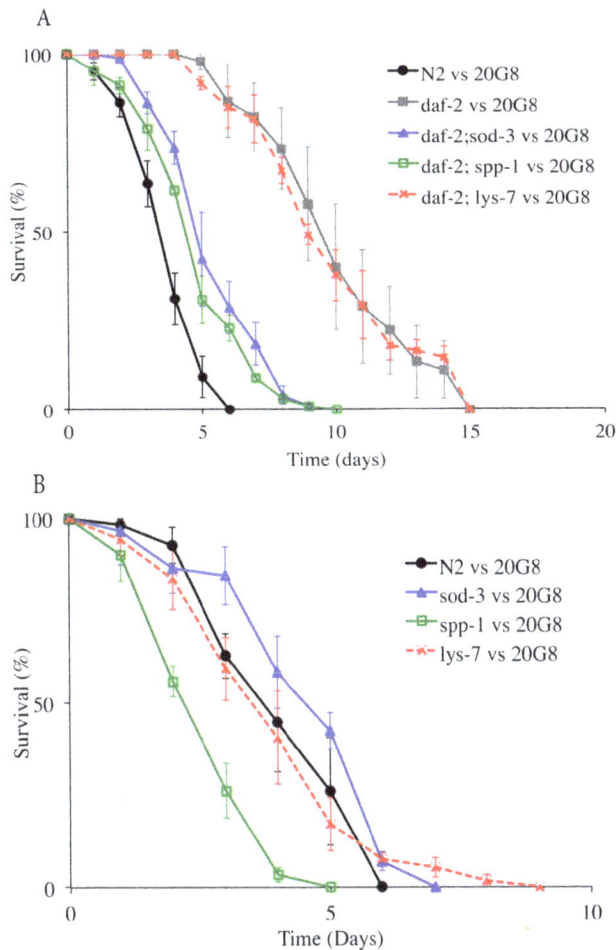

Figure 6. Nematode killing assay (wild type animals and *daf-2*, *daf-2;spp-1*, *daf-2;lys-7*, *daf-2;sod-3*, *spp-1*, *lys-7*, *sod-3* mutant nematodes vs the 20G8 clone). (A) The survival of the nematode was tested using *C. elegans* double mutants *daf-2;sod-3*, *daf-2;spp-1*, *daf-2;lys-7* and (B) the single mutant animals *sod-3*, *spp-1*, and *lys-7*. Each data point represents means ± the standard error of three replicate plates. *p* values were calculated on the pooled data of all of the plates done in each experiment by using the log-rank (Mantel–Cox) method and the values are provide in the text.

the IIS pathway is involved in the cellular defence of *C. elegans* against *Bacillus thuringensis* crystal pore-forming toxins (Cry PFT). It appears likely that *C. elegans* uses this conserved pathway as a general bacterial toxin defence mechanism, as, similar to our findings for violacein, reduction in DAF-2 signalling confers

resistance to Cry PFT, which is at least in part, dependent on DAF-16 function. However the newly described second branch of the IIS pathway mediated by WWP-1 appears to play a different role in defence against violacein, as in contrast to the hypersensitivity towards Cry PFT [22], *wwp-1* mutant animals were more resistant to violacein than wild type nematodes (Figure 5C). These contrasting observations may speak to differences in how the two toxins affect target cells and will be the subject of future studies investigating the cellular response of *C. elegans* to bacterial toxins.

Conclusion

In their natural habitat nematodes, including *C. elegans*, are major predators for bacteria that, in turn, have evolved a number of predatory defence mechanism including the production of inhibitory metabolites [61]. This work presents, for the first time, violacein as an antinematode compound and assesses toxicity and the cellular response to violacein exposure in a *C. elegans* animal model. We found that this small molecule facilitates intestinal accumulation of the *E. coli* host cells and stimulates apoptotic activity. Mutations in a number of genes in the IIS pathway (*daf-2* and *pdk-1*) and over-expression of another (*daf-16*) can confer significant resistance to violacein toxicity, providing evidence that *C. elegans* uses the DAF-2/DAF-16 innate immune signalling pathway to defend itself against this and potentially other noxious bacterial metabolites. Finally we demonstrate that defence against violacein requires, at least in part, the function of the antimicrobial peptide SPP-1 and possibly superoxide dismutase SOD-3. It is expected that the added knowledge will assist in the design of future mutational and/or gene expression studies aimed at determining the mechanisms of violacein toxicity in the nematode model including its direct molecular target/s. Such studies will not only prove important for further understanding the molecular basis for microbial antagonistic interactions but also for the development of microbial secondary metabolites such as violacein as novel chemotherapeutics.

Acknowledgments

We thank the IGB Open Laboratory for facilities and resources, the *C. elegans* gene knockout project and the *Caenorhabditis* Genetics Center (CGC) at the University of Minnesota for provision of *C. elegans* samples. We further thank Silvana Castro for technical assistance and A/Prof. Torsten Thomas for revision of the draft manuscript.

Author Contributions

Conceived and designed the experiments: FB PB EDS SE. Performed the experiments: FB MD AP JN DS. Analyzed the data: FB DS PB EDS SE. Contributed reagents/materials/analysis tools: PB EDS SE. Wrote the paper: FB AP DS PB EDS SE.

References

1. Holden-Dye L, Walker RJ (2007) Anthelmintic drugs. WormBook 2: 1–13.
2. Kaplan RM (2004) Drug resistance in nematodes of veterinary importance: a status report. Trends Parasitol 20: 477–481.
3. Prichard RK, Geary TG (2008) Drug discovery: Fresh hope to can the worms. Nature 452: 157–158.
4. Jones AK, Buckingham SD, Sattelle DB (2005) Chemistry-to-gene screens in *Caenorhabditis elegans*. Nat Rev Drug Discov 4: 321–330.
5. Sifri CD, Begun J, Ausubel FM (2005) The worm has turned–microbial virulence modeled in *Caenorhabditis elegans*. Trends Microbiol 13: 119–127.
6. Penesyan A, Ballestriero F, Daim M, Kjelleberg S, Thomas T, et al. (2012) Assessing the effectiveness of functional genetic screens for the identification of bioactive metabolites. Mar Drugs 11: 40–49. doi: 10.3390/md11010040.
7. Hoshino T (2011) Violacein and related tryptophan metabolites produced by *Chromobacterium violaceum*: biosynthetic mechanism and pathway for construction of violacein core. Appl Microbiol Biotechnol 91: 1463–1475.
8. Hakvag S, Fjarvik E, Klinkenberg G, Borgos S, Josefsen K, et al. (2009) Violacein producing *Collimonas* sp. from the sea surface microlayer of costal waters in trondelag, Norway. Mar Drugs 7: 576–588.
9. Aranda S, Montes-Borrego M, Landa BB (2011) Purple-pigmented violacein-producing *Duganella* spp. inhabit the rhizosphere of wild and cultivated olives in southern Spain. Microb Ecol 62: 446–459.
10. Pantanella F, Berlutti F, Passariello C, Sarli S, Morea C, et al. (2007) Violacein and biofilm production in *Janthinobacterium lividum*. J Appl Microbiol 102: 992–999.

11. Lu Y, Wang L, Xue Y, Zhang C, Xing X-H, et al. (2009) Production of violet pigment by a newly isolated psychrotrophic bacterium from a glacier in Xinjiang, China. Biochem Eng J 43: 135–141.

12. Matz C, Webb JS, Schupp PJ, Phang SY, Penesyan A, et al. (2008) Marine biofilm bacteria evade eukaryotic predation by targeted chemical defense. PLoS One 3: e2744.

13. Yang LH, Xiong H, Lee OO, Qi SH, Qian PY (2007) Effect of agitation on violacein production in *Pseudoalteromonas luteoviolacea* isolated from a marine sponge. Lett Appl Microbiol 44: 625–630.

14. Konzen M, De Marco D, Cordova CA, Vieira TO, Antonio RV, et al. (2006) Antioxidant properties of violacein: possible relation on its biological function. Bioorg Med Chem 14: 8307–8313.

15. Becker MH, Brucker RM, Schwantes CR, Harris RN, Minbiole KP (2009) The bacterially produced metabolite violacein is associated with survival of amphibians infected with a lethal fungus. Appl Environ Microbiol 75: 6635–6638.

16. Durán M, Ponezi A, Faljoni-Alario A, Teixeira MS, Justo G, et al. (2012) Potential applications of violacein: a microbial pigment. Med Chem Res 21: 1524–1532.

17. Marsh EK, May RC (2012) *Caenorhabditis elegans*, a model organism for investigating immunity. Appl Environ Microbiol 78: 2075–2081.

18. Tan MW, Shapira M (2011) Genetic and molecular analysis of nematode-microbe interactions. Cell Microbiol 13: 497–507.

19. Yanase S, Yasuda K, Ishii N (2002) Adaptive responses to oxidative damage in three mutants of *Caenorhabditis elegans* (age-1, mev-1 and daf-16) that affect life span. Mech Ageing Dev 123: 1579–1587.

20. Zhang X, Zhang Y (2009) Neural-immune communication in *Caenorhabditis elegans*. Cell Host Microbe 5: 425–429.

21. Portal-Celhay C, Bradley ER, Blaser MJ (2012) Control of intestinal bacterial proliferation in regulation of lifespan in *Caenorhabditis elegans*. BMC Microbiol 12: 49.

22. Chen CS, Bellier A, Kao CY, Yang YL, Chen HD, et al. (2010) WWP-1 is a novel modulator of the DAF-2 insulin-like signaling network involved in pore-forming toxin cellular defenses in *Caenorhabditis elegans*. PLoS One 5: e9494. doi: 9410.1371/journal.pone.0009494.

23. Mukhopadhyay A, Oh SW, Tissenbaum HA (2006) Worming pathways to and from DAF-16/FOXO. Exp Gerontol 41: 928–934.

24. Back P, Matthijssens F, Vlaeminck C, Braeckman BP, Vanfleteren JR (2010) Effects of sod gene overexpression and deletion mutation on the expression profiles of reporter genes of major detoxification pathways in *Caenorhabditis elegans*. Exp Gerontol 45: 603–610.

25. Durai S, Karutha Pandian S, Balamurugan K (2011) Changes in *Caenorhabditis elegans* exposed to Vibrio parahaemolyticus. J Microbiol Biotechnol 21: 1026–1035.

26. Roeder T, Stanisak M, Gelhaus C, Bruchhaus I, Grotzinger J, et al. (2010) Caenopores are antimicrobial peptides in the nematode *Caenorhabditis elegans* instrumental in nutrition and immunity. Dev Comp Immunol 34: 203–209.

27. Sulston J, Hodgkin J (1988) The nematode *Caenorhabditis elegans*. New York: Cold Spring Harbor Lab Press.

28. Marden P, Tunlid A, Malmcrona-Friberg K, Odham G, Kjelleberg S (1985) Physiological and morphological changes during short term starvation of marine bacteria isolates. Arch Microbiol 142: 326–332.

29. Brenner S (1974) The genetics of *Caenorhabditis elegans*. Genetics 77: 71–94.

30. Stiernagle T (2006) Maintenance of *C. elegans*. WormBook 1:11101895/wormbook11011. Available: http://www.wormbook.org/chapters/www_strainmaintain/strainmaintain.html. Accessed 2014 Sep 18.

31. Ballestriero F, Thomas T, Burke C, Egan S, Kjelleberg S (2010) Identification of compounds with bioactivity against the nematode *Caenorhabditis elegans* by a screen based on the functional genomics of the marine bacterium *Pseudoalteromonas tunicata* D2. Appl Environ Microbiol 76: 5710–5717.

32. Altschul SF, Gish W, Miller W, Myers EW, Lipman DJ (1990) Basic local alignment search tool. J Mol Biol 215: 403–410.

33. Egan S, James S, Holmström C, Kjelleberg S (2002) Correlation between pigmentation and antifouling compounds produced by *Pseudoalteromonas tunicata*. Environ Microbiol 4: 433–442.

34. Herrero M, Delorenzo V, Timmis K (1990) Transposon vectors containing non-antibiotic resistance selection markers for cloning and stable chromosomal insertion of foreign genes in gram-negative bacteria. J Bacteriol 172: 6557–6567.

35. Harrington D (2005) Linear Rank Tests in Survival Analysis. Encyclopedia of Biostatistics. Hoboken, NJ: John Wiley & Sons, Ltd.

36. Mantel N (1966) Evaluation of survival data and two new rank order statistics arising in its consideration. Cancer Chemo Rep Part 1 50: 163–170.

37. Balibar CJ, Walsh CT (2006) In vitro biosynthesis of violacein from L-tryptophan by the enzymes VioA-E from *Chromobacterium violaceum*. Biochemistry 45: 15444–15457.

38. Murphy CT, McCarroll SA, Bargmann CI, Fraser A, Kamath RS, et al. (2003) Genes that act downstream of DAF-16 to influence the lifespan of *Caenorhabditis elegans*. Nature 424: 277–283.

39. Mallo GV, Kurz CL, Couillault C, Pujol N, Granjeaud S, et al. (2002) Inducible antibacterial defense system in *C. elegans*. Curr Biol 12: 1209–1214.

40. Chavez V, Mohri-Shiomi A, Maadani A, Vega LA, Garsin DA (2007) Oxidative stress enzymes are required for DAF-16-mediated immunity due to generation of reactive oxygen species by *Caenorhabditis elegans*. Genetics 176: 1567–1577.

41. Alegado RA, Tan MW (2008) Resistance to antimicrobial peptides contributes to persistence of *Salmonella typhimurium* in the *C. elegans* intestine. Cell Microbiol 10: 1259–1273.

42. August PR, Grossman TH, Minor C, Draper MP, MacNeil IA, et al. (2000) Sequence analysis and functional characterization of the violacein biosynthetic pathway from *Chromobacterium violaceum*. J Mol Microbiol Biotechnol 2: 513–519.

43. Zhang X, Enomoto K (2011) Characterization of a gene cluster and its putative promoter region for violacein biosynthesis in *Pseudoalteromonas* sp. 520P1. Appl Microbiol Biotechnol 90: 1963–1971.

44. Jiang PX, Wang HS, Xiao S, Fang MY, Zhang RP, et al. (2012) Pathway redesign for deoxyviolacein biosynthesis in *Citrobacter freundii* and characterization of this pigment. Appl Microbiol Biotechnol 94: 1521–1532.

45. Ahmetagic A, Philip DS, Sarovich DS, Kluver DW, Pemberton JM (2011) Plasmid encoded antibiotics inhibit protozoan predation of *Escherichia coli* K12. Plasmid 66: 152–158.

46. Matz C, Deines P, Boenigk J, Arndt H, Eberl L, et al. (2004) Impact of violacein-producing bacteria on survival and feeding of bacterivorous nanoflagellates. Appl Environ Microbiol 70: 1593–1599.

47. Taylor CM, Martin J, Rao RU, Powell K, Abubucker S, et al. (2013) Using existing drugs as leads for broad spectrum anthelmintics targeting protein kinases. PLoS Pathog 9: e1003149.

48. Bromberg N, Dreyfuss JL, Regatieri CV, Palladino MV, Durán N, et al. (2010) Growth inhibition and pro-apoptotic activity of violacein in Ehrlich ascites tumor. Chem-Biol Interact 186: 43–52.

49. Garigan D, Hsu AL, Fraser AG, Kamath RS, Ahringer J, et al. (2002) Genetic analysis of tissue aging in *Caenorhabditis elegans*: a role for heat-shock factor and bacterial proliferation. Genetics 161: 1101–1112.

50. Kho MF, Bellier A, Balasubramani V, Hu Y, Hsu W, et al. (2011) The pore-forming protein Cry5B elicits the pathogenicity of *Bacillus* sp. against *Caenorhabditis elegans*. PLoS One 6: e29122.

51. Gems D, Riddle DL (2000) Genetic, behavioral and environmental determinants of male longevity in *Caenorhabditis elegans*. Genetics 154: 1597–1610.

52. Garsin DA, Villanueva JM, Begun J, Kim DH, Sifri CD, et al. (2003) Long-lived *C. elegans* daf-2 mutants are resistant to bacterial pathogens. Science 300: 1921.

53. Kenyon C, Chang J, Gensch E, Rudner A, Tabtiang R (1993) A *C. elegans* mutant that lives twice as long as wild type. Nature 366: 461–464.

54. Kawli T, He F, Tan MW (2010) It takes nerves to fight infections: insights on neuro-immune interactions from *C. elegans*. Dis Model Mech 3: 721–731.

55. McEwan DL, Kirienko NV, Ausubel FM (2012) Host translational inhibition by *Pseudomonas aeruginosa* Exotoxin A Triggers an immune response in *Caenorhabditis elegans*. Cell Host Microbe 11: 364–374.

56. Melo JA, Ruvkun G (2012) Inactivation of conserved *C. elegans* genes engages pathogen- and xenobiotic-associated defenses. Cell 149: 452–466.

57. Kleino A, Silverman N (2012) UnZIPping mechanisms of effector-triggered immunity in animals. Cell Host & Microbe 11: 320–322.

58. de Carvalho DD, Costa FT, Duran N, Haun M (2006) Cytotoxic activity of violacein in human colon cancer cells. Toxicol In Vitro 20: 1514–1521.

59. Konzen M, De Marco D, Cordova CA, Vieira TO, Antonio RV, et al. (2006) Antioxidant properties of violacein: possible relation on its biological function. Bioorg Med Chem 14: 8307–8313.

60. Evans EA, Kawli T, Tan MW (2008) *Pseudomonas aeruginosa* suppresses host immunity by activating the DAF-2 insulin-like signaling pathway in *Caenorhabditis elegans*. PLoS Pathog 4: e1000175. doi: 1000110.1001371/journal.ppat.1000175.

61. Jousset A (2012) Ecological and evolutive implications of bacterial defences against predators. Environ Microbiol 14: 1830–1843.

62. Penesyan A, Marshall-Jones Z, Holmström C, Kjelleberg S, Egan S (2009) Antimicrobial activity observed among cultured marine epiphytic bacteria reflects their potential as a source of new drugs. FEMS Microbiol Ecol 69: 113–124.

10

Sodium-Glucose Transporter-2 (SGLT2; SLC5A2) Enhances Cellular Uptake of Aminoglycosides

Meiyan Jiang[1], Qi Wang[1], Takatoshi Karasawa[1], Ja-Won Koo[1,2], Hongzhe Li[1], Peter S. Steyger[1]*

1 Oregon Hearing Research Center, Oregon Health & Science University, Portland, Oregon, United States of America, **2** Department of Otorhinolaryngology, Seoul National University College of Medicine, Bundang Hospital, Seongnam, Gyeonggi, Republic of Korea

Abstract

Aminoglycoside antibiotics, like gentamicin, continue to be clinically essential worldwide to treat life-threatening bacterial infections. Yet, the ototoxic and nephrotoxic side-effects of these drugs remain serious complications. A major site of gentamicin uptake and toxicity resides within kidney proximal tubules that also heavily express electrogenic sodium-glucose transporter-2 (SGLT2; SLC5A2) *in vivo*. We hypothesized that SGLT2 traffics gentamicin, and promotes cellular toxicity. We confirmed *in vitro* expression of SGLT2 in proximal tubule-derived KPT2 cells, and absence in distal tubule-derived KDT3 cells. D-glucose competitively decreased the uptake of 2-(N-(7-nitrobenz-2-oxa-1,3-diazol-4-yl)amino)-2-deoxyglucose (2-NBDG), a fluorescent analog of glucose, and fluorescently-tagged gentamicin (GTTR) by KPT2 cells. Phlorizin, an SGLT2 antagonist, strongly inhibited uptake of 2-NBDG and GTTR by KPT2 cells in a dose- and time-dependent manner. GTTR uptake was elevated in KDT3 cells transfected with SGLT2 (compared to controls); and this enhanced uptake was attenuated by phlorizin. Knock-down of SGLT2 expression by siRNA reduced gentamicin-induced cytotoxicity. *In vivo*, SGLT2 was robustly expressed in kidney proximal tubule cells of heterozygous, but not null, mice. Phlorizin decreased GTTR uptake by kidney proximal tubule cells in $Sglt2^{+/-}$ mice, but not in $Sglt2^{-/-}$ mice. However, serum GTTR levels were elevated in $Sglt2^{-/-}$ mice compared to $Sglt2^{+/-}$ mice, and in phlorizin-treated $Sglt2^{+/-}$ mice compared to vehicle-treated $Sglt2^{+/-}$ mice. Loss of SGLT2 function by antagonism or by gene deletion did not affect gentamicin cochlear loading or auditory function. Phlorizin did not protect wild-type mice from kanamycin-induced ototoxicity. We conclude that SGLT2 can traffic gentamicin and contribute to gentamicin-induced cytotoxicity.

Editor: Ines Armando, Universtiy of Maryland School of Medicine, United States of America

Funding: The colony-founding Sglt2+/− mice were a gift of the Wellcome Trust Sanger Institute (Hinxton, Cambridge, UK). This work was supported by NIH-NDCD grants R01 DC004555, R01 DC012588 (PSS), R03 DC011622 (HL), and P30 DC005983 [URL: https://www.nidcd.nih.gov/]. The funding agencies had no role in study design, data collection and analysis, preparation of the manuscript, or decision to publish.

Competing Interests: The authors have declared that no competing interests exist.

* Email: steygerp@ohsu.edu

Introduction

Aminoglycoside antibiotics, like gentamicin, are essential important clinically for treating critical gram-negative bacterial infections, and are frequently used worldwide [1,2]. Both infants and adults receive gentamicin for bacterial meningitis, endocarditis, septicemia and for prophylaxis in premature births and surgical cases. Unfortunately, the nephrotoxic and ototoxic side-effects of gentamicin therapy remain serious complications, limiting the clinical use of gentamicin [3]. Gentamicin-induced nephrotoxicity, characterized by proximal tubular necrosis without morphological changes in glomerular structures, can cause acute kidney failure and increased morbidity [4,5]. Acute renal toxicity is largely reversible because kidney tubule cells can proliferate to replace cells lost to aminoglycoside toxicity [6].

The mechanism of gentamicin-induced cytotoxicity is incompletely understood. Gentamicin can induce cell death mechanisms via mitochondrial damage and caspase activation [7–9], as well as the generation of toxic levels of reactive oxygen species [10,11]. Since it is difficult to inhibit the wide variety of cell death mechanisms that may be induced by gentamicin, an alternative strategy to prevent gentamicin-induced cytotoxicity is to block drug entry into cells. Gentamicin and other aminoglycosides are known to enter cells via at least two mechanisms: endocytosis and permeation through non-selective cation channels. In the kidney, the best characterized entry route for lumenal gentamicin is apical endocytosis and trafficking of gentamicin-laden endosomes to the Golgi complex and endoplasmic reticulum (ER) prior to release into the cytosol from the ER [12,13]. A non-endocytotic entry route for gentamicin into kidney cells has been demonstrated *in vitro* – via permeation of non-selective cation channels, presumptively transient receptor potential (TRP) channels [14,15]. Proximal tubule cells are presumed to be more pharmacologically sensitive to gentamicin because these cells take up and retain the drug. Distal tubule cells, however, are more resistant to gentamicin, most likely because they do not readily take up or retain gentamicin in the cytoplasm [14,16]. Another distinguishing feature is the abundant expression of sodium-glucose transporter-2 (SGLT2; a.k.a. SLC5A2) in proximal, but not distal, tubule cells [17,18].

SGLT2 is a low affinity, high capacity sodium-glucose electrogenic transporter of glycosides expressed in proximal tubules, and is responsible for ~90% of glucose resorption from the renal ultrafiltrate [18,19]. Antagonism of SGLT2 activity induces glycosuria [20,21] and aminoaciduria [22]. Aminoglycosides also induce glycosuria [23,24], and nephrotoxicity, predominantly within the proximal tubules [25]. The structure of SGLT2 resembles the major facilitator superfamily of transporters with a large, hydrophilic, elastic vestibule, an internal pore diameter of ~3 nm, and an exit pore (into cytosol) of ~1.5–2.5 nm [26,27], sufficiently large to potentially allow permeation by gentamicin. Non-lethal mutations in SGLT2 occur in humans, with little impact on kidney function besides glucosuria and aminoaciduria, with no reported loss of hearing acuity [22,28,29]. Several SGLT2 antagonists have been identified, including phlorizin, a hydrolyzable O-glucoside, several non-hydrolyzable antagonists including O-glycosides (sergliflozin [30], remogliflozin [20]) and C-glycosides (dapagliflozin [31], canagliflozin [21,32]). These non-hydrolyzable antagonists are being, or have been tested, to reverse Type II diabetes in mice [21,30,33] and humans [34].

We hypothesized that SGLT2 can traffic gentamicin into cells, and tested whether SGLT2 expression and was required for accelerated onset of gentamicin-induced toxicity in cell lines. If this hypothesis is correct, then loss of the SGLT2 function *in vivo* should reduce cellular uptake of gentamicin and protect against cytotoxicity. If so, this could potentially prevent nephrotoxicity and ototoxicity during gentamicin therapy.

Materials and Methods

Ethics Statement

The care and use of all animals reported in this study were approved by the Animal Care and Use Committee of Oregon Health & Science University (IACUC approval #IS00001801).

Conjugation and purification of GTTR

Gentamicin-Texas Red conjugate (GTTR) was produced as previously described [15,35–37]. Briefly, an excess of gentamicin (Sigma, MO, USA) in 0.1 M potassium carbonate (pH 10) was mixed with Texas Red (TR) succinimidyl esters (Invitrogen, CA) to minimize the possibility of over-labeling individual gentamicin molecules with more than one TR molecule, and to preserve the polycationic nature of the conjugate [38]. After conjugation, reversed phase chromatography, using C-18 columns (Grace Discovery Sciences, IL), was used to purify GTTR from unconjugated gentamicin, and potential contamination by un-reacted TR [39]. The purified GTTR conjugate was aliquoted, lyophilized, and stored desiccated, in the dark at −20°C until required.

Cell culture

The mouse kidney proximal tubule (KPT2) and distal tubule (KDT3) cell lines were generated and characterized as previously described [14,40]. These cell lines were maintained in DMEM with 10% FBS, without streptomycin or penicillin, at 37°C.

Competition and inhibition experiments

Cells plated on 8-well chambered coverslips were washed with DMEM twice and incubated as described below. To establish appropriate competition experiments, KPT2 cells were incubated with 0.4 mM 2-(N-(7-nitrobenz-2-oxa-1,3-diazol-4-yl)amino)-2-deoxyglucose (2-NBDG) (Life Technologies, NY, USA) without or with 1:1, 1:50 or 1:1000 molar ratios of D-glucose (Sigma, MO, USA) in DMEM. 2-NBDG (0.4 mM) was also incubated without or with 1:1, 1:10 or 1:50 molar ratios of phlorizin (Pfaltz & Baue, CT, USA) at 37°C for 20 mins. Phlorizin was solubilized in

Dimethyl sulfoxide (DMSO) (Final concentration of DMSO in buffer was <0.001%) prior to dilution in buffer to the required concentration. Cells were washed three times with PBS prior to fixation with 4% paraformaldehyde for 15 minutes.

To examine the effect of D-glucose on GTTR uptake by KPT2 cells, cells plated on chambered coverslips were washed with DMEM twice, co-incubated with GTTR (5 μg/mL, gentamicin base, gentamicin: ~450–477 g/mol; GTTR: ~1100 g/mol) and 1:40, 1:2000 or 1:40000 molar ratios of D-glucose for 20 minutes in DMEM at 37°C, with 5% CO$_2$, then washed with PBS three times to remove GTTR from extracellular media prior to fixation with 4% paraformaldehyde containing 0.5% Triton X-100 (FATX) for 15 minutes at room temperature. To examine the effect of phlorizin on the uptake of GTTR by KPT2 cells, cells were co-incubated with 5 μg/mL GTTR and 1:5, 1:10 or 1:20 molar ratios of phlorizin at 37°C for 20 mins respectively prior to washing and fixation.

To examine the effect of sodium on GTTR uptake by KPT2 cells, Na$^+$ free buffer was made up as follows: 140 mM choline chloride, 5 mM KCl, 2.5 mM CaCl$_2$, 1 mM MgSO4, 1 mM KH$_2$PO4, and 10 mM HEPES (pH 7.4); choline chloride was replaced 140 mM NaCl in Na$^+$ buffer. GTTR uptake experiments were performed described as above.

GTTR uptake and confocal microscopy

The cellular distribution of fluorescence was examined using a Bio-Rad 1024 ES scanning laser system. For each individual set of images to be compared, the same confocal settings were used, with two acquisition images per well, two wells per experimental condition, and each experiment performed at least three times to confirm consistency of experimental data. GTTR fluorescent pixel intensities were obtained by histogram function of the ImageJ software after removal of nuclei and intercellular pixels using Adobe Photoshop. Pixel intensities were statistically compared within each set of images per experiment, and not compared between replicate experiments due to varying acquisition settings to obtain the best dynamic range. To normalize data between experimental sets, the mean intensity was ratioed against the standard (e.g., GTTR only cells) and plotted [15].

Immunofluorescence

For immunolocalization of SGLT2, paraformaldehyde-fixed cells were washed in PBS, immunoblocked in 1% serum in PBS for 30 min and incubated with polyclonal anti-SGLT2 antisera (rabbit, Abcam, MA, USA; or goat, Santa Cruz Biotechnology, TX, USA) at room temperature for 1 hour. After washing with 1% serum in PBS, specimens were further incubated with 1:200 Alexa-488-conjugated goat-anti-rabbit or donkey anti-goat antisera (Invitrogen, CA) for 1 hour at room temperature, washed, post-fixed with 4% paraformaldehyde for 15 min, rinsed and mounted under coverslips with VectorShield (Vector Labs, CA). In vitro studies, when double-labeled for SGLT2 plus GTTR, cells were permeabilized by 0.5% Triton X-100 after immunolabeling.

Immunoblotting

Kidney and cochlear tissues were analyzed by immunoblot as described before [41–44]. Briefly, total protein extracts were prepared by homogenizing tissues in T-PER tissue protein extraction buffer (Thermo Scientific, IL, USA) with protease inhibitor (Sigma, MO, USA), and the total protein concentration determined using the bicinchoninic acid (BCA) assay. Protein samples (100 μg) were separated by 4–20% pre-cast polyacrylamide gel (Bio-Rad, CA, USA), transferred to polyvinylidene difluoride membranes (Millipore Corporation, MA, USA), blocked with 5% non-fat milk and then incubated at 4 C overnight with goat (1:50; Santa Cruz, CA, USA) or rabbit (1:50; Abcam, MA, USA) polyclonal antibodies against SGLT2 in 5% non-fat milk. Rabbit polyclonal antibodies against anti-actin (1:1000; Sigma, MO, USA) were also used as an internal standard. Peroxidase-conjugated anti-goat (1:1000) or anti-rabbit (1:2500) antisera were used to localize primary antisera and visualized using an ECL-Plus detection kit (Thermo Scientific, IL, USA), documented with a photoscanner and analyzed with the Fiji (freeware) program.

KDT3-SGLT2 cell line generation

Mouse SGLT2 cDNA from Open Biosystems (Clone ID: 4235707) was amplified by PCR, using primers 5′-TTT GAA TTC GCC ACC ATG GAG CAA CAC GTA GAG-3′ and 5′-CCC GTC GAC TTA TGC ATA GAA GCC CCA GAG-3′, digested with EcoRI/SalI, and subcloned into pBabe-puro vector. The resultant plasmid was transfected into Phoenix Eco packaging cell using Lipofectamine 2000. After 48 hours, the retrovirus-containing medium was collected, diluted (1:500) with growth medium and added to mouse kidney distal tubule KDT3 cells in DMEM with 10% FBS. Culture medium was changed again after 24 h and puromycin was added at 2.5 μg/ml to select for retrovirus-infected cells. From dozens of surviving cells after several days of puromycin treatment, several clones were selected, expanded and used for GTTR uptake experiments as described above. Puromycin was not applied during GTTR uptake experiments.

Transfection and cell viability measurement

Cell viability was determined by the reduction of 3-(4,5-dimethylthiazol-2-yl)-2,5-diphenyltetrazolium bromide (MTT), an indicator of mitochondrial dehydrogenase activity, as previously described [40,45]. Briefly, KPT2 cells were plated at 3000 cells per well in a 96-well plate. After incubation overnight to allow cells to attach to the plate, cells were treated with small interfering RNA (siRNA) and control for SGLT2 (Invitrogen, CA). Transfection of siRNA was performed using Lipofectamine RNAiMAX (Invitrogen, CA). After 48 hours, transfected cells were treated with gentamicin (5 or 10 mM) in DMEM (10% FBS) for 1, 2 or 3 days. Subsequently, 20 μl of 5 mg/ml MTT solution was added to each well, and cells incubated for 4 h at 37°C, 5% CO_2. Culture medium was then replaced with 200 μl DMSO in each well and the optical density recorded at 540 nm with background subtraction at 660 nm. Student's t-test was used for statistical analysis [40,45].

Mice

$Sglt2^{+/-}$ mice were obtained from the Wellcome Trust Sanger Institute (Hinxton, Cambridge, UK), and an in-house colony established from these founders. Homozygous mice were generated either by crossing heterozygotes together or by crossing heterozygotes with homozygotes. Littermates of wild-type and heterozygotes served as controls. A PCR-based genotyping method was used to identify mutant and wild-type alleles. The mutant allele was identified using primers Slc5a2_55706_F: 5′-AGC AGG AGG GTT CAG GCA GG -3′ and CAS_R1_term_x: 5′-TCG TGG TAT CGT TAT GCG CC -3′ (172-bp product). The wild type allele was identified using primers Slc5a2_55706_F and Slc5a2_55706_R: 5′-TTT TGC GCG TAC AGA CCA TC -3′ (412-bp product).

Mice (21–28 days old) received an intra-peritoneal (i.p.) injection of 800 mg/kg phlorizin (200 μg/μl phlorizin in 40% DMSO, pH 7.4; 4 μl/g), or the vehicle alone. Thirty minutes later, mice received an i.p. injection of 2 mg/kg GTTR (in sterile PBS, pH 7.4). After a further 30 minutes, cardiac serum was collected from deeply-anesthetized mice prior to cardiac perfusion with PBS, then 4% paraformaldehyde. Kidneys and cochleae were excised and post-fixed in FATX, and processed for immunofluorescence [39].

Determination of gentamicin levels in serum and in cells

Serum levels of gentamicin and the gentamicin epitope of GTTR were determined via enzyme-linked immunosorbent assay (ELISA). Serum supernatant was further diluted, centrifuged and protein extracted as needed for ELISA. Measurement of total gentamicin levels in serum was determined according to the manufacturers' instructions (EuroProxima, Arnhem, the Netherlands).

For cellular levels of gentamicin or the gentamicin epitope of GTTR, KPT2 cells were plated in 60 mm dishes and incubated at 37°C, with 5% CO_2 overnight. After washing with DPBS, cells were incubated in 5 μg/mL GTTR or 1 mM gentamicin and 1:5, 1:10 or 1:20 molar ratios of phlorizin respectively, as described above. After 20 minutes, cells were washed with DPBS three times and proteins extracted. The quantity of cell protein was measured by BCA protein assay kit. Gentamicin ELISAs were performed described as above.

Auditory testing

ABR thresholds to pure tones were obtained to evaluate hearing function. Wild-type, $Sglt2^{+/-}$ and $Sglt2^{-/-}$ mice were anesthetized and placed on a heating pad in a sound-proof, electrically isolated chamber. Needle electrodes were placed subcutaneously below the test ear, at the vertex, and with a ground on the claw. Each ear was stimulated separately with a closed tube sound delivery system sealed into the ear canal. The auditory brain-stem response to a 1-ms rise-time tone burst at 4, 8, 16, 24, and 32 kHz was recorded. Threshold was defined as an evoked response of 0.2 mV [46,47]. ABR thresholds were obtained both before and 30 minutes after phlorizin treatment in wild-type mice, and in $Sglt2^{+/-}$ and $Sglt2^{-/-}$ mice 6 and 12 weeks of age. In addition, ABRs were also obtained before and after aminoglycoside treatment.

Toxicity studies

Since immunofluorescence may not detect SGLT2 in the cochlea, toxicity studies with aminoglycosides in the presence or absence of phlorizin were conducted. Dosing with gentamicin to induce ototoxicity in vivo causes systemic toxicity in mice, therefore wild-type mice were treated with a similar aminoglycoside - kanamycin in the presence or absence of phlorizin [48]. ABR thresholds were obtained before kanamycin dosing. Four groups of mice were used: group 1, sterile Dulbecco's PBS (DPBS) only in the same delivery routes as for subsequent groups; group 2, 800 mg/kg kanamycin in DPBS twice daily, subcutaneously, for 14 days; group 3, 800 mg/kg kanamycin in DPBS twice daily, subcutaneously, plus DMSO vehicle only, i.p., for 14 days; group 4, 800 mg/kg kanamycin in DPBS twice daily, subcutaneously, plus phlorizin (100 mg/kg in DMSO, i.p.) for 14 days. Phlorizin was injected twice daily 15 minutes prior to each kanamycin injection. Subsequently, mice were allowed to recover for 3 weeks before final ABR thresholds were obtained to determine any permanent ABR threshold shift and mice euthanized.

Statistics

All in vitro experiments were performed multiple times to validate the observations, with the data expressed as means ± SEM. Statistical analysis was conducted using the nonparametric t test for comparison of 2 groups or ANOVA for comparisons of 3 groups (GraphPad Prism). For *in vivo* experiments, cytoplasmic GTTR fluorescence in kidney proximal tubules was compared between phlorizin treatment and control group. ABR thresholds (or threshold shifts) at each tested frequency were compared between $Sglt2^{-/-}$ mice and control mice, or between treatment groups in the kanamycin toxicity study, by nonparametric t-test (GraphPad Prism). A confidence level of 95% was considered statistically significant. $*p<0.05$ and $**p<0.01$.

Results

Uptake of a fluorescent glucose analog, 2-NBDG, was inhibited by phlorizin, an SGLT2 antagonist

We verified the presence (or absence) of SGLT2 immunoexpression in previously-characterized murine KPT2 and KDT3 cell [14]. SGLT2 was specifically immunolocalized at the periphery of KPT2 cells, but not KDT3 cells (Fig. 1A, B, respectively), presumptively at the cell membrane. Cellular uptake of the fluorescent glucose analog, 2-NBDG, is mediated by both SGLTs and also by facilitated glucose transporters (GLUTs) [49]. In KPT2 cells, 2-NBDG fluorescence was primarily localized at the cell periphery (Fig. 1C). Increasing concentrations of D-glucose (Fig. 1 C–F, K), or the SGLT2 antagonist phlorizin (Fig. 1 G–J, L), dose-dependently reduced 2-NBDG fluorescence in KPT2 cells. This demonstrated the presence of robust SGLT2 activity in KPT2 cells.

SGLT2-mediated uptake of GTTR by KPT2 cells can be competitively inhibited

GTTR is a fluorescently-tagged gentamicin conjugate used to visually test for gentamicin permeation of non-selective cation channels into cells [15,50–53]. We used phlorizin, an SGLT2 antagonist [54] or D-glucose to test whether SGLT2 was potentially GTTR-permeation. Increasing doses of phlorizin (Fig. 2 A–E) and D-glucose (Fig. 2 F, Fig. S1) significantly decreased GTTR fluorescence in KPT2 cells. We then used ELISA technology to verify the imaging data, and found that phlorizin reduced both GTTR and native gentamicin uptake by KPT2 cells in a dose-dependent manner (Fig. 2 G, H), validating GTTR as a tracer for gentamicin studies. Thus, SGLT2-mediated uptake of GTTR by KPT2 cell can be antagonized or competitively-inhibited.

SGLT2-mediated uptake of GTTR by KPT2 cells can be inhibited by Na$^+$ free buffer

SGLT2 is a Na$^+$-ligand symporter [55]. We examined whether GTTR uptake by KPT2 cells was attenuated in Na$^+$-free buffer after 5, 10 or 20 minutes at 37°C. GTTR fluorescence in KPT2 cells in Na$^+$-free buffer was significantly attenuated (~20%) after 20 minutes (Fig. 3 A). We used phlorizin to further verify this data over time. Phlorizin also significantly inhibited ~20% GTTR uptake of KPT2 cells, most consistently at the 20 minute timepoint (Fig. 3 B), and this timepoint was chosen for the majority of subsequent experiments. Thus, SGLT2 accounts for ~20% of total GTTR uptake in SGLT2-expressing KPT2 cells.

Enhanced GTTR uptake by KDT3 cells heterologously expressing SGLT2

To test if SGLT2 can enhance cellular uptake of GTTR, stable cell lines expressing SGLT2 were generated using KDT3 cells that do not endogenously express SGLT2 (Fig. 1 B). KDT3-derived cell lines expressing SGLT2 (KDT3-SGLT2) and empty vector

control cell lines (KDT3-pBabe) retained the parental KDT3 morphology (Fig. S2). Immunofluorescence revealed expression of SGLT2 in most KDT3-SGLT2 cells, with negligible immunofluorescence for SGLT2 in KDT3-pBabe cells (Fig. 4 A, G and D, J respectively).

Following a 20 minute incubation with GTTR, robust GTTR uptake was present in KDT3-SGLT2 cells immunolabeled for SGLT2, but not in control KDT3-pBabe cells lacking SGLT2 immunofluorescence (Fig. 4 B, E). Pixel intensity analysis revealed statistically significant increases in GTTR fluorescence within KDT3-SGLT2 cells compared to that of control KDT3-pBabe cells (Fig. 4 M). In the presence of phlorizin (100 µg/ml), GTTR fluorescence in KDT3-SGLT2 cells was significantly less than in KDT3-SGLT2 cells treated without phlorizin (Fig. 4 B, H, respectively). In addition, GTTR fluorescence in phlorizin-treated KDT3-SGLT2 cells was not significantly different to KDT3-pBabe cells with or without phlorizin treatment (Fig. 4 M), demonstrating the specificity of phlorizin for SGLT2 in these cells. Thus, exogenous expression of SGLT2 in KDT3 cells facilitated GTTR uptake.

Knock-down of SGLT2 reduced gentamicin-induced cytotoxicity

To test whether SGLT2 contributes to gentamicin-induced cytotoxicity, we transfected KPT2 cells with siRNA for SGLT2 to knock-down protein expression of SGLT2 prior to drug exposure. Immunofluorescence confirmed that SGLT2 siRNA reduced SGLT2 expression compared to control siRNA-transfected cells (Fig. 5 A–H). The effect of SGLT2 siRNA was apparent 1 day after transfection and further reduced SGLT2 expression 2 days after transfection (Fig. 5 A–D). This knock-down of SGLT2 expression lasted at least 5 days (Fig. 5 A–H). Two days after transfection with control or SGLT2 siRNA, KPT2 cells were treated with gentamicin (5 mM or 10 mM) for 1, 2 or 3 days, prior to MTT assay for cell viability [40,45]. Control and SGLT2 siRNA-transfected cells showed no difference in viability (Fig. S3), demonstrating that loss of SGLT2 did not affect cell viability. Although gentamicin reduced cell viability, SGLT2 knock-down attenuated the degree of gentamicin-induced toxicity, most significantly at 2 or 3 days of gentamicin treatment (Fig. 5 I). Thus, KPT2 cells with SGLT2 expression were more susceptible to gentamicin-induced cytotoxicity than KPT2 cells with SGLT2 knock-down, suggesting that SGLT2 trafficking of gentamicin contributed to gentamicin-induced cytotoxicity.

Immunoexpression of SGLT2 in renal and cochlear tissues *in vivo*

To determine if SGLT2 is appropriately located for gentamicin uptake *in vivo*, the immunoexpression of SGLT2 was characterized in fixed renal proximal tubules and cochleae *in situ* using two different antibodies. In the kidney, as previously described [17,18,56], SGLT2 was immunolocalized at the apical brush border membranes of wild-type renal proximal tubule cells, but not in adjacent distal tubule regions (Fig. 6 A, C), nor in the kidney of $Sglt2^{-/-}$ mice (Fig. 6 B, D). In the cochlea, the rabbit anti-SGLT2 antibody did not label wild-type marginal cells above background (Fig. 6 E), or exhibited non-specificity in marginal cells of $Sglt2^{-/-}$ mice (Fig. 6 F). The goat anti-SGLT2 antibody consistently exhibited non-specific fluorescence in marginal cells of $Sglt2^{-/-}$ mice (Fig. 6 H) similar to that observed in wild-type marginal cells (Fig. 6 G). Both SGLT2 antibodies consistently exhibited a punctate labeling pattern within the intra-stria vascularis (Fig. 6 I, K) that was not present in $Sglt2^{-/-}$ mice

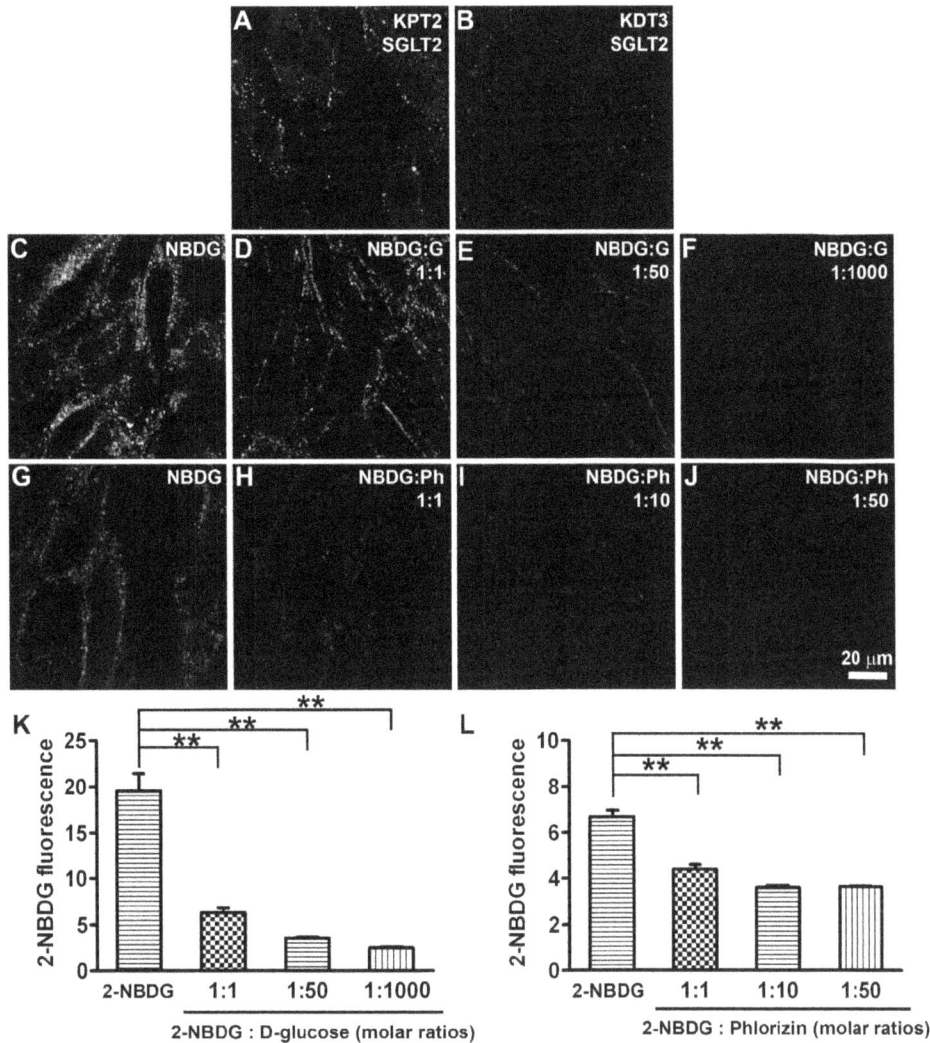

Figure 1. Uptake of the fluorescent glucose analog 2-NBDG is mediated by SGLT2 in KPT2 cells. KPT2 cells (A) had robust SGLT2 immunolabeling compared to KDT3 cells (B). Increasing doses of (C–F) D-glucose (molar ratios of 1:0, 1:1, 1:50 or 1:1000 [2-NBDG/D-glucose]), or (G–J) phlorizin (molar ratios of 1:0, 1:1, 1:10 or 1:50 [2-NBDG/phlorizin]) dose-dependently decreased 2-NBDG fluorescence in KPT2. Scale bar = 20 μm. (K, L). The fluorescence intensity of 2-NBDG in KPT2 cells was significantly decreased with increasing doses of D-glucose (K) or phlorizin (L; $**p < 0.01$).

(Fig. 6 G, L). Immunoblotting revealed SGLT2 protein expression in kidneys of wild-type and $Sglt2^{+/-}$ mice, but not in $Sglt2^{-/-}$ mice (Fig. 6 M). Immunoblotting of wild-type cochlear tissues detected actin, but not SGLT2 (data not shown), indicative of the low level expression of SGLT2 protein in cochlear tissues. PCR-based genotyping demonstrated the absence of wild-type alleles in $Sglt2^{-/-}$ mice (Fig. 6 N).

Phlorizin decreased renal uptake and increased serum levels of GTTR *in vivo*

Since phlorizin had no effect on the bactericidal activity of gentamicin on *E. coli* by disk diffusion assay (Table S1), the *in vitro* data suggested that phlorizin may decrease cellular uptake of GTTR *in vivo* and potentially reduce aminoglycoside-induced cytotoxicity *in vivo*. To test whether phlorizin decreased cellular uptake of GTTR *in vivo*, the intensity of cytoplasmic GTTR fluorescence was determined in proximal tubule cells of $Sglt2^{+/-}$ and $Sglt2^{-/-}$ mice. In $Sglt2^{+/-}$ mice, rabbit anti-SGLT2 immunolabeling was co-localized with GTTR fluorescence in

proximal, but not distal tubule cells (Fig. 7 A–C). GTTR fluorescence was diffusely distributed throughout the cytoplasm, and intensely localized at the brush border of proximal tubule cells of $Sglt2^{+/-}$ mice that received GTTR plus vehicle only *in vivo* (Fig. 7 D). Phlorizin pre-treatment visibly reduced cytoplasmic GTTR fluorescence within proximal tubule cells and at their brush border (Fig. 7 E, F). In $Sglt2^{-/-}$ mice, unexpectedly, GTTR fluorescence was diffusely distributed throughout the cytoplasm, and intense fluorescence at the brush border of proximal tubule cells (Fig. 7 G), as observed in untreated $Sglt2^{+/-}$ mice (Fig. 7 D). Phlorizin had no significant effect on the intensity (uptake) or distribution of GTTR fluorescence in proximal tubule cells of $Sglt2^{-/-}$ mice (Fig. 7 H, I). Thus, SGLT2 is not required for renal proximal tubule uptake of GTTR in $Sglt2^{-/-}$ mice, although GTTR uptake by these cells can be acutely inhibited by the SGLT2 antagonist, phlorizin, in $Sglt2^{+/-}$ mice.

In $Sglt2^{+/-}$ mice, phlorizin increased serum levels of both gentamicin and GTTR levels compared to vehicle-treated mice (Fig. 7 J, K). In $Sglt2^{-/-}$ mice, phlorizin had no effect on serum levels of gentamicin or GTTR (Fig. 7 J, K). Gentamicin and

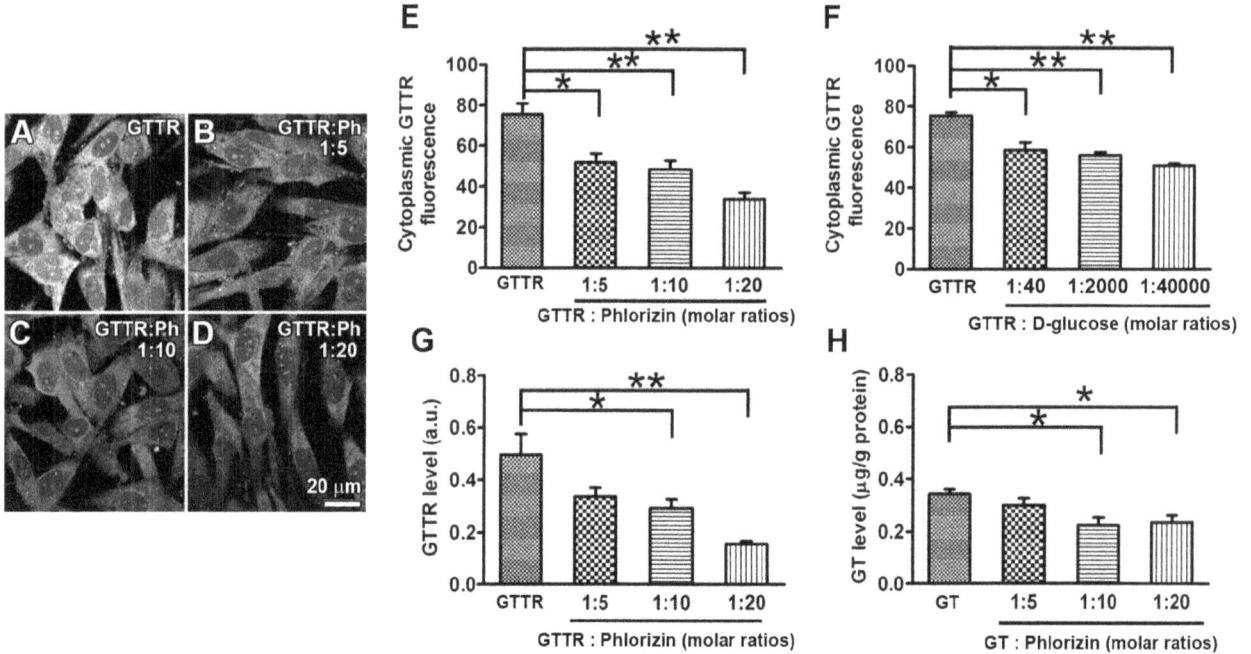

Figure 2. SGLT2-mediated uptake of GTTR can be competitively inhibited. (A–D) Cells were treated with 5 μg/ml GTTR for 20 minutes at 37°C with a dose-range of phlorizin (molar ratios of 1:0, 1:5, 1:10 or 1:20 [GTTR:phlorizin]) in DMEM buffer. Scale bar = 20 μm. Increasing doses of (E) phlorizin or (F) D-glucose (molar ratios of 1:0, 1:40, 1:2000 or 1:40000 [GTTR:D-glucose]) reduced GTTR fluorescence in KPT2 cells (*$p < 0.05$; **$p < 0.01$). Cell ELISAs demonstrated that (G) GTTR or (H) gentamicin levels in KPT2 cells are decreased by increasing doses of phlorizin.

GTTR serum levels in $Sglt2^{-/-}$ mice were significantly higher than in $Sglt2^{+/-}$ mice (Fig. 7 J, K). Thus, loss of SGLT2 function, by antagonism, or by gene deletion, increases serum levels of gentamicin.

Phlorizin did not affect cochlear uptake of GTTR or auditory function

To test whether the low levels of SGLT2 immunofluorescence in the cochlea (Fig. 6) were required for cochlear uptake of GTTR, we examined whether phlorizin modulated the distribution of GTTR in murine cochleae of $Sglt2^{+/-}$ and $Sglt2^{-/-}$ mice. Mice were injected with GTTR 30 minutes after phlorizin or vehicle injection. In the stria vascularis, GTTR was characteristically localized in marginal and intermediate cells (Fig. 8) as previously described [35]. The nucleoplasm of marginal cell nuclei

displayed negligible fluorescence (Fig. 8 A, C, E, G), as expected. There were no significant differences in the uptake or distribution of GTTR fluorescence between phlorizin- and vehicle-treated groups of $Sglt2^{+/-}$ or $Sglt2^{-/-}$ mice (Fig. 8 A–H). As observed in the kidney, SGLT2 was not required for cochlear uptake of GTTR in $Sglt2^{-/-}$ mice, and this uptake could not be inhibited by acute exposure to phlorizin in either $Sglt2^{+/-}$ or $Sglt2^{-/-}$ mice.

In wild-type mice, no statistically significant changes in auditory brainstem response (ABR) thresholds were observed after intraperitoneal (i.p.) injection with 800 mg/kg phlorizin or vehicle [DMSO; Fig. S4]. Furthermore, $Sglt2^{-/-}$ mice displayed no significant differences in ABR thresholds at 6 or 12 weeks of age compared to wild-type or $Sglt2^{+/-}$ (Fig. 8 I, J), and gender differences were minimal (Fig. S5). Thus, auditory function was not affected by phlorizin antagonism, or genomic loss of functional SGLT2.

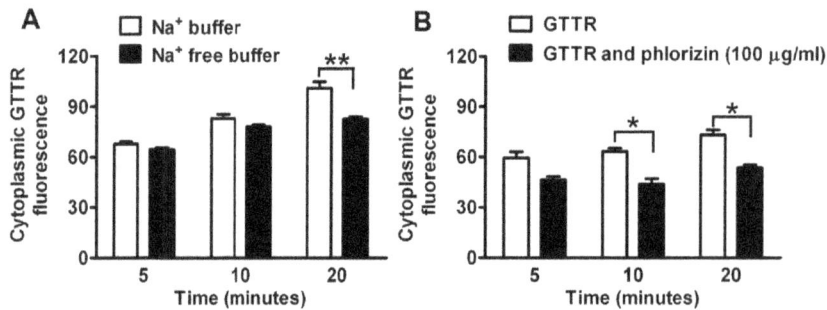

Figure 3. SGLT2-mediated uptake of GTTR by KPT2 cells was inhibited by Na⁺ free buffer. (A) KPT2 cells were incubated with GTTR for 5 minutes, 10 minutes or 20 minutes at 37°C in Na⁺ free buffer or Na⁺ buffer. GTTR fluorescence of KPT2 cell in Na⁺ buffer for 20 minutes was more intense than in Na⁺ free buffer (**$p < 0.01$). (B) KPT2 cells were treated with GTTR and phlorizin in DMEM buffer. GTTR uptake by KPT2 cells was also inhibited by phlorizin (100 μg/ml) over time (*$p < 0.05$).

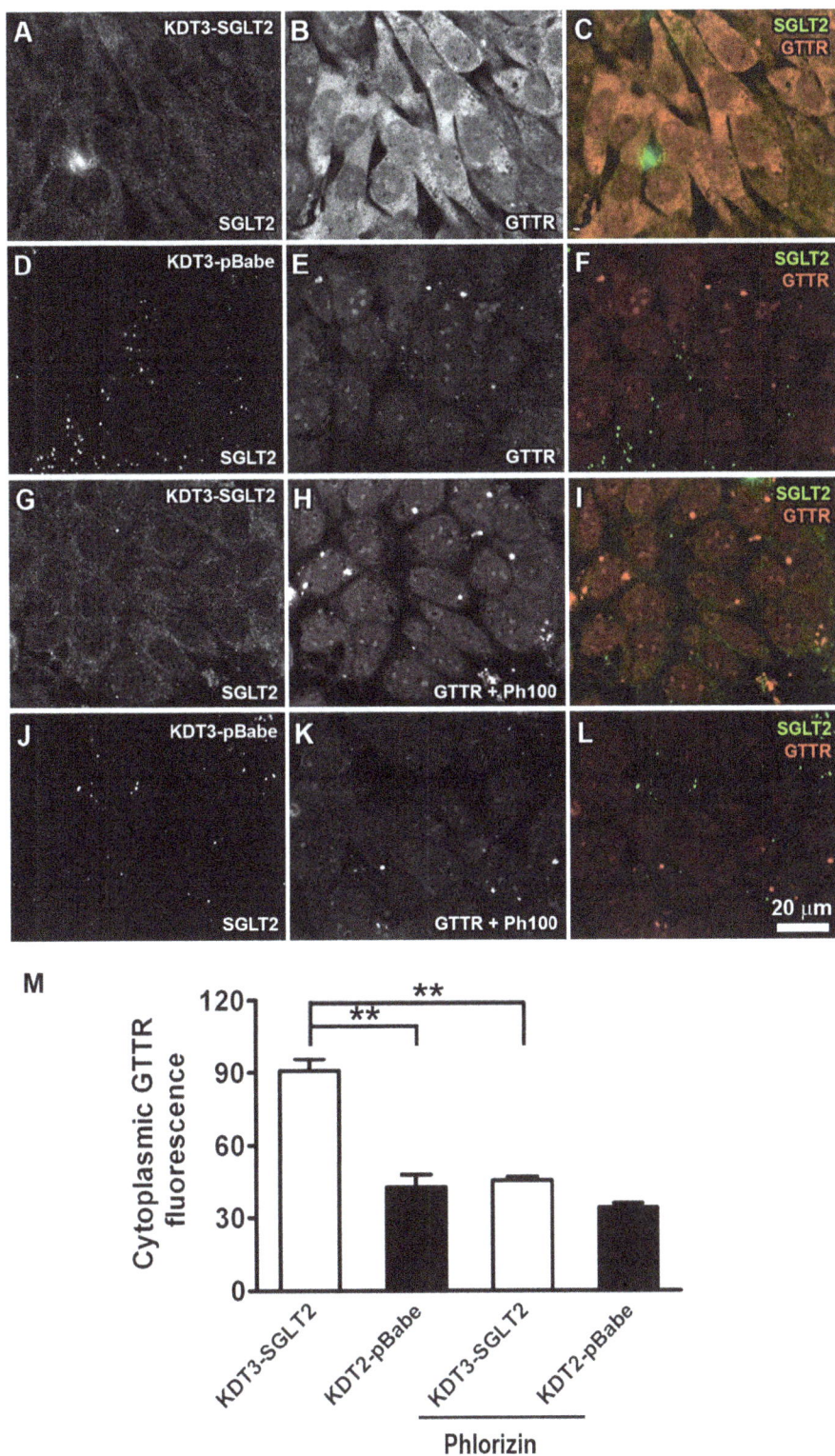

Figure 4. Heterologous expression of SGLT2 in KDT3 cells increased cellular uptake of GTTR. (A–C) KDT3-SGLT2 cells with positive SGLT2 immunofluorescence displayed robust GTTR uptake (B, C). (D–F) Empty vector control clones (KPT2-pBabe) showed negligible SGLT2 immunofluorescence (D) and weak, uniform levels of GTTR fluorescence (E, F) compared to (B, C). (H, I) GTTR fluorescence in KDT3-SGLT2 cells in the presence of phlorizin (100 μg/ml) was visibly less intense than in KDT3-SGLT2 cells without phlorizin treatment (B, C). (K, L) GTTR fluorescence in phlorizin-treated KDT3-pBabe cells showed weak levels of GTTR fluorescence as untreated in KDT3-pBabe cells (E, F). Scale bar = 20 μm. (M) Fluorescence intensities of GTTR in KDT3-SGLT2 or KDT3-pBabe cells in the presence or absence of phlorizin (100 μg/ml; **$p < 0.01$).

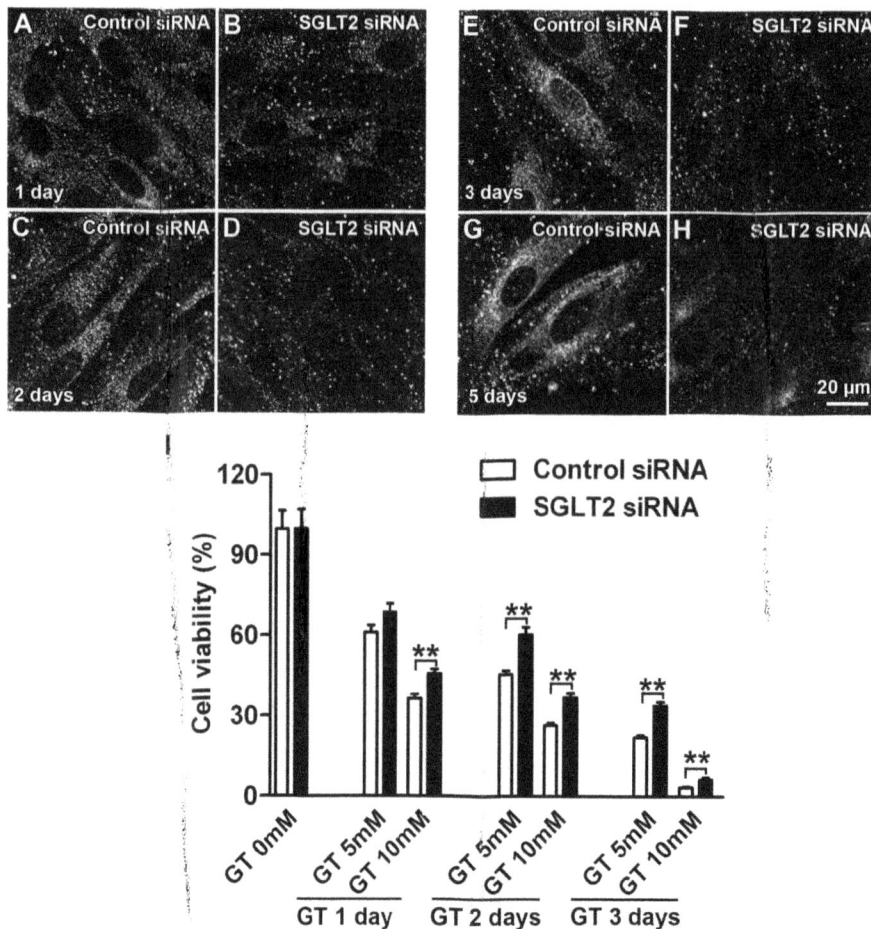

Figure 5. Knockdown of SGLT2 reduced gentamicin-induced cytotoxicity. (A–H) KPT2 cells transfected with siRNA for SGLT2 showed reduced immunoexpression of SGLT2 compared with cells transfected with control siRNA. (A–D) The effect of SGLT2 siRNA began within 1 day of transfection and was most apparent 2 days of transfection. (A–H) The effect SGLT2 siRNA tranfection lasted for at least 5 days. (I) MTT assay on cells (2-days post-transfection) treated with gentamicin for 1, 2 or 3 days revealed greater viability of SGLT2 siRNA-transfected KPT2 cells compared with KPT2 cells treated with control siRNA (**$p < 0.01$).

Toxicity studies

Dosing wild-type mice with gentamicin *in vivo* causes systemic toxicity prior to induction of ototoxicity [48]. To test whether chronic phlorizin exposure ameliorates aminoglycoside cochleotoxicity as assessed by ABR threshold shifts, toxicity studies with another aminoglycoside, kanamycin, using a well-established protocol [48], in the presence or absence of phlorizin were conducted in wild-type mice. ABR threshold shifts were obtained before and after kanamycin dosing. In mice treated with just DPBS, insignificant threshold shifts were observed 3 weeks after dosing (Fig. 9; Fig. S6). In mice treated with kanamycin in DPBS, threshold shifts were observed at 32 kHz that were statistically significant compared with the DPBS-only group (Fig. 9; Fig. S6). In mice treated with kanamycin plus DMSO (vehicle for phlorizin), threshold shifts were observed at 16 and 32 kHz compared with the DPBS-only group, however, these thresholds shifts were significantly different only at 32 kHz (Fig. 9; Fig. S6). Mice treated with kanamycin plus phlorizin had statistically significant threshold shifts at 4, 8, 16 and 32 kHz compared with the DPBS-only group (Fig. 9; Fig. S6). The kanamycin plus phlorizin group also had threshold shifts were significantly different at 4, 8 and 16 kHz compared to the kanamycin in DPBS group (Fig. 9; Fig. S6). No significant differences in

threshold shifts were observed between the kanamycin plus phlorizin and kanamycin plus DMSO groups (Fig. 9; Fig. S6). Thus, phlorizin did not protect auditory function from kanamycin-induced ototoxicity, and unexpectedly exacerbated drug-induced hearing loss at lower frequencies.

Discussion

Here we report evidence, for the first time, that the electrogenic sodium-glucose transporter SGLT2 contributes to the cellular uptake of aminoglycosides, particularly by proximal tubule cells that highly express SGLT2 [17,18,56]. *In vitro*, SGLT2-mediated uptake of 2-NBDG and GTTR was inhibited by phlorizin and D-glucose. Cellular uptake of GTTR was enhanced by heterologous expression of SGLT2 in KDT3 cells. Knock-down of SGLT2 expression by siRNA reduced gentamicin-induced cytotoxicity in KPT2 cells endogenously expressing SGLT2, further suggesting SGLT2 involvement in cellular uptake of gentamicin and subsequent cytotoxicity. *In vivo*, we observed SGLT2 immunoexpression at the apical brush border region of kidney proximal tubule cells, and phlorizin pre-treatment can acutely inhibit GTTR uptake by proximal tubules in *Sglt2*$^{+/-}$ mice. Loss of SGLT2 function increased serum levels of gentamicin and GTTR.

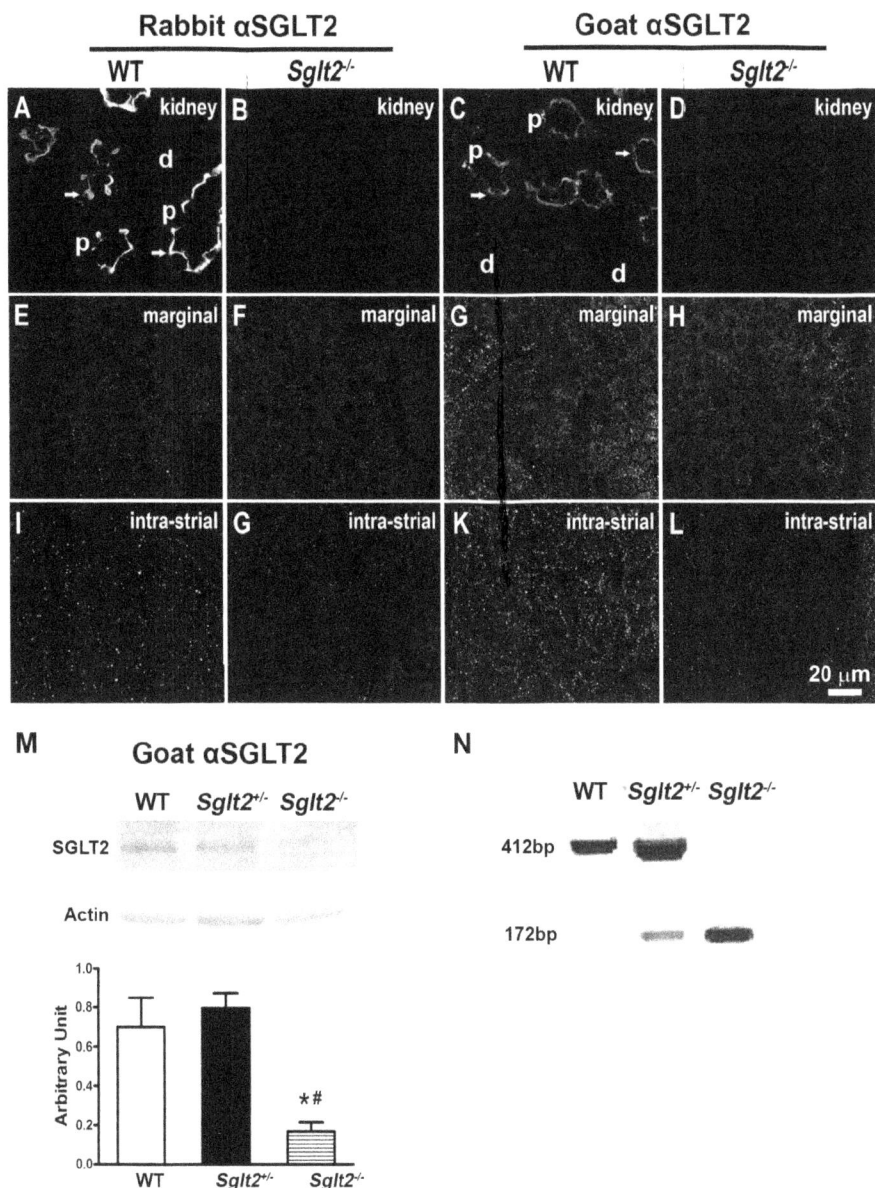

Figure 6. SGLT2 immunofluorescence in the kidney and cochlea. Two different SGLT2 antibodies were used, a rabbit polyclonal IgG to synthetic peptide derived from residues 250–350 of human SGLT2 and a goat polyclonal IgG against a murine peptide sequence within the N-terminal extracellular domain of SGLT2. (A, C) In wild-type mice, SGLT2 was immunolocalized at the apical membranes (arrows) of proximal tubules (p), but not in adjacent glomerular (not shown) or distal tubule (d) regions. (B, D) In $Sglt2^{-/-}$ mice, no immunoexpression for renal SGLT2 was observed with either antibody. (E, F) No labeling above background was observed in cochlear marginal cells of wild-type or $Sglt2^{-/-}$ mice with rabbit antisera for SGLT2. (G, H) Goat antisera for SGLT2 produced labeling patterns in cochlear marginal cells of both wild-type mice and $Sglt2^{-/-}$ mice, suggestive if substantial non-specificity in this cell type. (I, K) In the intra-strial layer of wild-type mice, predominantly composed of both marginal and intermediate cells, both antisera exhibited a punctate labeling pattern not observed in $Sglt2^{-/-}$ mice (J, L). Scale bar = 20 µm. (M) Immunoblotting with the goat antibody for SGLT2 revealed SGLT2 protein expression in wild-type and $Sglt2^{+/-}$ mice, but not $Sglt2^{-/-}$ mice. The ratio of SGLT2 to actin expression in kidney tissues of wild-type and $Sglt2^{+/-}$ mice were significantly higher than that in $Sglt2^{-/-}$ mice. There was no statistical difference in SGLT2 protein expression between wild-type and $Sglt2^{+/-}$ mice. (N) Genotyping demonstrated the absence of wild-type SGLT2 alleles in $Sglt2^{-/-}$ mice.

However, loss of SGLT2 function by phlorizin or gene knockout did not affect auditory function or cochlear uptake of GTTR. Phlorizin treatment exacerbated drug-induced hearing loss at lower frequencies.

Aminoglycosides enter cells, including kidney proximal tubule cells, via an endocytosis pathway [13,57–60]. However, aminoglycoside exposure also generates reactive oxygen species within

seconds in euthermic cells at room temperature, precluding endocytosis [61,62]. Aminoglycosides also enter cells via non-selective cation channels, including TRP channels and the TRP-like mechanoelectrical transduction channels of sensory hair cells in the inner ear [15,35,51,53,63]. Identifying the molecular mechanisms by which aminoglycosides can enter cells, and particularly kidney proximal tubule and cochlear cells, is crucial

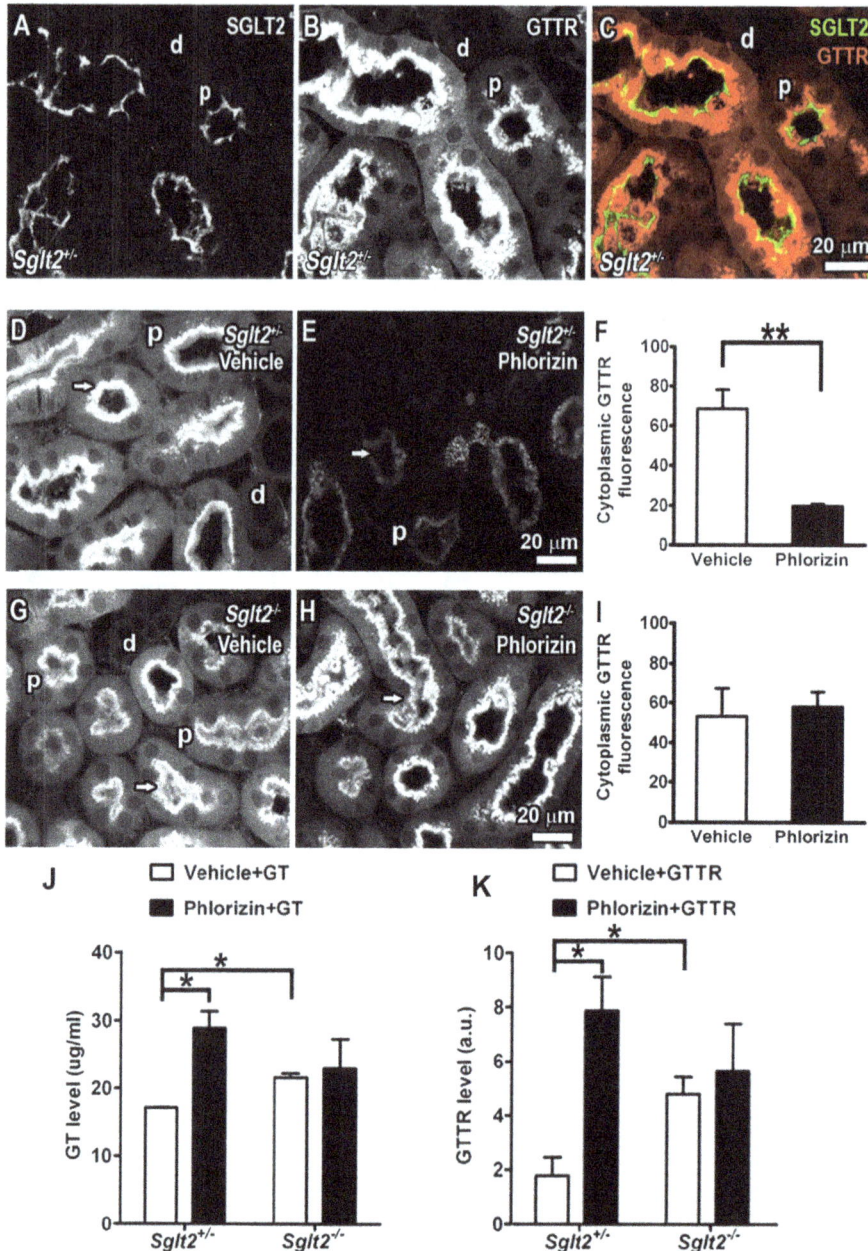

Figure 7. Phlorizin decreased renal GTTR uptake, and increased serum drug levels *in vivo*. (A) In *Sglt2*$^{+/-}$ mice, rabbit anti-SGLT2 immunolabeling was predominantly localized at the apical, lumenal region of proximal tubules (p), with negligible labeling in distal tubules (d). (B) GTTR fluorescence was most intense (as saturated puncta) in the apical region of proximal tubules (p), with less intense diffuse labeling in the cytoplasm of these same cells. Very weak and only diffuse GTTR fluorescence was observed in the cytoplasm of distal tubule cells (d). (C) Merged image showing colocalization of SGLT2 (green) and GTTR (red) in proximal tubules. (D–F) When *Sglt2*$^{+/-}$ mice were pre-treated with phlorizin, significantly reduced GTTR fluorescence was observed in the cytoplasm and apical brush border (arrows) of proximal tubule cells (E) compared to untreated mice (D, F; **$p<0.01$). (G, I) In *Sglt2*$^{-/-}$ mice, GTTR fluorescence was diffusely distributed throughout the cytoplasm of proximal tubule cells, with intense fluorescence at the apical brush border. (H, I) Phlorizin had no effect on the uptake, distribution or intensity of GTTR fluorescence in *Sglt2*$^{-/-}$ proximal tubule cells (**$p<0.01$). Scale bar = 20 μm. (J, K) In *Sglt2*$^{+/-}$ mice, phlorizin pre-treatment significantly increased both gentamicin and GTTR serum levels compared to vehicle treated control mice (*$p<0.05$). In *Sglt2*$^{-/-}$ mice, phlorizin did not significantly change gentamicin or GTTR serum levels. However, serum levels of gentamicin or GTTR serum level were significantly higher in *Sglt2*$^{-/-}$ mice than in *Sglt2*$^{+/-}$ mice in the absence of phlorizin treatment (*$p<0.05$).

to develop effective strategies to protect these pharmacologically-sensitive cells during clinically-essential gentamicin pharmacotherapy.

In the kidney, glomerular filtrate has ~150 mM Na$^+$ compared to intracellular levels of 12 mM, driving the inward electrogenic

activity of Na$^+$-ligand symporters present on the lumenal (apical) membrane of proximal tubule cells [55]. Proximal tubule cells also express high levels of SGLT2, a Na$^+$-ligand symporter that traffics glycosides like glucose, facilitating the renal resorption of 90% of lumenal glucose from glomerular filtrate in promixal tubules

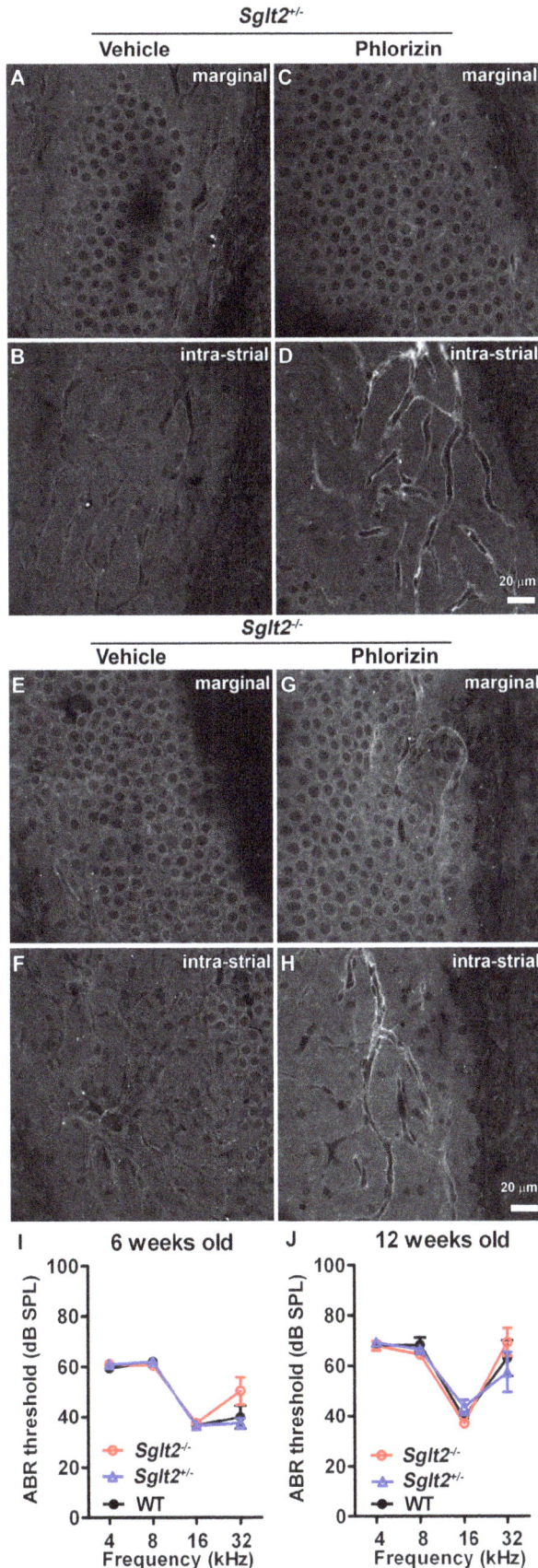

Figure 8. Loss of SGLT2 function had no effect on cochlear uptake of GTTR or auditory function. In the stria vascularis, GTTR was localized in marginal (A, E) and intermediate (B, F) cells of $Sglt2^{+/-}$ (A, B) and $Sglt2^{-/-}$ (E, F) mice. The nucleoplasm of marginal and intermediate cell nuclei displayed weak labeling. (C, D, G, H) Phlorizin had no effect on the uptake or distribution of GTTR fluorescence in the stria vascularis of $Sglt2^{+/-}$ or $Sglt2^{-/-}$ mice. Scale bar = 20 µm. (I, J) Wild-type, $Sglt2^{+/-}$ and $Sglt2^{-/-}$ mice, at 6 or 12 weeks of age, displayed no significant differences in ABR thresholds.

[18,19]. The kidney proximal tubule is also a primary location of aminoglycoside-induced nephrotoxicity [24,25,64].

In the cochlea, loop diuretics enhance the cochlear uptake of aminoglycosides [65]. Loop diuretics inhibit the $Na^+K^+Cl^-$ co-transporter (NKCC) [66], predominantly localized on the basolateral membrane of marginal cells [65,66], increasing the intra-strial concentration of Na^+ [65]. If SGLT2 is localized on the basolateral membrane of marginal cells, as implicated by the discrete, yet low level, of immunoexpression within the stria vascularis, SGLT2 may be appropriately located to traffic aminoglycosides into marginal cells, prior to clearance into endolymph, and uptake by hair cells as shown previously [36]. As noted above, SGLT2 has a large, hydrophilic, elastic vestibule, with an internal pore diameter of ~3 nm, and an exit pore (into cytosol) of ~1.5–2.5 nm [26,27] that is sufficiently large to potentially allow permeation by gentamicin. SGLT2 is blocked with high affinity by non-hydrolyzable glycoside derivatives that are generally well-tolerated acutely [30,33,67]. Thus, SGLT2 appeared to be rationally-identified candidate aminoglycoside transporter, and this hypothesis drove our experiments.

Fluorescently-conjugated gentamicin, GTTR, has been used to characterize the endocytotic trafficking of aminoglycosides to their intracellular domains in the kidney [12,13]. Although the relative molecular mass (g/mol) and minimum cross-sectional diameter (mcd) of GTTR is larger than that of untagged gentamicin (gentamicin, 440–470 g/mol, mcd, 0.81 nm, GTTR, ~1100 g/mol, mcd, ~1.47 nm), GTTR can also permeate non-selective cation channels, with a sufficiently large pore diameter, directly into cytoplasm [15,35,51]. Using GTTR, we have previously shown that cytoplasmic uptake of GTTR can occur rapidly at low temperatures, precluding endocytosis, and is regulated by cellular potential, pH, extracellular cations (Ca^{2+}, Gd^{3+}, La^{3+}), and non-specific cation channel blockers such as Ruthenium Red (RR), and verified using immunocytochemistry [15]. These properties are indicative of molecular permeation of ion channels, as for another fluorescent dye, FM 1–43 [68].

Using previously-described murine kidney cell lines [14], we observed specific SGLT2 immunofluorescence in proximal (KPT2), but not distal tubule (KDT3) cell lines, and in proximal tubules, but not distal tubules, *in vivo*. D-glucose is a primary substrate for SGLT2, which can also traffic the fluorescent glucose analog 2-NBDG. We found that a 1-fold molar excess of D-glucose competitively decreased 2-NBDG uptake by KPT2 cells. A 40-fold molar excess of D-glucose can also significantly decrease GTTR uptake of KPT2 cells. Thus, GTTR appears to have a greater affinity for SGLT2 than D-glucose and 2-NBDG. Uptake of 2-NBDG by KPT2 cells was also strongly inhibited by phlorizin, demonstrating robust SGLT2 activity in these cells, although not as efficaciously as D-glucose. Phlorizin also significantly decreased KPT2 uptake of GTTR and gentamicin in a dose-dependent manner by both immunofluorescence and ELISA. The uptake of GTTR was significantly attenuated ~20% by phlorizin or Na^+-free buffer, suggesting that this proportion of total GTTR uptake by KPT2 cells was mediated by SGLT2. Heterologous expression

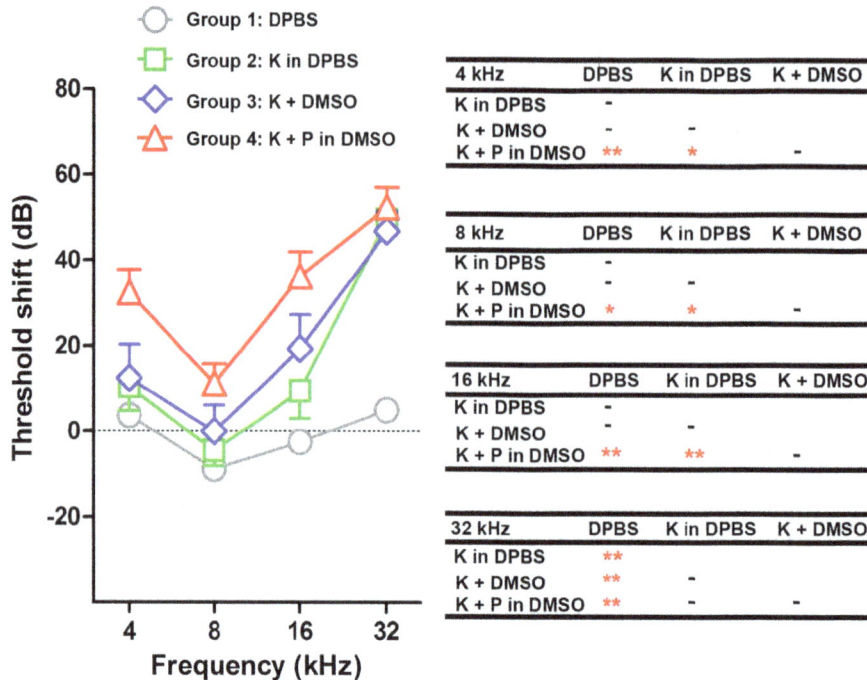

Figure 9. Phlorizin does not ameliorate kanamycin-induced ototoxicity. Three weeks after dosing, mice treated with kanamycin in DPBS or mice treated with kanamycin plus DMSO (vehicle for phlorizin) had significant ABR threshold shifts at 32 kHz only compared to mice treated with DPBS only (**$p < 0.01$). The kanamycin plus phlorizin group had significantly different threshold shifts at 4, 8 and 16 kHz compared to the kanamycin in DPBS group; and significantly different threshold shifts at 4, 8, 16 and 32 kHz compared DPBS only group (*$p < 0.05$; **$p < 0.01$). However, no significant threshold shifts were observed between kanamycin plus phlorizin (in DMSO) and kanamycin plus DMSO groups.

of SGLT2 in distal tubule-derived KDT3 cells significantly enhanced GTTR uptake, and this enhanced uptake can be abolished by phlorizin. The residual uptake of GTTR in these cells after phlorizin treatment likely represents GTTR uptake via previously-identified gentamicin-permeant cation channels, as demonstrated previously [14,15,69]. Furthermore, siRNA knockdown of SGLT2 expression in KPT2 cells reduced cellular susceptibility to gentamicin-induced cytotoxicity. Phlorizin had no apparent effect on the bactericidal activity of gentamicin on *E. coli* by disk diffusion assay. These *in vitro* data suggested that phlorizin, or other SGLT2 antagonists, may decrease cellular uptake of GTTR *in vivo*, and protect cells against aminoglycoside-induced cytotoxicity *in vivo*.

In vivo, we used two antisera for SGLT2 to determine if SGLT2 was appropriately located to contribute to aminoglycoside trafficking in the kidney and cochlea. Both antisera provided specific localization for SGLT2 in renal proximal tubules, with weak, less defined immunoexpression for SGLT2 in the cochlear stria vascularis (Fig. 5), and negligible SGLT2 immunolabeling in the organ of Corti. Although the SGLT2 inhibitor – phlorizin – acutely decreased GTTR uptake in proximal tubules of $Sglt2^{+/-}$ mice (Fig. 7), phlorizin did not inhibit cochlear uptake of GTTR of $Sglt2^{+/-}$ and $Sglt2^{-/-}$ mice (Fig. 8). It is not known whether phlorizin crosses the blood-brain barrier or the blood-labyrinth barrier, which may have limited its efficacy of blocking cochlear SGLT2. We also speculate (Fig. S7) that phlorizin inhibition, if any, of low cochlear levels of SGLT2-mediated GTTR trafficking would be compensated by phlorizin-induced elevation of GTTR serum levels (Fig. 7). Elevated serum levels of GTTR would increase trafficking into the cochlea via other residual cellular mechanisms of aminoglycoside uptake, e.g., endocytosis [12,13] or ion channel permeation [14,15]. However, genomic loss of

SGLT2 function, unexpectedly, did not reduce GTTR uptake by proximal tubules compared to $Sglt2^{+/-}$ mice. Phlorizin did not alter proximal tubule uptake of GTTR in $Sglt2^{-/-}$ mice (Fig. 7), indicating that phlorizin had negligible effects on the cellular uptake of aminoglycosides in $Sglt2^{-/-}$ mice, acting specifically on SGLT2 in $Sglt2^{+/-}$ mice. Thus, renal GTTR uptake by $Sglt2^{-/-}$ mice likely occurs via compensatory mechanisms such as endocytosis or aminoglycoside-permeant cation channels, as discussed above. In addition, genomic loss of SGLT2 function and phlorizin-pretreatment in $Sglt2^{+/-}$ mice elevated serum levels of gentamicin or GTTR (compared to control-treated heterozygous $Sglt2^{+/-}$ mice; Fig. 7). This may be the result of a reduced volume of distribution for these compounds, and/or by a reduced glomerular filtration rate (GFR) [70], providing an alternate explanation for the reduced GTTR cytoplasmic and punctate fluorescence in proximal tubule cells. However, phlorizin did not alter serum levels of gentamicin or GTTR in $Sglt2^{-/-}$ mice (Fig. 7) due to the absence of a binding partner (*i.e.*, SGLT2) that facilitates aminoglycoside trafficking, further indicating the specificity of phlorizin for SGLT2.

There is no demonstrable nephrotoxicity or renal damage following kanamycin treatment (at 700–900 mg/kg per dose twice daily for 14 days) to induce ototoxicity in mice [48], as used here. To test for permanent changes in auditory performance, a recovery period of three weeks is optimal [48]. In mice, gentamicin at doses to induce ototoxicity causes systemic toxicity and mortality [48,71]. Phlorizin or genomic loss of SGLT2 function did not affect auditory function, suggesting that SGLT2 activity is not required for cochlear function, and that glucose transport into the cochlea can be achieved by other transporters such as facilitated GLUTs [72,73]. Whether GLUTs are aminoglycoside-permeant remains uncertain, as crystal structures have yet to be determined. Phlorizin, and genomic loss of SGLT2, did not reduce cochlear uptake of GTTR. Phlorizin did not protect auditory function from kanamycin-induced ototoxicity in wild-type mice. Crucially, neither phlorizin nor genomic loss of SGLT2 increased serum levels of aminoglycosides. Thus, we did not attempt to repeat the ototoxicity studies with $Sglt2^{-/-}$ mice.

In summary, SGLT2 increased cellular uptake of gentamicin and exacerbated gentamicin-induced cytotoxicity *in vitro*. Acute inhibition of SGLT2 function reduced gentamicin-induced cytotoxicity in kidney proximal tubule cells, *in vitro*, and may reduce the risk of gentamicin-induced nephrotoxicity in the kidney *in vivo*. Acute inhibition, but not chronic loss, of SGLT2 function reduced GTTR uptake by kidney cells *in vivo*. Loss of SGLT2 function increased serum levels of gentamicin and GTTR, but did not prevent cochlear loading, and can increase the risk of aminoglycoside-induced ototoxicity. These data suggest that clinical antagonism of SGLT2 function by phlorizin, and phlorizin derivatives like O-glycosides and C-glycosides, may be contra-indicated if patients are undergoing aminoglycoside therapy.

Supporting Information

Figure S1 SGLT2-mediated uptake of GTTR is attenuated by D-glucose. Cells were treated with 5 μg/ml GTTR for 20 minutes at 37°C with increasing doses of D-glucose (molar ratios of 1:0, 1:40, 1:2000 or 1:40000 [GTTR/D-glucose]). Increasing doses of D-glucose reduced GTTR fluorescence in KPT2 cells. Scale bar = 20 μm.

Figure S2 KDT3-SGLT2 cell line generation. Parental KDT3, KDT3-SGLT2 and KDT3-pBabe cell lines have similar epitheloid morphology. Scale bar = 50 μm.

Figure S3 Cell growth of control siRNA and SGLT2 siRNA transfected KPT2 cell. MTT assay showed there was no difference for cell growth between control siRNA and SGLT2 siRNA transfected KPT2 cell at 1, 2 or 3 days after transfection.

Figure S4 Phlorizin did not affect auditory function. The ABR thresholds of wild-type mice 30 minutes after injection with

800 mg/kg phlorizin or vehicle (DMSO) control i.p., displayed no significant differences.

Figure S5 Auditory function of $Sglt2^{-/-}$ mice. Wild-type, $Sglt2^{+/-}$ or $Sglt2^{-/-}$ mice displayed no significant differences in ABR thresholds at 6 or 12 weeks of age. Male (blue) and female (red) displayed no significance differences.

Figure S6 Auditory function by ABR before or 3 weeks after kanamycin treatment with or without phlorizin in wild-type mice. In mice treated with kanamycin in DPBS, threshold shifts at 32 kHz were observed 1 day, post-treatment, 1.5 weeks post-treatment and 3 weeks post-treatment with kanamycin. In mice treated with kanamycin plus DMSO (vehicle for phlorizin), further threshold shifts were observed at 16 and 32 kHz at these 3 post-treatment time points. In mice treated with kanamycin plus phlorizin (in DMSO), threshold shifts were observed at 4, 16 and 32 kHz. Male (blue) and female (red) mice have little difference.

Figure S7 Schematic representation of the effect of phlorizin on SGLT2-mediated GTTR uptake by the kidney or cochlea, as suggested by our data and interpretation.

Table S1 Phlorizin had no effect on bactericidal activity of gentamicin. In *E. coli* disk diffusion assay, gentamicin (0.4 μg or 1 μg) alone induced a colony-free halo around the drug-impregnated disk, indicating baseline bactericidal effect. The colony-free diameter or halo thickness was not attenuated by increasing doses of phlorizin, indicating that phlorizin had no effect on the bactericidal activity of gentamicin.

Author Contributions

Conceived and designed the experiments: PSS MJ QW. Performed the experiments: MJ QW TK JK HL PSS. Analyzed the data: MJ PSS QW. Contributed reagents/materials/analysis tools: MJ QW. Wrote the paper: MJ PSS.

References

1. Forge A, Schacht J (2000) Aminoglycoside antibiotics. Audiol Neurootol 5: 3–22.
2. Mwengee W, Butler T, Mgema S, Mhina G, Almasi Y, et al. (2006) Treatment of plague with gentamicin or doxycycline in a randomized clinical trial in Tanzania. Clin Infect Dis 42: 614–621.
3. Mohr PE, Feldman JJ, Dunbar JL, McConkey-Robbins A, Niparko JK, et al. (2000) The societal costs of severe to profound hearing loss in the United States. Int J Technol Assess Health Care 16: 1120–1135.
4. Karahan I, Atessahin A, Yilmaz S, Ceribasi AO, Sakin F (2005) Protective effect of lycopene on gentamicin-induced oxidative stress and nephrotoxicity in rats. Toxicology 215: 198–204.
5. Nagai J, Takano M (2004) Molecular aspects of renal handling of aminoglycosides and strategies for preventing the nephrotoxicity. Drug Metab Pharmacokinet 19: 159–170.
6. Mingeot-Leclercq MP, Tulkens PM (1999) Aminoglycosides: nephrotoxicity. Antimicrobial Agents and Chemotherapy 43: 1003–1012.
7. Dzhagalov IL, Chen KG, Herzmark P, Robey EA (2013) Elimination of self-reactive T cells in the thymus: a timeline for negative selection. PLoS Biology 11: e1001566.
8. Servais H, Van Der Smissen P, Thirion G, Van der Essen G, Van Bambeke F, et al. (2005) Gentamicin-induced apoptosis in LLC-PK1 cells: involvement of lysosomes and mitochondria. Toxicology and Applied Pharmacology 206: 321–333.
9. Karasawa T, Steyger PS (2011) Intracellular mechanisms of aminoglycoside-induced cytotoxicity. Integrative Biology 3: 879–886.
10. Cuzzocrea S, Mazzon E, Dugo L, Serraino I, Di Paola R, et al. (2002) A role for superoxide in gentamicin-mediated nephropathy in rats. Eur J Pharmacol 450: 67–76.
11. Kolodkin-Gal I, Sat B, Keshet A, Engelberg-Kulka H (2008) The communication factor EDF and the toxin-antitoxin module mazEF determine the mode of action of antibiotics. PLoS Biology 6: e319.
12. Sandoval RM, Molitoris BA (2004) Gentamicin traffics retrograde through the secretory pathway and is released in the cytosol via the endoplasmic reticulum. American Journal of Physiology Renal Physiology 286: F617–624.
13. Sandoval RM, Dunn KW, Molitoris BA (2000) Gentamicin traffics rapidly and directly to the Golgi complex in LLC-PK(1) cells. American Journal of Physiology Renal Physiology 279: F884–890.
14. Karasawa T, Wang Q, Fu Y, Cohen DM, Steyger PS (2008) TRPV4 enhances the cellular uptake of aminoglycoside antibiotics. Journal of Cell Science 121: 2871–2879.
15. Myrdal SE, Steyger PS (2005) TRPV1 regulators mediate gentamicin penetration of cultured kidney cells. Hearing Research 204: 170–182.
16. Dai CF, Steyger PS (2008) A systemic gentamicin pathway across the stria vascularis. Hearing Research 235: 114–124.
17. Sabolic I, Vrhovac I, Eror DB, Gerasimova M, Rose M, et al. (2012) Expression of Na+-D-glucose cotransporter SGLT2 in rodents is kidney-specific and exhibits sex and species differences. American Journal of Physiology Cell Physiology 302: C1174–1188.
18. You G, Lee WS, Barros EJ, Kanai Y, Huo TL, et al. (1995) Molecular characteristics of Na(+)-coupled glucose transporters in adult and embryonic rat kidney. Journal of Biological Chemistry 270: 29365–29371.
19. Kanai Y, Lee WS, You G, Brown D, Hediger MA (1994) The human kidney low affinity Na+/glucose cotransporter SGLT2. Delineation of the major renal reabsorptive mechanism for D-glucose. Journal of Clinical Investigation 93: 397–404.

20. Fujimori Y, Katsuno K, Nakashima I, Ishikawa-Takemura Y, Fujikura H, et al. (2008) Remogliflozin etabonate, in a novel category of selective low-affinity sodium glucose cotransporter (SGLT2) inhibitors, exhibits antidiabetic efficacy in rodent models. J Pharmacol Exp Ther 327: 268–276.

21. Nomura S, Sakamaki S, Hongu M, Kawanishi E, Koga Y, et al. (2010) Discovery of canagliflozin, a novel C-glucoside with thiophene ring, as sodium-dependent glucose cotransporter 2 inhibitor for the treatment of type 2 diabetes mellitus. J Med Chem 53: 6355–6360.

22. Magen D, Sprecher E, Zelikovic I, Skorecki K (2005) A novel missense mutation in SLC5A2 encoding SGLT2 underlies autosomal-recessive renal glucosuria and aminoaciduria. Kidney Int 67: 34–41.

23. Garry F, Chew DJ, Hoffsis GF (1990) Urinary indices of renal function in sheep with induced aminoglycoside nephrotoxicosis. Am J Vet Res 51: 420–427.

24. Banday AA, Farooq N, Priyamvada S, Yusufi AN, Khan F (2008) Time dependent effects of gentamicin on the enzymes of carbohydrate metabolism, brush border membrane and oxidative stress in rat kidney tissues. Life Sciences 82: 450–459.

25. Nonclercq D, Wrona S, Toubeau G, Zanen J, Heuson-Stiennon JA, et al. (1992) Tubular injury and regeneration in the rat kidney following acute exposure to gentamicin: a time-course study. Renal Failure 14: 507–521.

26. Naftalin RJ (2008) Osmotic water transport with glucose in GLUT2 and SGLT. Biophys J 94: 3912–3923.

27. Liu T, Speight P, Silverman M (2009) Reanalysis of structure/function correlations in the region of transmembrane segments 4 and 5 of the rabbit sodium/glucose cotransporter. Biochemical and Biophysical Research Communications 378: 133–138.

28. Santer R, Kinner M, Lassen CL, Schneppenheim R, Eggert P, et al. (2003) Molecular analysis of the SGLT2 gene in patients with renal glucosuria. J Am Soc Nephrol 14: 2873–2882.

29. van den Heuvel LP, Assink K, Willemsen M, Monnens L (2002) Autosomal recessive renal glucosuria attributable to a mutation in the sodium glucose cotransporter (SGLT2). Hum Genet 111: 544–547.

30. Katsuno K, Fujimori Y, Takemura Y, Hiratochi M, Itoh F, et al. (2007) Sergliflozin, a novel selective inhibitor of low-affinity sodium glucose cotransporter (SGLT2), validates the critical role of SGLT2 in renal glucose reabsorption and modulates plasma glucose level. J Pharmacol Exp Ther 320: 323–330.

31. Obermeier MT, Yao M, Khanna A, Koplowitz B, Zhu M, et al. (2009) In Vitro Characterization and Pharmacokinetics of Dapagliflozin (BMS-512148), a Potent Sodium-Glucose Cotransporter Type II (SGLT2) Inhibitor, in Animals and Humans. Drug Metab Dispos.

32. Sha S, Devineni D, Ghosh A, Polidori D, Chien S, et al. (2011) Canagliflozin, a novel inhibitor of sodium glucose co-transporter 2, dose dependently reduces calculated renal threshold for glucose excretion and increases urinary glucose excretion in healthy subjects. Diabetes Obes Metab 13: 669–672.

33. Han S, Hagan DL, Taylor JR, Xin L, Meng W, et al. (2008) Dapagliflozin, a selective SGLT2 inhibitor, improves glucose homeostasis in normal and diabetic rats. Diabetes 57: 1723–1729.

34. Nauck MA, Del Prato S, Meier JJ, Duran-Garcia S, Rohwedder K, et al. (2011) Dapagliflozin Versus Glipizide as Add-on Therapy in Patients With Type 2 Diabetes Who Have Inadequate Glycemic Control With Metformin: A randomized, 52-week, double-blind, active-controlled noninferiority trial. Diabetes Care 34: 2015–2022.

35. Wang Q, Steyger PS (2009) Trafficking of systemic fluorescent gentamicin into the cochlea and hair cells. Journal of the Association for Research in Otolaryngology 10: 205–219.

36. Li H, Steyger PS (2011) Systemic aminoglycosides are trafficked via endolymph into cochlear hair cells. Sci Rep 1: 159.

37. Li H, Wang Q, Steyger PS (2011) Acoustic trauma increases cochlear and hair cell uptake of gentamicin. PLoS One 6: e19130.

38. Sandoval R, Leiser J, Molitoris BA (1998) Aminoglycoside antibiotics traffic to the Golgi complex in LLC-PK1 cells. J Am Soc Nephrol 9: 167–174.

39. Myrdal SE, Johnson KC, Steyger PS (2005) Cytoplasmic and intra-nuclear binding of gentamicin does not require endocytosis. Hear Res 204: 156–169.

40. Karasawa T, Wang Q, David LL, Steyger PS (2011) Calreticulin binds to gentamicin and reduces drug-induced ototoxicity. Toxicological Sciences 124: 378–387.

41. Xiao F, Jiang M, Du D, Xia C, Wang J, et al. (2013) Orexin A regulates cardiovascular responses in stress-induced hypertensive rats. Neuropharmacology 67: 16–24.

42. Jiang MY, Chen J, Wang J, Xiao F, Zhang HH, et al. (2011) Nitric oxide modulates cardiovascular function in the rat by activating adenosine A2A receptors and inhibiting acetylcholine release in the rostral ventrolateral medulla. Clin Exp Pharmacol Physiol 38: 380–386.

43. Jiang M, Zhang C, Wang J, Chen J, Xia C, et al. (2011) Adenosine A(2A)R modulates cardiovascular function by activating ERK1/2 signal in the rostral ventrolateral medulla of acute myocardial ischemic rats. Life Sci 89: 182–187.

44. Zhang CR, Xia CM, Jiang MY, Zhu MX, Zhu JM, et al. (2013) Repeated electroacupuncture attenuating of apelin expression and function in the rostral ventrolateral medulla in stress-induced hypertensive rats. Brain Res Bull 97: 53–62.

45. Karasawa T, Sibrian-Vazquez M, Strongin RM, Steyger PS (2013) Identification of cisplatin-binding proteins using agarose conjugates of platinum compounds. PLoS ONE 8: e66220.

46. Zhang F, Dai M, Neng L, Zhang JH, Zhi Z, et al. (2013) Perivascular macrophage-like melanocyte responsiveness to acoustic trauma–a salient feature of strial barrier associated hearing loss. FASEB Journal 27: 3730–3740.

47. Mitchell C, Kempton JB, Creedon T, Trune D (1996) Rapid acquisition of auditory brainstem responses with multiple frequency and intensity tone-bursts. Hearing Research 99: 38–46.

48. Wu WJ, Sha SH, McLaren JD, Kawamoto K, Raphael Y, et al. (2001) Aminoglycoside ototoxicity in adult CBA, C57BL and BALB mice and the Sprague-Dawley rat. Hearing Research 158: 165–178.

49. Blodgett AB, Kothinti RK, Kamyshko I, Petering DH, Kumar S, et al. (2011) A fluorescence method for measurement of glucose transport in kidney cells. Diabetes Technology and Therapeutics 13: 743–751.

50. Stepanyan RS, Indzhykulian AA, Velez-Ortega AC, Boger ET, Steyger PS, et al. (2011) TRPA1-mediated accumulation of aminoglycosides in mouse cochlear outer hair cells. J Assoc Res Otolaryngol 12: 729–740.

51. Alharazneh A, Luk L, Huth M, Monfared A, Steyger PS, et al. (2011) Functional hair cell mechanotransducer channels are required for aminoglycoside ototoxicity. PLoS One 6: e22347.

52. Vu AA, Nadaraja GS, Huth ME, Luk L, Kim J, et al. (2013) Integrity and regeneration of mechanotransduction machinery regulate aminoglycoside entry and sensory cell death. PLoS ONE 8: e54794.

53. Marcotti W, van Netten SM, Kros CJ (2005) The aminoglycoside antibiotic dihydrostreptomycin rapidly enters mouse outer hair cells through the mechano-electrical transducer channels. Journal of Physiology 567: 505–521.

54. Ehrenkranz JR, Lewis NG, Kahn CR, Roth J (2005) Phlorizin: a review. Diabetes/Metabolism Research and Reviews 21: 31–38.

55. Wright EM (2001) Renal Na(+)-glucose cotransporters. American Journal of Physiology Renal Physiology 280: F10–18.

56. Santer R, Calado J (2010) Familial renal glucosuria and SGLT2: from a mendelian trait to a therapeutic target. Clinical Journal of the American Society of Nephrology 5: 133–141.

57. Hashino E, Shero M (1995) Endocytosis of aminoglycoside antibiotics in sensory hair cells. Brain Res 704: 135–140.

58. Hiel H, Schamel A, Erre JP, Hayashida T, Dulon D, et al. (1992) Cellular and subcellular localization of tritiated gentamicin in the guinea pig cochlea following combined treatment with ethacrynic acid. Hear Res 57: 157–165.

59. Nagai J, Komeda T, Yumoto R, Takano M (2013) Effect of protamine on the accumulation of gentamicin in opossum kidney epithelial cells. J Pharm Pharmacol 65: 441–446.

60. Raggi C, Fujiwara K, Leal T, Jouret F, Devuyst O, et al. (2011) Decreased renal accumulation of aminoglycoside reflects defective receptor-mediated endocytosis in cystic fibrosis and Dent's disease. Pflugers Arch 462: 851–860.

61. Hirose K, Hockenbery DM, Rubel EW (1997) Reactive oxygen species in chick hair cells after gentamicin exposure in vitro. Hear Res 104: 1–14.

62. Mamdouh Z, Giocondi MC, Laprade R, Le Grimellec C (1996) Temperature dependence of endocytosis in renal epithelial cells in culture. Biochim Biophys Acta 1282: 171–173.

63. Tanaka R, Muraki K, Ohya S, Yamamura H, Hatano N, et al. (2008) TRPV4-like non-selective cation currents in cultured aortic myocytes. Journal of Pharmacological Sciences 108: 179–189.

64. Humes HD (1999) Insights into ototoxicity. Analogies to nephrotoxicity. Annals of the New York Academy of Sciences 884: 15–18.

65. Higashiyama K, Takeuchi S, Azuma H, Sawada S, Yamakawa K, et al. (2003) Bumetanide-induced enlargement of the intercellular space in the stria vascularis critically depends on Na+ transport. Hearing Research 186: 1–9.

66. Crouch JJ, Sakaguchi N, Lytle C, Schulte BA (1997) Immunohistochemical localization of the Na-K-Cl co-transporter (NKCC1) in the gerbil inner ear. Journal of Histochemistry and Cytochemistry 45: 773–778.

67. Pajor AM, Randolph KM, Kerner SA, Smith CD (2008) Inhibitor binding in the human renal low- and high-affinity Na+/glucose cotransporters. J Pharmacol Exp Ther 324: 985–991.

68. Meyers JR, MacDonald RB, Duggan A, Lenzi D, Standaert DG, et al. (2003) Lighting up the senses: FM1–43 loading of sensory cells through nonselective ion channels. Journal of Neuroscience 23: 4054–4065.

69. Wang T, Yang YQ, Karasawa T, Wang Q, Phillips A, et al. (2013) Bumetanide hyperpolarizes madin-darby canine kidney cells and enhances cellular gentamicin uptake by elevating cytosolic Ca(2+) thus facilitating intermediate conductance Ca(2+)–activated potassium channels. Cell Biochemistry and Biophysics 65: 381–398.

70. Vallon V, Gerasimova M, Rose M, Masuda T, Satriano J, et al. (2013) SGLT2 Inhibitor Empagliflozin Reduces Renal Growth and Albuminuria in Proportion to Hyperglycemia and Prevents Glomerular Hyperfiltration in Diabetic Akita Mice. American Journal of Physiology Renal Physiology.

71. Fetoni AR, Sergi B, Ferraresi A, Paludetti G, Troiani D (2004) alpha-Tocopherol protective effects on gentamicin ototoxicity: an experimental study. International Journal of Audiology 43: 166–171.

72. Takeuchi S, Ando M (1997) Marginal cells of the stria vascularis of gerbils take up glucose via the facilitated transporter GLUT: application of autofluorescence. Hearing Research 114: 69–74.

73. Ando M, Edamatsu M, Fukuizumi S, Takeuchi S (2008) Cellular localization of facilitated glucose transporter 1 (GLUT-1) in the cochlear stria vascularis: its possible contribution to the transcellular glucose pathway. Cell and Tissue Research 331: 763–769.

11

Characterization of a Novel Anti-Cancer Compound for Astrocytomas

Sang Y. Lee[1]*, Becky Slagle-Webb[1], Elias Rizk[1], Akshal Patel[1], Patti A. Miller[2], Shen-Shu Sung[3], James R. Connor[1]

1 Department of Neurosurgery, Pennsylvania State University College of Medicine, Penn State M.S. Hershey Medical Center, Hershey, Pennsylvania, United States of America, **2** Department of Radiology, Pennsylvania State University College of Medicine, Penn State M.S. Hershey Medical Center, Hershey, Pennsylvania, United States of America, **3** Department of Pharmacology, Pennsylvania State University College of Medicine, Penn State M.S. Hershey Medical Center, Hershey, Pennsylvania, United States of America

Abstract

The standard chemotherapy for brain tumors is temozolomide (TMZ), however, as many as 50% of brain tumors are reportedly TMZ resistant leaving patients without a chemotherapeutic option. We performed serial screening of TMZ resistant astrocytoma cell lines, and identified compounds that are cytotoxic to these cells. The most cytotoxic compound was an analog of thiobarbituric acid that we refer to as CC-I. There is a dose-dependent cytotoxic effect of CC-I in TMZ resistant astrocytoma cells. Cell death appears to occur via apoptosis. Following CC-I exposure, there was an increase in astrocytoma cells in the S and G2/M phases. In *in vivo* athymic (*nu/nu*) nude mice subcutaneous and intracranial tumor models, CC-I completely inhibited tumor growth without liver or kidney toxicity. Molecular modeling and enzyme activity assays indicate that CC-I selectively inhibits topoisomerase IIα similar to other drugs in its class, but its cytotoxic effects on astrocytoma cells are stronger than these compounds. The cytotoxic effect of CC-I is stronger in cells expressing unmethylated O^6-methylguanine methyltransferase (MGMT) but is still toxic to cells with methylated MGMT. CC-I can also enhance the toxic effect of TMZ on astrocytoma when the two compounds are combined. In conclusion, we have identified a compound that is effective against astrocytomas including TMZ resistant astrocytomas in both cell culture and *in vivo* brain tumor models. The enhanced cytotoxicity of CC-I and the safety profile of this family of drugs could provide an interesting tool for broader evaluation against brain tumors.

Editor: Javier S. Castresana, University of Navarra, Spain

Funding: This project is funded, in part, by a grant with the National Cancer Institute of the National Institutes of Health under Award Number R21CA167406 to SL. The content is solely the responsibility of the authors and does not necessarily represent the official views of the National Institutes of Health. This study also supported in part by the Elsa U. Pardee Foundation and the Tara Leah Witmer Endowment. The funders had no role in study design, data collection and analysis, decision to publish, or preparation of the manuscript

Competing Interests: Connor is partial owner of NuHope LLC which has a financial interest in development of compounds for treating brain tumors that were initially screened using cell lines chosen for HFE genotype. Lee has a royalty agreement with NuHope LLC.

* Email: SYL3@psu.edu

Introduction

Gliomas account for 28% of all primary brain and central nervous system (CNS) tumors, and 80% of gliomas are malignant [1]. Among gliomas, glioblastoma (glioblastoma multiforme, grade IV astrocytoma, GBM) is the most common malignant glioma. The mortality rate of primary malignant brain and CNS tumors is high; approximately 22,620 new adult cases of malignant brain and CNS cancers in 2013 [1] and 13,700 deaths occurred in 2012 [2]. The median survival for GBM patients was 14.6 months and the 2 year survival of patients with GBM was 10.4% for radiotherapy alone and only 26.5% undergoing combined therapy treatment of temozolomide (TMZ) and radiation [3].

The current standard treatment for GBM is total resection followed by radiotherapy alone or combination with TMZ chemotherapy [4,5]. TMZ is an oral alkylating agent used in the treatment of brain cancer, *e.g.*, GBM and oligodendroglioma [6]. It has also been used to treat melanoma, prostate cancer, pancreatic carcinoma, soft tissue sarcoma, and renal cell carcinoma [7–11]. TMZ inhibits cell reproduction by inhibiting DNA replication [12] and has unique characteristics compared with other alkylating agents. For example, it is administered orally, crosses the blood-brain barrier, is less toxic than other alkylating agents, and does not chemically cross-link DNA. However, although TMZ is the current chemotherapeutic standard for treating brain tumors and other cancers, as many as 50% of brain tumors are resistant to TMZ therapy [13,14]. In addition, almost all tumors eventually come back and the large majority of recurrent tumors are resistant to chemotherapy [15,16]. Therefore, the development of new treatment options including novel drugs for therapy resistant brain tumors is urgently needed.

In addition to the alkylation agents like TMZ, topoisomerase inhibitors are another group of anti-cancer drugs under evaluation. Topoisomerases are important nuclear enzymes that regulate the topology of DNA, maintain genomic integrity and are essential for DNA replication, recombination, transcription and chromosome segregation [17]. There are six human topoisomerase enzymes [18] and three of them, topoisomerase I, topoisomerase IIα and topoisomerase IIβ, have significant involvement in cancer and cancer chemotherapy [19]. The topoisomerase I enzyme nicks and rejoins one strand of the duplex DNA, and topoisomerase II enzyme transiently breaks and closes double-stranded DNA [20]. The topoisomerase I inhibitors (e.g., topotecan) have been used in patients with recurrent small-cell lung cancer, recurrent malignant gliomas, recurrent childhood brain tumors [21,22]. Although topoisomerase II inhibitors were studied in glioma cells [23–25], the topoisomerase II inhibitors haven't been widely used in adults with primary brain tumors due to their poor CNS penetrance. Therefore, small molecules with the capability to penetrate the brain would be highly desirable to treat gliomas *in vivo*.

We have previously reported that human neuroblastoma cells and human astrocytoma cells lines expressing commonly occurring polymorphisms in the HFE gene were resistant to chemotherapy and radiation [26]. The HFE gene product is involved in iron homeostasis and the common HFE polymorphisms, H63D and C282Y, lead to a number of changes in cells such as increased endoplasmic reticulum stress and increased oxidative stress [27–29]. In the present study, we used astrocytoma cell lines that we identified with the HFE gene variants and TMZ resistance to screen compounds from DIVERSet compound library from Chembridge (San Diego, CA) and found a number of effective compounds with a similar chemotype. We identified an analog of a thiobarbituric acid compound which has strong toxic effect on TMZ-resistant astrocytoma cells. We report here the characterization of the lead compound in *in vitro* cell culture and *in vivo* brain tumor models.

Materials and Methods

Materials

Dulbecco's Modified Eagle Medium (DMEM), fetal bovine serum (FBS) and other cell culture ingredients were purchased from Life Technologies (Grand Island, NY). All the PCR Array ingredients were supplied from SABiosciences (Frederick, MD). TMZ was purchased from Oakwood Products Inc. (West Columbia, SC) and was dissolved in cell culture medium or 100% DMSO. The lead chemotype compound–I (CC-I) was ordered from ChemBridge Corporation (San Diego, CA). The compound was dissolved in DMSO as a stock solution and diluted for the experiment. Topoisomerase enzymes I and IIα assay kits were ordered from TopoGen Inc. (Port Orange, FL). Merbarone was obtained from Calbiochem (San Diego, CA). All of the other chemicals used were purchased from Sigma Co. (St. Louis, MO).

Human astrocytoma cell culture, treatment and cytotoxicity assay

Human astrocytoma cells (SW1088-grade III, U87-MG-grade IV, CCF-STTG1-grade IV, T98G-grade IV, LN-18-grade IV) were ordered from American Type Culture Collection (ATCC, Manassas, VA) and maintained in DMEM (Gibco by Life Technologies, catalog 11885) supplemented with 100 U/mL penicillin, 100 μg/mL streptomycin, 0.29 mg/mL L-glutamine, and 10% FBS. All experiments were performed at 37°C in 5% CO_2 atmosphere cell culture conditions. For the cytotoxicity assays, the compounds tested were prepared by first diluting them from the stock solution in cell culture media. The compounds were exposed to the cells for 3–6 days. Cell cytotoxicity was performed by MTS [3-(4,5-dimethylthiazol-2-yl)-5-(3-carboxymethoxyphenyl)-2-(4-sulfophenyl)-2H-tetrazolium] cell proliferation assay (Pro-

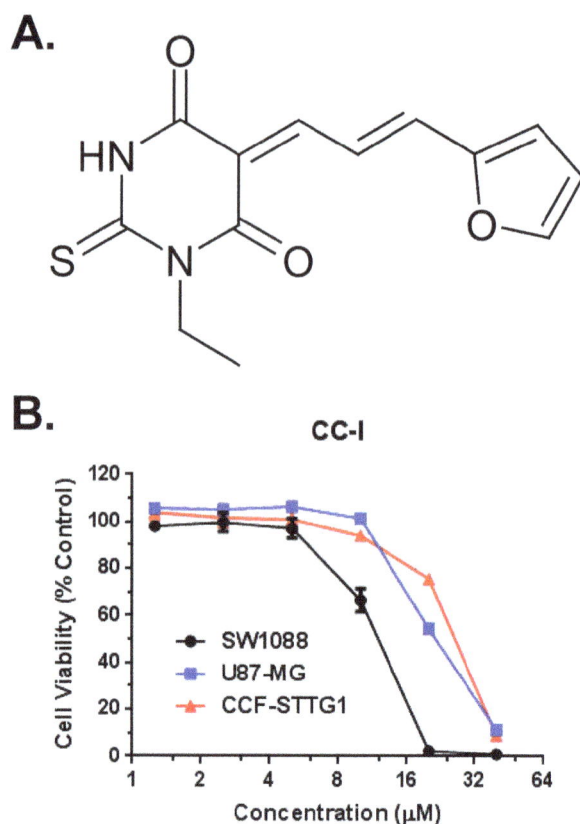

Figure 1. Chemical structure and cytotoxicity of CC-I in *in vitro*. (A) The structure of CC-I. (B) Cytotoxicity of CC-I in *in vitro*. Human astrocytoma cell lines were cultured with different doses of CC-I for 3 days and then the cytotoxicity was determined by SRB assay. The LC_{50} of CC-I to SW1088 cell lines (13.6 μM) are significantly different with the LC_{50} of CC-I to U87-MG and CCF-STTG1 cell lines (23.6 μM and 25.4 μM) (p<0.001).

mega, Madison, WI) or sulforhodamine B (SRB) assay at the end of the cell culture period.

Acute toxicity determination

Acute toxicity of CC-I was determined in athymic nude mice (strain 088 or 490, Charles River Laboratories, Wilmington, MA) according to the NIH drug development program's acute toxicity procedure with minor modification. To determine the acute toxicity, a total of six female mice (1–2 month old) were injected intraperitoneally with 3 different doses (e.g., 20 mg/kg, 37.5 mg/kg, 50 mg/kg) of CC-I or vehicle control once a week and then observed for a period of 7–14 days. The mice were observed daily for changes in body weight, visible and/or palpable dermal infection, presence of ascites, food consumption or nutrition status, and grooming or impaired mobility or death to determine acute toxicity. At 7–14 days after treatment, 0.5–1 ml of blood was collected through a cardiac heart puncture while the mice were under anesthesia (Ketamine 100 mg/kg body weight/xylazine 10 mg/kg body weight, intraperitoneally) for blood toxicity examination. All the animals in the study were housed in germ-free environmental rooms, and individual bubble systems. All the animal experiments were approved (IACUC #2011-062) by the Pennsylvania State University Institutional Animal Care and Use Committees.

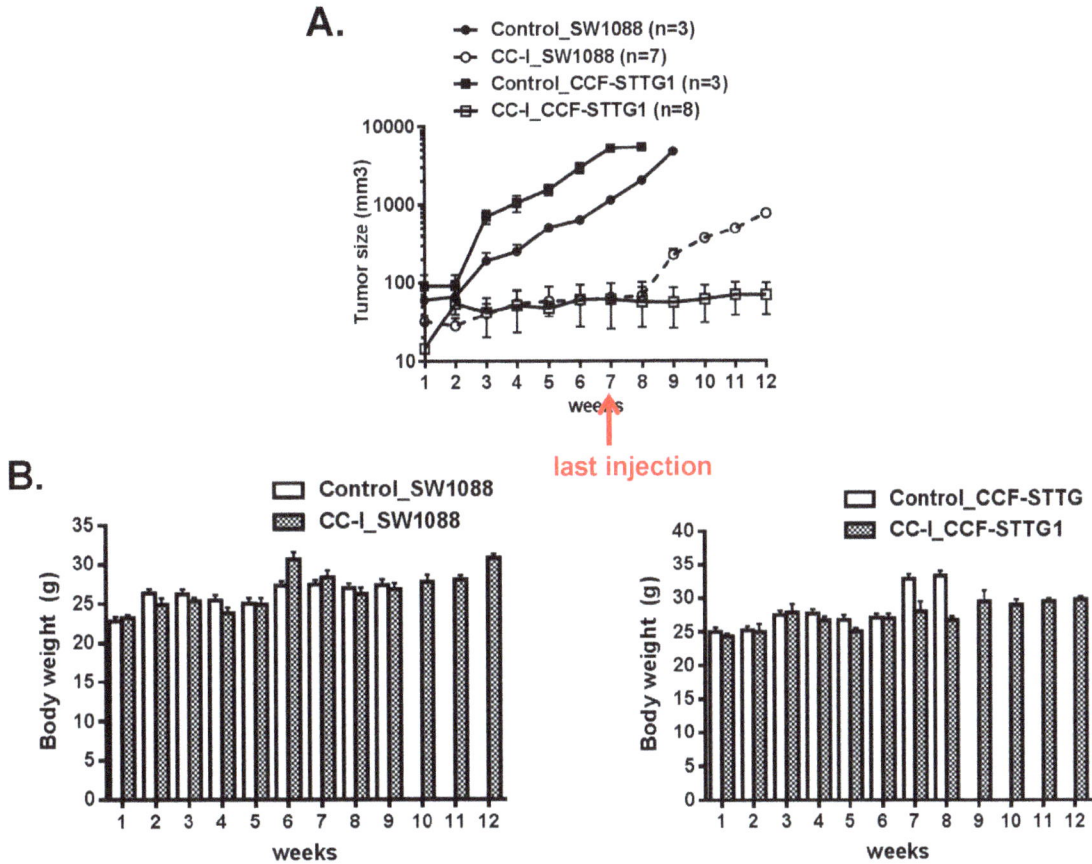

Figure 2. The anti-tumor effect of CC-I in a subcutaneous mouse tumor model. (A) Mice were implanted with ten million cells with the SW1088 or CCF-STTG1 cells. The starting tumor size for the CCF-STTG1 cells ranged from 80–100 mm³. The SW1088 cells grew more slowly so CC-I treatment was started when the tumors reached 30 mm³. CC-I was injected intraperitoneally at a concentration of 25 mg/kg body weight once a week for 7 weeks (n = 7~10). The control group was given PBS in the same volume and regimen (n = 3–8). The tumor slowly reoccurred in the TMZ-sensitive SW1088 astrocytoma injected nude mice but did not reoccur in the TMZ resistant CCF-STTG1 injected nude mice when CC-I was discontinued (beyond 7 weeks). CC-I inhibited the tumor growth and was not lethal in any of the treatment groups. Some error bars are too small to be visible. (B) Mean body weight of mice is presented in grams. Some error bars are too small to be visible.

Subcutaneous tumor model

To test the anti-tumor effect of CC-I against human astrocytoma tumor, one-two month old female immunodeficient (*nu/nu*) nude mice (strain 088, Charles River Laboratories, Wilmington, MA) were implanted 10×10^6 cells per mouse subcutaneously with TMZ sensitive SW1088 or TMZ resistant CCF-STTG1 astrocytoma cells. When the tumor reached approximately 32–100 mm³ in size, the mice (n = 10 or 11) were randomly divided into two groups. The CC-I was injected intraperitoneally at a concentration of 25 mg/kg body weight in a volume of 200–300 µL in 12.5% ethanol once a week for 7 weeks. The control group was given phosphate-buffered saline (PBS) in the same volume and regimen. Tumor size was measured weekly with a Vernier caliper for 7 weeks by an investigator blinded to experimental conditions. Tumor volume (V) was calculated according to the formula $V = a^2/2 \times b$, where a and b are minor and major axes of the tumor foci, respectively. The tumor size, health, and survival of the mice were visibly monitored daily and the tumor size measured weekly. We did not take pictures of the tumors. We will consider taking pictures for upcoming experiments. To monitor the toxicity of compounds, the animals were euthanized with ketamine/xylazine 100/10 mg/kg body weight intraperitoneally, and measured liver and kidney toxicity at the end of the experiment.

Intracranial xenograft model

Female immunodeficient nude mice (strain 088, Charles River Laboratories, Wilmington, MA) weighing 20–30 g were anesthetized by intraperitoneal injection of ketamine-xylazine 100 mg/kg–10 mg/kg body weight. Human U87-MG and CCF-STTG1 astrocytoma cell lines were implanted to create the brain tumor xenograft. In brief, the head was held in horizontal position and 1 million astrocytoma cells in a volume of 10 µL were injected slowly into the caudate putamen region using a small animal stereotactic apparatus. The stereotactic co-ordinates used for the xenografts are P = 0.5, L = 1.7, H = 3.8 mm. The astrocytoma cells were injected slowly for 10 minutes to avoid elevation in the intracranial pressure or upward cell suspension leakage through the track of the needle. The animals were given buprenorphine (0.05–0.1 mg/kg body weight subcutaneous) for pain during and after surgery. This was given every 8–12 hours for 24–48 hours after surgery. The animals were subjected to T1 weighted magnetic resonance imaging (MRI) twice; once to determine that a tumor is established in the brain (~3 weeks injection of astrocytoma cells) and at the end of the experiment. The animals were monitored on a daily basis and the body weight was recorded weekly. Once a tumor was observed, the mice (n = 12 or 15) were randomly divided into two groups. CC-I (25 mg/kg body weight) or PBS was injected once a week intraperitoneally. The overall

A.

B.

C.

D.

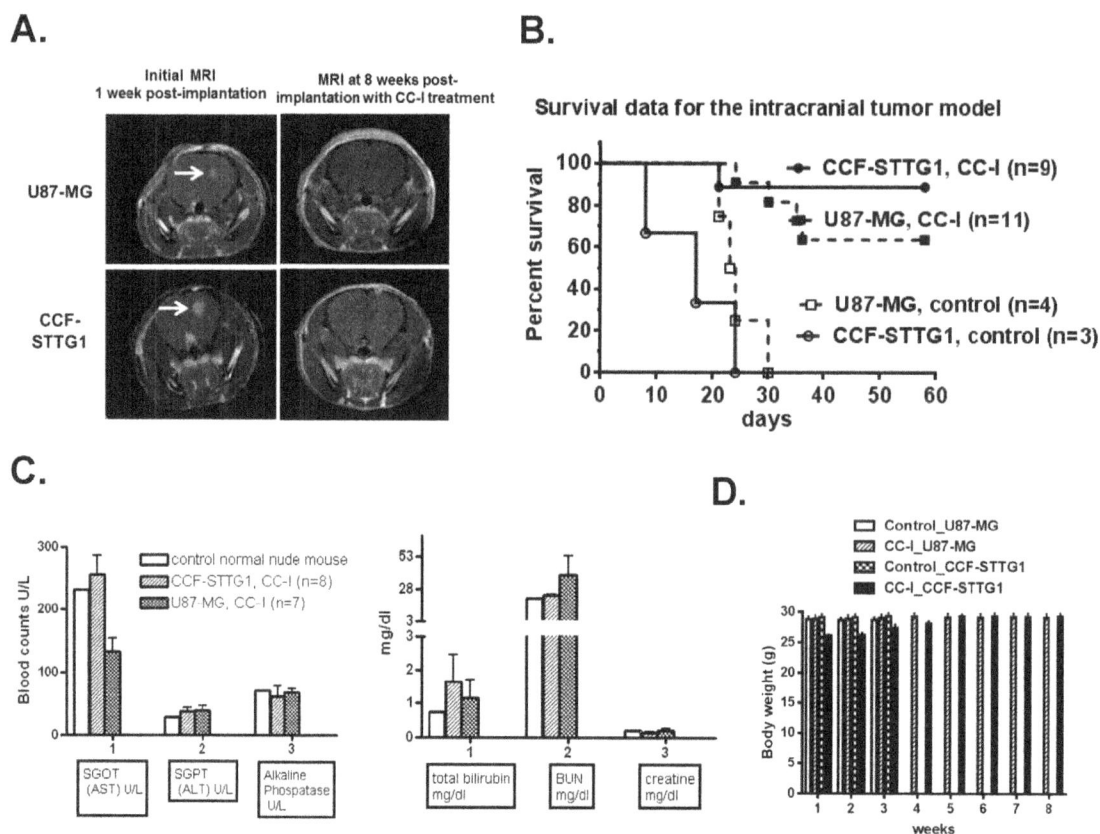

Figure 3. Anti-tumor effect of CC-I in an intracranial xenograft mouse model. (A) Representative MRI images taken with T1-weighted MRI contrast (7T MR imaging system) after intracranial tumor formation (one-three weeks post-implantation of astrocytoma cells) or after tumor formation followed by injection of CC-I (25 mg/kg body weight) for 7 weeks. CC-I completely inhibited tumor growth in both astrocytoma cell lines. (B) Kaplan-Meier survival graph of intracranial brain tumor mice after the administration of CC-I. CC-I extends the survival of the mice when compared to the untreated mice (n = 9 or 11) (p<0.0001). None of the mice which received PBS (control) survived after 30 days and median survival of all those animals was 20 days (n = 3 or 4). (C) Liver and kidney toxicity of CC-I. The liver and kidney toxicity (total bilirubin, blood urea nitrogen (BUN), creatine, aspartate aminotransferase (AST), alanine aminotransferase (ALT), and alkaline phosphatase) were determined using an automated chemistry analyzer machine (Roche Cobase MIRA) and kits manufactured by Thermo Electron. These data indicate no liver or kidney toxicity by CC-I in nude mice. Toxicity data displayed as means ± SEM. (D) Mean body weight of mice in grams.

survival of mice was performed by a Kaplan-Meier survival curve. The animals were euthanized according to acceptable method of euthanasia as defined by the American Veterinary Medical Association (AVMA) Guidelines on Euthanasia - Approved Euthanasia Methods, 2013. Once the animals receive a body condition score of less than 2, the animals were euthanized with ketamine/xylazine 100/10 mg/kg body weight intraperitoneally as well as a secondary method of cervical dislocation. At the termination of the experiment, plasma was collected for analysis of liver and kidney toxicity after euthanized with ketamine/xylazine 100/10 mg/kg body weight intraperitoneally.

T1 weighed MRI images

T1 weighted MRI contrast was used to visualize the tumor growth using 7T MRI system (Bruker, Biospec GmbH, Ettlingen, Germany). The imaging parameters of the T1 scan are TR/TE = 540 ms/11 ms, 8 averages, 192×192, 0.5 mm slice thickness, and 3.2 cm^2 FOV. The mice were anesthetized by inhalation of 1–2% isoflurane and placed in a position with brain located at the center of the coil. Intracranial tumor volume was estimated using Gadolinium enhanced T1 weighed multislice axial fast spin echo images. From these images the size of the tumor was

calculated using the Region-of-Interest tool available on the Paravision software (Bruker Biospec, Ettlingen, Germany).

Liver and kidney toxicity

The liver and kidney toxicity (total bilirubin, blood urea nitrogen (BUN), creatine, aspartate aminotransferase (AST), alanine aminotransferase (ALT), and alkaline phosphatase) was assessed for both subcutaneous tumor model and intracranial xenograft model using an automated chemistry analyzer (Roche Cobase MIRA) and kits manufactured by Thermo Electron (Louisville, CO). The blood was obtained from the control or CC-I injected mice with astrocytoma cells at the termination of the experiment.

Apoptosis assay

For apoptosis assay, the 3×10^6 of CCF-STTG1 cells were cultured for 48 hr with several concentrations (~36 μM) of CC-I or actinomycin D (~80 nM) as a positive control. The cells were harvested following trypsine-EDTA exposure and washed in cold PBS. Then 100 μL of the cell suspension (~1×10^6 cells) was incubated with 1 μL of 100 μg/mL red-fluorescent propidium iodide nucleic acid binding dye and 5 μL Annexin V-FITC (Molecular Probes, Carlsbad, CA) for 15 minutes at room

A. Apoptotic cell death

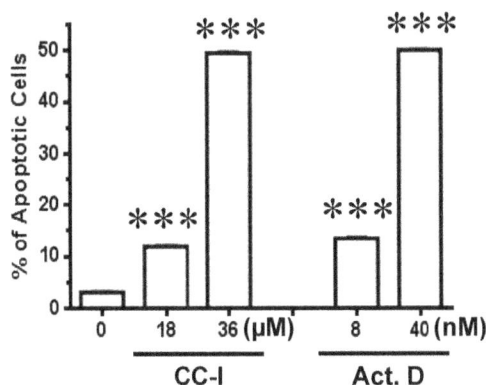

B. Necrotic cell death

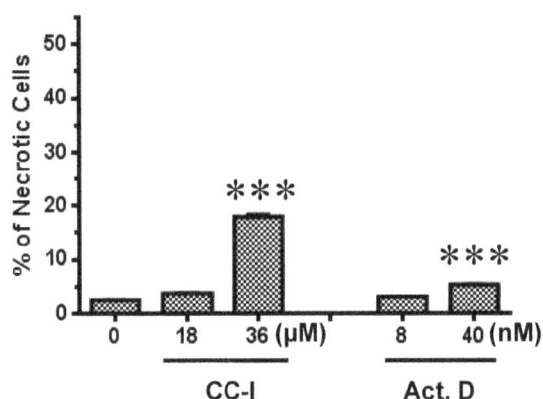

Figure 4. CC-I-induced cell death in CCF-STTG1 cells. Cell death was monitored with apoptotic and necrotic cell markers after 48 hours CC-I exposure in CCF-STTG1 cells. Cell death was determined with the recombinant annexin V conjugated to fluorescein, followed by flow cytometric analysis. Apoptotic cell death is shown in panel A. Panel B is necrotic cell death. Actinomycin D was used as a positive control to induce apoptotic cell death. The percentage of apoptotic cells following CC-I treatment was increased in a dose-dependent manner in CCF-STTG1 cells. There was not a pronounced dose dependent increase in necrotic cell death in the CCF-STTG1 cells until the higher concentration. Data assessed using Student t test and displayed as means ± SEM. Some error bars are too small to be visible. The symbols indicate a significant difference compared to the control. (***$p < 0.001$).

temperature in the dark. The cells were analyzed by flow cytometry (Becton Dickinson, Franklin Lakes, NJ) of emission at 530 nm (e.g. FL1) and >575 nm (e.g. FL3). The cells that are bound by Annexin V illustrate early apoptotic cells. Cells that are reactive for both Annexin V and propidium iodide are necrotic cells.

Gene expression profiling

We used Apoptosis PCR Array (SABiosciences, Frederick, MD) to determine which genes are altered by CC-I in TMZ resistant CCF-STTG1 cells. The PCR Array was performed according to the manufacturer's instructions. In brief, total RNA was extracted from vehicle (0.1% DMSO) treated or CC-I treated CCF-STTG1 cell lines using qPCR-Grade RNA Isolation kit. One μg of RNA was used for first strand cDNA synthesis by reverse transcription with MMLV reverse transcriptase. Then real-time PCR was performed with diluted cDNA and master mix with ROX filter.

For signal detection, the ABI Prism 7900 Sequence Detector System was programmed with an initial sterilization step of 2 minutes at 50°C, followed by 10 minutes denaturation at 95°C and then 40 cycles for 15 second at 95°C, 1 minute at 60°C and 30 second at 72°C. Each reaction sample was performed in triplicate. PCR Array data was calculated by the ΔΔcycle threshold (ΔΔCt) method, then normalized against multiple housekeeping genes and expressed as mean fold changes in CC-I treated samples relative to vehicle treated control samples.

Cell cycle analysis

For cell cycle analysis, CCF-STTG1 cells were cultured overnight at a density of 2–5×10⁶ cells per flask. The following day, the cells were treated with different concentrations of CC-I in fresh cell culture medium. After 24–48 hr later, the adherent cells were harvested and split (1×10⁶ cells per tube) for washing with HANK's buffer, then fixed in ice-cold 70% ethanol overnight at −20°C. For DNA staining day, the cells were incubated with propidium iodide (100 μg/ml) and RNase A (20 μg/ml) for 15 min at 4°C (protect from light). Samples were analyzed using BD FACS Calibur Flow Cytometry Analyzer.

Topoisomerase relaxation and decatenation assay

DNA relaxation and kinetoplast DNA (kDNA) decatenation assay was performed using topoisomerase I or II drug screening kit or Topopoisomerase II assay kit (TopoGEN, Inc., Port Orange, FL) according to the manufacturer's instructions [30]. Topoisomerase IIα decatenates kDNA which consists of highly catenated networks of circular DNA in an ATP-dependent reaction to yield individual minicircles of DNA. In brief, for topoisomerase IIα mediated kDNA decatenation assay, the 20 μL reaction mixture contains following components; 50 mM Tris-HCl, pH 8.0, 150 mM NaCl, 10 mM MgCl₂, 0.5 mM dithiothreitol, 30 μg/mL bovine serum albumin, 2 mM ATP, 260 ng of kDNA, several concentrations of compounds, and 4 U of human topoisomerase IIα. The final concentration of 0.5% (v/v) DMSO was used because this concentration does not affect activity of topoisomerase IIα. The incubation of assay mixture was carried out at 37°C for 30 minutes and terminated by the addition of 4 μL stop loading dye. The kDNA decatenation products from the reaction mixture was resolved on a 1% agarose gel at 100 V for 40 minutes, then stained with 0.5 μg/mL ethidium bromide in TAE buffer (4 mM Tris base/glacial acetic acid [0.11% (v/v)]/2 mM Na₂EDTA).

Molecular modeling study

The molecular modeling studies were based on the X-ray crystal structure of human topoisomerase IIα bound to L-peptide at 1.50 Å resolution (PDB identification code: 2q5a) [31]. The position of the L-peptide was used to specify the dimensions of the CC-I binding site for the docking study. Docking between topoisomerase IIα protein and CC-I was carried out using the GLIDE program (Grid Based Ligand Docking from Energetics, from Schrödinger, L.L.C.) [32,33]. The Jorgensen OPLS-2005 force field was employed in the GLIDE program. The optimal binding geometry for each model was obtained with GLIDE, which relies upon Monte Carlo sampling techniques coupled with energy minimization. GLIDE SP (Standard Precision mode) was used to dock the compound CC-I followed by GLIDE XP (Extra Precision mode). Schrödinger's LigPrep was used to generate the 3D conformations of CC-I

Table 1. Gene expression profile of human Apoptosis PCR Array in CC-I treated CCF-STTG1 cells.

Gene Name (Gene Symbol)	GenBank Accession Number	Description	Fold (Compare to control)
BAG-3/BIS (BAG3)	NM_004281	BCL2-associated athanogene 3	up 8.2
BCL-B/Boo (BCL2L10)	NM_020396	BCL2-like 10 (apoptosis facilitator)	up 29.4
BIP1/BP4 (BIK)	NM_001197	BCL2-interacting killer (apoptosis-inducing)	up 9.9
AIP1/API2 (BIRC3)	NM_001165	Baculoviral IAP repeat-containing 3	up 10.4
ILP-2/ILP2 (BIRC8)	NM_033341	Baculoviral IAP repeat-containing 8	up 8.0
ALPS2/FLICE2 (CASP10)	NM_001230	Caspase 10, apoptosis-related cysteine peptidase	up 16.8
MGC119078 (CASP14)	NM_012114	Caspase 14, apoptosis-related cysteine peptidase	up 45.5
Bp50/CDW40 (CD40)	NM_001250	CD40 molecule, TNF receptor superfamily member 5	up 95.1
CD154/CD40L (CD40LG)	NM_000074	CD40 ligand (TNF superfamily, member 5, hyper-IgM syndrome)	up 52.2
CIDE-A (CIDEA)	NM_001279	Cell death-inducing DFFA-like effector a	up 27.3
APT1LG1/CD178 (FASLG)	NM_000639	Fas ligand (TNF superfamily, member 6)	up 33.3
DP5/HARAKIRI (HRK)	NM_003806	Harakiri, BCL2 interacting protein (contains only BH3 domain)	up 66.6
LT/TNFB (LTA)	NM_000595	Lymphotoxin alpha (TNF superfamily, member 1)	up 110.6
ASC/CARD5 (PYCARD)	NM_013258	PYD and CARD domain containing	up 6.3
DIF/TNF-alpha (TNF)	NM_000594	Tumor necrosis factor (TNF superfamily, member 2)	up 703.4
APO2/CD261 (TNFRSF10A)	NM_003844	Tumor necrosis factor receptor superfamily, member 10a	up 22.8
S152/T14 (CD27)	NM_001242	CD27 molecule	up 8.1
4–1BB/CD137 (TNFRSF9)	NM_001561	Tumor necrosis factor receptor superfamily, member 9	up 302.6
CD27L/CD27LG (CD70)	NM_001252	CD70 molecule	up 38.2
CD153/CD30L (TNFSF8)	NM_001244	Tumor necrosis factor (ligand) superfamily, member 8	up 22.3

Statistical Analysis

All of the data was subjected to statistical analysis by the student t-test when comparing two groups. We used one-way ANOVA followed by Tukey-Kramer test for more than two group comparisons to determine if the differences are significant. For comparisons of time course or concentration data we performed repeated measures two-way ANOVA followed by Tukey-Kramer test. Differences among means are considered statistically significant when the p value is less than 0.05. The LC_{50} (50% lethal concentration) of compounds was determined using statistical software (GraphPad Prism 6) as a general indicator of a chemical's toxicity. In the *in vivo* brain tumor model, the tumor volume data was summarized as the mean values with standard errors. The mice survival was compared between the groups using Kaplan-Meier survival analysis with logrank test.

Results

Identification of a cytotoxic compound against TMZ resistant astrocytoma cells

Our screening approach identified a thiobarbituric acid analog and given the identification tag of chemotype compound–I (CC-I). The structure of CC-I is shown in **Figure 1A**. CC-I was cytotoxic to both the TMZ-resistant human astrocytoma cell lines CCF-STTG1 and to TMZ-sensitive SW1088 (**Figure 1B**). The LC_{50} of CC-I to SW1088, U87-MG and CCF-STTG1 cell lines is 13.6 μM, 23.6 μM and 25.4 μM respectively.

Acute toxicity of CC-I in nude mice

Injections of CC-I once a week at 50 or 75 mg/kg body weight were lethal within 7 days. A once a week injection at 35 mg/kg

body weight was tolerated. Therefore, we used approximately 70% of the tolerated dose (25 mg/kg body weight) of CC-I concentration for the *in vivo* tumor model study.

Anti-tumor effect of CC-I in the subcutaneous mouse tumor model

To establish the anti-tumor effect of CC-I on astrocytoma cells, we used the immunodeficient nude mouse subcutaneous tumor model injected with either TMZ sensitive SW1088 or TMZ resistant CCF-STTG1 cell lines. The mice with tumors from the CCF-STTG1 cell line showed no evidence of tumor progression following CC-I injections even after the injections ended (**Figure 2A**) whereas in the untreated control group the tumor volume dramatically increased over 7 weeks (p<0.0001). The tumors in mice from the SW1088 cell line also failed to progress during the injection period, but the tumor progressed when the CC-I injections were discontinued (**Figure 2A**). We did not take pictures of the tumors. We will consider taking pictures for upcoming experiments. The body weight for the control or CC-I treated mice did not decrease during course of the study (**Figure 2B**).

Anti-tumor effect of CC-I in intracranial brain tumor model

After establishing the *in vivo* efficacy and safety of CC-I against both TMZ sensitive and resistant cell lines in the subcutaneous brain tumor model, we examined the intracranial xenograft brain tumor model. U87-MG or CCF-STTG1 astrocytoma cells were injected into the mouse brain and formed tumors (verified by MRI) ~3 weeks post implantation (**Figure 3A**). None of the untreated control mice survived more than 30 days, and the

A. Cell cycle (day 1)

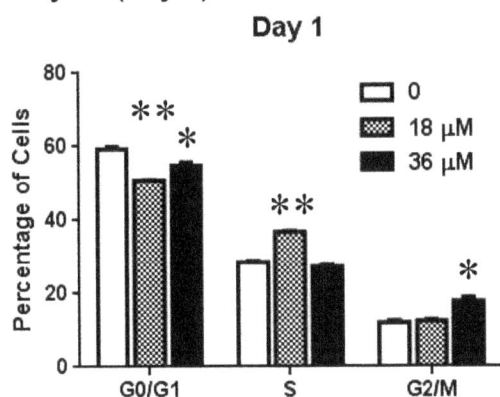

B. Cell cycle (day 2)

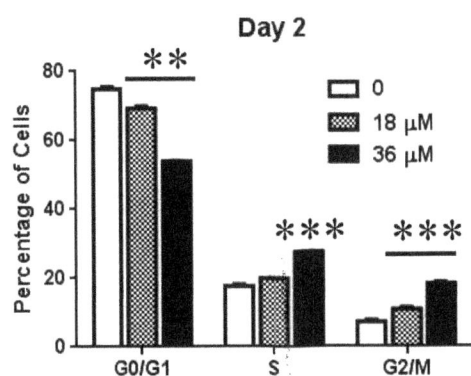

Figure 5. CC-I-induced cell cycle arrest in CCF-STTG1 cells. The CCF-STTG1 cells were treated with 18 or 36 μM of CC-I for 24 or 48 hours. The cells were stained with propidium iodide and then analyzed for cell cycle distribution using a FACScan analyzer. CC-I treatment significantly increased the S and G2/M cell population, but decreased in G0/G1 phase. The symbols indicate a significant difference compared to the control. (*p<0.05; **p<0.01; ***p<0.001).

median survival was 20 days. If the mice were being treated with CC-I, however 64% (7/11) of the U87-MG tumor bearing mice were still alive at 60 days and 89% (8/9) of the CCF-STTG1 tumor bearing mice were still live at 60 days (p<0.0001) (**Figure 3B**) and no tumor was visible on MRI (**Figure 3A**). Five mice in the U87-MG tumor group and six in the CCF-STTG1 tumor group receiving CC-I injections were alive 200 days after the tumor injection (137 days after the last CC-I injection). As with the systemic tumor model, there was no indication of liver or kidney toxicity from CC-I in intracranial xenograft mice (**Figure 3C**). The body weight of the animals did not decrease in the animals receiving CC-I (**Figure 3D**).

Apoptosis of CC-I in the TMZ resistant astrocytoma cells

Next we asked whether the cell death by CC-I to the TMZ resistant CCF-STTG1 astrocytoma cells is mediated through an apoptotic pathway. CC-I induced apoptosis in a dose dependent manner in CCF-STTG1 cell lines (**Figure 4A**). The amount of CCF-STTG1 apoptotic cell death at 36 μM was comparable to the positive control apoptosis inducer, actinomycin D. There is evidence of necrotic cell death in CCF-STTG1 following exposure to CC-I, but fewer cells were labeled and significance was not

achieved until twice the concentration at which apoptosis was first observed (**Figure 4B**).

Apoptosis gene array in CC-I treated TMZ resistant CCF-STTG1 cells

To determine which apoptotic pathway was activated by CC-I treatment, we performed gene expression profiles using targeted arrays for apoptosis. The Human Apoptosis Microarray revealed that tumor necrosis factor (TNF) pathway genes have the greatest changes in gene expression in the CC-I treated cells compared to the vehicle treated cells. CC-I (36 μM) increased TNF superfamily member 1, 2, 5, 6, and 9 as well as TNF receptor superfamily 5, 9, 10a from 30 to 700 fold. Among caspase pathway genes, only caspase 10 and caspase 14 were induced. The fold ratio of the altered genes is summarized in **Table 1**.

Effect of CC-I on the cell cycle of TMZ resistant astrocytoma cells

To better understand the cytotoxic effect of CC-I, we performed a cell cycle analysis in CCF-STTG1 cells after CC-I treatment. CC-I treatment of CCF-STTG1 cells resulted in a significant decrease in the G0/G1 phase, and an increase in the S and G2/M phase compared to untreated cells (**Figure 5A & B**).

Topoisomerase IIα inhibition by CC-I

We determined whether CC-I can bind human topoisomerase IIα in a molecular modeling study. The molecular modeling data between human topoisomerase IIα and CC-I suggested that CC-I fits into the cavity of human topoisomerase IIα where it could function as an inhibitor (**Figure 6A**). Therefore, we performed DNA relaxation and kDNA decatenation assays to determine the ability of CC-I to inhibit topoisomerase IIα enzyme activity. CC-I inhibited topoisomerase IIα activity in a dose dependent manner. At concentrations greater than 23 μM, CC-I inhibited topoisomerase IIα catalyzed kDNA decatenation (**Figure 6B**). Etoposide (VP16), a known topoisomerase II poison, inhibited topoisomerase IIα at 1 mM but not at 0.1 mM concentration (**Figure 6B**). Next, we determined whether CC-I is a specific inhibitor of topoisomerase IIα using a supercoiled DNA relaxation assay. CC-I did not enhance topoisomerase I-mediated relaxation of supercoiled pHOT1 DNA (**Figure 6C**). Camptothecin, a topoisomerase I inhibitor, was used as a positive control for the assay and showed the expected inhibition of topoisomerase I mediated DNA relaxation. In contrast, CC-I exhibited a strong inhibitory effect on topoisomerase IIα-mediated relaxation of supercoiled pHOT1 DNA (**Figure 6D**). The effective concentration of CC-I on topoisomerase IIα mediated DNA relaxation was first seen at 11 μM.

Comparison of cytotoxicity between CC-I and topoisomerase inhibitors on the astrocytoma cells

We compared the relative toxicity of structurally similar topoisomerase inhibitors using TMZ resistant CCF-STTG1 and T98G cells (**Figure 7A**). The LC$_{50}$ of CC-I for CCF-STTG1 and T98G astrocytomas was approximately 22.5 and 29.1 μM. The LC$_{50}$ concentration for CC-I is significantly lower than that found for merbarone (LC$_{50}$: >40 μM, p<0.01). We observed similar relative toxicity of these compounds on SW1088 and U87-MG cell lines.

A.

B. kDNA decatenation assay

C. DNA relaxation by Topo I

D. DNA relaxation by Topo IIα

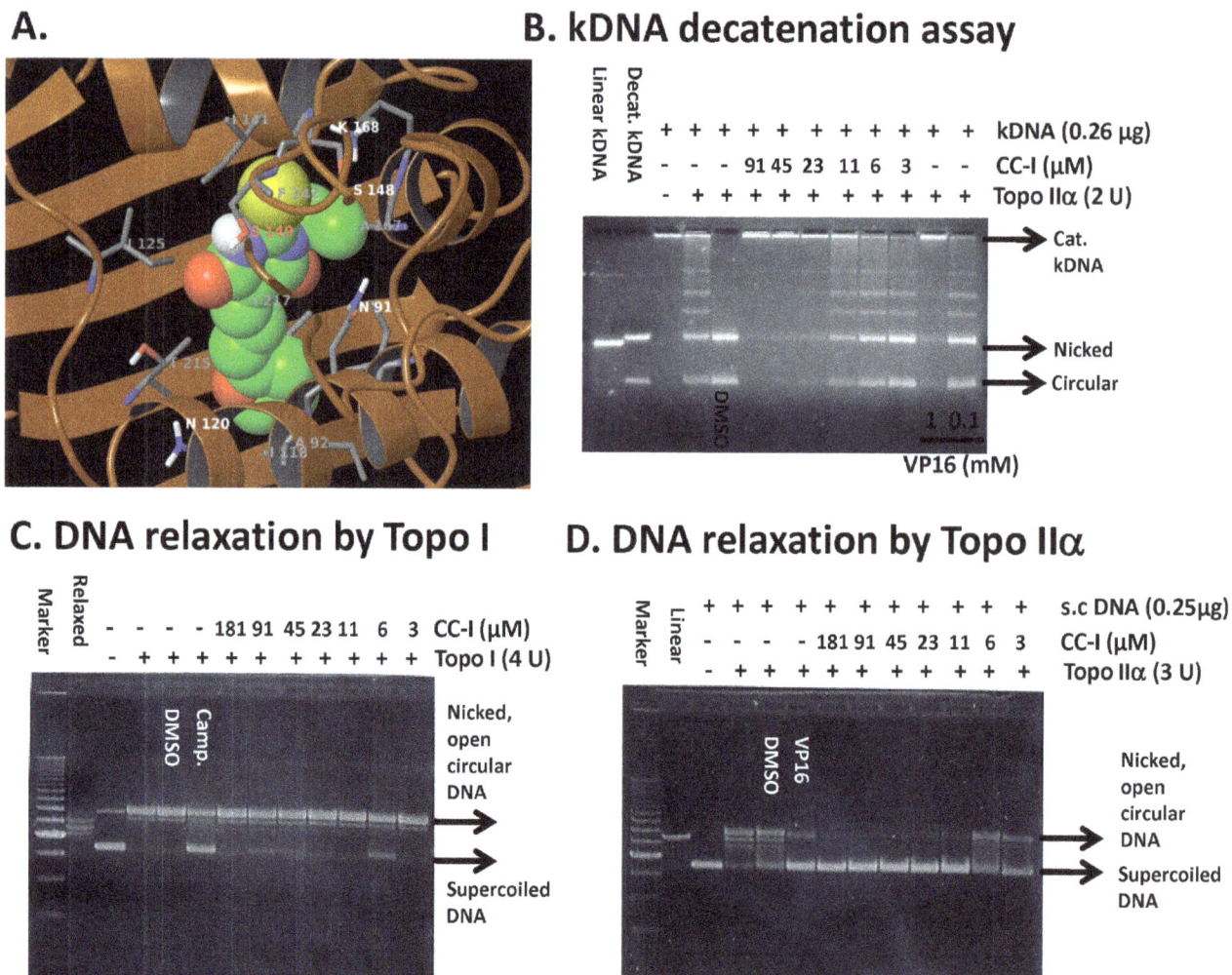

Figure 6. Topoisomerase IIα inhibition by CC-I. (A) Structure of CC-I docked into topoisomerase IIα (pdb code 1ZXM). Topoisomerase is shown as the brown-colored ribbon with residues on the binding site. Carbon atoms of CC-I are colored green, while those of topoisomerase is colored gray. Other atoms are colored according to atom types, i.e., nitrogen-blue, oxygen-red, sulfur-yellow, and polar hydrogen white. Non-polar hydrogen atoms are not shown. (B) The CC-I concentration-dependent inhibition of human topoisomerase IIα-mediated kDNA decatenation. All experiments were carried out according to instructions from the Topogen kit (Port Orange, FL). Reactions contained 4U of enzyme, 0.26 μg of DNA substrate, and different concentrations of the CC-I dissolved in DMSO (0.5% final concentration (v/v)). Different topological forms exhibited different mobility as indicated. Linear, linear kDNA; Decat., decatenated kDNA; Nicked, nicked decatenated kDNA; circular, circular decatenated kDNA; kDNA, kinetoplast DNA. VP16 was used as a positive control. (C) CC-I did not inhibit topo-I mediated supercoiled pHOT1 DNA relaxation. The procedures are described in method section. Camptothecin (camp.) was used as a positive control. (D) CC-I dose dependently inhibited topoisomerase IIα-mediated supercoiled pHOT1 DNA relaxation. VP16 was used as a positive control. s.c. DNA, super-coiled DNA.

Cytotoxicity of CC-I on the MGMT promoter methylated and unmethylated GBM cells

We determined the effect of CC-I using several GBM cell lines that have different MGMT promoter methylation status and MGMT protein expression levels. The LN-18 cell line, which has unmethylated MGMT promoter and MGMT protein expression [26,34], is more sensitive to CC-I than CCF-STTG1 or T98G cells (LC_{50}: 9.03 μM, 14.8 μM, and 13.5 μM respectively; p< 0.05) (**Figure 7B**). The latter cells have methylated MGMT promoter [26].

Combination effect of CC-I and TMZ on the TMZ resistant astrocytoma cell line

To test whether CC-I can enhance cytotoxicity of TMZ in astrocytoma cell lines, we determined effect of combination of both drugs (CC-I & TMZ) on the survival of CC-I resistant T98G

cell lines. Survival of cells was evaluated following treatment with concentrations of CC-I and TMZ around their respective the LC_{50}. There was an additive effect of both drugs. Cell survival which was significantly (p<0.001) reduced in the combined therapy group compared to single treatment in T98G cells after 3 days exposure (**Figure 8**).

Discussion

The present study investigated the development of anti-tumor compounds for TMZ resistant cancer cell lines. Using TMZ resistant cancer cell lines, we identified a lead compound CC-I which is an analog of thiobarbituric acid. The results of the *in vivo* study demonstrate that CC-I is a safe and effective anti-tumor compound against astrocytoma cell lines, including those shown to be resistant to chemotherapy and radiation. CC-I induced

A.

B.

CC-I

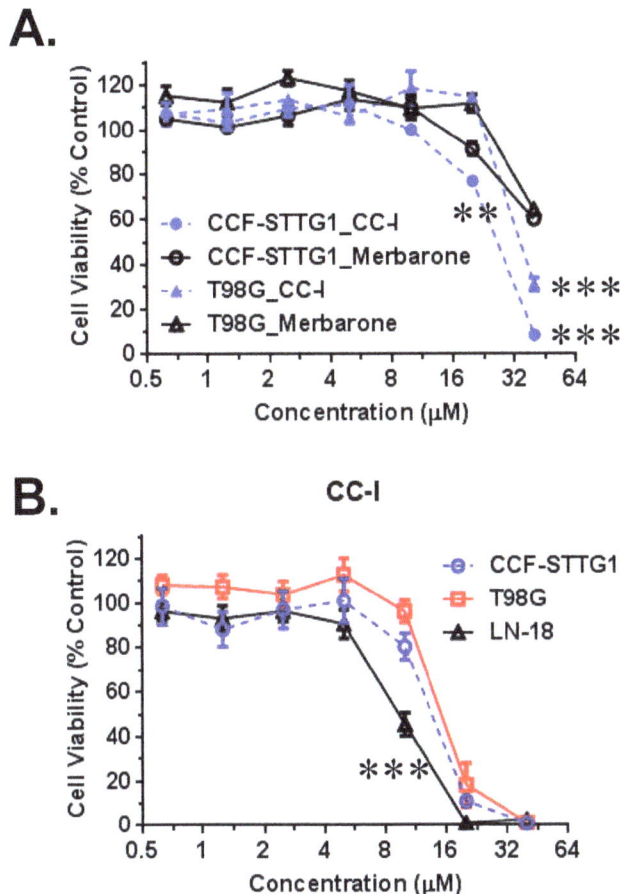

Figure 7. Cytotoxicity of CC-I, merbarone, and combination of CC-I and TMZ on the astrocytoma cells. (A) TMZ-resistant human CCF-STTG1 and T98G cell lines were cultured for 3 days with CC-I and other similar structure topoisomerase II inhibitor (merbarone) followed by cytotoxicity measurement by SRB assay. CC-I showed greater toxicity than merbarone on the astrocytomas. The symbols indicate a significant difference between the merbarone treated and CC-I treated groups (**$p < 0.01$; ***$p < 0.001$). (B) The MGMT methylated (T98G, CCF-STTG1) or un-methylated (LN-18) astrocytoma cell lines were cultured for 3 days with CC-I and determined cytotoxicity by SRB assay. T98G cells have methylated MGMT promoter, but show weak MGMT expression. CC-I is more cytotoxic to LN-18 cells which has un-methylated MGMT promoter and MGMT expression. The symbol (***) indicates the most difference between the cells ($p < 0.001$).

Figure 8. Combination effect of CC-I and TMZ on the T98G astrocytoma cells. T98G cells were cultured for 3 days with CC-I and TMZ, and cytotoxicity was evaluated by SRB assay. Both CC-I and TMZ treatment on the T98G cells showed much more cytotoxic effect than either single treatment. The symbol (***) indicates a significant difference between the control and single treatment groups ($p < 0.001$).

successful and identified a lead therapeutic agent, CC-I, with strong cytotoxicity to tumors, prevention of tumor recurrence, and an acceptable safety profile in *in vivo*. Tumors did not return in 45–66% (depending on cell line) of the mice for 151 days after the last injection and the mice were still alive at 200 days of age when the study was terminated.

CC-I belongs to the thiobarbituric acid family. Various barbituric acid derivatives have been studied as anti-inflammatory and anti-cancer compounds [37–39]. Thiobarbituric acid derivatives also have been studied as anti-tumor agents, uridine phosphorylase inhibitors, HIV-1 integrase inhibitors, and hepatitis C virus polymerase inhibitors [40–43]. An example of thiobarbituric acid derivative evaluated as a treatment for brain cancer is merbarone [5-(N-phenylcarboxamido)-2-thiobarbituric acid] which has a similar structure to CC-I. Merbarone is a non-sedating derivative of thiobarbituric acid and induces single strand breaks in DNA apparently without binding to DNA [44,45]. CC-I also shares structural similarity with ICRF-193 which is a bisdioxopiperazine derivative compound. It has been reported that merbarone and ICRF-193 inhibit topoisomerase [46]. The present study demonstrated that CC-I also inhibits topoisomerase activity within a similar concentration range to merbarone but CC-I is more cytotoxic to the TMZ resistant CCF-STTG1 astrocytoma cell lines than these two compounds. The reason for the differences in cytotoxicity may be due to a structural difference between CC-I which has diene motif linking the barbiturate C5 position with the terminal aromatic ring rather than a shorter amide linker as in merbarone. There is also a structure difference in the functional residue at N1 position; CC-I compound has N-ethyl group, but merbarone has a NH residue.

CC-I exposure resulted in S and G2/M arrest in CCF-STTG1 astrocytoma cell line. This observation is consistent with a number of anti-tumor agents such as 9-methoxycamptothecin, topoisomerase II poisons (doxorubicin, etoposide) [47,48]. For example, 9-methoxycamptothecin induced apoptosis through TNF and Fas/FasL pathway, oxidative stress, and G2/M cell cycle arrest in multiple cancer cell lines [47]. Camptothecin, a topoisomerase I poison, also triggers S and G2/M arrest in cancer cell lines [49]. Our PCR array data indicate that CC-I induces cell death through TNF signaling pathway and the Annexin V data indicate cells die

apoptosis and cell cycle arrest in astrocytoma cells. Because of its structural similarity to topoisomerase inhibitors, we examined CC-I for topoisomerase inhibition and found it selective for topoisomerase IIα. The cytotoxicity of CC-I is greater than other compounds of similar structure.

We have previously reported that human neuroblastoma cells and human astrocytoma cells lines expressing commonly occurring polymorphisms in the HFE gene were resistant to chemotherapy and radiation [26]. The CCF-STTG1 astrocytoma cell lines that carry the HFE C282Y gene variant were even more resistant to TMZ than T98G or U343-MG cell lines, which are considered standards for TMZ resistance [26,35,36]. The CCF-STTG1 cells are also resistant to geldanamycin, its derivatives, and radiation [26] and less sensitive to merbarone; a compound chemotypically similar to our CC-I compound that reached Phase II clinical trials. Our approach using TMZ resistant astrocytoma cells was

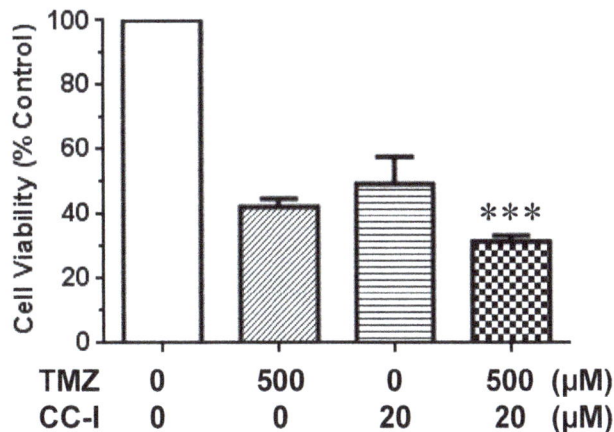

via apoptosis. Therefore our present cell cycle analysis study indicates that CC-I has a similar impact on cell cycle and subsequent apoptosis as many anti-cancer compounds.

CC-I was identified by screening against TMZ resistant astrocytoma cells. However, CC-I was also toxic to TMZ sensitive astrocytoma cells (SW1088, U87-MG). In vivo, CC-I showed greater efficacy against TMZ resistant CCF-STTG1 subcutaneous and intracranial tumors than TMZ sensitive astrocytoma cells (**Figure 2A & 3B**). MGMT methylation status influenced CC-I cytotoxicity, but CC-I has a lower LC_{50} than regardless of methylation status compared to TMZ [26]. This finding is important because there is a correlation between MGMT promoter methylation and GBM patient survival [50]. Because of the relative differences in effect based on methylation status (and HFE genotype) we investigate CC-I in combination with TMZ and found the addition of CC-I improves TMZ efficacy in TMZ resistant astrocytoma cell lines. These findings are consistent with several studies reporting a combination effect with an anti-tumor compound and TMZ in TMZ resistant astrocytoma cell lines [51,52]. The data suggest that CC-I could be considered an adjuvant therapy with TMZ. There are many limitations in translating studies, such as ours, that find compounds that show

efficacy in animal models to clinical application. Nonetheless, the results of the initial analyses of CC-I warrant further investigation.

In conclusion, we identified an anti-tumor compound for TMZ resistant and sensitive astrocytomas with strong *in vivo* efficacy and safety profiles in mouse tumor models. The cytotoxicity of CC-I is mediated by apoptosis, cell cycle arrest at S and G2/M phase. CC-I has a similar biological profile to other topoisomerase inhibitors but it is smaller and shows effects in orthotopic models, therefore we believe it has more attractive properties than most other topoisomerase inhibitors that allows it access the brain.

Acknowledgments

We thank to the Dr. Mohammed Alsaidi for technical help. We also thank to the Drug Discovery Core at Penn State Hershey for the compounds they provided. We further thank to Dr. Qing Yang for MRI imaging. We thank Dr. Mandy Snyder for her critical reading.

Author Contributions

Conceived and designed the experiments: SYL. Performed the experiments: SYL BS ER AP PAM SS. Analyzed the data: SYL JRC. Contributed to the writing of the manuscript: SYL BS PAM SS JRC.

References

1. Ostrom QT, Gittleman H, Farah P, Ondracek A, Chen Y, et al. (2013) CBTRUS Statistical Report: Primary brain and central nervous system tumors diagnosed in the United States in 2006–2010. Neuro Oncol 15 Suppl 2:ii1–56. doi:10.1093/neuonc/not151.

2. American Cancer Society (2012) Cancer Facts & Figures 2012. Atlanta: American Cancer Society. Available: http://www.cancer.org/research/cancerfactsstatistics/cancerfactsfigures2012/.

3. Stupp R, Mason WP, van den Bent MJ, Weller M, Fisher B, et al. (2005) Radiotherapy plus concomitant and adjuvant temozolomide for glioblastoma. N Engl J Med 352: 987–996.

4. Theeler BJ, Groves MD (2011) High-Grade Gliomas. Curr Treat Options Neurol 13: 386–399.

5. Nishikawa R (2010) Standard therapy for glioblastoma – a review of wehere we are. Neurol Med Chir (Tokyo) 50: 713–719.

6. Friedman HS, Kerby T, Calvert H (2000) TMZ and treatment of malignant glioma. Clin Cancer Res 6: 2585–2597.

7. Atallah E, Flaherty L (2005) Treatment of metastatic malignant melanoma. Curr Treat Options Oncol 6: 185–193.

8. van Brussel JP, Busstra MB, Lang MS, Catsburg T, Schröder FH, et al. (2000) A phase II study of TMZ in hormone-refractory prostate cancer. Cancer Chemother Pharmacol 45: 509–512.

9. Moore MJ, Feld R, Hedley D, Oza A, Siu LL (1998) A phase II study of TMZ in advanced untreated pancreatic cancer. Invest New drugs 16: 77–79.

10. Jakob J, Wenz F, Dinter DJ, Ströbel P, Hohenberger P (2009) Preoperative intensity-modulated radiotherapy combined with TMZ for locally advanced soft-tissue sarcoma. Int J Radiat Oncol Biol Phys 75: 810–816.

11. Park DK, Ryan CW, Dolan ME, Vogelzang NJ, Stadler WM (2007) A phase II trial of oral TMZ in patients with metastatic renal cell cancer. Cancer Chemother Pharmacol 50: 160–162.

12. Marchesi F (2007) Triazene compounds: mechanism of action and related DNA repair systems. Pharmacol Res 56: 275–287.

13. Friedman HS, McLendon RE, Kerby T, Dugan M, Bigner SH, et al. (1998) DNA mismatch repair and O6-alkylguanine-DNA alkyltransferase analysis and response to Temodal in newly diagnosed malignant glioma. J Clin Oncol 16: 3851–3857.

14. Hegi ME, Liu L, Herman JG, Stupp R, Wick W, et al. (2008) Correlation of O6-methylguanine methyltransferase (MGMT) promoter methylation with clinical outcomes in glioblastoma and clinical strategies to modulate MGMT activity. J Clin Oncol 26: 4189–4199.

15. Cahill DP, Levine KK, Betensky RA, Codd PJ, Romany CA, et al. (2007) Loss of the mismatch repair protein MSH6 in human glioblastomas is associated with tumor progression during temozolomide treatment. Clin Cancer Res 13: 2038–2045.

16. Yip S, Miao J, Cahill DP, Iafrate AJ, Aldape K, et al. (2009) MSH6 mutations arise in glioblastomas during temozolomide therapy and mediate temozolomide resistance. Clin Cancer Res 15: 4622–4629.

17. McClendon AK, Osheroff N (2007) DNA topoisomerase II, genotoxicity, and cancer. Mutat Res 623: 83–97.

18. Champoux JJ (2001) DNA topoisomerases: structure, function, and mechanism. Annu Rev Biochem 70: 369–413.

19. Beck WT (1996) DNA topoisomerases and tumor cell resistance to their inhibitors. In: Schilsky R, Milano G, Ratain M, editors. Principles of Antineoplastic Drug Development and Pharmacology (Basic and Clinical Oncology). CRC Press. 487–502.

20. Falaschi A, Abdurashidova G, Sandoval O, Radulescu S, Biamonti G, et al. (2007) Molecular and structural transactions at human DNA replication origins. Cell Cycle 6: 1705–1712.

21. Bruce JN, Fine RL, Canoll P, Yun J, Kennedy BC, et al. (2011) Regression of recurrent malignant gliomas with convection-enhanced delivery of topotecan. Neurosurgery 69: 1272–1279.

22. Minturn JE, Janss AJ, Fisher PG, Allen JC, Patti R, et al. (2011) A phase II study of metronomic oral topotecan for recurrent childhood brain tumors. Pediatr Blood Cancer 56: 39–44.

23. Matsumoto Y, Tamiya T, Nagao S (2005) Resistance to topoisomerase II inhibitors in human glioma cell lines overexpressing multidrug resistant associated protein (MRP) 2. J Med Invest 52: 41–48.

24. Chen Y, Su YH, Wang CH, Wu JM, Chen JC, et al. (2005) Induction of apoptosis and cell cycle arrest in glioma cells by GL331 (a topoisomerase II inhibitor). Anticancer Res 25: 4203–4208.

25. Schmidt F, Knobbe CB, Frank B, Wolburg H, Weller M (2008) The topoisomerase II inhibitor, genistein, induces G2/M arrest and apoptosis in human malignant glioma cell lines. Oncol Rep 19: 1061–1066.

26. Lee SY, Liu S, Mitchell RM, Slagle-Webb B, Hong Y-S, et al. (2011) HFE polymorphisms influence the response to chemotherapeutic agents via induction of p16INK4A. Int J Cancer 129: 2104–2114.

27. de Almeida SF, Picarote G, Fleming JV, Carmo-Fonseca M, Azevedo JE, et al. (2007) Chemical chaperones reduce endoplasmic reticulum stress and prevent mutant HFE aggregate formation. J Biol Chem 282: 27905–27912.

28. Liu Y, Lee SY, Neely E, Nandar W, Moyo M, et al. (2011) Mutant HFE H63D protein is associated with prolonged endoplasmic reticulum stress and increased neuronal vulnerability. J Biol Chem 286: 13161–13170.

29. Lee SY, Patton SM, Henderson RJ, Connor JR (2007) Consequences of expressing mutants of the hemochromatosis gene (HFE) into a human neuronal cell line lacking endogenous HFE. FASEB J 21: 564–576.

30. Gong Y, Firestone GL, Bjeldanes LF (2006) 3,3′-diindolylmethane is a novel topoisomerase IIalpha catalytic inhibitor that induces S-phase retardation and mitotic delay in human hepatoma HepG2 cells. Mol Pharmacol 69: 1320–1327.

31. Wendorff TJ, Schmidt BH, Heslop P, Austin CA, Berger JM (2012) The structure of DNA-bound human topoisomerase II alpha: conformational mechanisms for coordinating inter-subunit interactions with DNA cleavage. J Mol Biol 424: 109–124.

32. Friesner RA, Banks JL, Murphy RB, Halgren TA, Klicic JJ, et al. (2004) Glide: a new approach for rapid, accurate docking and scoring. 1. Method and assessment of docking accuracy. J Med Chem 47: 1739–1749.

33. Halgren TA, Murphy RB, Friesner RA, Beard HS, Frye LL, et al. (2004) Glide: a new approach for rapid, accurate docking and scoring. 2. Enrichment factors in database screening. J Med Chem 47: 1750–1759.

34. Mellai M, Monzeglio O, Piazzi A, Caldera V, Annovazzi L, et al. (2012) MGMT promoter hypermethylation and its associations with genetic alterations in a series of 350 brain tumors. J Neurooncol 107: 617–631.

35. Kanzawa T, Germano IM, Kondo Y, Ito H, Kyo S, et al. (2003) Inhibition of telomerase activity in malignant glioma cells correlates with their sensitivity to TMZ. Br J Cancer 89: 922–929.

36. Uzzaman M, Keller G, Germano IM (2007) Enhanced proapoptotic effects of tumor necrosis factor-related apoptosis-inducing ligand on TMZ-resistant glioma cells. J Neurosurg 106: 646–651.

37. Cebo B, Krupinska J, Mazur J, Piotrowicz J (1980) Antiinflammatory activity of the new aminomethyl derivatives of 1-cyclohexyl-5-alkyl and 1-cyclohexyl-5,5-dialklbarbituric acids. Farmaco Sci 35: 248–252.

38. Brewer AD, Minatelli JA, Plowman J, Paull KD, Narayanan VL (1985) 5-(N-phenylcarboxamido)-2-thiobarbituric acid (NSC 336628), a novel potential antitumor agent. Biochem Pharmacol 34: 2047–2050.

39. Singh P, Kaur M, Verma P (2009) Design, synthesis and anticancer activities of hybrids of indole and barbituric acids-identification of highly promising leads. Bioorg Med Chem Lett 19: 3054–3058.

40. Balas VI, Hadjikakou SK, Hadjiliadis N, Kourkoumelis N, Light ME, et al. (2008) Crystal structure and antitumor activity of the novel zwitterionic complex of tri-n-butyltin(IV) with 2-thiobarbituric acid. Bioinorg Chem Appl 654137. doi:10.1155/2008/654137.

41. Balas VI, Verginadis II, Geromichalos GD, Kourkoumelis N, Male L, et al. (2011) Synthesis, structural characterization and biological studies of the triphenyltin(IV) complex with 2-thiobarbituric acid. Eur J Med Chem 46: 2835–2844.

42. Rajamaki S, Innitzer A, Falciani C, Tintori C, Christ F, et al. (2009) Exploration of novel thiobarbituric acid-, rhodanine- and thiohydrantoin-based HIV-1 integrase inhibitors. Bioorg Med Chem Lett 19: 3615–3618.

43. Lee JH, Lee S, Park MY, Myung H (2011) Characterization of thiobarbituric acid derivatives as inhibitors of hepatitis C virus NS5B polymerase. Virol J 8: 18. doi:10.1186/1743-422X-8-18.

44. Warrell RP Jr, Muindi J, Stevens YW, Isaacs M, Young CW (1989) Induction of profound hypouricemia by a non-sedating thiobarbiturate. Metabolism 38: 550–554.

45. Glover A, Chun HG, Kleinman LM, Cooney DA, Plowman J, et al. (1987) Merbarone: an antitumor agent entering clinical trials. Invest New Drugs 5: 137–143.

46. Drake FH, Hofmann GA, Mong SM, Bartus JO, Hertzberg RP, et al. (1989) In vitro and intracellular inhibition of topoisomerase II by the antitumor agent merbarone. Cancer Res 49: 2578–2583.

47. Wang H, Ao M, Wu J, Yu L (2013) TNFα and Fas/FasL pathways are involved in 9-Methoxycamptothecin-induced apoptosis in cancer cells with oxidative stress and G2/M cell cycle arrest. Food Chem Toxicol 55: 396–410.

48. Kolb RH, Greer PM, Cao PT, Cowan KH, Yan Y (2012) ERK1/2 signaling plays an important role in topoisomerase II poison-induced G2/M checkpoint activation. PLoS One 7: e50281. doi:10.1371/journal.pone.0050281.

49. Bhonde MR, Hanski ML, Notter M, Gillissen BF, Daniel PT, et al. (2006) Equivalent effect of DNA damage-induced apoptotic cell death or long-term cell cycle arrest on colon carcinoma cell proliferation and tumour growth. Oncogene 25: 165–175.

50. Melguizo C, Prados J, González B, Ortiz R, Concha A, et al. (2012) MGMT promoter methylation status and MGMT and CD133 immunohistochemical expression as prognostic markers in glioblastoma patients treated with temozolomide plus radiotherapy. J Transl Med 10: 250. doi:10.1186/1479-5876-10-250.

51. Vlachostergios PJ, Hatzidaki E, Befani CD, Liakos P, Papandreou CN (2013) Bortezomib overcomes MGMT-related resistance of glioblastoma cell lines to temozolomide in a schedule-dependent manner. Invest New Drugs 31: 1169–1181.

52. Peigñan L, Garrido W, Segura R, Melo R, Rojas D, et al. (2011) Combined use of anticancer drugs and an inhibitor of multiple drug resistance-associated protein-1 increases sensitivity and decreases survival of glioblastoma multiforme cells in vitro. Neurochem Res 36: 1397–1406.

12

Comparison of Efficacy and Toxicity of Traditional Chinese Medicine (TCM) Herbal Mixture LQ and Conventional Chemotherapy on Lung Cancer Metastasis and Survival in Mouse Models

Lei Zhang[1,3], Chengyu Wu[2]*, Yong Zhang[1], Fang Liu[1,4], Xiaoen Wang[1], Ming Zhao[1], Robert M. Hoffman[1,3]*

1 AntiCancer, Inc., San Diego, California, United States of America, **2** Department of Traditional Chinese Medicine Diagnostics, Nanjing University of Chinese Medicine, Nanjing, China, **3** Department of Surgery, University of California San Diego, San Diego, California, United States of America, **4** Department of Anatomy, Second Military Medical University, Shanghai, China

Abstract

Unlike Western medicine that generally uses purified compounds and aims to target a single molecule or pathway, traditional Chinese medicine (TCM) compositions usually comprise multiple herbs and components that are necessary for efficacy. Despite the very long-time and wide-spread use of TCM, there are very few direct comparisons of TCM and standard cytotoxic chemotherapy. In the present report, we compared the efficacy of the TCM herbal mixture LQ against lung cancer in mouse models with doxorubicin (DOX) and cyclophosphamide (CTX). LQ inhibited tumor size and weight measured directly as well as by fluorescent-protein imaging in subcutaneous, orthotopic, spontaneous experimental metastasis and angiogenesis mouse models of lung cancer. LQ was efficacious against primary and metastatic lung cancer without weight loss and organ toxicity. In contrast, CTX and DOX, although efficacious in the lung cancer models caused significant weight loss, and organ toxicity. LQ also had anti-angiogenic activity as observed in lung tumors growing in nestin-driven green fluorescent protein (ND-GFP) transgenic nude mice, which selectively express GFP in nascent blood vessels. Survival of tumor-bearing mice was also prolonged by LQ, comparable to DOX. *In vitro*, lung cancer cells were killed by LQ as observed by time-lapse imaging, comparable to cisplatinum. LQ was more potent to induce cell death on cancer cell lines than normal cell lines unlike cytotoxic chemotherapy. The results indicate that LQ has non-toxic efficacy against metastatic lung cancer.

Editor: Beicheng Sun, The First Affiliated Hospital of Nanjing Medical University, China

Funding: This study was supported by National Cancer Institute grant CA132971 and grant BA2011019 from the Jiangsu Science and Technology Achievements Transformation Thematic Special Funds, China. The funders had no role in study design, data collection and analysis, decision to publish, or preparation of the manuscript.

Competing Interests: Lei Zhang, Yong Zhang, Xiaoen Wang and Ming Zhao are affiliates of AntiCancer Inc. Fang Liu was a former affiliate of AntiCancer Inc. Robert M. Hoffman is a non-salaried affiliate of AntiCancer Inc. AntiCancer Inc. markets animal models of cancer. There are no other competing interests. There are no patents, products in development or marketed products to declare.

* Email: chengyu720@163.com (CW); all@anticancer.com (RMH)

Introduction

Lung cancer is the leading cause of cancer death with non–small-cell lung cancer (NSCLC) accounting for approximately 80% of thoracic malignancies. The overall 5 years survival for NSCLC remains poor, approximately 16%. Lung cancer has shown little improvement in survival for the last 30 years [1]. Novel approaches to the treatment of lung cancer are urgently required. Nature products are important resources for drug development. For cancer disease, 60% of new drugs, originate from natural sources. There is currently increasing interest in traditional Chinese medicine (TCM) herbal mixtures which have been used to treat cancer for thousands of years in China. Unlike Western medicine that generally uses purified compounds and aims to target a single molecule or pathway, TCM compositions usually comprise multiple herbs and components. There is much

anecdotal evidence of the efficacy of TCM in the form of herbal mixtures in cancer patients [2–9].

In a previous study, we determined the inhibitory efficacy of the TCM herb Celastrus orbiculatus Thunb. (COT) on tumor growth, metastasis and antiogenesis of hepatocelluar carcinoma (HCC) Hep-G2 cells in an orthotopic nude mouse model using fluorescence imaging technology. Whole-body fluorescence imaging was performed to measure tumor growth and monitor metastasis development. High-dose, early treatment with COT demonstrated significant efficacy on controlling tumor volume and tumor weight in the human HCC Hep-G2 orthotopic tumor model [4].

The efficacy of the TCM herb tubeimu, extracted from the tuber of the plant *Bolbostemma paniculatum* was tested on MDA-MB-231 human breast cancer cells *in vitro*. The MDA-MB-231 cell line was engineered to express RFP in the cytoplasm and GFP linked to histone H2B in the nucleus, which allows real-time imaging of nuclear-cytoplasmic dynamics. Apoptosis was readily visualized in these cells by nuclear shape changes and fragmentation. The MDA-MB-231 RFP-GFP cells were cultured either in two-dimensions on plastic or in three dimensions on Gelfoam. Cells were treated with a dichloromethane extract of fresh tubeimu. Tubeimu induced apoptosis of MDA-MB-231 cells, as observed by fluorescence microscopy, as early as 24 hours of treatment in vitro in two-dimensional culture. By 48 hours' treatment, DNA fragmentation could be observed. Tubeimu also induced apoptosis of MDA-MB-231 cells in three-dimensional culture on Gelfoam, but to a lesser extent than in 2D culture [5].

Despite the very long-history and wide-spread use of TCM, there are very few direct comparisons of TCM and standard cytotoxic chemotherapy. We have previously developed a TCM formulation of herbs, LQ, that has been shown to have potent non-toxic therapeutic properties in clinically and in orthotopic mouse models of pancreatic cancer [6]. We compared LQ to gemcitabine, which is first-line therapy for pancreatic cancer for anti-metastatic and anti-tumor activity as well as safety. The therapeutic efficacy of LQ was comparable with gemcitabine but with less toxicity [6].

In the present report, we used state-of-art fluorescence imaging technology and animal models to compare efficacy and toxicity of LQ to doxorubicin (DOX) and cyclophosphamide (CTX) on lung cancer. With the use of GFP and/or RFP stably expressed in human cancer cells or mice, the efficacy of LQ was compared to standard chemotherapy on tumor and metastatic growth as well as angiogenesis.

We demonstrate here that LQ has antitumor, anti-metastatic and anti-angiogenic efficacy in clinically-relevant mouse models of lung cancer comparable to doxorubicin and cyclophosphamide, without their toxicity.

Materials and Methods

Ethics statement

All animal studies were conducted with an AntiCancer Institutional Animal Care and Use Committee (IACUC)-protocol specifically approved for this study and in accordance with the principals and procedures outlined in the National Institute of Health Guide for the Care and Use of Animals under Assurance Number A3873-1. In order to minimize any suffering of the animals the use of anesthesia and analgesics were used for all surgical experiments. Animals were anesthetized with a 20 μL mixture of Ketamine (22–44 mg/kg), Acepromazine (0.75 mg/kg), and Xylazine (2–5 mg/kg) by intramuscular injection 10 minutes before surgery. The response of animals during surgery was monitored to ensure adequate depth of anesthesia. Ibuprofen (7.5 mg/kg orally in drinking water every 24 hours for 7 days post-surgery) was used in order to provide analgesia post-operatively in the surgically-treated animals. The animals were observed on a daily basis and humanely sacrificed by CO_2 inhalation when they met the following humane endpoint criteria: prostration, skin lesions, significant body weight loss, difficulty breathing, epistaxis, rotational motion and body temperature drop. The use of animals was necessary to understand the in vivo efficacy, in particular, anti-metastatic efficacy of the agents tested. Animals were housed with no more than 5 per cage. Animals were housed in a barrier facility on a high efficiency particulate air (HEPA)-filtered rack under standard conditions of 12-hour light/dark cycles. The animals were fed an autoclaved laboratory rodent diet.

Cell lines and cell culture

Human lung cancer cell lines H460 [12], A549 [13], mouse Lewis lung carcinoma (LLC) [14], and monkey normal kidney endothelial cell line (VERO) [15] were maintained in RPMI-1640 (HyClone, South Logan, UT) with 10% fetal bovine serum (Gemini Bio-Products, Calabasas, CA). Human umbilical vein endothelial cells (HUVEC) [16] was maintained in EGM-2 Bulletkit medium (Lonza, Anaheim, CA).

LLC cells stably expressed red fluorescent protein (RFP), as previously described [10,11]. H460 cells stably expressed green fluorescent protein (GFP), as previously described [12,17–19]. H460 dual-color cells expressed GFP linked to histone H2B in the nucleus as previously described [20–22].

Mice

Athymic nude mice (nu/nu) and C57BL/6 mice (AntiCancer Inc., San Diego, CA), 6–8 weeks old, were used in this study. Nestin-driven-GFP (ND-GFP) transgenic C57/B6 nude mice (AntiCancer, Inc.) expressing GFP under control of the nestin promoter were also used [23–26]. Non-transgenic C57 B/6 mice (AntiCancer, Inc.) were also used.

Subcutaneous tumor growth

A549, H460, LLC, and H460-GFP cells were harvested by trypzinization and washed two times with phosphate-buffered saline (PBS) (HyClone, South Logan, UT). Cells (5×10^6) were injected subcutaneously into the right flank of mice in a total volume of 100 μl PBS within 30 min of harvesting. The subcutaneous tumors were also used as the source of tissue for orthotopic implantation into the lung.

Surgical orthotopic implantation (SOI)

Tumor pieces (1 mm³) derived from H460-GFP subcutaneous tumors growing in the nude mouse were implanted by surgical orthotopic implantation (SOI) [27] onto the left visceral pleura in cohorts of additional nude mice [12,28–30]. The mice were anesthetized by Isofluran inhalation. A small 1 cm transverse incision was made on the left-lateral chest of the nude mice via the fourth intercostal space. A small incision (0.4–0.5 cm) between the third and fourth rib on the chest wall provided access to the pleural space and resulted in total lung collapse. Tumor fragments (1 mm³) were sewn together with an 8-0 surgical suture and fixed by making one knot. The lung was taken up by forceps and the tumor fragment sewn into the lower part of the lung with one suture. The lung tissue was then returned into the chest cavity. The chest muscles and skin were closed with a 6-0 surgical suture. The closed condition of the chest wall was examined immediately and, if a leak existed, it was closed by additional sutures. After closing the chest wall, the lung was un-inflated by withdrawing air from the chest cavity with a 25-gauge ½ needle. After the withdrawal of air, a completely inflated lung can be seen through the thin chest wall of the mouse. Then, the skin and chest muscle were closed with a 6-0 surgical suture in one layer. All procedures of the operation described above were performed with a 7× microscope [12].

Experimental metastasis syngeneic mouse model

LLC-RFP cells (2×10^6) were harvested by trypsinization and washed with cold PBS, then injected in the tail vein of nude or C57B/6 mice in a total volume of 100 μl with a 1 ml 29-gauge, latex-free syringe within 30 minutes of harvesting. The seeding and arrest of single cancer cells on the lung, accumulation of cancer-cell emboli, cancer-cell viability, and metastatic colony

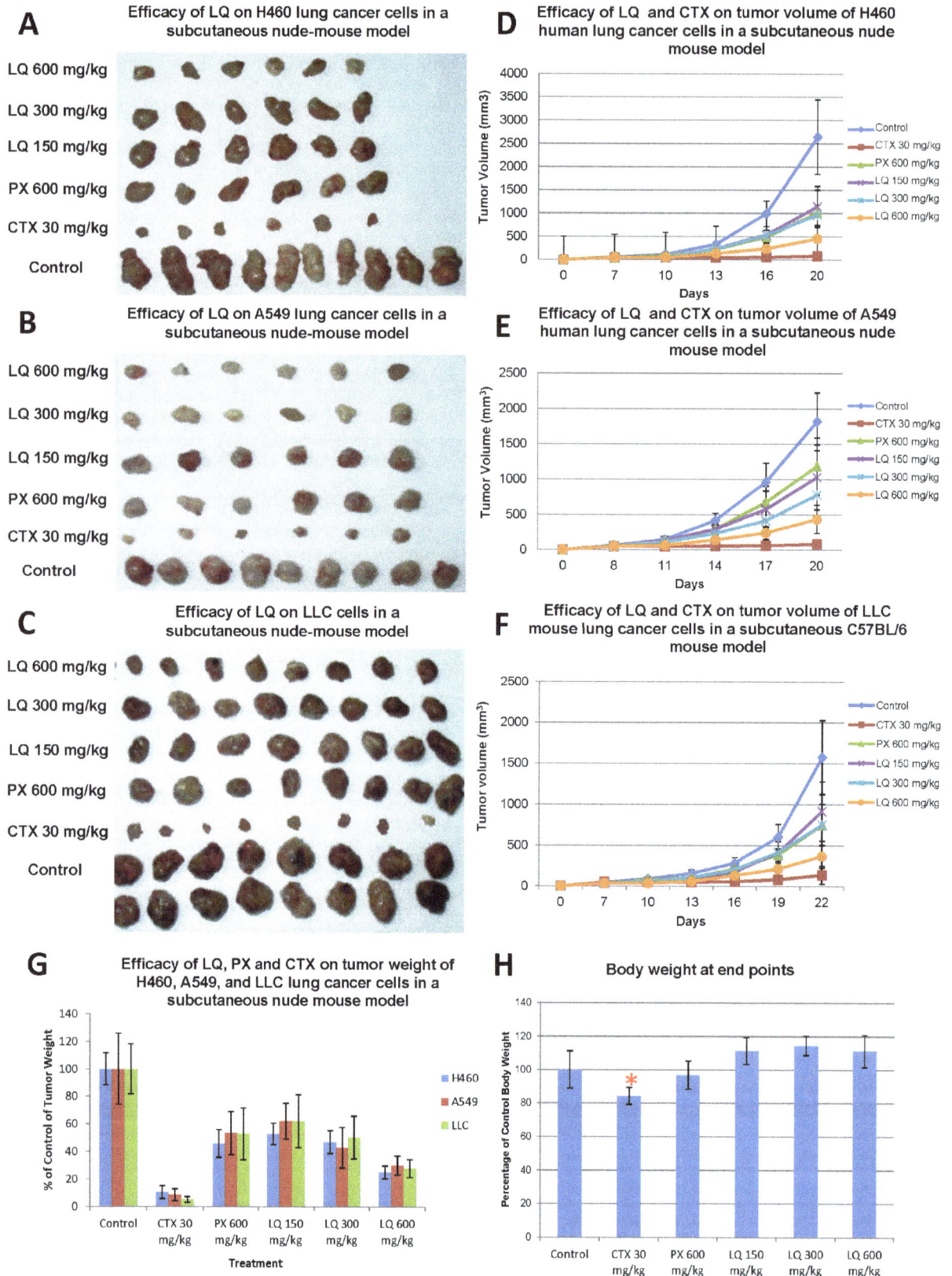

A Efficacy of LQ on H460 lung cancer cells in a subcutaneous nude-mouse model

B Efficacy of LQ on A549 lung cancer cells in a subcutaneous nude-mouse model

C Efficacy of LQ on LLC cells in a subcutaneous nude-mouse model

D Efficacy of LQ and CTX on tumor volume of H460 human lung cancer cells in a subcutaneous nude mouse model

E Efficacy of LQ and CTX on tumor volume of A549 human lung cancer cells in a subcutaneous nude mouse model

F Efficacy of LQ and CTX on tumor volume of LLC mouse lung cancer cells in a subcutaneous C57BL/6 mouse model

G Efficacy of LQ, PX and CTX on tumor weight of H460, A549, and LLC lung cancer cells in a subcutaneous nude mouse model

H Body weight at end points

Figure 1. Efficacy of LQ on tumor size, growth and weight in subcutaneous nude mouse tumor models. For the H460 and A549 cell lines, each treatment group contained 6 mice and 10 mice were used for the untreated control groups. For the LLC cell line, each treatment group contained 8 mice and 16 mice were used for the untreated control group. After the subcutaneous tumors grew, the nude mice were given either CTX (30 mg/kg/day i.p.) for 7 days or PX (600 mg/kg/day p.o) for 10 days. PBS (po) was used in the control group. Mice were treated at 150, 300, and 600 mg/kg/day of LQ (po) for 10 days. Tumor size was measured twice a week. Tumor weight was measured at the endpoint. Statistical significance between groups was determined with the Student's t-test. H460 (**A&D**), A549 (**B&E**), and LLC (**C&F**) had growth inhibition and tumor size inhibition after treatment with all agents. The tumor weight was also significantly inhibited by all agents (**G**) ($p<0.01$). After LQ treatment, lung tumor growth and tumor weight were significantly inhibited, compared to the untreated control group in all models ($p<0.01$). CTX and PX treatment resulted in a significant inhibition of tumor weight ($P<0.01$) compared to the control group for all cell lines. LQ had more efficacy than PX ($p<0.05$) on all cell lines (**G**). CTX induced loss of body weight ($p<0.05$), but not LQ or PX (**H**).

formation were imaged at 14 days after cell injection [31]. Individual images of excised lungs from the mice in the experimental metastasis model were obtained with the OV100 Small Animal Imaging System (Olympus Corp., Tokyo, Japan). The total red fluorescence area and lung weight were measured for metastatic burden. The images were analyzed using Imaging-Pro plus 6.0 software (MediaCybernetics Inc., Rockville, MD).

Color-coded angiogenesis model

ND-GFP transgenic nude mice, 6 to 8 weeks old, were used. The mice were anesthetized and RFP-expressing LLC cells (5×10^5 cells in 25 μl) were injected into the skin of the ear and footpad of the ND-GFP nude mice with a 1 ml 27G ½ latex-free syringe [32]. The total length of GFP-expressing blood vessels and fluorescence intensity were quantitated. The images were analyzed with Imaging-Pro plus 6.0 software (MediaCybernetics Inc.).

Assessment of efficacy in vivo

Efficacy of treatment was determined by standard measurements of tumor volume and tumor weight in the subcutaneous models. Tumor volume was calculated using the formula (long diameter×short diameter^2)/2. In the orthotopic models, mice were sacrificed and explored when they appeared pre-morbid. After euthanasia, each mouse underwent laparotomy and median sternotomy and was then imaged in order to identify primary and metastatic tumors by imaging RFP or GFP fluorescence. The OV100 Small Animal Imaging System was used [33]. After performing full-body, open imaging, the solid organs were thoroughly examined for any evidence of metastasis. Body weight, organ structure observation, and general appearance of each mouse were monitored and recorded as evidence of systemic toxicity.

Haematoxylin and Eosin (H&E) staining

The heart, liver, spleen, lung, kidney, and intestine of mice from each group were collected and embedded in frozen tissue matrix (OCT) and frozen immediately with liquid nitrogen. All the tissues were prepared as frozen sections and cut into 5 μm-thick slides for H&E staining. The histopathological structures of different organs were examined using an Olympus BH-2 microscope.

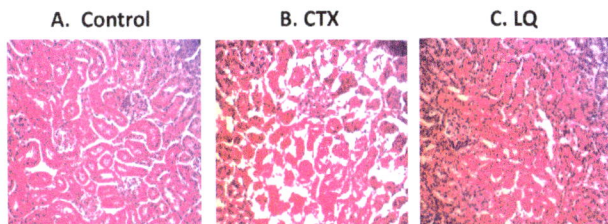

A. Control　　**B. CTX**　　**C. LQ**

Figure 2. Toxicity of treatment on the renal cortex. (A) Renal cortex of untreated mouse. (B) The renal cortex of mice treated with CTX demonstrated renal toxicity. Toxicity was mainly in the proximal tubular cells, including degeneration, obstruction, necrosis and swelling. (C) The renal cortex of mice treated with LQ had no toxicity.

Preparation of crude extracts of Chinese herbs

LQ is a mixture of Chinese medicinal herbs, comprising *Sinapis alba*, *Atractylodes macrocephala*, *Coix lacryma-jobi*, and *Polyporus adusta* and prepared at the School of Pharmacy, Nanjing University of Traditional Chinese Medicine as previously reported [6]. The ratio used (2:3:4:3) is from dried plants. The plant parts used were *Sinapis albe:* seed; *Atractylodes macrocephala:* root; *Coix lacryma-jobi:* kernel; *Polyporus adusta:* whole mushroom. We followed US Pharmacopoeias UPS231 for heavy metal testing. All the herbs were tested and the levels of heavy metal were below the US Pharmacopoeias minimum daily dose. Each herb was extracted in boiling water for 20 mins, the solution was filtered. The residue was extracted with 75% ethanol, and the extract was filtered. Both the water extract and ethanolic extract were combined and concentrated by lyophilization. To obtain 100 mg of LQ required 402 mg of dried herbs. The lyophilized powder was suspended in PBS. The mixture was vortexed for 1 min and incubated at 80°C for 30 min. The sample was cooled to room temperature and was then centrifuged at 2000 rpm for 10 min. The supernatant was collected at a final concentration to 90 mg/ml. LQ was then diluted for dose-ranging experiments in the mouse models or filtered through a 0.2 μm membrane for *in vitro* use. *In vivo* dosing was by gavage.

Drugs

Doxorobucin (DOX) (Bedford Laboratories, Bedford, OH), cyclophosphonate (CTX) (SIGMA, St. Louis, MO), Pingxiao (PX) (C.P. Pharmaceutical Co., Ltd., Xian, China), cisplatinum (CDDP) (DONG-A Pharmaceutical Co., Seoul, Korea), and paclitaxel (Taxol) (Bedford Laboratories, Bedford, OH) were used.

Time-lapse imaging of H460 dual-color cells

A FluoView FV1000 confocal laser microscope (Olympus Corp., Tokyo, Japan) was used to image H460 dual-color cells treated with LQ and CDDP for time-lapse imaging. High-resolution images were captured directly for a 72-hour period with 30 min intervals at 37°C.

MTS assay

The cells were exposed to various concentrations of LQ and chemotherapeutic agents for 72 hours. The number of viable cells was subsequently determined using the Cell Titer 96 Aqueous One Solution Cell Proliferation assay (Promega, Madison, WI) as described in the instructions.

Statistical analysis

Data were assessed using the Student's t-test. Kaplan-Meier analysis with a log-rank test was used to determine survival and differences between control and treatment groups. A p value of ≤ 0.05 was defined as statistically significant.

Results and Discussion

Comparison of efficacy of LQ and CTX on subcutaneous mouse models of lung cancer

We tested the efficacy of different dosages of LQ and CTX on different lung cancer cell lines *in vivo*. H460, A549, and LLC cells

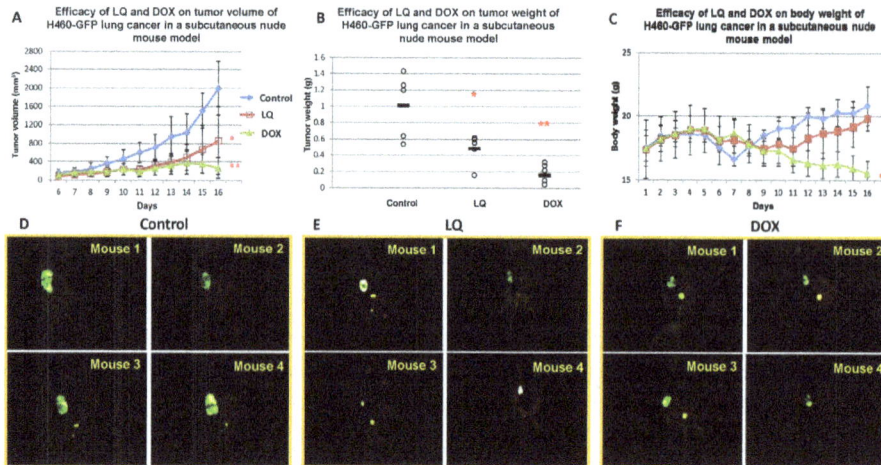

Figure 3. Efficacy of LQ on tumor volume (A), tumor weight (B) and body weight (C) in subcutaneous mouse tumor models of H460-GFP. Five mice were used in each group. After the subcutaneous tumors grew to 100 mm², the nude mice were given either DOX (7.5 mg/kg i.v.) (twice a week for 3 times). PBS (po), LQ (600 mg/kg daily po). Tumor size and body weight were measured every day. Tumor weight was measured at the endpoint. Statistical significance between groups was determined with the Student's t-test. After LQ treatment, lung tumor growth and tumor weight were significantly inhibited ($P<0.01$), compared to the untreated control group. DOX treatment also resulted in a significant tumor growth and weight inhibition compared to the untreated controls ($P<0.01$). The DOX-treated mice had significant body-weight loss compared to the LQ-treated and untreated controls ($P<0.01$). $*p<0.05$; $**p<0.01$. Images of H460-GFP tumors in live mice from the subcutaneous nude-mouse model are shown for the different treatment groups, and untreated control (**D**); LQ (**E**); and DOX (**F**).

growing subcutaneously for 7 days in nude mice were treated with either LQ (150, 300, and 600 mg/kg) for 10 days, the TCM herbal mixture PX [34] (600 mg/kg, daily gavage) for 10 days, or CTX (30 mg/kg/dose daily i.p) for 7 days. In the control group, mice were given PBS by oral administration. Tumor growth measured by size was significantly inhibited by LQ at doses of 150 mg/kg, 300 mg/kg, and 600 mg/kg in a dose-dependent manner (Figure 1A–F) (all p<0.01). Final tumor weight was also

significantly decreased after LQ treatment at all dosages (all p<0.01) (Figure 1G). CTX and PX also inhibited tumor size and tumor weight significantly, compared to the control mice (all p<0.01) (Fig. 1A–G). LQ (600 mg/kg) was more effective than PX (600 mg/kg) based on tumor size and weight (all p<0.01) (Fig. 1D, E, F and G).

The body weight and general appearance of each mouse was observed during the experiments and at the end point. The body

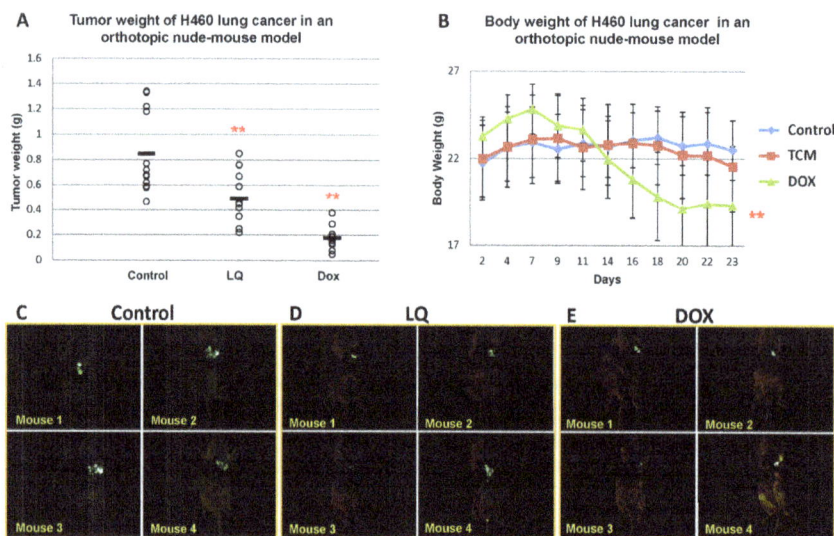

Figure 4. Efficacy of LQ on tumor volume, tumor weight (A) and body weight (B) in orthotopic H460-GFP lung cancer nude-mouse models. Ten mice were used in each group. H460-GFP fragments from tumors grown subcutaneously (1 mm in diameter) were harvested and implanted by surgical orthotopic implantation (SOI) into the left lungs of nude mice. The nude mice were given either PBS (po); DOX (7.5 mg/kg i.v. twice a week for 3 times from day 7) or LQ (600 mg/kg/day po daily. from day 7). Body weights were measured every day. Tumor weights were measured at the endpoint. Statistical significance between groups was determined with the Student's t-test. Lung tumor growth was significantly inhibited by LQ ($p<0.01$), without body weight loss. DOX inhibited lung tumor growth ($p<0.001$) but induced significant body weight loss compared to control mice ($p<0.001$). Open images of H460-GFP tumors in the orthotopic nude mouse model are shown for the different treatment groups and untreated control (**C**); LQ (**D**); and DOX (**E**).

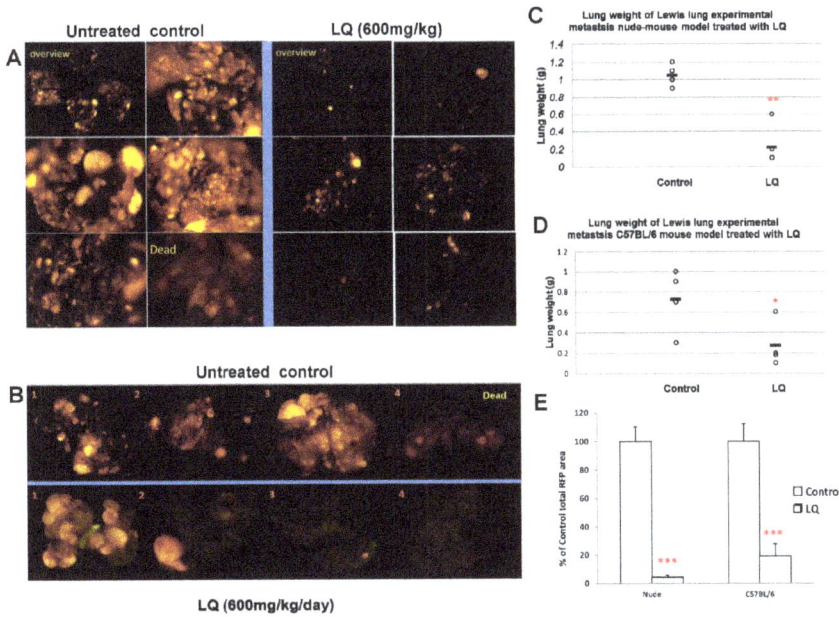

Figure 5. Efficacy of LQ on experimental lung metastasis. Six mice were used in each group. Individual images of excised lungs from the mice in the experimental metastasis model were obtained. LLC-RFP cancer cells (2×10^6) were injected into the tail vein of nude mice or C57BL/6 mice. From day 2, the mice were given PBS (po) every day as the untreated control. Mice were treated with LQ (600 mg/kg/day po) for 10 days. The LQ-treated mice had significantly reduced lung weight ($p < 0.001$) and reduced red fluorescence area in the lung ($p < 0.001$) compared to the untreated PBS controls. **A** and **B** are images of the lungs. **C–D** are lung weights. **E** is the total red fluorescence area.

weight in CTX-treated mice significantly decreased after treatment ($p < 0.05$) and recovered slowly after treatment termination. However, there was no decrease on body weight in LQ groups with all dosages. (Fig. 1H). The heart, liver, spleen, lung, kidney, and intestine of a mouse from each group (untreated control, LQ and CTX) were stained with H&E to visualize the toxicity after treatment. The histological structure of each organ was observed

and compared microscopically. Figure 2 shows morphological changes in the renal tissue after CTX administration. After CTX treatment, the proximal tubule cells were injured, including swelling, degeneration, and necrosis and sloughing of tubular epithelial cells (Fig. 2B). However, the changes in renal tissue were not observed in LQ treated mice (Fig. 2C). Those results indicated that renal toxicity was caused by CTX, but not LQ.

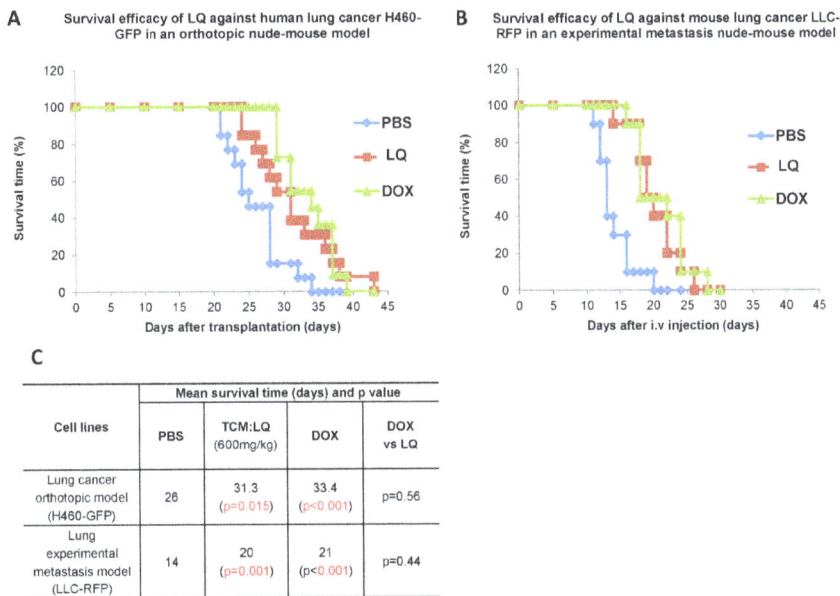

Figure 6. Survival curve of LQ-treated mice with H460-GFP orthotopic human lung cancer (A), and mice with LLC experimental lung metastasis (B). Ten mice were used in each group. LQ significantly increased survival in both the orthotopic ($p = 0.015$) and experimental metastasis ($p = 0.001$) models. DOX also significantly increased ($p < 0.001$) survival in both models.

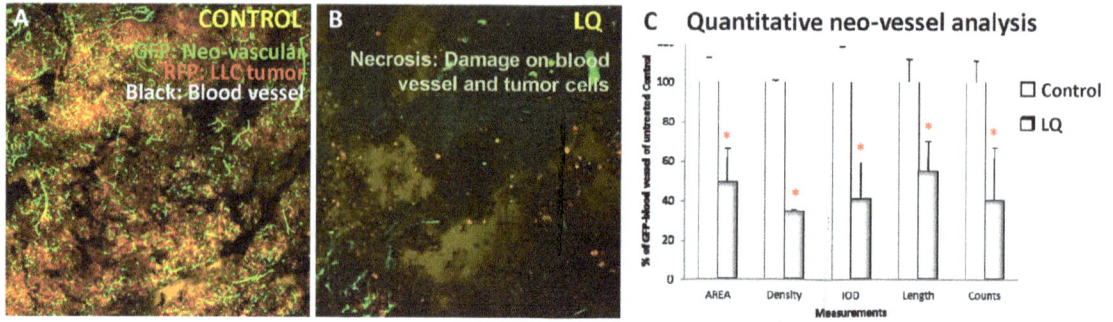

Figure 7. Effect of LQ on tumor blood vessels. LLC-RFP cells were grown in nestin-driven GFP (ND-GFP) transgenic nude mice in which nascent blood vessel expressed GFP. Three mice were used in each group. In the LQ-treated mice, tumor blood vessels appeared to be destroyed. The GFP florescence of blood vessels was quantitated for area, density, integrated optical density (IOD), length and counts. The measurements indicated that LQ decreased blood vessels by 50%.

The purpose of this experiment was to compare LQ to conventional chemotherapy at published therapeutic effective doses (ED) [35]. A low dose of doxorubicin was also tested which did not have apparent toxicity but had less efficacy compared to LQ (data not shown).

Comparison of LQ and DOX on the H460-GFP human lung cancer subcutaneous mouse model

From the data last section; we used 600 mg/kg of LQ (daily gavage) on the H460 human lung cancer cell line. We tested the H460-GFP cells growing subcutaneously compared to the H460 parental cell line in nude mice treated with LQ. We used DOX

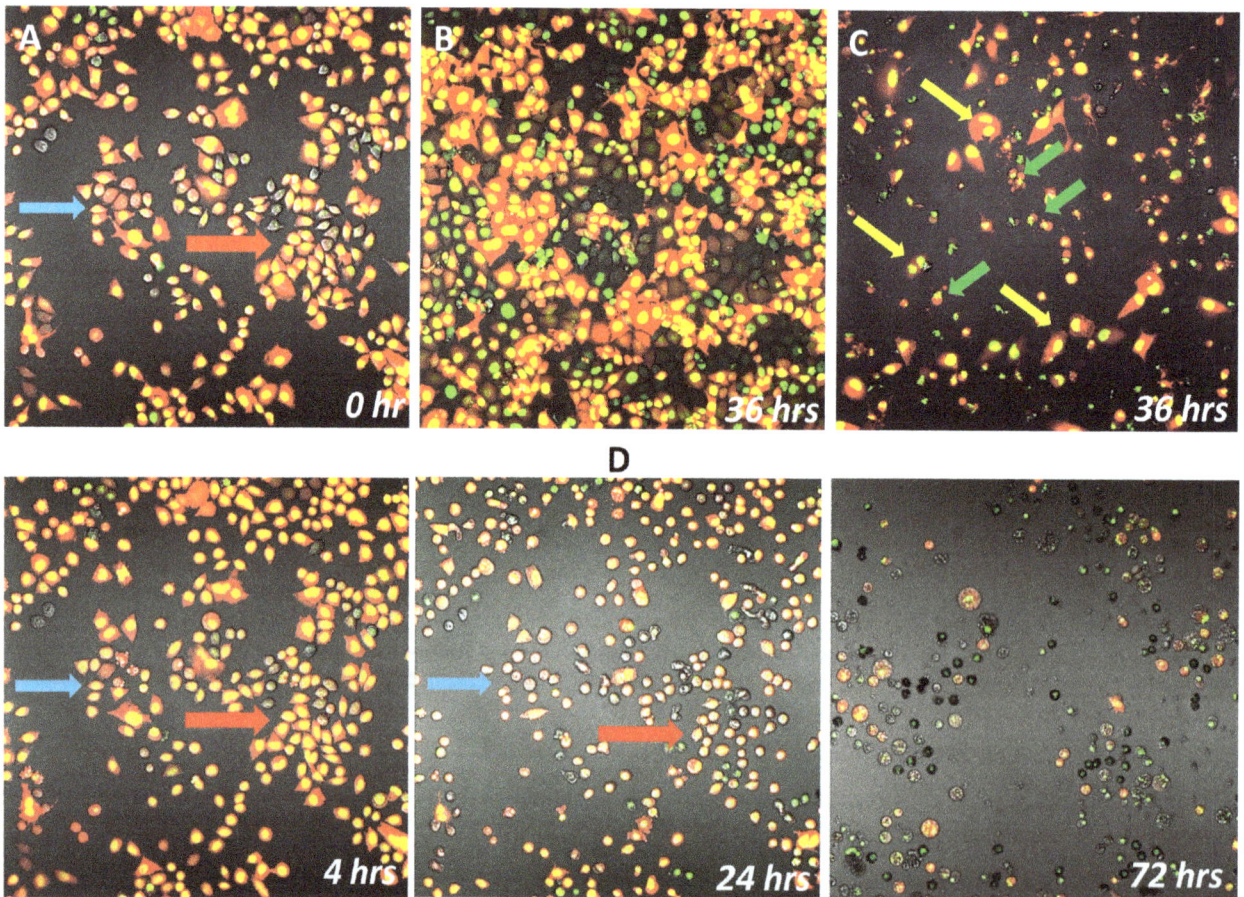

Figure 8. Time-Lapse imaging of H460 dual-color cells expressing GFP in the nucleus and RFP in the cytoplasm treated with LQ and CDDP. A and B) untreated control cells at 0 hour and 36 hours. **C**) CDDP-treated cells at 36 hours. **D**) time lapse imaging of LQ-treated H460 cells at 4, 24 and 72 hours. Cell proliferation was inhibited. Cell proliferation indicated by yellow arrows. Apoptotic cells indicated by green arrows. Blue and red arrows indicate changes in the same field.

(7.5 mg/kg i.v., twice a week for three times) as the positive control to compare with LQ. In the control group, mice were given PBS for oral administration. Tumor imaging (Figure 3D, E & F) and caliper measurement (Fig. 3A) showed that tumor growth was inhibited after LQ and DOX treatment (the P value = 0.029, P value = 0.005, respectively), compared to the control mice. Final tumor weight was also significantly inhibited after LQ treatment and DOX treatment compared to the PBS control ($p = 0.037$, $p = 0.01$ respectively.) (Fig. 3B). However, DOX resulted in a significant loss in body weight compared to the LQ-treated animals ($p = 0.002$) and to the untreated controls (p = 0.001) (Fig. 3C). In contrast, no significant body weight loss was found in the LQ group compared to the control (Fig. 3C). The data also suggested that H460-GFP and the H460 parental cell line showed no difference on growth in mice and response to LQ.

Comparison of LQ and DOX efficacy on the H460-GFP human lung cancer orthotopic model

To determine the anti-metastatic efficacy of LQ, an orthotopic nude-mouse model of human lung cancer, H460-GFP was used [12]. Open images of tumors in the lung from treated and untreated mice are shown in Fig. 4C–E. Tumor size was inhibited by LQ and DOX (all p<0.01) (Fig. 4A). Tumor weight was significantly decreased in the LQ-treated mice (600 mg/kg daily gavage) compared to the PBS control group ($p = 0.0041$) (Fig. 4A). DOX (7.5 mg/kg i.v., twice a week for three times) also inhibited tumor weight ($p<0.0001$) (Fig. 4A). DOX resulted in significant

body weight loss compared to the untreated control ($p<0.001$) (Fig. 4B). In contrast, no significant body weight loss was found in the LQ-treated animals compared to the untreated control (Fig. 4B). There was no metastasis found in the control group.

The results above indicated LQ can significantly inhibit orthotopic lung tumor growth without toxicity unlike DOX which caused significant body weight loss.

Efficacy of LQ on experimental lung cancer metastasis in nude mice and in immune-competent C57BL/6 mice

We tested the anti-metastatic efficacy of LQ in experimental mouse models. RFP-expressing Lewis lung carcinoma cells (LLC-RFP) cells were injected i.v. in both nude and C57BL/6 immune-competent mice in order to obtain experimental metastasis. Individual images of excised lungs from the mice were obtained. The lungs in the untreated control group were enlarged, and the color was dark red due to the size of the tumor. The LQ-treated animals had pinkish lungs with significantly reduced size compared to untreated controls. The red fluorescence intensity and area due to the presence of the LLC-RFP tumor was reduced in the LQ-treated mice compared to the untreated control mice ($p<0.01$) (Fig. 5A–B). LQ similarly inhibited tumor colonization in the lung in both nude and C57BL/6 mice (Fig. 5A–B). Reduced total lung weight in the LQ group, compared to the untreated control, also indicated reduced tumor colonization in both LQ-treated nude and C57BL/6 mice ($p = 0.001$ and $p = 0.01$, respectively) (Fig. 5C–D). The total area of red florescence of

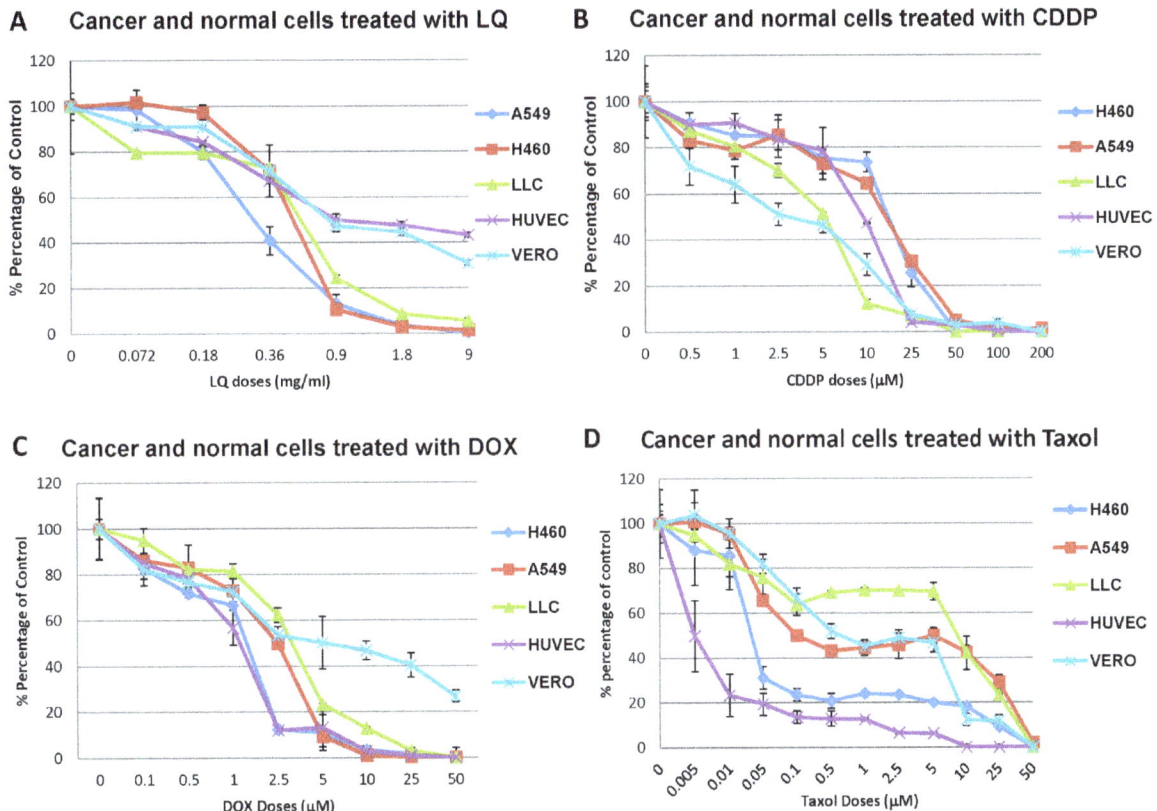

Figure 9. Efficacy of LQ on proliferation of cancer and normal cells *in vitro*. Cells were seeded in 96-well plates and incubated overnight. Cells were then exposed to LQ. H460, LLC and A549 were sensitive to LQ at 72 h (**A**). IC$_{50}$ values were between 0.36~0.45 mg/ml. All experiments showed a dose-dependent inhibition by LQ. However, LQ was less active on the normal cell lines HUVEC and VERO (**A**). The cancer and normal cells were all sensitive to DOX and CDDP (**B. C**). However, HUVEC and VERO cells were more sensitive to paclitaxol than the cancer cells (**D**). The graphs show combined values from two independent experiments, with each data point repeated in triplicate.

the lung was calculated. The fluorescent area correlated with lung weight. The total RFP tumor area in LQ-treated mice was 4% and 20%, compared to their nude or C57BL/6 control, respectively (all p value<0.001) (Fig. 5E). The results indicated that LQ can significantly inhibit the tumor colonization in the lung thereby indicating anti-metastatic efficacy.

LQ prolonged survival of tumor-bearing animals

LQ significantly prolonged survival of mice with orthotopically-implanted H460 lung cancer as well as in mice with LLC experimental lung metastasis. Median-survival increased from 26 days in the untreated control animals to 30.3 days in mice treated with LQ ($p = 0.005$), to 33.4 days in mice treated with DOX in the orthotopic lung cancer model ($p = 0.0001$) (Fig. 6A, C). In the experimental metastasis model, median survival was increased from 14 days in the untreated control animals to 20 days in the LQ-treated mice (p = 0.001) and 22 days in the DOX-treated mice ($p<0.001$) (Fig. 6B, C). LQ prolonged survival of tumor-bearing animals comparable to DOX.

Destruction of tumor blood vessels by LQ

In the ND-GFP mouse model, only new blood vessels induced by the tumor express nestin-driven GFP. Observation with the FV-1000 imaging system showed that blood vessels in the tumor were very dense and interwoven to form a blood vessel network (Fig. 7A). After LQ treatment, GFP blood vessels in LLC-RFP tumors in the ear and footpad were severely damaged (Fig. B). The GFP blood vessels fragmented in different areas of the tumor, and tumor necrosis was observed. Total fluorescent area, integrated optical intensity and length and density of GFP blood vessels were quantitated. The results showed that tumor blood vessels were reduced to ~50% of control ($p<0.05$) after LQ administration (Fig. 7C).

LQ induced cancer cell death *in vitro*

H460 dual-color cells, expressing GFP in the nucleus and RFP in the cytoplasm were cultured in 35 mm peri-dish treated with LQ or CDDP. Cell proliferation was imaged every 4 hours for 72 hours. Cell proliferation was totally inhibited by LQ. After treatment, there was no mitosis. H460 dual-color cells treated with LQ decreased in size. The cytoplasm shrank at an early time point and disappeared, but the nuclei remained (Fig. 8A, B, D). At late-stages, nuclei fragmented. In the CDDP-treated H460 dual-color cells, apoptotic cells and few proliferating cells were observed (Fig. 8C). Approximately 60% of cells were apoptotic at the 36-hour time point. Fragmented nuclei were observed but had a different appearance than in cells treated with LQ. LQ may trigger a different cell death mechanism than apoptosis induced by CDDP. Further investigations will be conducted in the future.

Comparison of efficacy of LQ and chemotherapy on cancer cell proliferation *in vitro*

H460, A549, and LLC proliferation was tested *in vitro* with LQ and compared with CDDP, DOX, and paciltaxel (Fig. 9A–D). All cell lines were sensitive to LQ in a similar pattern. The IC$_{50}$ were between 0.36 mg/ml to 0.45 mg/ml (Fig. 9A). H460, A549, and

LLC proliferation was also inhibited by CDDP, paciltaxel, and DOX (Fig. 9B–D) with IC$_{50}$ between 5~25 μM, 1~5 μM and 5~25 μM, respectively. However, normal cell lines HUVEC and VERO were resistant to LQ with an IC$_{50}$ approximately 9 mg/ml (more than 20-fold less sensitive than the cancer cells). In contrast, HUVEC and VERO were sensitive to CDDP comparable with cancer cells. HUVEC was sensitive to DOX comparable to cancer cells. However, VERO was 5-fold less sensitive to DOX than the cancer cells. HUVEC and VERO were more sensitive to paclitaxel than the cancer cells with the IC$_{50}$ between 0.005~0.01 μM (more than 100-fold more sensitive than the cancer cells). In order to rule out apparent efficacy due to pH change with LQ in the cell culture medium, the pH value at different concentrations of LQ was determined. The pH of RPMI 1640 medium with 10% FBS (no LQ), 9 mg/ml LQ or 1.8 mg/ml LQ was 7.61, 7.51, and 7.6, respectively. We also tested H460-dual color lung cancer cell proliferation in RPMI 1640 medium with 10% FBS at pH 7.51 using the MTT assay or live time imaging. The growth curve and growth pattern of cells at pH 7.51 was not significantly different from growth in RPMI 1640 medium with 10% FBS (pH 7.61) (Figure S1).

In the present report, we compared the efficacy of LQ against lung cancer in mouse models with DOX and CTX. LQ inhibited tumor size and weight comparable to cytotoxic chemotherapy but very importantly, without apparent toxicity. Survival of tumor-bearing mice was also prolonged by LQ, comparable to DOX. LQ had anti-metastatic efficacy observed by decreased cancer colonies in the lung. LQ also had anti-angiogenic activity as observed in lung tumors growing in ND-GFP transgenic nude mice, which selectively express GFP in nascent blood vessels. *In vitro*, death of lung cancer cells are induced by LQ comparable to cytotoxic chemotherapy. LQ was more potent to induce death in cancer cells than normal cells, unlike cytotoxic chemotherapy. The results of the present study indicate that LQ has non-toxic efficacy against metastatic lung cancer. Future studies should test LQ and other TCM in combination with cytotoxic chemotherapy in appropriate mouse models as a prelude to clinical studies. In addition, further experiments will test the antitumor and anti-metastatic efficacy of each herb separately in appropriate animal models.

Supporting Information

Figure S1 (**A–C**) The growth curve of lung cancer cells growth in complete medium (CM) with different pH values. (**D**) The images of H460-Dual color cells growth in CM in pH 7.61 and pH 7.51.

Acknowledgments

This paper is dedicated to the memory of A. R. Moossa, M.D.

Author Contributions

Conceived and designed the experiments: LZ RMH. Performed the experiments: LZ RMH. Analyzed the data: LZ CW YZ FL XW MZ RMH. Contributed reagents/materials/analysis tools: CW RMH. Wrote the paper: LZ RMH.

References

1. Siegel R, Naishadham D, Jemal A (2012) Cancer statistics, 2012. CA: A Cancer J Clin 62: 10–29.
2. Newman DJ, Cragg GM (2012) Natural products as sources of new drugs over the 30 years from 1981 to 2010. J Natural Prod 75: 311–335.
3. Zou YH, Liu XM (2003) Effect of astragalus injection combined with chemotherapy on quality of life in patients with advanced non-small cell lung cancer. Chinese J Int Trad Western Med 23: 733–735.
4. Wang M, Zhang X, Xiong X, Yang Z, Sun Y, et al. (2012) Efficacy of the Chinese traditional medicinal herb Celastrus orbiculatus Thunb on human

hepatocellular carcinoma in an orthothopic fluorescent nude mouse model. Anticancer Res 32: 1213–1220.

5. Hu M, Zhao M, An C, Yang M, Li Q, et al. (2012) Real-time imaging of apoptosis induction of human breast cancer cells by the traditional Chinese medicinal herb tubeimu. Anticancer Res 32: 2509–2514.

6. Zhang L, Wu C, Zhang Y, Liu F, Zhao M, et al. (2013) Efficacy comparison of traditional Chinese medicine LQ versus gemcitabine in a mouse model of pancreatic cancer. J Cell Biochem 114: 2131–2137.

7. Qi F, Li A, Inagaki Y, Gao J, Li J, et al. (2010) Chinese herbal medicines as adjuvant treatment during chemo- or radio-therapy for cancer. Bioscience Trends 4: 297–307.

8. Gai RY, Xu HL, Qu XJ, Wang FS, Lou HX, et al. (2008) Dynamic of modernizing traditional Chinese medicine and the standards system for its development. Drug Discoveries & Therapeutics 2(1): 2–4.

9. Wong R, Sagar CM, Sagar SM (2001) Integration of Chinese medicine into supportive cancer care: a modern role for an ancient tradition. Cancer Treatment Reviews 27: 235–246.

10. Katz MH, Takimoto S, Spivack D, Moossa AR, Hoffman RM, et al. (2003) A novel red fluorescent protein orthotopic pancreatic cancer model for the preclinical evaluation of chemotherapeutics. J Surg Res 113: 151–160.

11. Zhao M, Suetsugu A, Ma H, Zhang L, Liu F, et al. (2012) Efficacy against lung metastasis with a tumor-targeting mutant of Salmonella typhimurium in immunocompetent mice. Cell Cycle 11: 187–193.

12. Yang M, Hasegawa S, Jiang P, Wang X, Tan Y, et al. (1998) Widespread skeletal metastatic potential of human lung cancer revealed by green fluorescent protein expression. Cancer Res 58: 4217–4221.

13. Wang X, Fu X, Hoffman RM (1992) A patient-like metastasizing model of human lung adenocarcinoma constructed via thoracotomy in nude mice. Anticancer Res 12: 1399–1402.

14. Rashidi B, Yang M, Jiang P, Baranov E, An Z, et al. (2000) A highly metastatic Lewis lung carcinoma orthotopic green fluorescent protein model. Clin Exp Metastasis 18: 57–60.

15. Martínez-Gutierrez M, Castellanos JE, Gallego-Gómez JC (2011) Statins reduce dengue virus production via decreased virion assembly. Intervirology 54: 202–216.

16. Unger RE, Peters K, Sartoris A, Freese C, Kirkpatrick CJ (2014) Human endothelial cell-based assay for endotoxin as sensitive as the conventional Limulus Amebocyte Lysate assay. Biomaterials. pii: S0142-9612(13)01543-3. doi:10.1016/j.biomaterials.2013.12.059. [Epub ahead of print].

17. Hoffman RM, Yang M (2006) Subcellular imaging in the live mouse. Nature Protoc 1: 775–782.

18. Hoffman RM, Yang M (2006) Color-coded fluorescence imaging of tumor-host interactions. Nature Protoc 1: 928–935.

19. Hoffman RM, Yang M (2006) Whole-body imaging with fluorescent proteins. Nature Protoc 1: 1429–1438.

20. Yamamoto N, Jiang P, Yang M, Xu M, Yamauchi K, et al. (2004) Cellular dynamics visualized in live cells in vitro and in vivo by differential dual-color nuclear-cytoplasmic fluorescent-protein expression. Cancer Res 64: 4251–4256.

21. Jiang P, Yamauchi K, Yang M, Tsuji K, Xu M, et al. (2006) Tumor cells genetically labeled with GFP in the nucleus and RFP in the cytoplasm for imaging cellular dynamics. Cell Cycle 5: 1198–1201.

22. Hoffman RM (2005) The multiple uses of fluorescent proteins to visualize cancer in vivo. Nature Reviews Cancer 5: 796–806.

23. Li L, Mignone J, Yang M, Matic M, Penman S, et al. (2003) Nestin expression in hair follicle sheath progenitor cells. Proc Natl Acad Sci USA 100: 9958–9961.

24. Amoh Y, Li L, Yang M, Moossa AR, Katsuoka K, et al. (2004) Nascent blood vessels in the skin arise from nestin-expressing hair-follicle cells. Proc Natl Acad Sci USA 101: 13291–13295.

25. Amoh Y, Yang M, Li L, Reynoso J, Bouvet M, et al. (2005) Nestin-linked green fluorescent protein transgenic nude mouse for imaging human tumor angiogenesis. Cancer Res 65: 5352–5357.

26. Hayashi K, Yamauchi K, Yamamoto N, Tsuchiya H, Tomita K, et al. (2009) A color-coded orthotopic nude-mouse treatment model of brain-metastatic paralyzing spinal cord cancer that induces angiogenesis and neurogenesis. Cell Prof 42: 75–82.

27. Hoffman RM. (1999) Orthotopic metastatic mouse models for anticancer drug discovery and evaluation: a bridge to the clinic. Invest New Drugs 17: 343–359.

28. Astoul P, Colt HG, Wang X, Hoffman RM (1993) Metastatic human pleural ovarian cancer model constructed by orthotopic implantation of fresh histologically-intact patient carcinoma in nude mice. Anticancer Res 13: 1999–2002.

29. Astoul P, Colt HG, Wang X, Hoffman RM (1994) A "patient-like" nude mouse model of parietal pleural human lung adenocarcinoma. Anticancer Res 14: 85–91.

30. Astoul P, Colt HG, Wang X, Boutin C, Hoffman RM (1994) "Patient-like" nude mouse metastatic model of advanced human pleural cancer. J Cell Biochem 56: 9–15.

31. Kimura H, Hayashi K, Yamauchi K, Yamamoto N, Tsuchiya H, et al. (2010) Real-time imaging of single cancer-cell dynamics of lung metastasis. J Cell Biochem 109: 58–64.

32. Liu F, Zhang L, Hoffman RM, Zhao M (2010) Vessel destruction by tumor-targeting Salmonella typhimurium A1-R is enhanced by high tumor vascularity. Cell Cycle 9: 4518–4524.

33. Yamauchi K, Yang M, Jiang P, Xu M, Yamamoto N, et al. (2006) Development of real-time subcellular dynamic multicolor imaging of cancer-cell trafficking in live mice with a variable-magnification whole-mouse imaging system. Cancer Res 66: 4208–4214.

34. Zhang QY, Zhao WH, Lai YJ (2005) Effect on late-stage mammary cancer treated by endocrinotherapy or chemotherapy combined with pingxiao capsule. Chinese J Integrated Trad Western Med 25: 1074–1076.

35. Ottewell PD, Mönkkönen H, Jones M, Lefley DV, Coleman RE, et al. (2008) Antitumor effects of doxorubicin followed by Zoledronic acid in a mouse model of breast cancer. J Natl Cancer Inst 100: 1167–1178.

13

Association of ITPA Genotype with Event-Free Survival and Relapse Rates in Children with Acute Lymphoblastic Leukemia Undergoing Maintenance Therapy

Alenka Smid[1], **Natasa Karas-Kuzelicki**[1], **Miha Milek**[1¤], **Janez Jazbec**[2], **Irena Mlinaric-Rascan**[1]*

1 Faculty of Pharmacy, University of Ljubljana, Ljubljana, Slovenia, **2** University Children's Hospital, University Medical Centre Ljubljana, Ljubljana, Slovenia

Abstract

Although the treatment of acute lymphoblastic leukemia (ALL) has improved significantly over recent decades, failure due to treatment-related toxicities and relapse of the disease still occur in about 20% of patients. This retrospective study included 308 pediatric ALL patients undergoing maintenance therapy and investigated the effects of genetic variants of enzymes involved in the 6-mercaptopurine (6-MP) metabolism and folate pathway on survival and relapse rates. The presence of at least one of the non-functional *ITPA* alleles (94C>A and/or IVS2+21A>C variant) was associated with longer event-free survival compared to patients with the wild-type *ITPA* genotype (p = 0.033). Furthermore, patients carrying at least one non-functional *ITPA* allele were shown to be at a lower risk of suffering early (p = 0.003) and/or bone marrow relapse (p = 0.017). In conclusion, the *ITPA* genotype may serve as a genetic marker for the improvement of risk stratification and therapy individualization for patients with ALL.

Editor: David R. Booth, University of Sydney, Australia

Funding: This work was supported by Slovenian Research Agency grant No. J3-3615. The funder had no role in study design, data collection and analysis, decision to publish, or preparation of the manuscript.

Competing Interests: The authors have declared that no competing interests exist.

* Email: irena.mlinaric@ffa.uni-lj.si

¤ Current address: Max-Delbrück-Center for Molecular Medicine, Berlin Institute for Medical Systems Biology, Berlin, Germany

Introduction

Treatment of pediatric acute lymphoblastic leukemia (ALL) has improved drastically over the last forty years, with the long term-survival rate for children diagnosed with ALL now such that approximately 85% survive for 5 years or more after diagnosis [1]. This great improvement can be attributed both to risk-adjusted treatment based on the identification of several biological and clinical prognostic factors which facilitated the definition of patient subgroups with different relapse risks, and the implementation of rationally designed phases in the treatment backbone of ALL [2]. Despite this remarkable progress, treatment failure due to treatment-related toxicities, which can be life threatening, and drug resistance, which leads to relapse, still occur in about 20% of patients [3], for whom there is a very low likelihood of long-term survival [1,4,5,6].

Genetic polymorphisms in thiopurine-S-methyltransferase (*TPMT*) influence the toxicity of 6-mercaptopurine (6-MP), used as the backbone of the maintenance therapy of ALL, and thus represent one of the best examples of clinically important pharmacogenetic markers [7,8]. Patients on 6-MP therapy exhibiting decreased TPMT activity have been shown to be at greater risk of developing toxic effects, due to undesirably high thioguanine nucleotides (TGN) accumulation in cells [9]. On the other hand, ultra-high enzyme activity can lead to superior 6-MP tolerability but also an increased risk of relapse [10]. Despite the

very well characterized influences of genetic polymorphisms on TPMT activity, it is known that there are other genetic and non-genetic factors influencing it. Our previous study in ALL patients revealed that the presence of low-activity methylenetetrahydrofolate reductase (*MTHFR*) alleles contributes to 6-MP toxicity, presumably by limiting S-adenosylmethionine (SAM) synthesis [11]. Further molecular and functional studies carried out *in vitro* on several cancer cell lines have demonstrated that SAM is responsible for direct post-translational TPMT stabilization, resulting in increased TPMT activity [12]. Since folate and methionine pathways are crucial for SAM synthesis, other polymorphisms in enzyme-coding genes (such as 5-methyltetrahydrofolate-homocysteine methyltransferase reductase (*MTRR*), methylenetetrahydrofolate dehydrogenase 1 (*MTHFD1*), betaine–homocysteine S-methyltransferase (*BHMT*), and glycine N-methyltransferase (*GNMT*)) could influence TPMT activity and thus affect the safety and efficacy of 6-MP therapy. A recent study in ALL patients conducted by Stocco et al. [13] has shown that genetic variation in *PACSIN2* also influences TPMT activity and is significantly associated with the incidence of 6-MP-related gastrointestinal toxicity [13].

Inosine triphosphate pyrophosphatase (ITPA) is an enzyme involved in 6-MP metabolism whose activity is determined by polymorphisms in the *ITPA* gene and is currently being extensively studied in relation to 6-MP treatment outcome. ITPA catalyses the pyrophosphohydrolysis of inosine triphosphate (ITP) into inosine monophosphate (Figure 1), and prevents the accumulation of potentially toxic compounds, such as ITP and deoxy-ITP or xanthosine triphosphate (XTP), that may be incorporated into RNA and DNA; in this way, it rescues the cell from apoptosis [14]. Although *ITPA* has been linked to 6-MP toxicity, different studies on patients with inflammatory bowel disease, liver transplant recipients and ALL patients showed inconsistent results [7,15,16,17,18,19,20,21,22,23,24]. Two SNPs in the *ITPA* gene, namely a non-synonymous C94A transition and the intronic IVS2+21A>C variant, with a frequency of approximately 6% and 13%, respectively, in Caucasians [25], have been correlated with defective enzyme activity [14,26] leading to a higher risk of myelotoxicity and hepatotoxicity in pediatric ALL patients [7,22,27]. Furthermore, a recent Korean study suggested the *ITPA* 94C>A polymorphism might be associated with survival rate in pediatric ALL patients [21].

Despite intense efforts devoted to precise risk classification, patient characteristics at relapse reveal that more than half of relapsed patients were originally classified as non-high risk [28]. This highlights the need for the identification of additional prognostic markers.

In the present retrospective study we aimed to investigate the association of selected polymorphisms in genes involved in the folate and methionine pathway (*MTHFR* 677C>A, *MTHFR* 1298A>C, *MTRR* 66A>G, *MTHFD1* 1958G>A, *BHMT* 742G>A, *GNMT* 1298C>T), as well as polymorphisms of *PACSIN2* (rs2413739) and *ITPA* (94C>A and IVS2+21A>C), and long-term outcome and relapse rates in pediatric ALL patients undergoing maintenance therapy with 6-MP.

Patients and Methods

Study subjects

ALL patients diagnosed and treated at University Children's Hospital, University Medical Centre, Ljubljana, Slovenia from 1970 to 2006 were identified through the National Cancer Registry. Of 408 registered patients with childhood ALL, adequate documentation and intact genetic material was obtained from 308, which represented the final study group. Ethical approval for this study was obtained from the National Medical Ethics Committee of Slovenia (59/07/10) which waived the need for written informed consent, since the data was collected retrospectively and was analyzed anonymously.

The following therapy protocols were applied: from 1970 to 1983 the USA Pediatric Oncology Group (POG) protocols were used; from 1983 the therapy was changed to the German Berlin-Frankfurt-Muenster (BFM) protocols (BFM-83, -86, -90, -95 and IC2002). Maintenance therapy was included in all protocols and lasted from 1 to 3 years. The maintenance phase consisted of 6-MP taken daily per os (50 mg/m^2) and weekly oral low-dose methotrexate (MTX) (20 mg/m^2).

Patients' characteristics, such as gender, age at diagnosis, therapy protocol and risk group allocation, were obtained from their medical records. Patients were grouped according to age at diagnosis in 4 groups: (a) <1 year, (b) 1–6 years, (c) 6–12 years, and (d) >12 years, and according to treatment protocols to 3 groups: (a) POG protocol, (b) BFM83 or BFM86 protocol, and (c) BFM90, BFM95 or IC BFM2002 protocol.

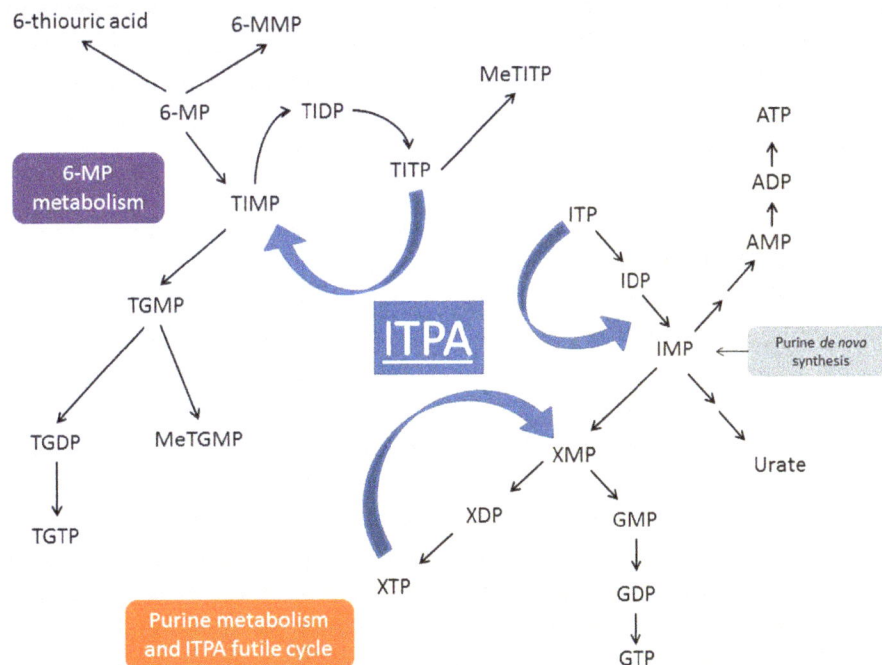

Figure 1. Schematic representation of the *ITPA* involvement in 6-MP metabolism and endogenous purine metabolism. 6-MP, 6-mercaptopurine; 6-MMP, 6-methylmercaptopurine; TIMP, thioinosine monophosphate; TIDP, thioinosine diphosphate; TITP, thioinosine triphosphate; MeTITP, methylthioinosine triphosphate; TGMP, thioguanosine monophosphate; TGDP, thioguanosine diphosphate; TGTP, thioguanosine triphosphate; MeTGMP, methylthioguanosine monophosphate; ITPA, inosine triphosphate pyrophosphatase; IMP, inosine monophosphate; XMP, xanthosine monophosphate; XDP, xanthosine diphosphate; XTP, xanthosine triphosphate; GMP, guanosine monophosphate; GDP, guanosine diphosphate; GTP, guanosine triphosphate; ITP, inosine triphosphate; IDP, inosine diphosphate; AMP, adenosine monophosphate; ADP, adenosine diphosphate; ATP, adenosine triphosphate.

Event and risk group definition

The event in the calculation of event-free survival (EFS) probability was defined as any relapse, death, or secondary malignant neoplasm.

Patients were stratified into risk groups according to criteria described previously [29,30,31,32]. The standard risk group included the low-risk group of POG, SR-L and SR-H groups of BFM-83 and SR groups of protocols BFM-86, -90 and -95. Intermediate or high-risk groups included the high-risk group of POG, R-group of BFM-86, MR groups of BFM-83, -90, -95 and HR groups of BFM-83, -90, -95 protocols.

DNA extraction and genotyping

DNA extraction was performed as previously described [11,33]. All the analyzed polymorphisms were determined by means of TaqMan chemistry using either the ABI Prism 7000 Sequence Detection system or the Roche LightCycler 480 system, in accordance with the manufacturers' instructions. The amount of DNA used in each individual assay was 10 ng. The TaqMan SNP Genotyping Assay part numbers are available in Table S1.

Statistical analysis

The distributions of genotypes and possible deviations from the Hardy-Weinberg equilibrium were assessed by Fisher's exact test. Odds ratios, 95% confidence intervals (95% CI) and P values, showing the association of occurrence of relapse with studied genotypes, were calculated using logistic regression analysis. Covariates considered to be potential confounders, such as gender, age at diagnosis, ALL therapy protocol and risk group, were also placed in the model.

Kaplan-Meier event-free survival curves (EFS) for ALL patients were created, and the survival differences according to different genetic polymorphisms and prognostic variables were analyzed by log-rank test. The starting point for the observation time was date of diagnosis. Multivariate analysis was conducted using the Cox proportional-hazards regression model to analyze predictive factors. Cumulative incidence curves of the different types of relapses were estimated, adjusting for competing risks of the other events, and compared with the Gray's test. In both analyses, the observation time was censored at the last follow-up date if no event was observed.

To explore the independent and synergistic effects of the studied SNPs on the occurrence of relapse we performed a multifactor dimensionality reduction (MDR) analysis [34].

Statistical analyses were performed using R, IBM SPSS for Windows, version 21.0 and MDR software. For all tests, a 2-sided p<0.05 was considered statistically significant.

Results

Patients and genotyping

In total, 308 patients with childhood ALL were included in this study. Their clinical characteristics are presented in Table 1.

Genotyping was performed once the clinical data were extracted from patients' medical files and analyzed by researchers who were blinded to patients' medical data. All genotype distributions were in Hardy-Weinberg equilibrium (p>0.05). The variant allele frequencies of the analyzed polymorphisms are presented in Table 2 and summary of genotype data provided in Table S2.

Analysis of patient survival and characteristics of relapsed patients

All the included patients were followed up for at least 5 years after diagnosis. The most significant improvement in overall, as well as event-free, survival rate was observed when patients were treated in accordance with later protocols, increasing from 58% (5-year OS) and 43% (5-year EFS) in patients treated by POG protocol to 91% (5-year OS) and 83% (5-year EFS) in patients treated by protocols BFM90/95/2002, as described in Table 3. Overall and event-free survival rates were not different in patients stratified to different risk groups (Table 3).

A total of 102 (33.1%) patients relapsed during the median follow-up time of 155 months, 5 patients (1.6%) died because of therapy-related complications and 6 patients (1.9%) developed secondary malignancy.

Of all relapsed patients, 36 (35.3%) were stratified to a standard risk group and only 21 (20.6%) to an intermediate or high risk group; 45 relapsed patients treated using POG protocols were not allocated to any risk group since stratification according to risk factors was not well established then. Clinical characteristics of relapsed patients are shown in Table 4.

Association of selected genotypes with relapse rate, overall and event-free survival

In search of novel biomarkers for the prediction of long-term outcomes and the occurrence of relapse, we addressed the involvement of selected polymorphisms in genes involved in 6-MP metabolism and folate and methionine pathways.

Multiple logistic regression analysis revealed that among all genes tested, only *ITPA* had a statistically significant effect on the occurrence of relapse, when the model was adjusted to treatment protocol group, age group and gender; namely, that *ITPA* wild-type patients were at higher risk of relapse than patients carrying at least one variant *ITPA* allele (odds ratio (OR) = 1.9; 95% CI = 1.1–3.4; p = 0.026).

In the univariate analysis using the Kaplan-Meier curve, the event-free survival (EFS) was affected by treatment protocol group, gender and *ITPA* genotype. EFS was lower in earlier protocols (43% in POG, 63% in BFM83/86 and 83% in BFM90/96/2002, p<0.001), in male patients as compared to female patients (60% and 72%, respectively; p = 0.022) and in the wild-type *ITPA* patients as compared to variant *ITPA* patients (62% and 74%, respectively; p = 0.030) (Figure 2A).

Multivariate analysis conducted with the Cox proportional-hazards regression model adjusted for treatment protocol group, gender and age group confirmed that among the genotypes tested, the *ITPA* 94CC/IVS2+21AA genotype was the only risk factor for lower EFS (HR 1.6, 95% CI 1.0–2.4; p = 0.033) (Table 5).

Influence of *ITPA* genotype on the site of relapse

Next, we aimed to explore whether the *ITPA* genotype has different effects in accordance with the relapse site. Patients experiencing relapse were divided into subgroups based on the site of the first relapse; namely, into those experiencing isolated bone marrow relapse (50 patients), combined bone marrow and extramedullary relapse (21 patients) and isolated extramedullary relapse (31 patients). The results of the multinomial logistic regression analysis adjusted to protocol treatment group, age group and gender revealed that wild-type *ITPA* patients were at higher risk of suffering medullary relapse (HR 2.5, 95% CI 1.2–5.2; p = 0.017) compared to variant *ITPA* patients, whereas the risk of experiencing a combined or isolated extramedullary relapse was not significantly different in patients with different *ITPA*

Table 1. Clinical characteristics of patients with childhood ALL (N = 308).

Characteristic	No	%
Gender		
Female	143	46.4
Male	165	53.6
Mean age at diagnosis ± standard deviation		5.9±4.3 years
<1 year	10	3.2
1<6 years	192	62.3
6<12 years	68	22.1
>12 years	38	12.3
ALL therapy protocol, n (%)		
POG	92	29.9
BFM 83/BFM 86	89	28.9
BFM 90/BFM 95/IC-BFM 2002	127	41.2
Risk group, n (%)		
Standard risk	134	43.5
Intermediate or high risk	94	30.5
undetermined	80	26

Abbreviations: POG, Pediatric Oncology Group; BFM, Berlin-Frankfurt-Muenster.

genotypes. Furthermore, in competing risk regression analysis the risk of suffering a bone marrow relapse was two times higher for wild-type *ITPA* patients compared to variant *ITPA* patients (p = 0.040). The model was adjusted for age, gender and treatment protocol group. The cumulative incidence of bone marrow relapse in wild-type *ITPA* patients was 19.6%, while it was 10.7% in variant *ITPA* patients (Figure 2B).

Influence of the *ITPA* genotypes on the time to relapse

Patients suffering relapse were divided into 3 subgroups in accordance with time to relapse criteria: very early (less than18 months from diagnosis), early (more than 18 months after

diagnosis and less than 6 months after the end of treatment) and late relapse (more than 6 months after the end of treatment).

The frequency of *ITPA* wild-type patients was significantly higher in the group of patients suffering early relapse (84%) compared to the group having no relapse (60%), very early relapse (59%) or late relapse (67%) (Fisher's exact test, p = 0.020). After adjustment to treatment protocol group, age and gender, the multinomial regression model confirmed that the effect of *ITPA* genotype was significant only for the occurrence of early relapse, such that *ITPA* wild-type patients are at higher risk of experiencing an early relapse than *ITPA* variant patients (HR 3.9, 95% CI 1.6–9.6; p = 0.003) (Figure 3).

Table 2. Frequency of the analyzed polymorphisms in patients with childhood ALL.

Gene	Variant	Wild type (N)	Heterozygous (N)	Homozygous variant (N)	Variant allele frequency (%)
TPMT	rs1142345 and rs1800460 (*3A)	290	18	0	3
TPMT	rs1800460 (*3B)	308	0	0	0
TPMT	rs1142345 (*3C)	305	3	0	0
MTHFR	rs1801133 (677C>A)	134	134	40	35
MTHFR	rs1801131 (1298 A>C)	143	131	34	33
MTRR	rs1801394 (66A>G)	59	140	109	58
MTHFD1	rs2236225 (1958G>A)	94	149	65	45
BHMT	rs3733890 (742G>A)	144	122	42	33
GNMT	rs10948059 (1289C>T)	89	139	80	49
PACSIN2	rs2413739[†]	111	139	55	41
ITPA	rs1127354 (94C>A)	268	40	0	6
ITPA	rs7270101 (IVS2+21A>C)	235	68	5	13

[†]PACSIN2 rs2413739 could not be determined in 3 patients.
Abbreviations: TPMT, Thiopurine S-methyltransferase; MTHFR, methylenetetrahydrofolate reductase; MTRR, 5-methyltetrahydrofolate-homocysteine methyltransferase reductase; MTHFD1, methylenetetrahydrofolate dehydrogenase 1; BHMT, betaine—homocysteine S-methyltransferase; GNMT, glycine N-methyltransferase; PACSIN2, protein kinase C and casein kinase substrate in neurons protein; ITPA, Inosine triphosphate pyrophosphatase.

Table 3. Overall and event-free survival rates in patients with childhood ALL, according to treatment protocol and risk group stratification.

	5-year OS/EFS	Log-rank p	10-year OS/EFS	Log-rank p
Overall survival rate ± SE %				
Treatment protocol (n)				
POG (92)	60±5	<0.001	51±5	<0.001
BFM 83/86 (89)	74±5		71±5	
BFM 90/95/2002 (127)	91±2		91±3	
Risk group (n)[1]				
Standard (94)	83±3	0.983	81±3	0.983
Intermediate or high (134)	84±4		80±4	
Event-free survival rate ± SE %				
Treatment protocol (n)				
POG (92)	43±5	<0.001	40±5	<0.001
BFM 83/86 (89)	64±5		61±5	
BFM 90/95/2002 (127)	83±3		82±3	
Risk group (n)[1]				
Standard (94)	72±4	0.645	70±4	0.645
Intermediate or high (134)	76±4		74±5	

[1]Risk group was not determined for 80 patients (those who were treated under POG protocol).
Abbreviations: POG, Pediatric Oncology Group; BFM, Berlin-Frankfurt-Muenster.

Gene-gene interactions

Gene-gene interactions between the variants and event-free survival or relapse were analysed using multifactor dimensionality reduction. There was no statistically significant multi-gene interaction model for event or relapse (data not shown).

Discussion

The remarkable improvements in risk-directed treatments of childhood ALL observed over recent decades have resulted in an impressive increase in long term survival rates, currently approaching 85% in the developed world [35]. In contrast to the steadily improving outcome for patients with newly diagnosed ALL, little progress has been made in the treatment of relapsed ALL patients whose long-term survival rates range only from about 35% to 40% [1,36,37]. Relapse of the disease, therefore, remains a major concern in the treatment of ALL.

Due to the well-known importance of maintenance therapy for event-free and overall survival, this retrospective study, including all patients treated in Slovenia from 1970 to 2006 receiving at least one cycle of maintenance therapy, was conducted. We selected 8 candidate genes and identified 12 polymorphisms that could influence the metabolism of 6-MP, either by being directly involved in the metabolism or indirectly by altering TPMT activity, and evaluated their effects on survival and relapse rates.

We found that the inheritance of at least one of the non-functional *ITPA* alleles (94C>A and/or IVS2+21A>C) is associated with improved event-free survival and lower relapse rates in patients undergoing maintenance treatment for ALL.

ITPA is a human ITPase, a 'house-cleaning' enzyme, which protects cells from endogenous non-canonical nucleotides and, in the case of thiopurine therapy, from exogenous nucleotides, such as thio-ITP and methyl-thio-ITP, derived from thiopurines [38]. One possible explanation for the protective role of non-functional *ITPA* polymorphisms in relation to the relapse of ALL could be

that the absence of functional ITPA activity can lead to the accumulation of non-canonical nucleotides that may cause DNA damage; such has also been demonstrated in an *in vitro* study on human HeLa cells with a knockdown *ITPA* gene [39]. Leukaemic blasts of the relapsed disease are more drug-resistant than blasts at initial diagnosis, which is probably the result of drug-resistant subclones being present at diagnosis or developing during therapy through the acquisition of additional genomic lesions [35]. It is possible that the action of 6-MP during the maintenance is partially reversed by the fully functional ITPA activity; this may then lead to the increased survival of leukemic blasts, which can, in turn, progress to relapse of the disease.

A more in-depth analysis of the impact of the *ITPA* genotype on relapse revealed that wild-type *ITPA* patients were at higher risk of suffering early relapse (HR 3.9, 95% CI 1.6–9.6; p = 0.003), whereas its effect on very early or late relapse was not statistically significant.

A study by Henderson et al. investigating the mechanism of relapse in ALL demonstrated that a shorter time interval to first relapse correlated with a higher quantity of the relapsing clone at diagnosis [40]. It is, therefore, possible that in the event of very early relapse occurring less than 18 months after the diagnosis, the amount of therapy-resistant sub-clones of leukemic blasts is too high for maintenance therapy to be effective in eradicating them. However, since the number of leukemic blasts in early relapse occurring more than 18 months after the diagnosis and less than 6 months after the end of treatment is lower, the influence of the *ITPA* genotype emerges by counteracting the 6-MP action. In the case of the wild-type *ITPA* genotype, the leukemic blasts are better protected against the therapy and are able to survive longer, resulting in a later expansion and onset of the relapse. On the other hand, the influence of the *ITPA* genotype diminishes again in late relapse occurring more than 6 months after the end of treatment. A possible explanation for this could lie in observed differences in gene expression between early and late relapse.

Table 4. Clinical characteristics of relapsed patients (N = 102).

Characteristic	No	%
Gender		
Female	39	38.2
Male	63	61.8
Time to relapse		
Median time (range) 24 (4–163) months		
Very early relapse[1]	32	31.4
Early relapse[2]	43	42.1
Late relapse[3]	27	26.5
Site of first relapse		
Bone marrow	50	49.0
Combined[4]	21	20.6
Isolated EMD[5]	31	30.4
Site of any relapse		
Bone marrow (with or without EMD)	89	87.3
Only isolated EMD[5]	13	12.7
Treatment protocol		
POG	53	52.0
BFM 83/86	30	29.4
BFM 90/95/2002	19	18.6
Initially assigned risk group		
Standard risk	36	35.3
Intermediate or high risk	21	20.6
Undetermined	45	44.1

Abbreviations: POG, Pediatric Oncology Group; BFM, Berlin-Frankfurt-Muenster; EMD, extramedullary disease.
[1]less than18 months from diagnosis.
[2]more than 18 months after diagnosis and less than 6 months after the end of treatment.
[3]more than 6 months after the end of treatment.
[4]Bone marrow with or without extramedullary disease.
[5]CNS/testes/Other (spinal channel, liver, iris, mesenterium, lymph nodes neck, labia major).

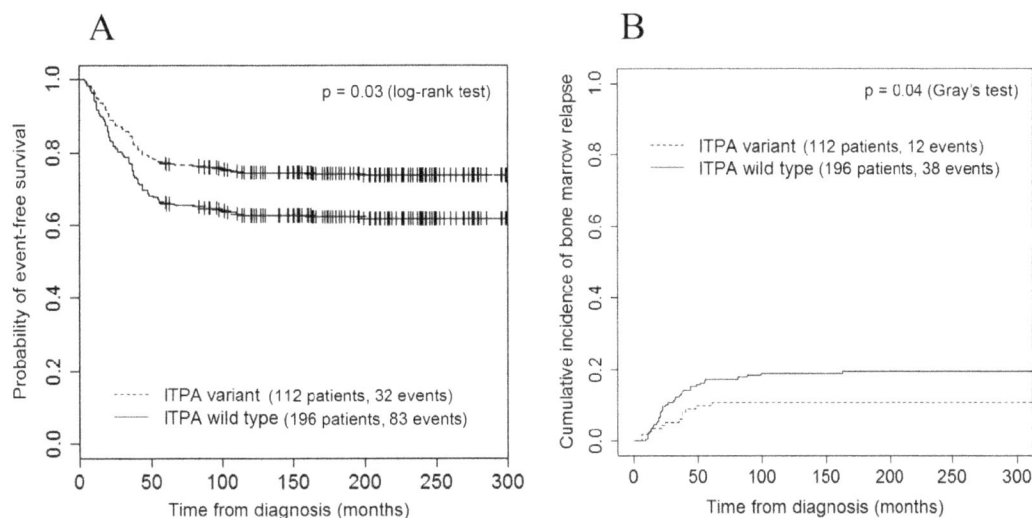

Figure 2. Event-free survival and cumulative incidence of bone marrow relapse by *ITPA* genotype (N = 308). A) *ITPA* wild-type genotype was shown to be an independent risk factor for lower event-free survival rate. 5-year EFS rates were 62% in wild-type patients and 74% in patients with at least 1 variant allele (log-rank test, p = 0.03). **B)** Cumulative incidences of bone marrow relapse in patients with wild-type *ITPA* and patients with at least one variant *ITPA* allele were 19.6% and 10.7%, respectively (Gray's test, p = 0.04).

Table 5. Results of analysis of genotypes' influence on survival rates.

Genotype (No. of patients)	No. of events (% patients)	Univariate (Log rank)		Multivariate (Cox regression)	
		5-year EFS (%)	p	Hazard ratio (95% CI)	p
Overall survival					
ITPA **genotype (94C>A/IVS2+21A>C)**					
CC/AA (196)	59 (30%)	71%	0.164		
at least 1 variant allele[1] (112)	25 (22%)	77%			
Event-free survival					
ITPA **genotype (94C>A/IVS2+21A>C)**					
CC/AA (196)	83 (42%)	62%	**0.030***	1.6 (1.04–2.37)	**0.033***
at least 1 variant allele[1] (112)	32 (29%)	74%			
TPMT **genotype**					
1*/1* (287)	107 (37%)	79%	0.919		
1*/3*[2] (21)	8 (38%)	69%			
MTHFR **genotype (677C>A/1298 A>C)**					
CC/AA (31)	15 (48%)	58%	0.298		
at least 1 variant allele[3] (277)	100 (36%)	67%			
MTRR **66A>G**					
AA (59)	21 (36%)	66%	0.757		
AG/GG (249)	94 (38%)	66%			
MTHFD **1958G>A**					
GG (94)	34 (36%)	68%	0.865		
GA/AA (214)	81 (38%)	65%			
BHMT **742G>A**					
GG (144)	55 (38%)	67%	0.843		
GA/AA (164)	60 (37%)	67%			
GNMT **1298C>T**					
CC (89)	32 (36%)	67%	0.841		
CT/TT (219)	83 (38%)	66%			
PACSIN2 **rs2413739**					
CC (111)	40 (36%)	68%	0.727		
CT/TT (194)	73 (38%)	64%			

*denotes statistically significant difference.
[1]*ITPA* genotype combinations with at least one variant allele (94CA/IVS2+21AA, 94CA/IVS2+21AC, 94CA/IVS2+21CC, 94CC/IVS2+21AC, 94CC/IVS2+21CC).
[2]either a *1/*3A or *1/*3C *TPMT* genotype.
[3]*MTHFR* genotype combinations with at least variant allele (677CC/1298AC, 677CT/1298AA, 677CC/1298CC, 677TT/1298AA, 677CT/1298AC, 677TT/1298AC).
Abbreviations: *TPMT*, Thiopurine S-methyltransferase; *MTHFR*, methylenetetrahydrofolate reductase; *MTRR*, 5-methyltetrahydrofolate-homocysteine methyltransferase reductase; *MTHFD1*, methylenetetrahydrofolate dehydrogenase 1; *BHMT*, betaine–homocysteine S-methyltransferase; *GNMT*, glycine N-methyltransferase; *PACSIN2*, protein kinase C and casein kinase substrate in neurons protein; *ITPA*, Inosine triphosphate pyrophosphatase.

Hogan et al. found that a set of genes involved in nucleotide biosynthesis and folate metabolism were specifically up-regulated in late relapse [41]. Another previous study by Zaza et al. has shown that the down-regulation of genes involved in purine metabolism correlates with the decreased *de novo* purine synthesis in EVT6/RUNX1 ALL and it has been postulated that this accounts for the chemosensitivity of this genetic subtype [42]. Overexpression in the blasts of patients who experience a late relapse offers an explanation for the resistance to antimetabolites and the lack of influence of the *ITPA* genotype in these patients.

The observation that *ITPA* genotype was associated only with bone marrow relapse could be explained by the role of the blood-brain and blood-testes barriers; these protect the leukemic blasts hidden in the CNS or testes (the so-called sanctuary sites) from the action of chemotherapeutics [43] and so the genes involved in the metabolism of anticancer drugs do not play such an important role in extramedullary relapse cases.

Several studies have previously been conducted studying the effects of *ITPA* polymorphisms on treatment outcome in ALL patients [7,21,22,24,25]; however, most of them focused on evaluating the effect of *ITPA* genotype on treatment-related toxicities and only a few on long-term survival. Stocco *et al.* reported that on maintenance therapy dosages, once the adjustment for *TPMT* had been made, the additional influence of *ITPA* genotype on 6-MP toxicity emerged [7]. No effect on long-term survival was seen therein. The effects of the combined *TPMT* and *ITPA* genotype on the mercaptopurine pharmacokinetics were also shown in a study by de Beaumais *et al.* [25]

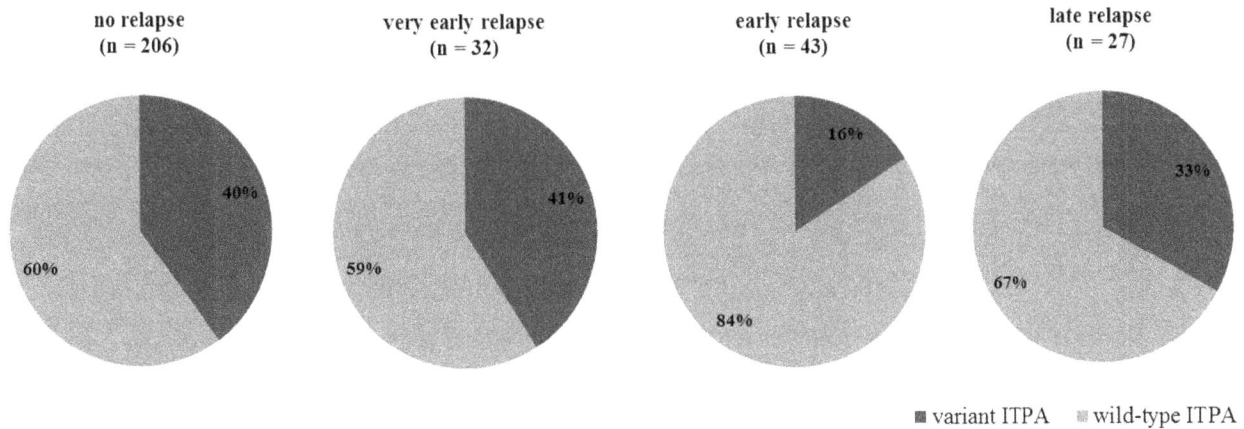

Figure 3. Influence of *ITPA* genotype on the incidence of relapses grouped according to time to relapse. Pie chart slices represent the percent of patients with different *ITPA* genotypes; wild-type *ITPA*: 94CC/IVS2+21AA ITPA genotype combination, variant *ITPA*: 94CA/IVS2+21AA, 94CA/IVS2+21AC, 94CA/IVS2+21CC, 94CC/IVS2+21AC, 94CC/IVS2+21CC ITPA genotype combinations. Very early relapse, <18 months after diagnosis; early relapse,>18 months after diagnosis and <6 months after the end of treatment; late relapse,>6 months after discontinuation of therapy (multinomial regression model adjusted to treatment protocol group, age group at diagnosis and gender; reference category = no relapse; p = 0.003).

Furthermore, a study conducted in Indian ALL patients showed an independent role for both *TPMT* and *ITPA* in terms of association with the incidence of hematological toxicity [24]. The only study that associated *ITPA* genotype with survival rate was that conducted by Kim *et al.* in 100 Korean patients with pediatric ALL. They evaluated 18 loci in 16 candidate genes of pharmacogenetic interest, finding that the *ITPA* genotype, though not *TPMT*, had a significant effect on the event-free survival rate, which was lower in *ITPA* variants [21]. Like Kim *et al.*, we have not seen any effect of the *TPMT* genotype on event-free survival in pediatric ALL patients. However, and in contrast to their study, we found a different effect of the *ITPA* genotype on the event-free survival rate, such that it was lower in wild-type *ITPA* patients. A significant effect was observed most particularly on the occurrence of relapse, with patients carrying at least one *ITPA* variant allele being at lower risk of suffering early and/or bone marrow relapse. One possible explanation for the observed differences may be the fact that the study by Kim *et al.* was evaluating only one of the two non-functional polymorphisms of the *ITPA* gene, which is probably due to ethnic differences in *ITPA* genotype frequencies between Asian and Caucasian populations. Although the *ITPA* 94C>A polymorphism has been more frequently associated with 6-MP related toxicities, we believe that genotyping for *ITPA* IVS2+21A>C is also important due to its higher frequency in Caucasian populations. Indeed, when the two polymorphisms were evaluated independently, no significant correlation with event-free survival and/or relapse rate could be shown in our study. However, since both of these variants lower ITPA activity, we considered it rational to evaluate them together, thus producing only two variables: (a) wild-type *ITPA* and (b) variant *ITPA*. This could also explain why other studies, such as that by Stocco *et al.* evaluating only the *ITPA* 94C>A polymorphism, did not detect a significant effect of the *ITPA* genotype on survival or relapse rate [7]. Furthermore, since no association between *ITPA* genotype and treatment-related toxicity (especially febrile neutropenia) could be observed in the Korean study, they postulated that the lower survival rate in the variant group might not be influenced by accumulated toxic metabolites but by different factors, such as other genetic polymorphisms that were linked with *ITPA* 94C>A [21]. In contrast, our results support the hypothesis

that variant *ITPA* alleles potentiate the effects of maintenance therapy by increasing the levels of toxic metabolites such that leukemic blasts are not protected and, therefore, unable to survive the attack of 6-MP. The effect is more pronounced in leukemic blasts in bone marrow and less so in blasts hidden in the sanctuary sites. Unfortunately, this hypothesis cannot be supported by measurements of toxic metabolites levels, since this was a retrospective study.

Our study evaluated a limited number of patients and, although we have taken into account - and adjusted our statistical models to - other factors influencing survival rates (the most important being treatment protocol), the effect of *ITPA* genotype should be evaluated in a larger cohort of patients treated according to the recent treatment protocols. In addition, we would like to point out that our conclusions are drawn from the analysis of DNA samples isolated from archival bone marrow smears prepared at the time of diagnosis of ALL. Since they contained a mixture of cells with different ratios of leukemic blasts and non-leukemic cells, the presence of hyperdiploid blasts and the potentially confounding effect of the acquisition of additional copies of chromosomes containing polymorphic genes on the concordance between the genotype and inherited phenotype could not be taken into account. Furthermore, additional molecular studies should be performed to elucidate the possible role of ITPA in relation to 6-MP therapy for the survival and expansion of leukemic blasts.

In summary, this study of pediatric ALL patients undergoing maintenance therapy shows that the *ITPA* genotype might be associated with event-free survival, with patients carrying at least one non-functional *ITPA* allele having higher event-free survival rates and a lower risk of suffering early and/or bone marrow relapse than wild-type *ITPA* patients.

Supporting Information

Table S1 List of TaqMan assays used for genotyping. Footnotes: Abbreviations: TPMT, Thiopurine S-methyltransferase; MTHFR, methylenetetrahydrofolate reductase; MTRR, 5-methyltetrahydrofolate-homocysteine methyltransferase reductase; MTHFD1, methylenetetrahydrofolate dehydrogenase 1; BHMT, betaine–homocysteine S-methyltransferase; GNMT, glycine N-methyltransferase; PACSIN2, protein kinase C and casein kinase substrate in neurons protein; ITPA, Inosine triphosphate pyrophosphatase.

Table S2 Summary of genotype data of ALL patients analysed in the study. Footnotes: [1] less than 18 months from diagnosis. [2] more than 18 months after diagnosis and less than 6 months after the end of treatment. [3] more than 6 months after the end of treatment. [4] percentage of patients within genotype group. Abbreviations: TPMT, Thiopurine S-methyltransferase; MTHFR, methylenetetrahydrofolate reductase; MTRR, 5-methyltetrahydrofolate-homocysteine methyltransferase reductase; MTHFD1, methylenetetrahydrofolate dehydrogenase 1; BHMT, betaine–homocysteine S-methyltransferase; GNMT, glycine N-methyl-

References

1. Freyer DR, Devidas M, La M, Carroll WL, Gaynon PS, et al. (2011) Postrelapse survival in childhood acute lymphoblastic leukemia is independent of initial treatment intensity: a report from the Children's Oncology Group. Blood 117: 3010–3015.
2. van der Veer A, Waanders E, Pieters R, Willemse ME, Van Reijmersdal SV, et al. (2013) Independent prognostic value of BCR-ABL1-like signature and IKZF1 deletion, but not high CRLF2 expression, in children with B-cell precursor ALL. Blood.
3. Pui CH, Robison LL, Look AT (2008) Acute lymphoblastic leukaemia. Lancet 371: 1030–1043.
4. Malempati S, Gaynon PS, Sather H, La MK, Stork LC (2007) Outcome after relapse among children with standard-risk acute lymphoblastic leukemia: Children's Oncology Group study CCG-1952. J Clin Oncol 25: 5800–5807.
5. Reismuller B, Attarbaschi A, Peters C, Dworzak MN, Potschger U, et al. (2009) Long-term outcome of initially homogenously treated and relapsed childhood acute lymphoblastic leukaemia in Austria–a population-based report of the Austrian Berlin-Frankfurt-Munster (BFM) Study Group. Br J Haematol 144: 559–570.
6. Nguyen K, Devidas M, Cheng SC, La M, Raetz EA, et al. (2008) Factors influencing survival after relapse from acute lymphoblastic leukemia: a Children's Oncology Group study. Leukemia 22: 2142–2150.
7. Stocco G, Cheok MH, Crews KR, Dervieux T, French D, et al. (2009) Genetic polymorphism of inosine triphosphate pyrophosphatase is a determinant of mercaptopurine metabolism and toxicity during treatment for acute lymphoblastic leukemia. Clin Pharmacol Ther 85: 164–172.
8. Krynetski EY, Evans WE (1998) Pharmacogenetics of cancer therapy: getting personal. Am J Hum Genet 63: 11–16.
9. Karas-Kuzelicki N, Mlinaric-Rascan I (2009) Individualization of thiopurine therapy: thiopurine S-methyltransferase and beyond. Pharmacogenomics 10: 1309–1322.
10. Schmiegelow K, Forestier E, Kristinsson J, Soderhall S, Vettenranta K, et al. (2009) Thiopurine methyltransferase activity is related to the risk of relapse of childhood acute lymphoblastic leukemia: results from the NOPHO ALL-92 study. Leukemia 23: 557–564.
11. Karas-Kuzelicki N, Jazbec J, Milek M, Mlinaric-Rascan I (2009) Heterozygosity at the TPMT gene locus, augmented by mutated MTHFR gene, predisposes to 6-MP related toxicities in childhood ALL patients. Leukemia 23: 971–974.
12. Milek M, Smid A, Tamm R, Kuzelicki NK, Metspalu A, et al. (2012) Post-translational stabilization of thiopurine S-methyltransferase by S-adenosyl-L-methionine reveals regulation of TPMT*1 and *3C allozymes. Biochem Pharmacol 83: 969–976.
13. Stocco G, Yang W, Crews KR, Thierfelder WE, Decorti G, et al. (2012) PACSIN2 polymorphism influences TPMT activity and mercaptopurine-related gastrointestinal toxicity. Hum Mol Genet 21: 4793–4804.
14. Sumi S, Marinaki AM, Arenas M, Fairbanks L, Shobowale-Bakre M, et al. (2002) Genetic basis of inosine triphosphate pyrophosphohydrolase deficiency. Hum Genet 111: 360–367.
15. Zabala-Fernandez W, Barreiro-de Acosta M, Echarri A, Carpio D, Lorenzo A, et al. (2011) A pharmacogenetics study of TPMT and ITPA genes detects a relationship with side effects and clinical response in patients with inflammatory bowel disease receiving Azathioprine. J Gastrointestin Liver Dis 20: 247–253.
16. Shipkova M, Franz J, Abe M, Klett C, Wieland E, et al. (2011) Association between adverse effects under azathioprine therapy and inosine triphosphate pyrophosphatase activity in patients with chronic inflammatory bowel disease. Ther Drug Monit 33: 321–328.
17. Kim JH, Cheon JH, Hong SS, Eun CS, Byeon JS, et al. (2010) Influences of thiopurine methyltransferase genotype and activity on thiopurine-induced leukopenia in Korean patients with inflammatory bowel disease: a retrospective cohort study. J Clin Gastroenterol 44: e242–248.
18. Jung YS, Cheon JH, Park JJ, Moon CM, Kim ES, et al. (2010) Correlation of genotypes for thiopurine methyltransferase and inosine triphosphate pyrophosphatase with long-term clinical outcomes in Korean patients with inflammatory

bowel diseases during treatment with thiopurine drugs. J Hum Genet 55: 121–123.
19. Uchiyama K, Nakamura M, Kubota T, Yamane T, Fujise K, et al. (2009) Thiopurine S-methyltransferase and inosine triphosphate pyrophosphohydrolase genes in Japanese patients with inflammatory bowel disease in whom adverse drug reactions were induced by azathioprine/6-mercaptopurine treatment. J Gastroenterol 44: 197–203.
20. Breen DP, Marinaki AM, Arenas M, Hayes PC (2005) Pharmacogenetic association with adverse drug reactions to azathioprine immunosuppressive therapy following liver transplantation. Liver Transpl 11: 826–833.
21. Kim H, Kang HJ, Kim HJ, Jang MK, Kim NH, et al. (2012) Pharmacogenetic analysis of pediatric patients with acute lymphoblastic leukemia: a possible association between survival rate and ITPA polymorphism. PLoS One 7: e45558.
22. Wan Rosalina WR, Teh LK, Mohamad N, Nasir A, Yusoff R, et al. (2012) Polymorphism of ITPA 94C>A and risk of adverse effects among patients with acute lymphoblastic leukaemia treated with 6-mercaptopurine. J Clin Pharm Ther 37: 237–241.
23. Adam de Beaumais T, Fakhoury M, Medard Y, Azougagh S, Zhang D, et al. (2011) Determinants of mercaptopurine toxicity in paediatric acute lymphoblastic leukemia maintenance therapy. Br J Clin Pharmacol 71: 575–584.
24. Dorababu P, Nagesh N, Linga VG, Gundeti S, Kutala VK, et al. (2012) Epistatic interactions between thiopurine methyltransferase (TPMT) and inosine triphosphate pyrophosphatase (ITPA) variations determine 6-mercaptopurine toxicity in Indian children with acute lymphoblastic leukemia. Eur J Clin Pharmacol 68: 379–387.
25. Adam de Beaumais T, Jacqz-Aigrain E (2012) Pharmacogenetic determinants of mercaptopurine disposition in children with acute lymphoblastic leukemia. Eur J Clin Pharmacol 68: 1233–1242.
26. Heller T, Oellerich M, Armstrong VW, von Ahsen N (2004) Rapid detection of ITPA 94C>A and IVS2 + 21A>C gene mutations by real-time fluorescence PCR and in vitro demonstration of effect of ITPA IVS2 + 21A>C polymorphism on splicing efficiency. Clin Chem 50: 2182–2184.
27. Hawwa AF, Millership JS, Collier PS, Vandenbroeck K, McCarthy A, et al. (2008) Pharmacogenomic studies of the anticancer and immunosuppressive thiopurines mercaptopurine and azathioprine. Br J Clin Pharmacol 66: 517–528.
28. Moricke A, Zimmermann M, Reiter A, Henze G, Schrauder A, et al. (2010) Long-term results of five consecutive trials in childhood acute lymphoblastic leukemia performed by the ALL-BFM study group from 1981 to 2000. Leukemia 24: 265–284.
29. Reiter A, Schrappe M, Ludwig WD, Hiddemann W, Sauter S, et al. (1994) Chemotherapy in 998 unselected childhood acute lymphoblastic leukemia patients. Results and conclusions of the multicenter trial ALL-BFM 86. Blood 84: 3122–3133.
30. Schrappe M, Reiter A, Ludwig WD, Harbott J, Zimmermann M, et al. (2000) Improved outcome in childhood acute lymphoblastic leukemia despite reduced use of anthracyclines and cranial radiotherapy: results of trial ALL-BFM 90. German-Austrian-Swiss ALL-BFM Study Group. Blood 95: 3310–3322.
31. Laks D, Longhi F, Wagner MB, Garcia PC (2003) [Survival evaluation of children with acute lymphoblastic leukemia treated with Berlin-Frankfurt-Munich trial]. J Pediatr (Rio J) 79: 149–158.
32. Riehm H, Reiter A, Schrappe M, Berthold F, Dopfer R, et al. (1987) [Corticosteroid-dependent reduction of leukocyte count in blood as a prognostic factor in acute lymphoblastic leukemia in childhood (therapy study ALL-BFM 83)]. Klin Padiatr 199: 151–160.
33. Karas Kuzelicki N, Milek M, Jazbec J, Mlinaric-Rascan I (2009) 5,10-Methylenetetrahydrofolate reductase (MTHFR) low activity genotypes reduce the risk of relapse-related acute lymphoblastic leukemia (ALL). Leuk Res 33: 1344–1348.

transferase; PACSIN2, protein kinase C and casein kinase substrate in neurons protein; ITPA, Inosine triphosphate pyrophosphatase.

Acknowledgments

We thank Ms. Klavdija Stare for her dedicated work and Mr. Benedict Dries-Jenkins for proofreading the manuscript.

Author Contributions

Conceived and designed the experiments: IMR AS. Performed the experiments: AS MM. Analyzed the data: AS NKK MM JJ. Contributed reagents/materials/analysis tools: IMR JJ. Wrote the paper: AS IMR.

34. Hahn LW, Ritchie MD, Moore JH (2003) Multifactor dimensionality reduction software for detecting gene-gene and gene-environment interactions. Bioinformatics 19: 376–382.

35. Bhojwani D, Pui CH (2013) Relapsed childhood acute lymphoblastic leukaemia. Lancet Oncol 14: e205–217.

36. Tallen G, Ratei R, Mann G, Kaspers G, Niggli F, et al. (2010) Long-term outcome in children with relapsed acute lymphoblastic leukemia after time-point and site-of-relapse stratification and intensified short-course multidrug chemotherapy: results of trial ALL-REZ BFM 90. J Clin Oncol 28: 2339–2347.

37. Raetz EA, Bhatla T (2012) Where do we stand in the treatment of relapsed acute lymphoblastic leukemia? Hematology Am Soc Hematol Educ Program 2012: 129–136.

38. Marinaki AM, Ansari A, Duley JA, Arenas M, Sumi S, et al. (2004) Adverse drug reactions to azathioprine therapy are associated with polymorphism in the gene encoding inosine triphosphate pyrophosphatase (ITPase). Pharmacogenetics 14: 181–187.

39. Menezes MR, Waisertreiger IS, Lopez-Bertoni H, Luo X, Pavlov YI (2012) Pivotal role of inosine triphosphate pyrophosphatase in maintaining genome stability and the prevention of apoptosis in human cells. PLoS One 7: e32313.

40. Henderson MJ, Choi S, Beesley AH, Sutton R, Venn NC, et al. (2008) Mechanism of relapse in pediatric acute lymphoblastic leukemia. Cell Cycle 7: 1315–1320. Epub 2008 Feb 1329.

41. Hogan LE, Meyer JA, Yang J, Wang J, Wong N, et al. (2011) Integrated genomic analysis of relapsed childhood acute lymphoblastic leukemia reveals therapeutic strategies. Blood 118: 5218–5226.

42. Zaza G, Yang W, Kager L, Cheok M, Downing J, et al. (2004) Acute lymphoblastic leukemia with TEL-AML1 fusion has lower expression of genes involved in purine metabolism and lower de novo purine synthesis. Blood 104: 1435–1441.

43. Buhrer C, Hartmann R, Fengler R, Dopfer R, Gadner H, et al. (1993) Superior prognosis in combined compared to isolated bone marrow relapses in salvage therapy of childhood acute lymphoblastic leukemia. Med Pediatr Oncol 21: 470–476.

14

Andrographis paniculata Leaf Extract Prevents Thioacetamide-Induced Liver Cirrhosis in Rats

Daleya Abdulaziz Bardi[1]**, Mohammed Farouq Halabi**[1]**, Pouya Hassandarvish**[2]**, Elham Rouhollahi**[3]**, Mohammadjavad Paydar**[3]**, Soheil Zorofchian Moghadamtousi**[4]**, Nahla Saeed Al-Wajeeh**[1]**, Abdulwali Ablat**[5]**, Nor Azizan Abdullah**[3]**, Mahmood Ameen Abdulla**[1]*****

1 Department of Biomedical Science, Faculty of Medicine, University of Malaya, Kuala Lumpur, Malaysia, 2 Department of Medical Microbiology, Faculty of Medicine, University of Malaya, Kuala Lumpur, Malaysia, 3 Department of Pharmacology, Faculty of Medicine, University of Malaya, Kuala Lumpur, Malaysia, 4 Biomolecular Research Group, Biochemistry Program, Institute of Biological Sciences, Faculty of Science, University of Malaya, Kuala Lumpur, Malaysia, 5 Institute of Biological Science, Faculty of Science, University of Malaya, Kuala Lumpur, Malaysia

Abstract

This study investigated the hepatoprotective effects of ethanolic *Andrographis paniculata* leaf extract (ELAP) on thioacetamide-induced hepatotoxicity in rats. An acute toxicity study proved that ELAP is not toxic in rats. To examine the effects of ELAP *in vivo*, male *Sprague Dawley* rats were given intraperitoneal injections of vehicle 10% Tween-20, 5 mL/kg (normal control) or 200 mg/kg TAA thioacetamide (to induce liver cirrhosis) three times per week. Three additional groups were treated with thioacetamide plus daily oral silymarin (50 mg/kg) or ELAP (250 or 500 mg/kg). Liver injury was assessed using biochemical tests, macroscopic and microscopic tissue analysis, histopathology, and immunohistochemistry. In addition, HepG2 and WRL-68 cells were treated *in vitro* with ELAP fractions to test cytotoxicity. Rats treated with ELAP exhibited significantly lower liver/body weight ratios and smoother, more normal liver surfaces compared with the cirrhosis group. Histopathology using Hematoxylin and Eosin along with Masson's Trichrome stain showed minimal disruption of hepatic cellular structure, minor fibrotic septa, a low degree of lymphocyte infiltration, and minimal collagen deposition after ELAP treatment. Immunohistochemistry indicated that ELAP induced down regulation of proliferating cell nuclear antigen. Also, hepatic antioxidant enzymes and oxidative stress parameters in ELAP-treated rats were comparable to silymarin-treated rats. ELAP administration reduced levels of altered serum liver biomarkers. ELAP fractions were non-cytotoxic to WRL-68 cells, but possessed anti-proliferative activity on HepG2 cells, which was confirmed by a significant elevation of lactate dehydrogenase, reactive oxygen species, cell membrane permeability, cytochrome *c*, and caspase-8,-9, and, -3/7 activity in HepG2 cells. A reduction of mitochondrial membrane potential was also detected in ELAP-treated HepG2 cells. The hepatoprotective effect of 500 mg/kg of ELAP is proposed to result from the reduction of thioacetamide-induced toxicity, normalizing reactive oxygen species levels, inhibiting cellular proliferation, and inducing apoptosis in HepG2 cells.

Editor: Aamir Ahmad, Wayne State University School of Medicine, United States of America

Funding: This research is supported by University of Malaya grant PV046/2012A, and High Impact Research Grant UM-MOHE UM.C/625/1/HIR/MOHE/SC/09 from the Ministry of Higher Education Malaysia. The funders had no role in study design, data collection and analysis, decision to publish, or preparation of the manuscript.

Competing Interests: The authors have declared that no competing interests exist.

* Email: ammeen@um.edu.my

Introduction

The liver is the largest internal organ in vertebrates, including humans. It is vital to survival, as it is the key organ of metabolism and excretion, and supports every other organ. It is continuously exposed to xenobiotics, and hence is prone to many diseases [1]. In chronically damaged liver, hepatic fibrosis arises from perpetual wound-healing, which results in abnormal production and accumulation of connective tissue and ultimately leads to liver cirrhosis [2]. The rate of fibrosis progression is reported to depend on the cause of liver disease, host, and environmental factors. Liver cirrhosis has a high global prevalence. It is the end-stage of most liver pathologies and leads to chronic liver dysfunction, altered metabolism, and death [1]. Reactive oxygen species (ROS) play an important role in liver pathology, particularly in cases of alcohol- and toxicity-induced liver disease [3]. Because conventional and synthetic drugs used to treat liver disease are often inadequate and can have serious side effects, many people, even in developed countries, turn to complementary and alternative medicine (CAM). Medicinal plants reported to possess hepatoprotective activity include *Vitex negundo* [4], *Boesenbergia rotunda* [5], *Phyllanthus niruri* [6], *Ipomoea aquatic* [7], *Orthosiphon stamineus* [8], *Zingiber officinale* [9], and *Caesalpinia sappan* [10].

Andrographis paniculata (Malay: Hempedu Bumi) is a herbaceous annual belonging to the family Acanthaceae, native to Southeast Asia, especially China, India, and Sri Lanka. This herb has been traditionally used in Ayurvedic medicine (Indian traditional medicine). It is most used to treat and prevent infectious diseases, as it thought to strengthen the immune system [11]. In addition, *A. paniculata* has analgesic [12], antioxidant [13], antibiofilm [14], gastroprotective [15], wound-healing [16], antiflarial [17], antimicrobial [18,19], anticancer [20], and antimalarial effects [21]. Phytochemical analyses have revealed that it is a rich source of diterpenoids and 2'-oxygenated flavonoids, including andrographolide, neoandrographolide, 14-deoxy-11, 12-didehydroandrographolide, 14-deoxyandrographolide, isoandrographolide, 14-deoxyandrographolide-19-β-d-glucoside, homoandrographolide, andrographan, andrographosterin, and stigmasterol [22]. Andrographolide is the primary bioactive phytochemical of *A. paniculata* [23], where it is found in the leaves with a concentration of >2% [22]. When ingested orally, andrographolide accumulates in the visceral organs such as liver. It is absorbed quickly and P-glycoprotein was suggested to be participants in the intestinal absorption. Andrographolide metabolized extensively in rats and humans, structural illucidation of metabolites have shown andrographolid analogues, sulfonates and sulfate ester compounds which isolated from rat feces, urine and small intestine, achieving ~90% elimination within 48 h [22,24]. Andrographolide exhibited significant cytotoxic activity against cancer cells, induces apoptosis, possess anti-inflammatory and Anti-angiogenic activity and has chemo-protective potential towards normal cells [24].

Thioacetamide (TAA) is hepatotoxic, with a single dose able to produce centrilobular necrosis followed by a regenerative response in animals [25,26]. High dosages lead to liver cirrhosis and hepatocarcinoma [27]. The aim of the present study was to evaluate the possible hepatoprotective effects of *A. paniculata* leaf extract (ELAP) both *in vivo* (using TAA-induced liver cirrhosis in the rat as a model system) and *in vitro* (using cultured hepatic cell lines). Results were compared to the effects of silymarin, a drug commonly used as a liver support during treatment of liver cirrhosis.

Materials and Methods

Ethics Statement

The study was approved by the Ethics Committee for Animal Experimentation, Faculty of Medicine, University of Malaya, Malaysia (Ethic No. PM/05/08/2012/MMA (a) (R)). All animals received humane care according to the criteria outlined in the Guide for the Care and Use of Laboratory Animals prepared by the United States National Academy of Sciences and published by the National Institutes of Health [28].

Chemicals and Consumables

TAA was purchased from Sigma-Aldrich (Germany) and silymarin was from International Laboratory (USA). HepG2 human hepatocarcinoma cells and WRL-68 normal human liver cells were purchased from the American Type Culture Collection

(ATCC, Manassas, USA) and cultured in supplemented Dulbecco's modified Eagle medium (DMEM, Invitrogen, Carlsbad, CA) at 37°C in a humidified atmosphere of carbon dioxide and air (5:95). DMEM was supplemented with 10% heat-inactivated fetal bovine serum (FBS), 100 mg/mL streptomycin, and 100 U/mL penicillin (all from Invitrogen). The MTT assay [3-(4,5-dimethylthiazol-2-yl)-2,5-diphenyltetrazolium bromide] was purchased from Invitrogen.

Preparation of Plant Extracts (ELAP)

Air-dried *A. paniculata* leaves were obtained from Ethno Resources (Selangor, Malaysia), and their identity was confirmed by comparison with voucher specimen no. 43261 deposited at the Herbarium of Rimba Ilmu, Institute of Science Biology, University of Malaya, Kuala Lumpur. Leaves were ground to a fine powder using an electric blender, and 100 g powder was suspended in 500 mL of 95% ethanol for 3 days. This mixture was filtered using fine muslin cloth followed by Whatman no. 1 filter paper, then concentrated under reduced pressure (Eyela rotary evaporator, Sigma-Aldrich, USA), frozen, and lyophilized (yield: 4.7 g crude extract). For *in vivo* experiments, extract was dissolved in 10% Tween-20 and administered orally to rats at 250 or 500 mg/kg body weight (5 mL/kg body weight). For *in vitro* experiments, liquid-liquid partitioning was performed on crude extract to yield *n*-hexane (HF), chloroform (CF), butanol (BF), and aqueous (AF) soluble ELAP fractions [9], which were stored at −20°C.

Acute Toxicity Test

The study was carried out using the "fix dose" method of the Organisation for Economic Co-operation and Development guideline no. 420 [29] and following the Animal Research: Reporting *In Vivo* Experiments (ARRIVE) guidelines [24]. Healthy adult female *Sprague Dawley* (SD) rats (6–8 weeks old, 180–200 g) obtained from the Animal House (Faculty of Medicine, University of Malaya) were randomly divided into two groups of 6 rats each. The treatment group received 2500 mg/kg ELAP, and the control group received only vehicle (10% Tween-20). To eliminate food from the gastrointestinal tract that might intervene with ELAP absorption, food was withheld overnight prior to ELAP administration and for another 3 to 4 h after administration. Water was provided throughout. Animals were observed for 0.5, 2, 4, 24, and 48 h for the onset of behavioral changes, clinical or toxicological symptoms, and death. Mortality was further observed over a period of 2 weeks every morning and animals were sacrificed on day 15. Gross necropsies were performed, histopathology was performed on livers and kidneys, and serum biochemistry was assessed following standard methods [30].

Table 1. Effect of ELAP on body weight, liver weight, and liver index after 8 weeks.

Treatment	Body weight (g)	Liver weight (g)	Liver index (LW/BW %)
Normal control	347±7	10.33±1.15	2.95±0.22
TAA control	172±6[#]	12.61±0.41[#]	7.38±0.36[#]
Silymarin (50 mg/kg)	374±8*	10.66±1.69	2.99±0.32*
ELAP (250 mg/kg)	236±8*[#]	10.92±0.59[#]	4.66±0.14*[#]
ELAP (500 mg/kg)	231±4*[#]	10.85±0.31[#]	4.75±0.24*[#]

Data expressed as mean ± SEM (*n* = 6 rats/group).
*P<0.05 compared with TAA control,
[#]P<0.05 compared with normal control.

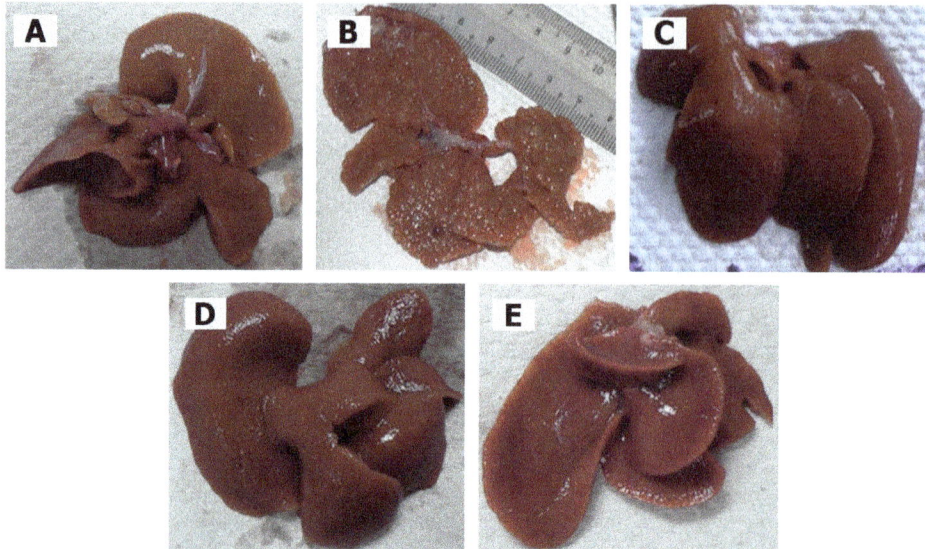

Figure 1. Macroscopic liver appearance. (A) Normal control liver has regular smooth surface. (B) TAA control liver has rough nodular surface, with uniform distribution of micronodules (<0.3 cm) and macronodules (≥0.3 cm). (C) TAA+silymarin liver has normal smooth surface. (D) TAA+ 250 mg/kg ELAP liver has nearly smooth surface and few micronodules. (E) TAA+500 mg/kg ELAP liver has normal smooth surface and nearly normal anatomical shape and appearance. Livers shown are representative samples ($n = 6$/group).

In Vivo Experimental Design

Healthy adult male SD rats (6–8 weeks old, 150–180 g) were obtained from the Animal House. Rats were housed individually in cages with wide-mesh wire bottoms in an animal room at $25°±2°$C, 50–60% humidity, and exposed to a 12-h light/dark cycle. Animals were maintained on standard pellet diet and tap water. Animals were acclimatized under standard laboratory conditions for a period of 2 weeks before the experiment.

Rats were divided randomly into five groups of six rats each. The normal control group was injected intraperitoneally with vehicle (10% Tween-20, 5 mL/kg) thrice weekly and received daily oral of distilled water (5 mL/kg). A TAA control group (to establish the liver cirrhosis model) was injected intraperitoneally with 200 mg/kg TAA thrice weekly for eight weeks [31] and received daily oral administration of vehicle (10% Tween-20, 5 mL/kg). A positive control group was injected intraperitoneally

with 200 mg/kg TAA thrice weekly and given daily oral silymarin at 50 mg/kg. The two experimental groups were injected intraperitoneally with 200 mg/kg TAA thrice weekly and given daily oral ELAP at 250 or 500 mg/kg.

The experiment was carried out for 8 weeks [5]. The animals were given water ad libitum. The body weights of the animals were recorded weekly. At the end of the 8th week, rats were fasted for 24 h after the last treatment and then anesthetized under ketamine (30 mg/kg, 100 mg/mL) and xylazil (3 mg/kg, 100 mg/ mL) anesthesia. Blood was withdrawn through the jugular vein and collected in gel-activated tubes, which were allowed to clot, centrifuged, and analyzed at the Clinical Diagnosis Laboratory of the University of Malaya Hospital for liver function test.

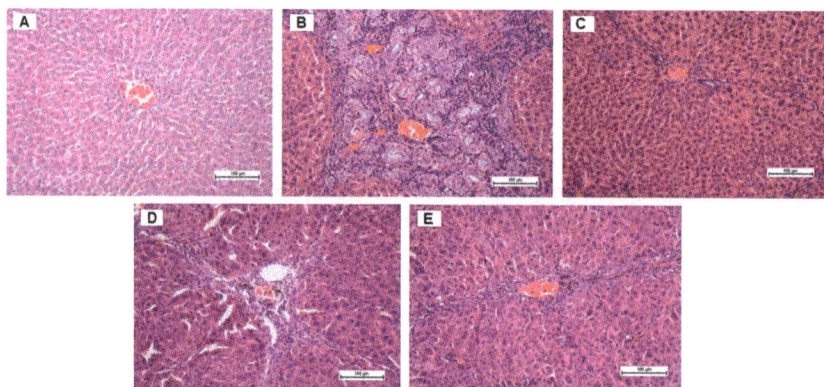

Figure 2. H&E staining of histopathological liver sections. (A) Normal control liver has normal histological structure and architecture. (B) TAA control liver has structural damage, irregular regenerating pseudolobules with dense fibrotic septa, proliferation of bile duct, and presence of centrilobular and inflammatory cells. (C) TAA+silymarin liver has mild inflammation but no fibrotic septa. (D) TAA+250 mg/kg ELAP liver has partially preserved hepatocytes, small area of necrosis, narrow fibrotic septa. (E) TAA+500 mg/kg ELAP liver has partially preserved hepatocytes and small areas of mild necrosis. Sections shown are from representative samples ($n = 6$/group).

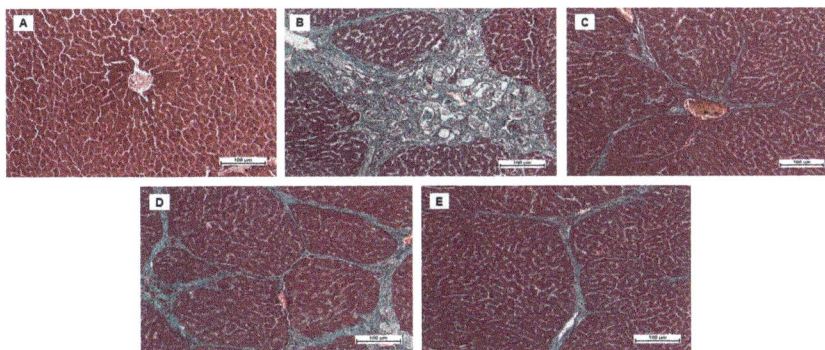

Figure 3. Masson's Trichrome staining of histopathological liver sections. (A) Normal control liver has normal architecture. (B) TAA control liver shows proliferation of bile duct, dense fibrous septa, and collagen fibers. (C) TAA+silymarin liver shows minimal fibrous septa and collagen fibers. (D) TAA+250 mg/kg ELAP liver shows moderate fibrous septa and irregular regenerating nodules. (E) TAA+500 mg/kg ELAP liver shows mild fibrous septa and collagen fibers. [Sections shown are from representative samples ($n = 6$/group)].

Macroscopic Liver Assessment

The abdominal and thoracic cavities were opened and rats displaying macroscopic evidence of pathology in organs other than the liver were excluded from the study. The livers were weighed and carefully examined for any gross pathology, then washed in ice-cold saline, blotted on filter paper, and weighed. Histopathology and immunohistochemistry was performed on liver tissues. Additionally, antioxidant activity and levels of oxidative stress were measured.

Histopathology

Liver specimens were fixed in 10% buffered formalin and processed by an automated tissue processing machine (Leica, Germany). Sections were stained with hematoxylin and eosin (H&E) and Masson's trichrome. Stained liver slices were evaluated under a Nikon microscope (Y-THS, Japan).

Immunohistochemistry

Liver tissue sections were heated at 60°C for 30–60 min in an oven (Venticell, MMM, Einrichtungen, Germany) and deparaffinized in xylene (2×3 min). Tissues were rehydrated using absolute, 95%, and 70% alcohol (2×3 min each) followed by running water. Antigen retrieval was performed in 10 mM sodium citrate buffer boiled in a microwave (Sanyo, Super Showe wave, Japan) for 10 min. The tissue slides were then cooled and placed in Tris Buffered Saline (TBS) with 0.05% Tween-20. Immunohistochemistry was performed following manufacturer's instructions using ARK (Animal Research Kit), Peroxidase (kit #K3955) (DakoCytomation, USA). Endogenous peroxidase was quenched

using peroxidase blocking solution (0.03% hydrogen peroxide sodium azide) for 5 min. The slides were rinsed gently with distilled water for 3 min and then incubated with biotinylated primary antibodies against (PCNA; 1:200) for 15 min. Tissue sections were gently washed twice with wash distilled water 3 min and kept in the buffer bath in a humid chamber. Streptavidin-HRP was added and sections were incubated for 15 min, then washed twice with distilled water for 3 min. Tissue sections were incubated with diaminobenzidine (DAB) substrate Chromagen for 5 min, then washed and counterstained with hematoxylin for 5 sec. Finally, sections were dipped ten times in weak ammonia (0.037 M/L), rinsed in distilled or deionized water for 2 to 5 min, and mounted with mounting medium. Under a light microscope, positive antigens stained brown against a blue hematoxylin background. The proliferation index of PCNA-stained liver sections was assessed by counting the percentage of labeled cells per 1000 liver cells, and the number of mitotic cells was expressed as mitotic index [9].

Antioxidant Activity and Oxidative Stress

Liver samples were washed, placed in ice-cold 10% (w/v) phosphate buffer solution (PBS), pH 7.4, and then homogenized on ice using a Teflon homogenizer (Polytron, Heidolph RZR 1, Germany). Cell debris was removed by centrifugation at 4500 rpm for 15 min at 4°C. The supernatant was used to determine antioxidant activity using assay kits for superoxide dismutase and catalase (SOD and CAT; Cayman Chemical Company, USA), and to analyze oxidative stress using assay kits for thiobarbituric

Table 2. Effect of ELAP on PCNA labeling and mitotic index.

Treatment	PCNA labeling	Mitotic index[a]
Normal control	0	0
TAA control	24.83±0.92[#]	72.5±1.22[#]
Silymarin (50 mg/kg)	0.33±0.19*[#]	16.83±1.10*[#]
ELAP (250 mg/kg)	11.66±1.64*[#]	42.33±1.28*[#]
ELAP (500 mg/kg)	0.66±0.19*[#]	20.83±1.53*[#]

[a]Percentage of labeled cells per 1000 liver cells. Data expressed as mean ± SEM (6 = rats/group).
*$P < 0.05$ compared with TAA control,
[#]$P < 0.05$ compared with normal control.

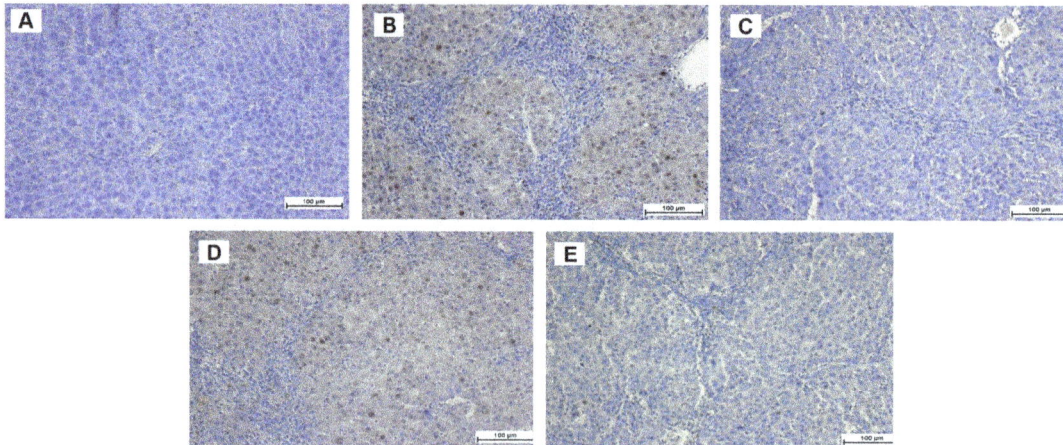

Figure 4. PCNA labeling in histopathological liver sections. (A) Normal controls stained without primary antibody show normal liver architecture and no PCNA labeling. (B) TAA controls have many PCNA-positive hepatocyte nuclei. (C) TAA+silymarin rats have no PCNA-positive hepatocytes. (D) TAA+250 mg/kg ELAP rats have moderate hepatocyte regeneration, as indicated by moderate presence of PCNA-positive hepatocyte nuclei. (E) TAA+500 mg/kg ELAP rats have mild PCNA expression with few regenerative hepatocytes.

acid reactive substance (TBARS; Cayman Chemical Company) and nitric oxide (QuantiChrom, USA).

Biochemical Parameters

Blood was collected into clot-activator tubes and serum was separated by centrifuging at 2500 rpm for 15 min. Aspartate aminotransferase (AST), alanine aminotransferase (ALT), total protein, albumin, globulin, total bilirubin, conjugated bilirubin, alkaline phosphatase (ALP), and gamma glutamyl transferase (GGT) were assayed spectrophotometrically at Clinical Diagnostic Laboratory of the University Malaya Medical Centre.

In Vitro Cell Viability Assay

Cell viability was measured using a standard colorimetric MTT reduction assay. HepG2 and WRL-68 cells were cultured in 96-wells plates at a density of 5×10^5 cells/well. After 24 h, cells were treated with increasing concentrations of the four ELAP fractions (CF, HF, BF, and AF) or 0.2% DMSO (vehicle control) for 24 h. Next, 2 mg/mL MTT (Invitrogen) was added and the cells were incubated for 4 h. After incubation, the supernatants were aspirated, and 100 μL DMSO were added to dissolve formazan crystals. Optical density was measured at 570 nm using a multiwell plate reader (Asys UVM340, Eugendorf, Austria) and IC_{50} values of the isolated fractions were determined from the linear portion of dose-response curves. Three ELAP fractions (CF,

HF, and BF) demonstrated significant antiproliferative effects in HepG2 cells and were used for further investigation.

Lactate Dehydrogenase (LDH) Release Assay

To confirm the cytotoxicity of the three ELAP fractions, we measured LDH released from cells into the medium using a Pierce LDH cytotoxicity assay kit (Thermo Scientific, Pittsburgh, PA). HepG2 cells were incubated with different concentrations of isolated ELAP fractions for 48 h and then LDH activity was determined in the supernatant. As a positive control, cells were completely lysed using Triton X-100 (2%) to give maximum LDH release. LDH reaction solution (100 μL) was added to the supernatants (100 μL) for 30 min and red color intensity measured at 490 nm. The level of LDH released from treated cells was expressed as a percentage of the positive control.

Reactive Oxygen Species (ROS) Assay

To determine the effect of ELAP fractions on ROS formation in HepG2 cells, 1×10^4 HepG2 cells per well were seeded in 96-well plates and incubated at 37°C for 24 h. Cells were then treated with ELAP fractions at different concentrations for 24 h. Treated cells were stained with 50 μL dihydroethidium (2.5 μg/mL) for 30 min and then washed twice with PBS. ROS formation was measured at 490 nm using a fluorescence microplate reader (Tecan Infinite M 200 PRO, Männedorf, Switzerland).

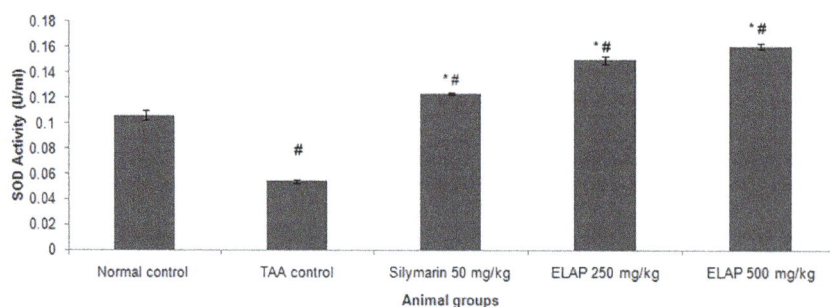

Figure 5. Effect of ELAP on SOD activity in liver tissue. Data expressed as mean ± SEM (*n* = 6 rats/group). *P<0.05 compared with TAA control, #P<0.05 compared with normal control.

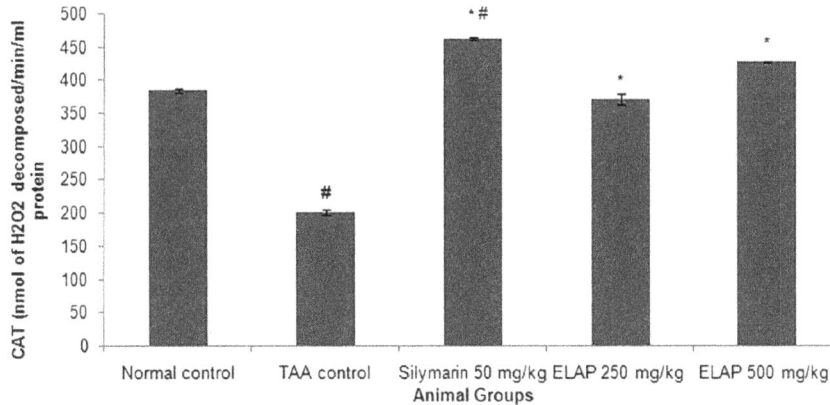

Figure 6. Effect of ELAP on CAT activity in liver tissue. Data expressed as mean ± SEM (n = 6 rats/group). *P<0.05 compared with TAA control, #P<0.05 compared with normal control.

Multiple Cytotoxicity Assay

A Cellomics high-content screening (HCS) system and multi-parameter Cytotoxicity 3 kit (Thermo Scientific) were used to simultaneously analyze cell membrane permeability, cytochrome c release, and mitochondrial membrane potential (MMP) in ELAP-treated HepG2 cells. Briefly, 1×10^4 liver cancer HepG2 cells were seeded into 96-well plates and incubated at 37°C for 24 h. Cells were then incubated with different concentrations of ELAP fractions for 24 h. Treated cells were stained with cell permeability and MMP dyes, fixed with 3.5% paraformaldehyde, blocked with 1×blocking buffer, and labeled by cytochrome c immunohisto-chemistry according to the vendor's protocol. Hoechst 33342 dye was used to counterstain the nucleus and cells were analyzed using the ArrayScan HCS system with ArrayScan II Data Acquisition and Data Viewer version 3.0 (Cellomics).

Caspase Activity Assay

Caspase activity in ELAP-treated HepG2 cells was determined using Caspase-Glo-3/7, -8, and -9 assay kits (Promega, Madison, WI). Briefly, HepG2 cells (1×10^4 cells/well) were cultured in a white 96-well microplate overnight. Cells were treated with different concentrations of isolated ELAP fractions for 24 h. Next, Caspase-Glo reagent (100 µL) was added, and cells were incubated for 30 min. Caspase activities were measured by luminescence measurement using a microplate reader (Infinite M200 PRO Tecan, Austria).

Statistical Analysis

Data from the rat study were expressed as mean ± SEM using one-way analysis of variance (ANOVA) followed by Bonferroni's post hoc test using SPSS software for Windows, version 20 (SPSS, Chicago, IL, USA). In vitro study results were expressed as mean ± standard deviation (SD) of at least three independent experiments. ANOVA was performed using GraphPad Prism software (GraphPad Software, San Diego, CA). The level of statistical significance was set at 0.05 for all experiments.

Results

Acute Toxicity Analysis

Animals pretreated with 2500 mg/kg ELAP or vehicle were observed for 14 days. All animals were alive and there were no signs of toxicity-related macroscopic or behavioral abnormalities. Biochemical and histopathological analyses showed no significant difference between control and ELAP-treated groups (Figure S1 and Tables S1, S2). These findings verify that 2500 mg/kg ELAP was safe and non-toxic in rats.

Effects of ELAP on TAA-Induced Liver Cirrhosis

Body Weight, Liver Weight, and Liver Index. Body weight of all animals was measured weekly and prior to scarification. Body and liver weights after two months of ELAP treatment are shown in Table 1. The control group had normal weight gains from 194 to 347 g over 8 weeks. The TAA control group weighed significantly (P<0.05) less than all other groups. The highest liver index (liver/body weight ratio) was observed in the TAA control

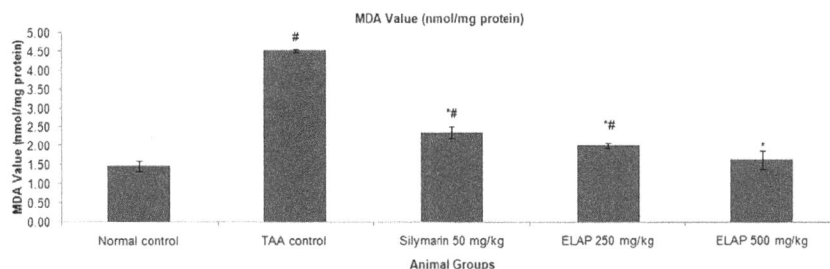

Figure 7. Effect of ELAP on MDA levels in liver tissue. Data expressed as mean ± SEM (n = 6 rats/group). *P<0.05 compared with TAA control, #P<0.05 compared with normal control.

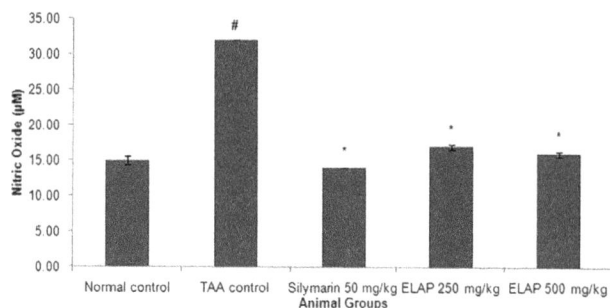

Figure 8. Effect of ELAP on NO levels in liver tissue. Data expressed as mean ± SEM ($n = 6$ rats/group). *$P<0.05$ compared with TAA control, #$P<0.05$ compared with normal control.

group. Administration of 250 or 500 mg/kg ELAP significantly ($P<0.05$) lowered the liver index, an effect comparable to that in the silymarin administered group.

Macroscopic Appearance of Liver. In this study, TAA treatment caused liver cirrhosis in rats. In comparison to the normal liver topology with a regular and smooth surface, the livers in the TAA control group were rough and nodular, with uniform micronodules (<0.3 cm) and macronodules (≥0.3 cm) throughout. Treatment with either silymarin or ELAP remarkably enhanced the recovery of TAA-induced liver structure damage as shown in Figure 1.

Histopathological Analysis of Liver Sections. Normal control livers were clear of any pathological abnormality, having distinct plates of hepatic cells, sinusoidal spaces, and a central vein (Figure 2A). Hepatocytes were polygonal with well-preserved cytoplasm and prominent nuclei. In contrast, TAA control livers (Figure 2B) showed a loss of normal architecture, with signs of inflammation and congestion with cytoplasmic vacuolation, fatty changes, sinusoidal dilatation, and centrilobular necrosis. The presence of regenerating micro- and macronodules, with bundles of collagen surrounding the lobules, resulted in large fibrous septa accompanied by severe proliferation of bile duct, heavy invasion of inflammatory cells, and distorted tissue architecture. Degenerative cell changes (such as cloudy swelling, hydropic degeneration, loss of nucleus and nucleolus, and necrosis) were also observed.

Rat livers treated with 250 mg/kg ELAP (Figure 2D) showed fewer macronodules and fibrotic nodules compared with TAA controls, as well as reduced inflammation and necrosis of hepatocytes with mild cytoplasmic vacuolation, and almost no visible changes compared to the reference group, with the exception of regenerative parenchyma nodules surrounded by septa of fibrous tissue. Livers from animals fed with 500 mg/kg

ELAP (Figure 2E) showed remarkable histological differences compared with those treated with 250 mg/kg ELAP. They had regular hepatic architecture, minimal disruption of hepatic cellular structure, well-preserved cytoplasm, very minor fibrotic septae, and low lymphocyte infiltration. Similar effects were found in silymarin-treated rats (Figure 2C). These results indicate that 500 mg/kg ELAP administration was as effective as silymarin in protecting rat liver against cirrhosis.

Masson's Trichrome Staining of Liver Sections. Masson's Trichrome staining of control and ELAP-treated liver sections is shown in Figure 3. Liver tissues from normal controls showed no collagen deposition, whereas those from TAA controls showed bile duct proliferation with dense fibrous septa and increased deposition of collagen fibers around the congested central vein, indicating severe fibrosis. Liver tissues from silymarin-treated rats showed minimal collagen deposition, indicating minimal fibrosis. Livers from rats treated with 250 mg/kg ELAP showed moderate deposition of collagen fibers and moderate congestion around the central vein, and those from rats treated with 500 mg/kg ELAP showed only mild collagen deposition and mild congestion around the central vein. This supports the protective effects of ELAP against TAA-induced hepatic toxicity.

Immunohistochemical Staining of Liver Sections. The effect of ELAP on cell proliferation following TAA-induced liver damage was examined by immunohistochemical analysis of PCNA expression in the liver parenchyma using anti-PCNA antibody (Figure 4 and Table 2). Hepatocytes of the normal control group showed no PCNA staining, indicating that no cell regeneration was occurring. In contrast, hepatocytes of the TAA control group had upregulated PCNA expression and an elevated mitotic index, indicating proliferation to repair the severe liver tissue damage induced by TAA. Liver tissues treated with 250 mg/kg ELAP, 500 mg/kg ELAP, or silymarin had reduced hepatocyte regeneration compared to the TAA controls, as indicated by reduced PCNA expression and a significant reduction of the mitotic index. ELAP had an outstanding effect on PCNA labeling and mitotic index in a dose-dependent manner.

Effect of ELAP on Liver Antioxidant Enzyme Levels. Antioxidant enzymatic activities of SOD and CAT in liver homogenates are shown in Figures 5 and 6. SOD activity was decreased in TAA controls compared with normal controls (0.054 ± 0.00 vs. 0.106 ± 0.01 U/mL SOD, respectively). SOD activity was significantly higher in ELAP-treated groups (0.149 ± 0.01 and 0.161 ± 0.01 U/mL SOD for 250 and 500 mg/kg ELAP, respectively) compared with TAA controls ($P<0.05$ for each), and was comparable to the silymarin-treated group (0.123 ± 0.01 U/mL SOD). Similarly, CAT activity was

Table 3. Effect of ELAP on biochemical parameters in TAA-induced liver cirrhosis.

Treatment	Total protein (g/L)	Albumin (g/L)	Globulin (g/L)	Total bilirubin (μM)	Conjugated bilirubin (μM)
Normal control	68.67±1.4	12.83±0.70	54.5±0.22	2.7±0.21	1±0.00
TAA control	60.83±0.47#	7.83±0.16#	68.66±1.4#	9±0.21#	5.3±0.00#
Silymarin (50 mg/kg)	67.33±0.95*	11.83±0.74*	54.33±1.47*	5.6±0.76	3±0.36*#
ELAP (250 mg/kg)	61.5±0.67*#	11.66±0.21*	48±0.85*#	7±0.36*#	3.5±0.61*#
ELAP (500 mg/kg)	70±1.46*	12.16±0.30*	54.5±0.5*	5±0.77*#	3±0.22*#

Data expressed as mean ± SEM ($n = 6$ rats/group).
*$P<0.05$ compared with TAA control,
#$P<0.05$ compared with normal control.

Table 4. Effect of ELAP on serum liver biomarkers in TAA-induced liver cirrhosis.

Treatment	ALP (IU/L)	ALT (IU/L)	AST (IU/L)	GGT (IU/L)
Normal control	100.66±2.02	64±0.516	174.5±3.31	5±0.00
TAA control	243.83±7.44#	209.83±2.13#	322.16±2.52#	12±0.00#
Silymarin (50 mg/kg)	132.66±10.44*#	70.3±2.01*	184.33±3.25*	7±0.00*
ELAP (250 mg/kg)	179.5±2.34*#	76.5±7.45*	246.5±31.7	6.83±0.16*#
ELAP 500 mg/kg	125.5±2.18*#	52.67±4.65*	190.67±21.7*	6.67±0.55*

Data expressed as mean ± SEM (n = 6 rats/group).
*P<0.05 compared with TAA control,
#P<0.05 compared with normal control.

reduced in TAA controls compared with normal controls (200±2.98 vs. 384.00±1.93 nmol H_2O_2 decomposed/min/mL protein, respectively). As for SOD, CAT activity in the ELAP-treated groups was significantly higher than in TAA controls (370.42±5.50 and 426.24±9.92 nmol H_2O_2 decomposed/min/mL protein for 250 and 500 mg/kg ELAP, respectively; P<0.05 for each), and comparable results were found in the silymarin-treated group (472.01±5.80 nmol H_2O_2 decomposed/min/mL protein).

Effect of ELAP on Oxidative Stress. Oxidative stress in liver tissue was assessed by measuring MDA and NO levels in liver homogenates, as shown in Figures 7 and 8. MDA levels increased significantly in TAA-treated rats compared with normal control rats (4.52±0.01 vs. 1.47±0.11 nmol/mg protein, respectively; P< 0.05). Compared to TAA-treated animals, administration of 500 mg/kg ELAP significantly lowered the MDA level (1.62±0.00 nmol/mg protein; P<0.05), restoring a level comparable to normal controls. NO levels, which indicate severe cell damage in cirrhotic livers, also increased significantly in TAA-treated rats (32.00±0.68 μM) compared with the other groups (P<0.05). ELAP significantly restored the altered NO level (17.00±0.42 and 16.00±0.93 μM for 250 and 500 mg/kg ELAP, respectively; P<0.05 for each), and the same effect was found in the silymarin group (14.00±0.48 μM).

Effect of ELAP on Biochemical Parameters. Administration of TAA led to significant increases of ALT, AST, ALP, globulin, total bilirubin, conjugated bilirubin, and GGT, which are indicators of liver injury (Tables 3 and 4). Significant decreases in total protein and albumin were also observed, indicating acute hepatocellular damage. ELAP and silymarin treatment significantly reduced enzyme activities of ALT, AST, ALP, globulin, total bilirubin, conjugated bilirubin, and GGT. Additionally, total protein and albumin levels were increased in ELAP- or silymarin-treated groups compared with the TAA controls. Thus, ELAP counteracted the toxic effect of TAA by restoring normal liver function. A marginal effect was observed

at a dose of 250 mg/kg, whereas ELAP effectively prevented TAA-induced liver damage at a dose of 500 mg/kg.

In Vitro Evaluation of ELAP Fractions

Effect of ELAP fractions on HepG2 Cancer Cells Proliferation. To determine the anti-proliferative effect of the four isolated ELAP fractions in HepG2 and WRL-68 cells, the MTT assay was carried out. CF, HF, and BF had a significant anti-proliferative effect on HepG2 cells (Table 5). CF had the strongest cytotoxic effect, with an IC_{50} value of 17.43±0.93 μg/mL after 24 h, whereas AF did not affect HepG2 cell proliferation. Normal human hepatic WRL-68 cells were not affected by treatment with any of the four fractions, indicating a selective cytotoxicity in HepG2 cells.

Effect of ELAP fractions on LDH Leakage. To test the effects of ELAP fractions on membrane integrity, we measured the levels of LDH released into the medium of HepG2 cells following treatment with CF, HF, or BF. CF, HF, and BF significantly increased LDH release in a dose-dependent manner at concentrations of 12.5 to 100 μg/mL (Figure 9).

Effect of ELAP fractions on ROS Generation. ROS generation can lead to metabolic impairment and cell death. We measured the formation of ROS in HepG2 cells after treatment with CF, HF, and BF. We found a significant dose-dependent generation of ROS in HepG2 cells treated with CF and HF at 25 to 100 μg/mL, whereas BF did not cause ROS generation (Figure 10).

Effect of ELAP fractions on Mitochondria-Initiated Events. To investigate the effects of fractions on apoptosis, multiple parameters were measured in treated and control HepG2 cells. Untreated HepG2 cells were strongly stained with MMP dye, whereas cells treated for 24 h with CF, HF, or BF (12.5 to 100 μg/mL) were not (Figure 11). Also, cytochrome c was released from mitochondria at these concentrations, and cell membrane permeability was significantly elevated (Figure 11).

Table 5. IC_{50} values of isolated fractions on HepG2 and WRL-68 cells after 24 h.

Cell line	IC_{50} (μg/mL)			
	CF	BF	HF	AF
HepG2	17.43±0.93	19.541±1.113	22.787±1.812	>100
WRL-68	52.124±1.356	58.853±2.45	61.43±2.88	>100

Data expressed as mean ± standard deviation (SD) of at least three independent experiments.

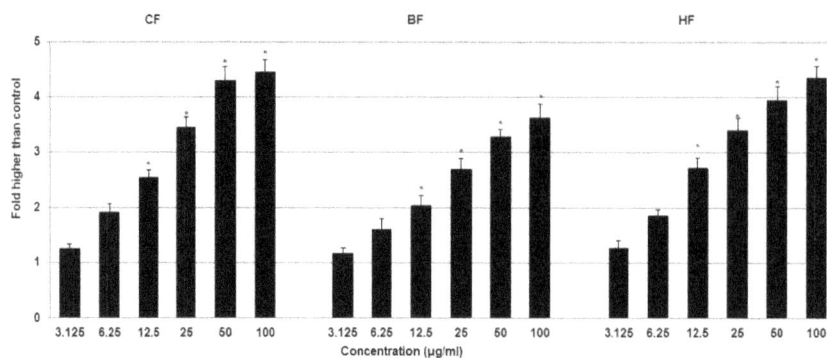

Figure 9. Effect of ELAP fractions on LDH leakage in HepG2 cells. LDH assay was used to assess the loss of membrane integrity in HepG2 cells treated with CF, HF, or BF. Significant cytotoxicity was observed at 12.5 to 100 μg/mL. Data represent mean ± SD of three independent experiments. *P<0.05 compared with no treatment.

Effect of ELAP fractions on Caspase Activation. Since caspases are key mediators of apoptotic pathways, we measured caspase activation in HepG2 cells. After 24 h treatment, CF, HF, and BF all activated caspase-3/7 and -8 at concentrations of 12.5 to 100 μg/mL caspases (Figure 12). CF and HF also activated caspase-9 at these concentrations, whereas BF activated caspase-9 only at higher concentrations (50 and 100 μg/mL). At these higher concentrations (50 and 100 μg/mL), executioner caspase-3/7 was induced all fractions more than 4-fold compared with control. High caspase-8 and -9 levels were also induced by treatment with CF, HF, and BF.

Discussion

The liver is a sensitive organ and is more prone to toxic injuries than other organs. Several phytochemicals exhibit hepatoprotective effects in liver injury [32]; however many are toxic, which limits their clinical use [33]. We evaluated the acute toxicity of ELAP in female rats and found no toxic effects. This is in accordance with another study that reported the extract to be genotoxically safe [11]. Our observations, however, are in contrast to findings reported by Oyewo et al. that administration of *A. paniculate* aqueous extract induced chronic inflammatory responses in tissues that may caused by disruption of the plasma membrane at high doses of the extract. [34]. In this study, we observed a significance reduction in body weight in the TAA control group and ELAP-treated groups (500 mg/kg and 250 mg/kg) compared to normal control. In contrast, liver weight increased

significantly in the TAA control group. ELAP treatment, however, reduced the liver weight to almost normal values. This effect might be due to reduced inflammation [35].

In this study, TAA-induced hepatotoxicity was correlated with a marked increase in serum liver biomarkers ALP, ALT, AST, and GGT. These values were significantly reduced upon ELAP administration. Similar observations of serum liver biomarkers improvement with ELAP treatment were previously reported [36–38]. TAA was reported to interfere with RNA movement from the nucleus to the cytoplasm, causing membrane damage that results in increased release of serum liver markers [39,40]. In this study, TAA injection at a dose (200 mg/kg, 3 times/week caused liver cirrhosis in rats. The severe damage was ameliorated by ELAP treatment in a dose - dependent manner. These observations are in agreement with a study by Rajalakshmi et al. [41]. Blocking normal liver cells regeneration will result in hepatic fibrosis. Brenner [42] reported that telomerase, b-cell lymphoma-extra large (Bcl-xL) and adiponectin genes are sensitive to fibrosis while Fas and cathepsin B genes are resistant to hepatocyte apoptosis and, thus, are resistant to hepatic fibrosis. When hepatocytes undergo apoptosis they produce chemokines and cytokines, including macrophage inflammatory protein 2 (MIP-2), keratinocyte-derived chemokine (KC) and transforming growth factor β (TGF-β), which in turn activate hepatic stellate cells (HSCs) to lose their retinoid and express α-smooth muscle actin (α-SMA) and produce extracellular matrix (ECM) proteins. This is result in excessive production of collagens and degradation of the normal ECM [42]. Reduced collagen synthesis with ELAP administration

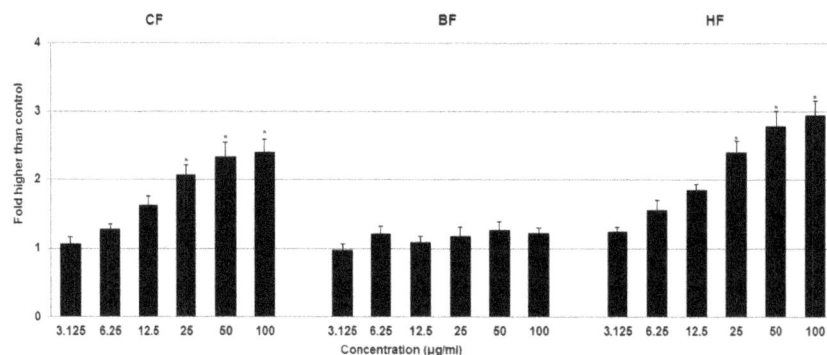

Figure 10. Effect of ELAP fractions on ROS generation in HepG2 cells. The level of ROS increased significantly after CF and HF treatment at concentrations from 25 to 100 μg/mL. Data represent mean ± SD of three independent experiments. *P<0.05 compared with no treatment.

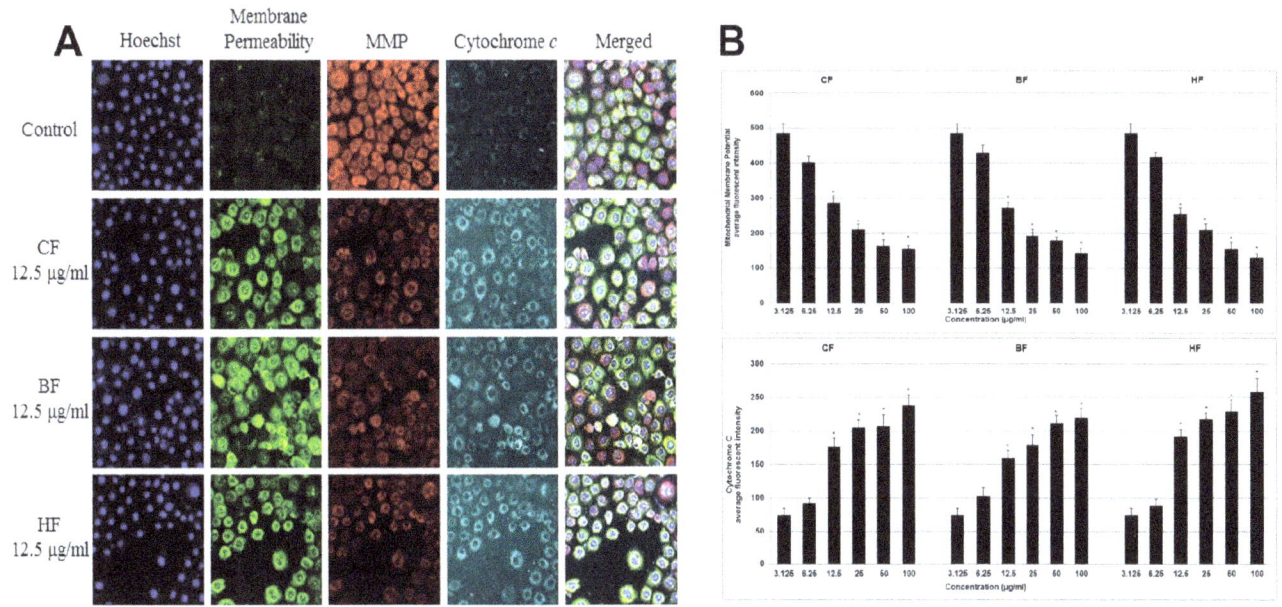

Figure 11. Effect of ELAP fractions on apoptotic markers in HepG2 cells. (A) HepG2 cells were treated with medium alone and 12.5 μg/mL CF, HF, or BF, then stained with Hoechst 33342, cell membrane permeability, MMP, and cytochrome *c* dyes. Isolated fractions caused a marked elevation in cytochrome *c* release and cell membrane permeability, and a noticeable decrease in MMP. (B) Dose-dependent reduction of MMP and increase in cell permeability in treated HepG2 cells (12.5 to 100 μg/mL). Cytochrome *c* was significantly released at CF, HF, or BF concentrations of 12.5 to 100 μg/mL. Data represent mean ± SD of three independent experiments. *$P<0.05$ compared with no treatment.

as seen in Masson's trichrome stained slides indicates the antiapoptotic properties of ELAP, which might result from andrographolide ability to decrease α-SMA and TGF-β [43], which are implicated in apoptosis of activated hepatic stellate cells [44].

PCNA has recently been identified as the polymerase S accessory protein [45]. Ng et al. [46] reported a significant relationship between PCNA and tumor invasiveness. In this study, normal liver controls and silymarin-treated samples showed no significant PCNA staining, indicating the absence of cellular regeneration. An upregulation of PCNA expression was observed in TAA-treated controls, indicating extensive proliferation, likely in an attempt to repair TAA-induced tissue damage [5]. On the other hand, treatment using silymarin or ELAP decreased cellular proliferation levels due to a reduction in PCNA expression. Luo et al. [47] found that andrographolide is able to inhibit the

expression and activity of matrix metalloproteinase (MMP)-9 by inhibiting (NF)-κB-mediated MMP-9 expression, thus preventing the proliferation of tumor cells. Decreased SOD and CAT activities in the TAA control group may be due to excessive ROS generation in cirrhotic tissue as a result of over expression of NADPH oxidase [48]. On the other hand, the increase in enzyme activities upon ELAP or silymarin treatment may stem from a decrease in ROS and free radicals, due to scavenging by andrographolides. This finding is in accordance with similar reports on the effects of *A. paniculata* on liver damage [36,38]. The decreased hepatic antioxidant enzyme activities of the TAA control group could explain the elevated MDA lipid peroxidation (as assessed by the TBARS assay) and the increase in NO concentration. ROS and NO are known to be involved in apoptosis of hepatocytes [49]. Administration of both ELAP and silymarin reduced MDA activity and reduced TAA-induced liver

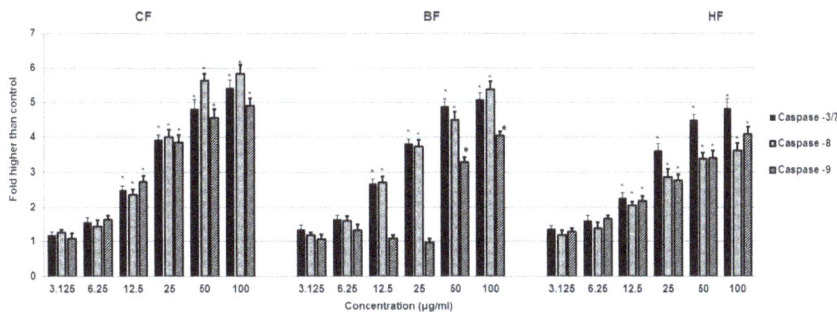

Figure 12. Effect of ELAP fractions on Caspases activation in HepG2 cells. Relative luminescence dose-dependent expression of caspase-3/7, -8, and -9 in HepG2 cells treated with CF and HF caused significant activation at 12.5 to 100 μg/mL concentrations. In HepG2 cells treated with CF and HF, caspase-3/7, -8, and -9 expression increased significantly and in a dose-dependent manner at concentrations from 12.5 to 100 μg/mL. BF activated caspase-3/7 and -8 at the same concentrations, but caspase-9 was activated only at higher concentrations (50 and 100 μg/mL). Activation was measured by relative luminescence. Data represent mean ± SD of three independent experiments. *$P<0.05$ compared with no treatment.

toxicity. Similar trends were observed in *A. paniculata's* protective effect against ethanol-induced hepatotoxicity [50]. Compared with the TAA control group, both ELAP and silymarin treatment significantly reduced NO generation in liver cells. However, the presence of 14-deoxyandrographolide and 14-deoxy-11,12-dide-hydroandrographolide, major diterpenoids, reported to stimulate NO release in tissue endothelial cells [51]. The authors reported that 14-deoxy-11,12-didehydroandrographolide caused NO stim-ulation via activation of constitutive nitric oxide synthase (cNOS), which was followed by upregulation of γ-glutamylcysteine synthetase activity and resulted in reduced oxidative stress [52]. In contrary, the NO inhibitory property of *A. paniculata* was associated with its bioactive diterpenoid andrographolide, which caused the reduction of inducible NO synthase (iNOS) mRNA and protein expression [53,54].

A. paniculata is reported to suppress growth of different human cancer cells, including Jurkat, PC-3, Colon 205, and HepG2 cells. Extracts of *A. paniculata* aerial parts were found to induce cell cycle arrest and mitochondrial-dependent apoptosis in human acute myeloid leukemia cells [22,55]. The current study reports that three isolated fractions of ELAP inhibit HepG2 proliferation via mitochondrial-dependent apoptosis. The IC_{50} values of isolated fractions in HepG2 cells were estimated as 17 to 22 μg/mL, which is comparable to findings in previous studies [56,57]. The significant LDH release, which is a marker of irreversible cell injury, also confirmed the cytotoxic effect of CF, HF, and BF on HepG2 cells. Yen et al. [58] have found that andrographolide markedly inhibits cerebral endothelial cell growth. ROS have been reported to initiate apoptotic signaling [59]. We show excessive ROS formation in HepG2 cells after treatment with CF and HF. Andrographolide is the main active compound in *A. paniculata* that triggers ROS formation in lymphoma cells and induces apoptosis via mitochondrial-mediated pathways [60]. Our exper-iment showed that BF did not induce significant ROS generation in HepG2 cells. CF, HF, and BF activated caspase-9 and -3/7, suggesting the induction of apoptosis via intrinsic pathways. Previous studies also showed the induction of mitochondrial-mediated apoptosis by *A. paniculata* and andrographolide on lymphoma and human leukemia HL-60 cells [56,60]. Activation of caspase-8 by CF, HF, and BF suggests that extrinsic pathways induce apoptosis. Andrographolide from *A. paniculata* has been shown to induce cell death and cell cycle arrest at G2/M phase in HepG2 cells [57]. Andrographolide also significantly enhances tumor necrosis factor–related apoptosis-inducing ligand (TRAIL)-induced apoptosis in various human cancer cell lines [42]. Concurrent activation of both extrinsic and intrinsic apoptotic pathways by plant-mediated constituents such as curcumin has been shown previously [61]. In the present study, significant cytotoxic effect of three different isolated fractions suggested the presence of more than one active compound (andrographolide) against HepG2 cells, which remains to be determined.

Conclusion

Based on the results from both *in vivo* and *in vitro* studies, 500 mg/kg of ELAP significantly protected against TAA-induced

liver damage. The acute toxicity study demonstrated that rats treated with ELAP at 2500 mg/kg did not show signs of toxic damage. However, treatment using ELAP reduced the liver weight to almost normal. Our histopathology finding and Masson's Trichrome staining showed the inhibitory effect of treatment with ELAP, which may be due to its ability to inhibit hepatocyte proliferation, as indicated by PCNA staining. ELAP significantly elevated the concentration of serum CAT and SOD, while it significantly decreased hepatic MDA and NO compared to the TAA control group. The pathological increases in serum levels of liver biomarkers caused by TAA toxicity were restored signifi-cantly upon ELAP treatment. ELAP fractions were free of toxic effects on WRL-68 cells, but inhibited HepG2 cell proliferation in both a dose-dependent manner. Furthermore, treatment of HepG2 cells with ELAP fractions significantly increased LDH release and ROS generation in a dose-dependent manner. A multiparameter cytotoxicity assay showed a dose-dependent reduction of MMP, a significant increase in cell membrane permeability, and a concentration-dependent increase in cyto-chrome *c* release in HepG2 cells treated with different concentra-tions of ELAP fractions. Marked dose-dependent increases in caspase-9 and -3/7 activity and a gradual increase in caspase-8 activity were detected in the treated HepG2 cells. ELAP fractions activated downstream caspase molecules and consequently trig-gered apoptosis in HepG2 cells that could be mediated via the intrinsic, mitochondrial-caspase-9 pathway and the extrinsic, death receptor-linked caspase-8 pathway. ELAP has thus been demonstrated to accelerate the recovery of TAA-induced liver damage and ELAP fraction possesses a significant cytotoxic effect in HepG2 cells, suggesting the involvement of several active compounds towards cytotoxic effects against HepG2 cells.

Supporting Information

Figure S1   Effect of ELAP on liver and kidney histology in acute toxicity study. Histological sections of liver (A, B) and kidney (C, D) from rat treated with vehicle (10% Tween-20; A, C) or ELAP (2500 mg/kg; B, D). H&E staining demonstrates the normal structural appearance of liver and kidney parenchyma.

Table S1   Effect of ELAP on liver function biochemical parameters in acute toxicity study.

Table S2   Effect of ELAP on renal function biochemical parameters in acute toxicity study.

Author Contributions

Conceived and designed the experiments: DAB MFH MP NAA MAA. Performed the experiments: DAB MFH PH ER MP NSA AA. Analyzed the data: DAB PH MP SZM. Contributed reagents/materials/analysis tools: DAB MAA. Wrote the paper: DAB SZM.

References

1. Schuppan D, Afdhal NH (2008) Liver cirrhosis. The Lancet 371: 838–851.
2. Hsiao TJ CLH, Hsieh PS, Wong RH (2007) Risk of betel quid chewing on the development of liver cirrhosis: A community-based Case-Control Study. Epidemiol Elsevier 17: 479–485.
3. Poli GPM (1997) Oxidative damage and fibrogenesis. Free Radic Biol Med 22: 287–305.
4. Kadir FA, Othman F, Abdulla MA, Hussan F, Hassandarvish P (2011) Effect of *Tinospora crispa* on thioacetamide-induced liver cirrhosis in rats. Indian Journal of Pharmacology 43: 64.
5. Salama SM, Bilgen M, Al Rashdi AS, Abdulla MA (2012) Efficacy of *Boesenbergia rotunda* treatment against thioacetamide-induced liver cirrhosis in a rat model. Evidence-based complementary and alternative medicine 2012: 10.

6. Amin ZA, Bilgen M, Alshawsh MA, Ali HM, Hadi AHA, et al. (2012) Protective role of *Phyllanthus niruri* extract against thioacetamide-induced liver cirrhosis in rat model. Evidence-Based Complementary and Alternative Medicine 2012: 9.

7. Alkiyumi SS, Abdullah MA, Alrashdi AS, Salama SM, Abdelwahab SI, et al. (2012) *Ipomoea aquatica* extract shows protective action against thioacetamide-induced hepatotoxicity. Molecules 17: 6146–6155.

8. Alshawsh MA, Abdulla MA, Ismail S, Amin ZA (2011) Hepatoprotective effects of *Orthosiphon stamineus* extract on thioacetamide-induced liver cirrhosis in rats. Evidence-Based Complementary and Alternative Medicine 2011: 6.

9. Abdulaziz Bardi D, Halabi MF, Abdullah NA, Rouhollahi E, Hajrezaie M, et al. (2013) *In vivo* evaluation of ethanolic extract of *zingiber officinale* rhizomes for its protective effect against liver cirrhosis. BioMed Research International 2013: 10.

10. Kadir FA, Kassim NM, Abdulla MA, Kamalidehghan B, Ahmadipour F, et al. (2014) Pass-predicted hepatoprotective activity of *Caesalpinia sappan* in thioacetamide-induced liver fibrosis in rats. The Scientific World Journal 2014: 12.

11. Chandrasekaran C, Thiyagarajan P, Sundarajan K, Goudar KS, Deepak M, et al. (2009) Evaluation of the genotoxic potential and acute oral toxicity of standardized extract of *Andrographis paniculata* (KalmCold™). Food and Chemical Toxicology 47: 1892–1902.

12. Shivaprakash G, Gopalakrishna H, Padbidri DS, Sadanand S, Sekhar SS, et al. (2011) Evaluation of *Andrographis paniculata* leaves extract for analgesic activity. Journal of Pharmacy Research 4: 3375–3377.

13. Qader SW, Abdulla MA, Chua LS, Najim N, Zain MM, et al. (2011) Antioxidant, total phenolic content and cytotoxicity evaluation of selected Malaysian plants. Molecules 16: 3433–3443.

14. Murugan K, Selvanayaki K, Al-Sohaibani S (2011) Antibiofilm activity of *Andrographis paniculata* against cystic fibrosis clinical isolate *Pseudomonas aeruginosa*. World Journal of Microbiology and Biotechnology 27: 1661–1668.

15. Wasman S, Mahmood A, Suan Chua L, Alshawsh MA, Hamdan S (2011) Antioxidant and gastroprotective activities of *Andrographis paniculata* (Hempedu Bumi) in *Sprague Dawley* rats. Indian Journal of Experimental Biology 49: 767.

16. Al-Bayaty FH, Abdulla MA, Hassan MIA, Ali HM (2012) Effect of *Andrographis paniculata* leaf extract on wound healing in rats. Natural Product Research 26: 423–429.

17. Sheeja BD, Sindhu D, Ebanasar J, Jeeva S (2012) Larvicidal activity of *Andrographis paniculata* (Burm.f) Nees against *Culex quinquefasciatus* Say (Insecta: Diptera-Culicidae), a filarial vector. Asian Pacific Journal of Tropical Disease 2: S574–S578.

18. Arunadevi R, Sudhakar S, Lipton AP (2010) Assessment of antibacterial activity and detection of small molecules in different parts of *Andrographis paniculata*. Journal of Theoretical and Experimental Biology 6: Article 192.

19. Singha PK, Roy S, Dey S (2003) Antimicrobial activity of *Andrographis paniculata*. Fitoterapia 74: 692–694.

20. Ajaya Kumar R, Sridevi K, Vijaya Kumar N, Nanduri S, Rajagopal S (2004) Anticancer and immunostimulatory compounds from *Andrographis paniculata*. Journal of Ethnopharmacology 92: 291–295.

21. Govindarajan M (2011) Evaluation of *Andrographis paniculata* Burm.f. (Family:Acanthaceae) extracts against *Culex quinquefasciatus* (Say.) and *Aedes aegypti* (Linn.) (Diptera:Culicidae). Asian Pacific Journal of Tropical Medicine 4: 176–181.

22. Jarukamjorn K, Nemoto N (2008) Pharmacological aspects of *Andrographis paniculata* on health and its major diterpenoid constituent andrographolide. Journal of health science 54: 370–381.

23. Ramya M, Kumar CKA, Sree MT, Revathi K (2012) A review on hepatoprotective plants. International Journal of Phytopharmacy Research 3: 83–86.

24. Varma A, Padh H, Shrivastava N (2011) Andrographolide: a new plant-derived antineoplastic entity on horizon. Evidence-Based Complementary and Alternative Medicine 2011: 9.

25. Staňková P, Kučera O, Lotková H, Roušar T, Endlicher R, et al. (2010) The toxic effect of thioacetamide on rat liver *in vitro*. Toxicology *in Vitro* 24: 2097–2103.

26. Wong WL, Abdulla MA, Chua KH, Kuppusamy UR, Tan YS, et al. (2012) Hepatoprotective effects of *panus giganteus* (berk.) corner against thioacetamide-(taa-) induced liver injury in rats. Evidence-Based Complementary and Alternative Medicine 2012: 10.

27. Natarajan SK, Thomas S, Ramamoorthy P, Basivireddy J, Pulimood AB, et al. (2006) Oxidative stress in the development of liver cirrhosis: a comparison of two different experimental models. Journal of Gastroenterology and Hepatology 21: 947–957.

28. Garber J, Barbee R, Bielitzki J, Clayton L, Donovan J, et al. (2010) Guide for the care and use of laboratory animals. The National Academic Press, Washington DC 8: 220.

29. OECD (2002) Guidance Document on Acute Oral Toxicity. Environmental health and safety monograph series on testing and assessment: 01–24.

30. Ghosh M (2005) Fundamentals of Experimental Pharmacology Hilton and Company. Kolkata, India.

31. Ljubuncic P, Song H, Cogan U, Azaizeh H, Bomzon A (2005) The effects of *Pistacia lentiscus* in experimental liver disease.. Journal of Ethnopharmacology 100: 198–204.

32. Mittal DK, Joshi D, Shukla S (2012) Hepatoprotective Role of Herbal Plants–A Review. International Journal of Pharmaceutical Sciences 3: 150–157.

33. Lagarto Parra A, Silva Yhebra R, Guerra Sardiñas I, Iglesias Buela L (2001) Comparative study of the assay of *Artemia salina* L. and the estimate of the medium lethal dose (LD50 value) in mice, to determine oral acute toxicity of plant extracts. Phytomedicine 8: 395–400.

34. Oyewo EB, Akanji MA, Iniaghe MO, Fakunle PB (2012) Toxicological Implications of Aqueous Leaf Extract of *Andrographis paniculata* in Wistar Rat. Nature and Science 10: 91–108.

35. Aoyama T, Inokuchi S, Brenner DA, Seki E (2010) CX3CL1-CX3CR1 interaction prevents carbon tetrachloride-induced liver inflammation and fibrosis in mice. Hepatology 52: 1390–1400.

36. Sivaraj A, Vinothkumar P, Sathiyaraj K, Sundaresan S, Devi K, et al. (2011) Hepatoprotective potential of Andrographis paniculata aqueous leaf extract on ethanol. Journal of Applied Pharmaceutical Science 1: 204–208.

37. Chao WW, Lin BF (2012) Hepatoprotective Diterpenoids Isolated from *Andrographis paniculata*. Chinese Medicine 3: 136–143.

38. Nagalekshmi R, Menon A, Chandrasekharan DK, Nair CKK (2011) Hepatoprotective activity of *Andrographis Paniculata* and *Swertia Chirayita*. Food and Chemical Toxicology 49: 3367–3373.

39. Alshawsh MA, Abdulla MA, Ismail S, Amin ZA (2011) Hepatoprotective effects of *Orthosiphon stamineus* extract on thioacetamide-induced liver cirrhosis in rats. Evidence-Based Complementary and Alternative Medicine 2011: doi:10.1155/2011/103039.

40. Binduja S, Visen P, Dayal R, Agarwal D, Patnaik G (1996) Protective action of ursolic acid against chemical induced hepato-toxicity in rats. Indian Journal of Pharmacology 28: 232.

41. Rajalakshmi G, Jothi KA, Venkatesan R, Jegatheesan K (2012) Hepatoprotective activity of *andrographis paniculata* on paracetamol induced liver damage in rats. Journal of Pharmacy Research 5: 2983–2986.

42. Brenner DA (2009) Molecular pathogenesis of liver fibrosis. Transactions of the American Clinical and Climatological Association 120: 361.

43. Lee T-Y, Lee K-C, Chang H-H (2010) Modulation of the cannabinoid receptors by andrographolide attenuates hepatic apoptosis following bile duct ligation in rats with fibrosis. Apoptosis 15: 904–914.

44. Kisseleva T, Brenner DA (2011) Anti-fibrogenic strategies and the regression of fibrosis. Best Practice & Research Clinical Gastroenterology 25: 305–317.

45. Takasaki Y (2010) Anti-proliferating cell nuclear antigen (PCNA) antibody. Nihon Rinsho Japanese Journal of Clinical Medicine 68: 578.

46. Ng IOL, Lai E, Fan ST, Ng M, Chan ASY, et al. (2006) Prognostic significance of proliferating cell nuclear antigen expression in hepatocellular carcinoma. Cancer 73: 2268–2274.

47. Luo W, Liu Y, Zhang J, Luo X, Lin C, et al. (2013) Andrographolide inhibits the activation of NFκB and MMP9 activity in H3255 lung cancer cells. Experimental and Therapeutic Medicine 6: 743–746.

48. Paik Y-H, Brenner DA (2011) NADPH oxidase mediated oxidative stress in hepatic fibrogenesis. The Korean Journal of Hepatology 17: 251–257.

49. Wang JH, Redmond HP, Di Wu Q, Bouchier-Hayes D (1998) Nitric oxide mediates hepatocyte injury. American Journal of Physiology-Gastrointestinal and Liver Physiology 275: G1117–G1126.

50. Singha PK, Roy S, Dey S (2007) Protective activity of andrographolide and arabinogalactan proteins from *Andrographis paniculata* Nees. against ethanol-induced toxicity in mice. Journal of Ethnopharmacology 111: 13–21.

51. Zhang C, Tan B (1999) Effects of 14-deoxyandrographolide and 14-deoxy-11, 12-didehydroandrographolide on nitric oxide production in cultured human endothelial cells. Phytotherapy Research 13: 157–159.

52. Mandal S, Nelson VK, Mukhopadhyay S, Bandhopadhyay S, Maganti L, et al. (2013) 14-deoxyandrographolide targets adenylate cyclase and prevents ethanol-induced liver injury through constitutive NOS dependent reduced redox signaling in rats. Food and Chemical Toxicology.

53. Chiou WF, Chen CF, Lin JJ (2000) Mechanisms of suppression of inducible nitric oxide synthase (iNOS) expression in RAW 264.7 cells by andrographolide. British Journal Of Pharmacology 129: 1553–1560.

54. Chiou WF, Lin JJ, Chen CF (1998) Andrographolide suppresses the expression of inducible nitric oxide synthase in macrophage and restores the vasoconstriction in rat aorta treated with lipopolysaccharide. British Journal Of Pharmacology 125: 327–334.

55. Geethangili M, Rao YK, Fang S-H, Tzeng Y-M (2008) Cytotoxic constituents from *Andrographis paniculata* induce cell cycle arrest in Jurkat cells. Phytotherapy Research 22: 1336–1341.

56. Cheung H-Y, Cheung S-H, Li J, Cheung C-S, Lai W-P, et al. (2005) Andrographolide isolated from *Andrographis paniculata* induces cell cycle arrest and mitochondrial-mediated apoptosis in human leukemic HL-60 cells. Planta Medica 71: 1106–1111.

57. Li J, Cheung H-Y, Zhang Z, Chan GK, Fong W-F (2007) Andrographolide induces cell cycle arrest at G2/M phase and cell death in HepG2 cells via alteration of reactive oxygen species. European Journal Of Pharmacology 568: 31–44.

58. Yen T-L, Hsu W-H, Huang SK-H, Lu W-J, Chang C-C, et al. (2013) A novel bioactivity of andrographolide from *Andrographis paniculata* on cerebral ischemia/reperfusion-induced brain injury through induction of cerebral endothelial cell apoptosis. Pharmaceutical Biology 51: 1150–1157.

59. Tan S, Sagara Y, Liu Y, Maher P, Schubert D (1998) The regulation of reactive oxygen species production during programmed cell death. The Journal of Cell Biology 141: 1423–1432.

60. Yang S, Evens AM, Prachand S, Singh AT, Bhalla S, et al. (2010) Mitochondrial-mediated apoptosis in lymphoma cells by the diterpenoid lactone andrographolide, the active component of *Andrographis paniculata*. Clinical Cancer Research 16: 4755–4768.

61. Karunagaran D, Rashmi R, Kumar T (2005) Induction of apoptosis by curcumin and its implications for cancer therapy. Curr Cancer Drug Targets 5: 117–129.

15

Functional Dissection of the *Clostridium botulinum* Type B Hemagglutinin Complex: Identification of the Carbohydrate and E-Cadherin Binding Sites

Yo Sugawara, Masahiro Yutani, Sho Amatsu, Takuhiro Matsumura, Yukako Fujinaga*

Laboratory of Infection Cell Biology, International Research Center for Infectious Diseases, Research Institute for Microbial Diseases, Osaka University, Yamada-oka, Suita, Osaka, Japan

Abstract

Botulinum neurotoxin (BoNT) inhibits neurotransmitter release in motor nerve endings, causing botulism, a condition often resulting from ingestion of the toxin or toxin-producing bacteria. BoNTs are always produced as large protein complexes by associating with a non-toxic protein, non-toxic non-hemagglutinin (NTNH), and some toxin complexes contain another non-toxic protein, hemagglutinin (HA), in addition to NTNH. These accessory proteins are known to increase the oral toxicity of the toxin dramatically. NTNH has a protective role against the harsh conditions in the digestive tract, while HA is considered to facilitate intestinal absorption of the toxin by intestinal binding and disruption of the epithelial barrier. Two specific activities of HA, carbohydrate and E-cadherin binding, appear to be involved in these processes; however, the exact roles of these activities in the pathogenesis of botulism remain unclear. The toxin is conventionally divided into seven serotypes, designated A through G. In this study, we identified the amino acid residues critical for carbohydrate and E-cadherin binding in serotype B HA. We constructed mutants defective in each of these two activities and examined the relationship of these activities using an *in vitro* intestinal cell culture model. Our results show that the carbohydrate and E-cadherin binding activities are functionally and structurally independent. Carbohydrate binding potentiates the epithelial barrier-disrupting activity by enhancing cell surface binding, while E-cadherin binding is essential for the barrier disruption.

Editor: Eric A. Johnson, University of Wisconsin, Food Research Institute, United States of America

Funding: Funding provided by the Funding Program for Next Generation World-Leading Researchers (NEXT Program) from the Japan Society for the Promotion of Science (JSPS) and grants from the Ministry of Education, Culture, Sports, Science and Technology of Japan (20790336, 21390128). The funders had no role in study design, data collection and analysis, decision to publish, or preparation of the manuscript.

Competing Interests: The authors have declared that no competing interests exist.

* Email: yukafuji@biken.osaka-u.ac.jp

Introduction

Botulinum neurotoxin (BoNT) is the etiological agent of the disease botulism. Different types of this toxin are produced by various strains of the spore-forming anaerobic bacteria *Clostridium botulinum*, *C. butyricum*, and *C. barati*, and are conventionally classified into seven serotypes, designated A through G [1]. Recently, a novel serotype H was reported, although it is still controversial whether this designation is appropriate [2], [3]. BoNTs exist as large protein complexes by associating with non-toxic proteins termed neurotoxin-associated proteins (NAPs), which consist of non-toxic non-hemagglutinin (NTNH) and hemagglutinin (HA) [4]. BoNT and NTNH form a protein complex termed 12S toxin or M toxin. In serotypes A–D and G, another complex termed 16S toxin or L toxin is also formed, which consists of BoNT, NTNH, and HA. Type A toxin is reported to form 19S toxin, which is considered to be a dimer of 16S toxin [5].

BoNT complex is a unique toxin in that it can give rise to food-borne disease, although the target site is nerve endings. Orally ingested toxin withstands the harsh conditions of the gastrointestinal tract, enters the general circulation through the intestinal epithelial barrier, and reaches peripheral nerve endings where

BoNT cleaves SNARE proteins, resulting in the inhibition of neurotransmitter release [6], [7]. NAPs are known to potentiate oral toxicity of the toxin complex. It was shown that the larger the molecular size of the toxin complex, the higher the oral toxicity [8]. NTNH stabilizes BoNT in the harsh conditions in the digestive tract, thereby potentiating the oral toxicity. HA is also reported to contribute to the stability of BoNT [9]. However, a direct association between BoNT and HA was not observed in a structural model of 16S toxin, and it remains controversial whether HA contributes to the stability of BoNT [10]. Besides the protection by NAPs, two specific activities of HA appear to enhance the oral toxicity: carbohydrate binding and the epithelial barrier disruption. It has been observed that the toxin complex containing HA interacts with epithelial cells through carbohydrate binding [11], [12]. In addition, mice that orally ingested with the toxin complex were protected by the inclusion of some kinds of monosaccharide that are recognized by HA [13]. These observations indicate that carbohydrate binding activity greatly contributes to the oral toxicity of BoNT. Meanwhile, we recently found that HA disrupts the epithelial barrier [14]. Upon the addition of HA, the barrier function of the epithelial cell monolayer, which is mediated by tight junctions, is compromised, and an influx of macromolecules through the gaps between the cells is promoted. We identified E-cadherin, a major constituent of adherens junction, as the target molecule of type A and B HAs [15]. These HAs directly bind to E-cadherin and inhibit its function, thereby disrupting tight junctions. We hypothesized that HA disrupts the epithelial barrier and allows further influx of the toxin complex that resides on the intestinal lumen, which, in turn, potentiates the oral toxicity of the toxin [16].

HA consists of three proteins termed HA1, HA2, and HA3. Alternatively, they are also referred to as HA33, HA17, and HA70, respectively. Two HA1, one HA2, and one HA3 molecules constitute the HA monomers, which trimerize via HA3 to form a whole HA complex; twelve HA molecules constitute the large whole HA complex [13], [17], [18]. Among three HA proteins, HA1 and HA3 show carbohydrate-binding activity. HA1 binds to galactose in types A through C, while HA3 proteins of these serotypes interact with sialylated molecules [12], [19], [20], [21]. Type C HA1 has an additional binding site for sialic acid [22]. Meanwhile, the E-cadherin binding site is assumed to be located in the HA2-HA3 connecting region [18]. Here, we identified amino acid residues critical for carbohydrate binding and E-cadherin binding in type B HA. Then, we assessed how these two activities are related.

Materials and Methods

Plasmid construction

DNA fragments encoding botulinum hemagglutinins were amplified by PCR from genomic DNA isolated from *C. botulinum* type B strain Okra or type C strain Stockholm using gene-specific primers containing restriction sites at their 5′ ends. The type B HA1 and HA2 fragments were inserted into the HindIII-KpnI site of the pT7-FLAG-1 vector (Sigma Aldrich). The type C HA1 fragment was inserted into the NotI-BglII site of the pT7-FLAG-1 vector. The HA3 fragments were inserted into the KpnI-SalI site of the pET52b(+) expression vector (Merck Millipore). The type B HA1 and HA3 fragments were also inserted into the BamHI-SalI site of the pGEX-5X-3 vector (GE Healthcare). Site-directed mutagenesis was performed using the PrimeSTAR Max polymerase (Takara Bio). The inserted regions of these vectors were confirmed by DNA sequencing.

Protein expression and purification

Protein expression was conducted as previously described using Overnight Express Autoinduction System 1 (Merck Millipore) [18]. FLAG-tagged proteins, FLAG-HA1 and FLAG-HA2, were purified using anti-FLAG M2 agarose (Sigma Aldrich). HA3 proteins were expressed and purified as the uncleaved forms, although they are cleaved to yield two fragments, designated

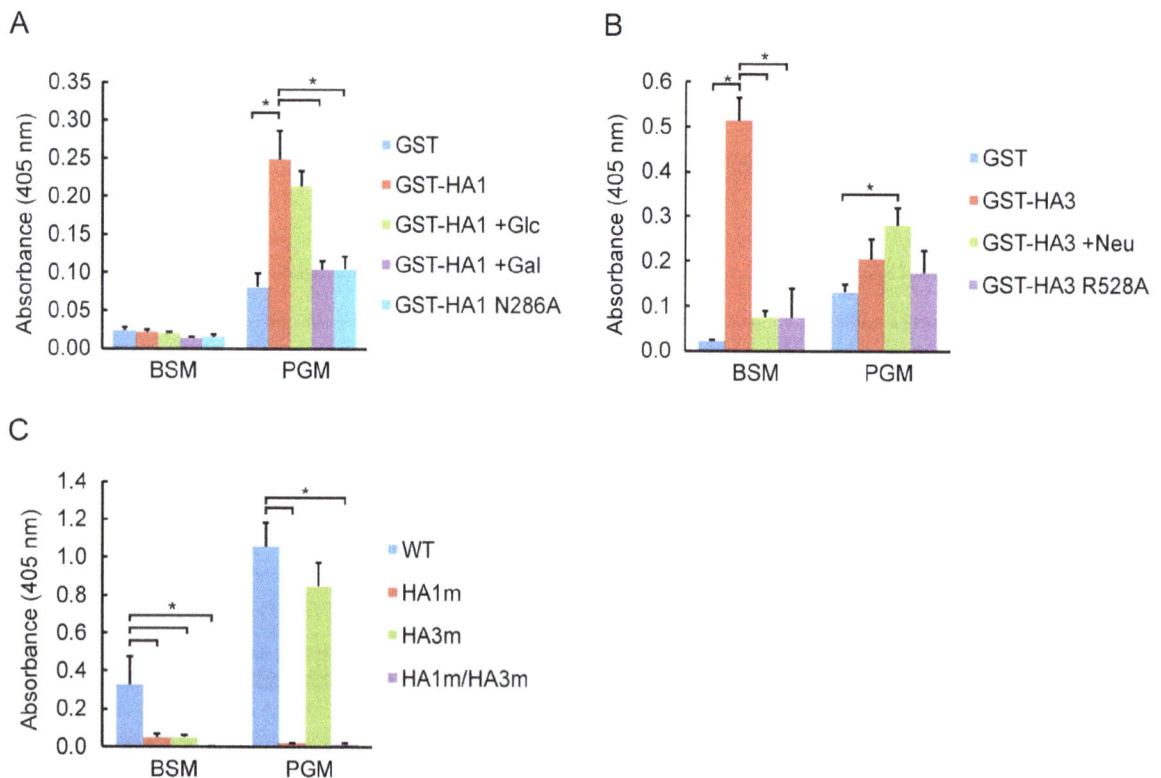

Figure 1. HA/B complex binds to mucins through carbohydrate recognition by HA1/B and HA3/B. (A) Mucin binding of GST-HA1/B. Mucin binding of GST, GST-HA1/B, and GST-HA1/B harboring Asn286 to Ala mutation (N286A) was assessed as described in Materials and Methods. Binding of GST-HA1/B was measured in the presence or absence of 100 mM glucose (Glc) or galactose (Gal). HA1/B preferentially bound to PGM in a galactose-binding-dependent manner. (B) Mucin binding of GST, GST-HA3/B, and GST-HA3/B harboring Arg528 to Ala mutation (R528A). Binding of GST-HA3/B was assessed with the coated mucin pretreated with (+Neu) or without 5 mU/ml neuraminidase. HA3/B preferentially bound to BSM in a sialic-acid-dependent manner. (C) Mucin binding of HA/B complex (WT) and those harboring either HA1/B N286A or HA3/B R528A, or both (HA/B harboring HA1 N286A mutation, HA1 m; HA/B harboring HA3 R528A mutation, HA3 m; HA/B harboring HA1N286A and HA3 R528A mutations, HA1 m/HA3 m). The results show that carbohydrate binding activities of HA1 and HA3 are fully functional in the HA/B complex. Values are means ± S.E.M. from three independent experiments. *$p<0.05$, unpaired t test.

A

B

C

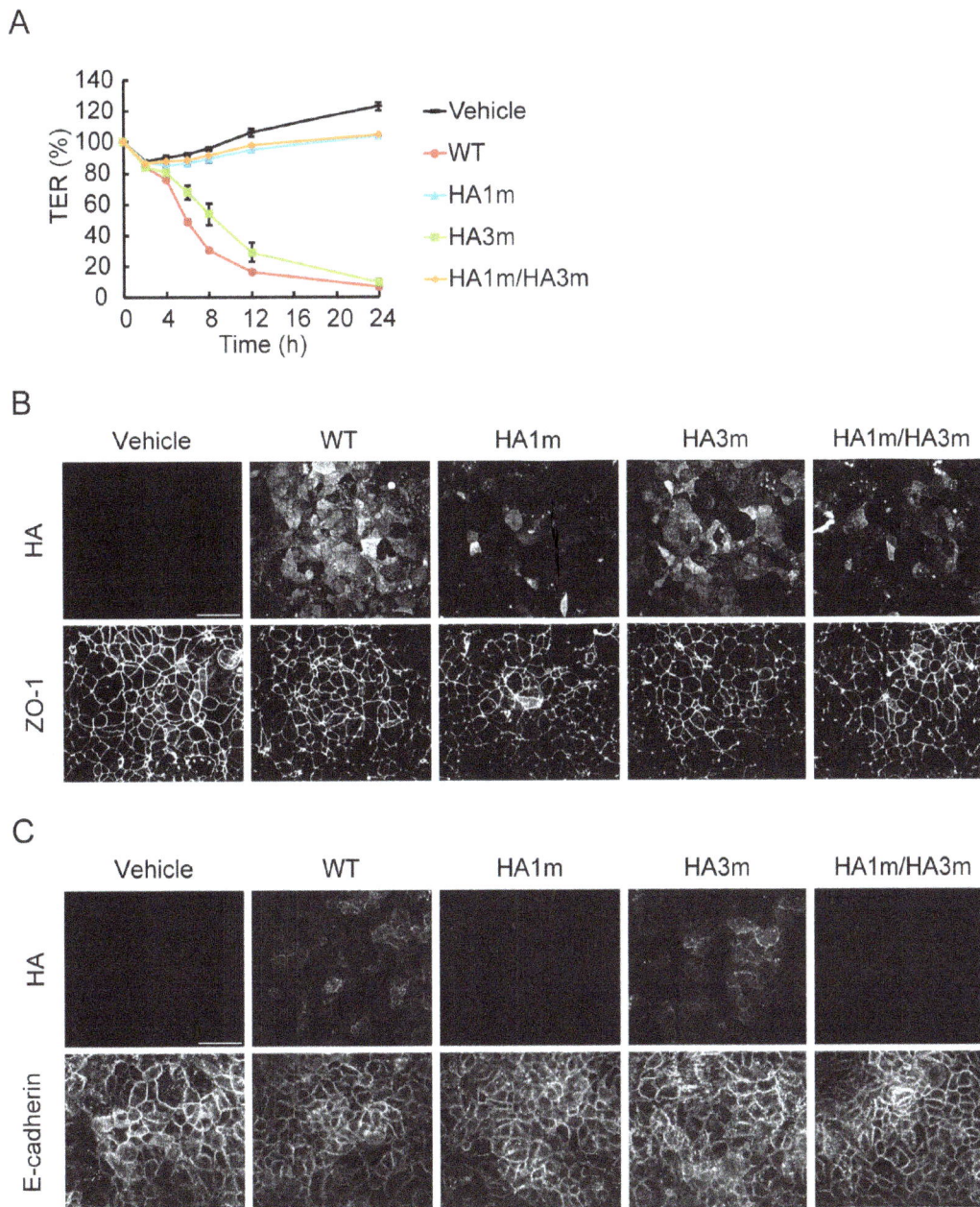

Figure 2. Carbohydrate binding of HA1/B is required for apical to basolateral translocation through Caco-2 cell monolayers. (A) Caco-2 cells were grown on Transwell chambers, and treated with 500 nM HA/B (WT) or those harboring carbohydrate-binding-defective mutations (HA/B harboring HA1 N286A mutation, HA1 m; HA/B harboring HA3 R528A mutation, HA3 m; HA/B harboring HA1 N286A and HA3 R528A mutations, HA1 m/HA3 m) from the upper side of the chambers. TER was measured at time points up to 24 h. The barrier-disrupting activity was inhibited by HA1 mutation. Values are means ± S.E.M. of triplicate wells. (B, C) Caco-2 cells were treated with HA the same as in A. After 1 h of treatment, cells were fixed and the apical (B) or basolateral (C) surface of the cell was stained with the anti-B16S antibody. ZO-1 and E-cadherin were also stained to show the apical and basolateral positions of the epithelial cells, respectively. HA1 mutation greatly affected the apical cell surface binding and subseqent translocation to the basolateral surface of the HA complex. Scale bars: 50 μm.

HA3a and HA3b (alternatively termed HA20 and HA50, respectively), in native toxin complex [4]. We previously showed that the barrier-disrupting activity of recombinant HA complex is comparable to that of native HA complex, indicating that the cleavage of HA3 is dispensable for the activity [18]. The Strep-HA3 proteins were purified using StrepTrap HP (GE Healthcare) according to the manufacturer's instructions. GST-fused proteins, GST-HA1 and GST-HA3, were purified using GSTrap HP (GE

Healthcare). For reconstitution of the HA complex, FLAG-HA1, FLAG-HA2, and Strep-HA3 proteins were mixed at a molar ratio of 4:4:1 in PBS, pH 7.4, and incubated for 3 h at 37°C. The HA complex was captured with a StrepTrap HP followed by a brief wash with PBS, and then eluted with PBS containing 3 mM D-desthiobiotin (Merck Millipore). Purified proteins were dialyzed against PBS, pH 7.4. E-cadherin ectodomain protein was expressed in HEK293 cells, and purified as previously described

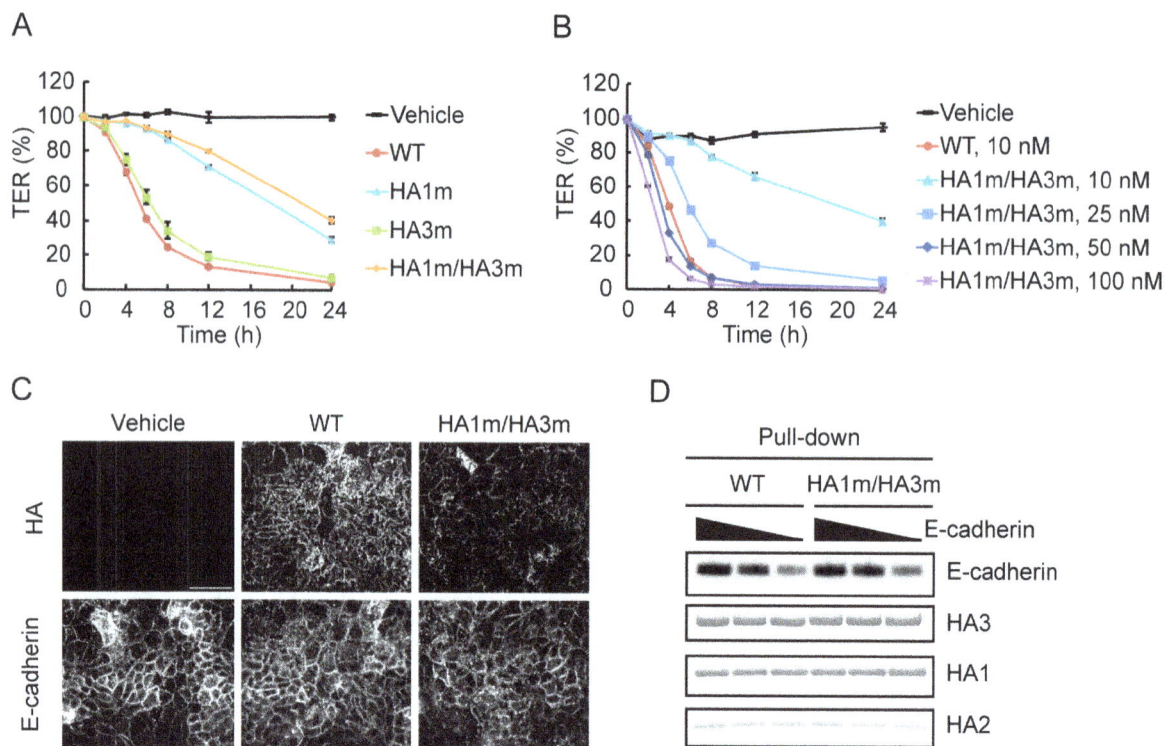

Figure 3. Carbohydrate binding of HA/B is not required for E-cadherin binding. (A) Caco-2 cells were grown on Transwell chambers, and treated with 10 nM HA/B (WT) or those harboring carbohydrate-binding-defective mutations (HA/B harboring HA1 N286A mutation, HA1 m; HA/B harboring HA3 R528A mutation, HA3 m; HA/B harboring HA1 N286A and HA3 R528A mutations, HA1 m/HA3 m) from the lower side of the chambers. HA1 mutation affected the barrier-disrupting activity, whereas HA3 mutation largely did not. (B) Caco-2 cell monolayers were treated with 10 nM HA/B or the indicated concentrations of HA/B carbohydrate-binding-defective mutant (HA1m/HA3m). These HAs were treated with the lower side of the chamber. Values are means ± S.E.M. of triplicate wells. The mutant showed sufficient activity to reduce the TER value completely when applied at higher concentrations (50 nM and 100 nM). (C) Caco-2 cells were treated the same as in A. After 1 h of treatment, cells were fixed and the basolateral surface of the cell was stained with the anti-B16S antibody. E-cadherin staining is also indicated. Binding to the basolateral surface of the cells was reduced in the carbohydrate-binding-defective mutant. Scale bar: 50 μm. (D) Varying concentrations of the recombinant E-cadherin ectodomain protein (300, 100, and 30 nM) was pulled-down with resin-bound HA/B complex or its carbohydrate-binding-defective mutant. The mutations did not affect E-cadherin binding.

[15]. Protein concentrations were determined using the BCA Protein Assay Reagent (Thermo Scientific).

Mucin binding assay

Mucin binding assay was performed as described by Nakamura et al. with some modifications [22]. The wells of 96-well microtiter plates were coated with 10 μg of bovine submaxillary mucin (BSM) or 1 μg of porcine gastric mucin (PGM) (Sigma Aldrich). After washing with PBS containing 0.2% Tween20 and blocking with 1% BSA, mucin-coated wells were incubated with 1 μM GST-tagged HA1, HA3, or 50 nM HA complex. For the removal of sialic acid, the wells were pretreated with 5 mU/ml *Arthrobacter ureafaciens* neuraminidase (Nacalai Tesque) for 1 h at 37°C before the addition of HA. Bound proteins were probed with rabbit anti-GST antibody (Sigma Aldrich) or rabbit anti-B16S antibody [15], followed by peroxidase-labeled goat anti-rabbit IgG (Jackson Immunoresearch). The plates were developed using ABTS (Roche Diagnostics), and absorbance was measured at 405 nm using a microplate photometer (Thermo Scientific).

Cell culture

Caco-2 cells, the human colon adenocarcinoma-derived cell line, were obtained from Riken Cell Bank, and were grown in a 5% CO_2 humidified incubator with Eagle's MEM (Life Technologies) supplemented with 20% heat-inactivated fetal bovine serum.

Immunofluorescence

Cells grown on Transwell pore filters were fixed with 4% paraformaldehyde in PBS for 20 min. Cells were incubated with rabbit anti-B16S antibody before permeabilization. After permeabilization with 0.5% Triton X-100 in PBS for 5 min, monolayers were incubated with primary antibodies, mouse anti-E-cadherin (BD Biosciences) or anti-ZO-1 (Zymed) monoclonal antibodies, probed with secondary antibodies coupled to Alexa488 (Invitrogen) or Cy3 (Jackson Immunoresearch), and mounted with ProLong Antifade kit (Invitrogen). All procedures were carried out at room temperature. Images were captured with an Olympus IX71 microscope equipped with a LCPlanFl 40×/0.60 objective (Olympus) and a CSU-X1 confocal scanner unit (Yokogawa). Images were analyzed with MetaMorph imaging software (version 7.7.10.0, Universal Imaging).

Measurement of transepithelial electrical resistance (TER)

Measurement of TER was carried out using a Millicell-ERS (Merck Millipore) as described previously [23].

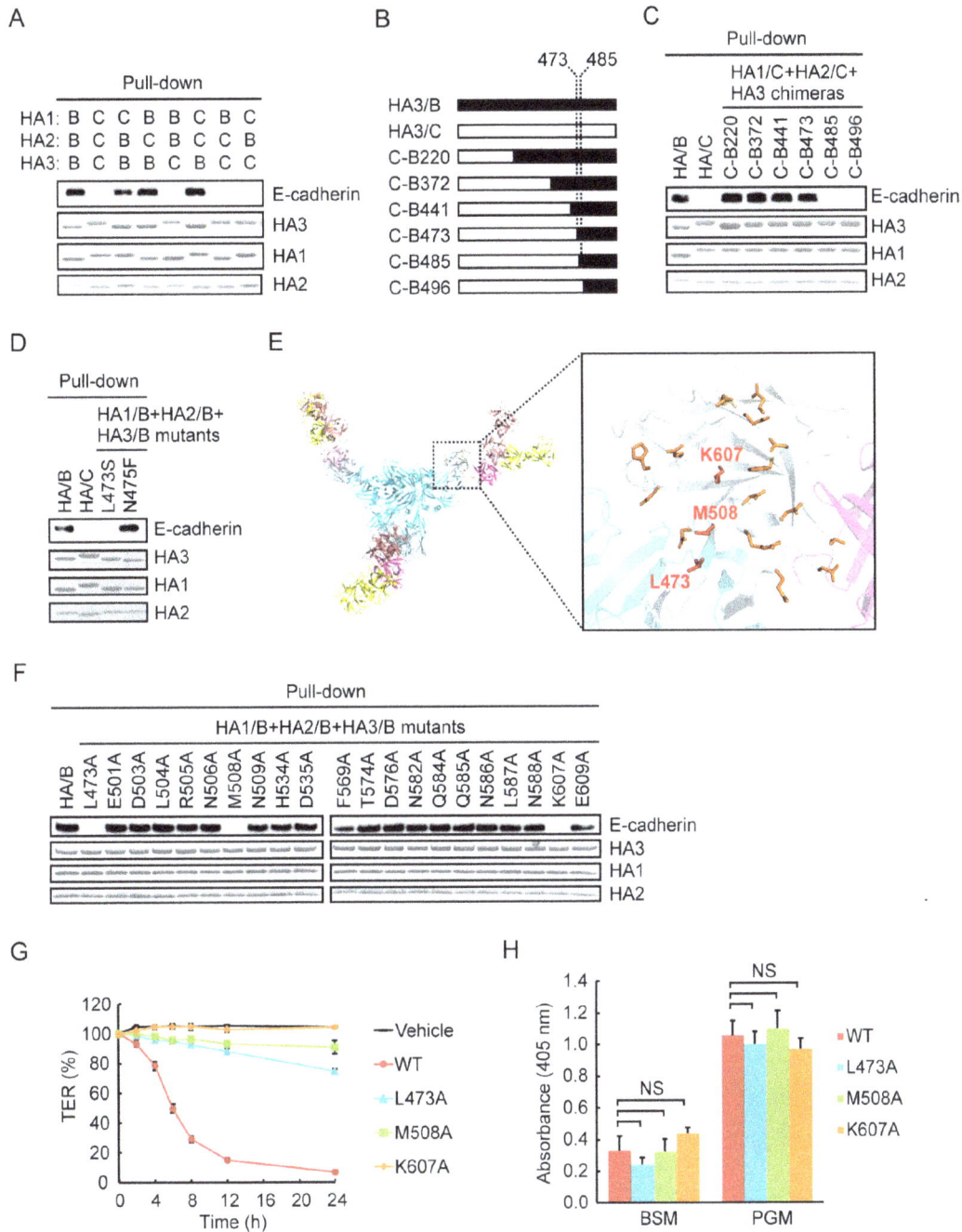

Figure 4. Identification of the E-cadherin binding sites in HA/B. (A) Chimeric HA complexes in which each subcomponent of the complex is replaced between types B and C were prepared, and E-cadherin binding was assessed by HA pull-down assay using the recombinant E-cadherin ectodomain protein. The results indicate that HA3/B is required for E-cadherin binding. The molecular sizes of HA proteins differ between types B and C. (B) Schematic representation of chimeric HA3 constructs between types B (black) and C (white). (C) E-cadherin binding of the HA/B, HA/C, and HA/C complexes containing each of the chimeric HA3 constructs was assessed by HA pull-down assay. (D) E-cadherin binding of the HA/B, HA/C, and HA/B complex harboring HA3 point mutations (Leu473 to Ser, L473S; Asn475 to Phe, N475F) was assessed by HA pull-down assay. E-cadherin binding was inhibited by mutating Leu473, suggesting that this amino acid residue is involved in this binding. (E) Crystal structure of the whole HA/B complex ([18], left) and a higher-magnification image of the boxed region (right). Leu473, Met508, and Lys607 are shown as a stick model and colored red. Other mutated residues (Glu501, Asp503, Leu504, Arg505, Asn506, Asn509, His534, Asp535, Phe569, Thr574, Asp576, Asn582, Gln584, Gln585, Asn586, Leu587, Asn588, Glu609) are shown as a stick model and colored orange. HA1, salmon and yellow; HA2, magenta; HA3, cyan; C-terminal region from Leu473 of HA3, pale cyan. (F) E-cadherin binding of HA/B and those containing HA3 point mutants in which each of the amino acid residues indicated in E was substituted with alanine was assessed by HA pull-down assay. The results show that Leu473, Met508, and Lys607 are critically involved in E-cadherin binding. (G) Caco-2 cells were grown on Transwell chambers, and treated with 10 nM HA/B complex or those harboring E-cadherin-binding-defective mutations (Leu473 to Ala, L473A; Met508 to Ala, M508A; Lys607 to Ala, K607A) from the lower side of the chambers. Values are means ± S.E.M. of triplicate wells. (H) Mucin binding of the HA/B complex or those harboring E-cadherin-binding-defective mutations. These mutations did not affect the carbohydrate binding. Values are means ± S.E.M. of three independent experiments. NS, not significant.

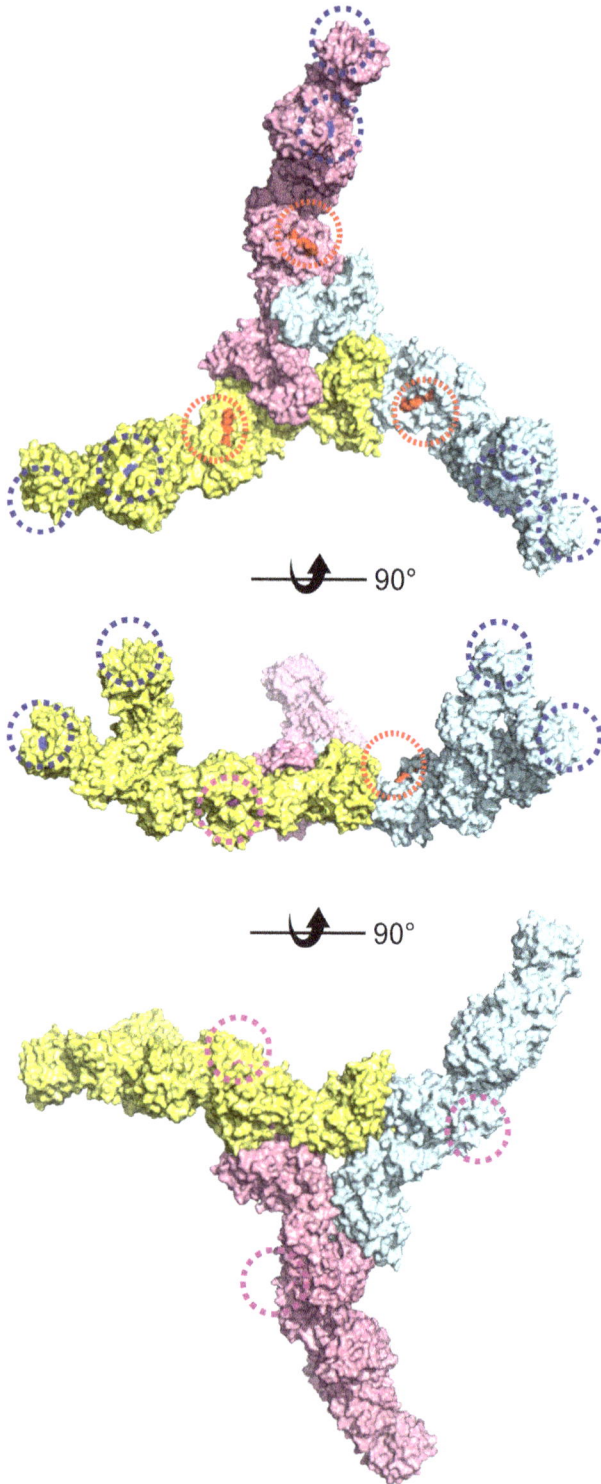

Figure 5. Crystal structure and the carbohydrate and E-cadherin binding sites of the HA/B complex. The galactose binding sites of HA1 (Asn286) are colored blue and marked with dotted blue circles. The sialic acid-binding sites of HA3 (Arg528) are colored magenta and marked with dotted magenta circles. The E-cadherin binding sites of HA3 (Leu473, M508, and Lys607) are colored red and marked with dotted red circles. These sites are structurally independent. Each of the three HA monomers is colored pale cyan, pale yellow, or pink.

HA pull-down assay

The concentration of the E-cadherin ectodomain protein was adjusted to 100 nM in Hepes buffer (20 mM Hepes-NaOH, pH 7.35, 2 mM CaCl$_2$, 150 mM NaCl, and 0.01% Triton X-100). A 200 µl aliquot of E-cadherin solution was mixed with StrepTactin Superflow agarose gel (Merck Millipore) coupled with HA proteins, and rotated for 1 h at 4°C. The gels were washed twice with 1 ml of Hepes buffer and bound proteins were eluted with SDS-PAGE sample buffer. Samples were separated by SDS-PAGE and transferred to a PVDF membrane. Western blot was performed with an anti-mouse E-cadherin antibody, ECCD-2 (Takara Bio). HA proteins were stained with Coomassie blue to show that comparable amounts of HA proteins were coupled to the affinity resin among the compared samples.

Results

The galactose binding site of HA1 and the sialic acid binding site of HA3 have already been identified in serotypes A and C (HA/A and HA/C) [13], [20], [24]. The amino acid residues involved in binding are conserved among types A, B and C, suggesting that HA1/B and HA3/B could recognize galactose and sialic acid, respectively, in a similar manner to types A and C. A recent study, in which the crystal structure of HA1/B bound with its ligand lactose was solved, supports this assumption as to HA1 [25]. It was reported that the galactose binding of HA1/C is abolished by alanine substitution of Asn278, and sialic acid binding of HA3/A is abrogated by alanine substitution of Arg528 [13], [20]. We constructed mutants of HA1/B and HA3/B in which amino acids equivalent to these residues, Asn286 in HA1/B and Arg528 in HA3, were substituted with alanine (HA1 N286A and HA3 R528A). We assessed the effects of these mutations on carbohydrate binding by mucin binding assay [22], [26]. As shown in Fig. 1A, HA1/B bound to porcine gastric mucin (PGM), whose carbohydrate moiety is devoid of sialic acid, but not to bovine submaxillary mucin (BSM), which is rich in sialic acid. This binding was inhibited in the presence of galactose, showing that the binding is dependent on the galactose binding of HA1/B. HA1 N286A did not bind to PGM, indicating that Asn286 is critically involved in carbohydrate binding, as observed in other serotypes. In contrast to HA1/B, HA3/B preferentially bound to BSM (Fig. 1B). This binding was significantly decreased when the coated mucin was pretreated with neuraminidase, indicating that the binding is dependent on sialic acid. BSM binding of HA3/B was abolished by alanine substitution of Arg528, confirming that this residue is critically involved in sialic acid binding. Then, we assessed whether these carbohydrate binding sites function in the HA complex. HA complexes containing carbohydrate-binding-defective mutations were prepared as described in the Materials and Methods section, and carbohydrate binding activity was evaluated by mucin binding assay. As shown in Fig. 1C, BSM binding was greatly inhibited by either HA1 or HA3 mutation (HA1 m, HA3 m), and completely abolished in the presence of both mutations (HA1 m/HA3 m) compared to the wild-type HA/B (WT). PGM binding was abolished when the galactose binding of HA1 was defective. Collectively, these results show that Asn286 of HA1/B and Arg528 of HA3/B are critically involved in carbohydrate binding, through galactose and sialic acid recognition, respectively, and these types of binding are fully functional in the whole HA complex.

HA/B binds to E-cadherin, and in turn, disrupts the epithelial barrier [15]. We considered whether carbohydrate binding affects the epithelial-barrier-disrupting activity of HA. To address this issue, the barrier-disrupting activity of HA complex and those harboring carbohydrate-binding-defective mutations was assessed by measuring TER using Caco-2 cells. The effects of the carbohydrate binding mutations differed depending on whether the HA complex was applied from the apical or basolateral side of the cells. Since E-cadherins reside on the basolateral surface, apically applied HA needs to translocate through the epithelial cells to affect the epithelial barrier [15]. When the complex was applied from the apical side, the activity was completely inhibited by the HA1 mutation, and attenuated by the HA3 mutation (Fig. 2A). Binding to the apical surface of the cells was significantly reduced in the complex containing HA1 mutation, and apical to basolateral translocation was largely inhibited by this mutation (Fig. 2B, C). These results are consistent with our previous notion that HA1 is essential for apical to basolateral translocation and suggest that carbohydrate binding by HA1 is responsible for this activity [18].

When the complex was applied from the basolateral side, the barrier-disrupting activity was reduced but not abolished in the complex containing the HA1 mutant (Fig. 3A). Meanwhile, the activity was not largely affected by HA3 mutation. The barrier-disrupting activity of the complex harboring mutations in both HA1 and HA3 was reduced to approximately one-fifth of that of the wild type; however, it shows sufficient activity to disrupt the epithelial barrier completely when applied at higher concentrations, suggesting that carbohydrate binding is not essential for the activity (Fig. 3B, HA1 m/HA3 m 50 nM and 100 nM). Attachment of the double mutant to the basolateral surface of the cells is less efficient than that of the wild type, and its localization was relatively specific to cell-to-cell junctions (Fig. 3C). The E-cadherin extracellular domain is known to be modified with glycans, such as by N-glycosylation, O-glycosylation, and O-mannosylation [27], [28], [29]. Therefore, it is conceivable that these modifications enhance E-cadherin binding of HA by carbohydrate recognition. However, direct binding to E-cadherin was comparable between the wild type and the double mutant in a pull-down assay (Fig. 3D). These lines of evidence suggest that carbohydrate binding is not required for E-cadherin binding, but potentiates the barrier-disrupting activity by promoting cell surface attachment of HA.

We previously showed that HA/A and HA/B interact with E-cadherin, but HA/C does not [15], [23]. By comparing HA/B and HA/C, we sought to identify the E-cadherin binding site. First, we constructed chimera HA complexes in which each HA subcomponent is replaced between HA/B and HA/C. E-cadherin binding was assessed by pull-down assay. As shown in Fig. 4A, all the chimeras that contained HA3/B interacted with E-cadherin, indicating that HA3 is the determinant of E-cadherin binding. Next, we constructed chimera HA3 between HA/B and HA/C to narrow down the determinant region further (Fig. 4B). HA3 is divided into four domains termed I, II, III, and IV [18]. We found that the C-terminal part of HA3/B from residue 473, which encompasses a C-terminal small portion of domain III and the entirety of domain IV, is required for E-cadherin binding (Fig. 4C). The binding was lost when the HA3/B region was further narrowed down to the C-terminal part from residue 485. Between residues 473 and 484, only two residues, Leu473 and Asn475, differ between HA3/B and HA3/C. Thus, we substituted each of these two residues in HA3/B to that of HA3/C, and examined E-cadherin binding. In a pull-down assay, mutation in Leu473 abolished the binding, whereas that in Asn475 did not affect it (Fig. 4D). Thus, Leu473 was found to be involved in E-cadherin binding.

To identify further the residues involved in E-cadherin binding, we selected the residues of HA3/B that fulfill the following criteria:

residues located within the C-terminal part from Leu473, those differing between types B and C, and those located at the molecular surface adjacent to Leu473 (Fig. 4E). We mutated each of these residues to alanine, and E-cadherin binding was assessed by pull-down assay. As a result, binding was remarkably abrogated by mutating Met508 and Lys607 in addition to Leu473 (Fig. 4F).

Finally, we examined the influence of E-cadherin-binding-defective mutations on the barrier-disrupting activity by TER measurement. As shown in Fig. 4G, HA complexes containing each of these mutations in HA3 showed reduced activity compared with the wild-type complex. Remarkably, the complex containing Lys607Ala mutation completely lost its activity. Reduction in TER was not observed even when cells were treated with 100 nM of the mutant (data not shown). Carbohydrate binding activity was not reduced by this mutation (Fig. 4H). These results support our previous finding that the barrier-disrupting activity is completely dependent on E-cadherin binding [15].

Discussion

Botulism often arises as a food-borne disease; however, BoNT alone is not sufficient to cause it and NTNH and HA are required for oral toxicity. In the case of serotype B, NTNH potentiates the oral toxicity of BoNT by about twenty-fold compared with BoNT alone. HA further potentiates the oral toxicity of BoNT. The oral toxicity of type B 16S toxin, which consists of BoNT, NTNH, and HA, is about seven hundred times higher than that of 12S toxin, which is comprised of BoNT and NTNH [8]. Thus, HA is crucial for oral poisoning and renders the toxin complex a unique oral toxin.

HA possesses two specific activities, namely, carbohydrate binding and epithelial barrier disruption, the latter of which is dependent on E-cadherin binding [16]. It has been noted that carbohydrate binding of 16S toxin is attributable mainly to that of HA1 [12], [19]. This notion is readily understandable from the structures of HA complex and 16S toxin; the galactose binding sites of HA1 are positioned at the outer most parts of the complexes (Fig. 5) [10], [13]. The carbohydrate binding activity of HA1 was also prevalent in our experimental settings. In mucin binding assay, both BSM and PGM binding of the HA complex was greatly affected by galactose-binding-defective mutation of HA1. The inhibition of BSM binding by the HA1 mutation is apparently inconsistent with the result that GST-HA1 did not bind to BSM. Although BSM is rich in sialic acids, it also contains other carbohydrates including galactose; therefore, it is probable that HA1 proteins that constitute the HA complex recognize galactose moieties in BSM. There are three possible reasons why BSM binding of the HA complex was greatly inhibited by HA1 mutation, even though GST-HA1 did not bind to BSM. First, HA1 binding to BSM could be facilitated by sialic acid binding of HA3 present in the same complex. The observation that BSM binding of HA complex was greatly affected also by the sialic acid-binding-defective mutation of HA3 supports this notion (Fig. 1C). Second, HA1 binding to BSM could be enhanced in the HA complex due to multivalency effects. GST-HA1 is considered to behave as a divalent molecule due to dimerization through GST, whereas six galactose binding sites are present in the HA complex. We detected mucin binding of GST-HA1 but not Strep- or FLAG-tagged HA1, indicating profound effects of multivalency in carbohydrate binding (data not shown). Last, as described above, HA1 is positioned at the outer most part of the HA complex, in which the galactose binding sites are arranged in almost the same orientation, enabling cooperative ligand binding of these sites (Fig. 5), while this might not be the case in GST-HA1. Galactose binding of HA1 potentiated the barrier-disrupting activity by promoting cell surface binding and, in the case when HA was applied from the apical surface of cell monolayer, apical to basolateral translocation. In contrast to the galactose binding sites of HA1, sialic acid binding sites of HA3 are located at the inner part of the complex (Fig. 5) [10], [13]. Since this sialic acid binding site of HA3 is conserved among serotypes, it appears to have an important role in the pathogenesis of botulism. As shown in Fig. 1C, it is apparent that the sialic acid binding of HA3 renders the HA complex a more potent lectin with dual carbohydrate binding specificity. Meanwhile, although the sialic acid binding of HA3 enhances the barrier-disrupting activity, this enhancing effect was subtle and much more limited than that of galactose binding of HA1. It remains to be elucidated whether the sialic acid binding plays a critical role in the pathogenesis of botulism.

In this study, we identified E-cadherin binding sites in HA/B complex. We identified the critical residues by comparing HA/B and HA/C, the latter of which is unable to bind to E-cadherin [15]. These are Leu473, Met508, and Lys607, among which Lys607 appears to be critically involved in the interaction. E-cadherin binding was reduced or abolished by mutating these residues to alanine. Meanwhile, carbohydrate binding was not largely affected by these mutations. Reciprocally, HA complex harboring carbohydrate-binding-defective mutations showed comparable E-cadherin binding to that of the wild type. Consistent with these results, the residues involved in E-cadherin binding are located on the HA3 molecular surface that is opposite the sialic acid binding site of HA3 (Fig. 5). Collectively, we concluded that E-cadherin binding is structurally and functionally independent of carbohydrate binding.

We previously reported that HA/A also binds to E-cadherin and disrupts the epithelial barrier [15], [23]. The amino acid sequence identity between HA3/A (strain 62A) and HA3/B (strain

Okra) is 98%, and the three residues that we identified as E-cadherin binding sites are completely conserved in HA3/A. While we were preparing this manuscript, the co-crystal structures of two distal extracellular cadherin domains of E-cadherin and a part of HA/A that includes HA1, HA2, and the C-terminal two domains, domains III and IV, of HA3 were solved [30]. The authors showed that this distal part of E-cadherin recognizes a molecular surface formed by HA2 and HA3, in which Leu473, Met508, and Lys607 are involved. This result is consistent with our previous notion that the HA2-3 connecting region of HA/B appears to be the E-cadherin binding site [18]. Thus, it is likely that HA/B binds to E-cadherin in the same manner as HA/A.

In conclusion, we identified carbohydrate and E-cadherin interaction sites in HA/B complex and showed that these two binding activities are functionally and structurally independent. Our result and the mutant constructs that are deficient in each of these functions would be useful to examine the roles of HA, especially those of sialic acid binding of HA3 and E-cadherin binding, in the pathogenesis of type B botulism.

Acknowledgments

We are grateful to Masatoshi Takeichi and Yoshikazu Tsukasaki (Riken CDB) for providing the E-cadherin expressing cells. We also thank Takayuki Yoshimura, Asami Sano, Yuki Sano, and Chiyoko Aoki for technical assistance.

Author Contributions

Conceived and designed the experiments: YS YF. Performed the experiments: YS MY SA TM. Analyzed the data: YS MY SA TM YF. Contributed reagents/materials/analysis tools: MY SA TM. Contributed to the writing of the manuscript: YS YF.

References

1. Poulain B, Popoff MR, Molgo J (2008) How do the Botulinum Neurotoxins block neurotransmitter release: from botulism to the molecular mechanism of action. The Botulinum J 1: 14–87.
2. Barash JR, Arnon SS (2014) A Novel Strain of *Clostridium botulinum* That Produces Type B and Type H Botulinum Toxins. J Infect Dis 209: 183–191.
3. Johnson EA (2014) Validity of Botulinum Neurotoxin Serotype H. J Infect Dis 210: 992–993.
4. Oguma K, Inoue K, Fujinaga Y, Yokota K, Watanabe T, et al. (1999) Structure and function of *Clostridium botulinum* progenitor toxin. J.Toxicol-Toxin Reviews 18: 17–34.
5. Inoue K, Fujinaga Y, Watanabe T, Ohyama T, Takeshi K, et al. (1996) Molecular composition of *Clostridium botulinum* type A progenitor toxins. Infect Immun 64: 1589–1594.
6. Schiavo G, Matteoli M, Montecucco C (2000) Neurotoxins affecting neuroexocytosis. Physiol Rev 80: 717–766.
7. Simpson LL (2004) Identification of the major steps in botulinum toxin action. Annu Rev Pharmacol Toxicol 44: 167–193.
8. Sakaguchi G (1982) *Clostridium botulinum* toxins. Pharmacol Ther 19: 165–194.
9. Sugii S, Ohishi I, Sakaguchi G (1977) Correlation between oral toxicity and in vitro stability of *Clostridium botulinum* type A and B toxins of different molecular sizes. Infect Immun 16: 910–914.
10. Benefield DA, Dessain SK, Shine N, Ohi MD, Lacy DB (2013) Molecular assembly of botulinum neurotoxin progenitor complexes. Proc Natl Acad Sci U S A 110: 5630–5635.
11. Fujinaga Y, Inoue K, Watanabe S, Yokota K, Hirai Y, et al. (1997) The haemagglutinin of *Clostridium botulinum* type C progenitor toxin plays an essential role in binding of toxin to the epithelial cells of guinea pig small intestine, leading to the efficient absorption of the toxin. Microbiology 143: 3841–3847.
12. Fujinaga Y, Inoue K, Nomura T, Sasaki J, Marvaud JC, et al. (2000) Identification and characterization of functional subunits of *Clostridium botulinum* type A progenitor toxin involved in binding to intestinal microvilli and erythrocytes. FEBS Lett 467: 179–183.
13. Lee K, Gu S, Jin L, Le TT, Cheng LW, et al. (2013) Structure of a bimodular botulinum neurotoxin complex provides insights into its oral toxicity. PLoS Pathog 9: e1003690.
14. Matsumura T, Jin Y, Kabumoto Y, Takegahara Y, Oguma K, et al. (2008) The HA proteins of botulinum toxin disrupt intestinal epithelial intercellular junctions to increase toxin absorption. Cell Microbiol 10: 355–364.
15. Sugawara Y, Matsumura T, Takegahara Y, Jin Y, Tsukasaki Y, et al. (2010) Botulinum hemagglutinin disrupts the intercellular epithelial barrier by directly binding E-cadherin. J Cell Biol 189: 691–700.
16. Fujinaga Y, Sugawara Y, Matsumura T (2013) Uptake of botulinum neurotoxin in the intestine. Curr Top Microbiol Immunol 364: 45–59.
17. Hasegawa K, Watanabe T, Suzuki T, Yamano A, Oikawa T, et al. (2007) A novel subunit structure of *Clostridium botulinum* serotype D toxin complex with three extended arms. J Biol Chem 282: 24777–24783.
18. Amatsu S, Sugawara Y, Matsumura T, Kitadokoro K, Fujinaga Y (2013) Crystal structure of *Clostridium botulinum* whole hemagglutinin reveals a huge triskelion-shaped molecular complex. J Biol Chem 288: 35617–35625.
19. Arimitsu H, Sakaguchi Y, Lee JC, Ochi S, Tsukamoto K, et al. (2008) Molecular properties of each subcomponent in *Clostridium botulinum* type B haemagglutinin complex. Microb Pathog 45: 142–149.
20. Inoue K, Sobhany M, Transue TR, Oguma K, Pedersen LC, et al. (2003) Structural analysis by X-ray crystallography and calorimetry of a haemagglutinin component (HA1) of the progenitor toxin from *Clostridium botulinum*. Microbiology 149: 3361–3370.
21. Fujinaga Y, Inoue K, Watarai S, Sakaguchi Y, Arimitsu H, et al. (2004) Molecular characterization of binding subcomponents of *Clostridium botulinum* type C progenitor toxin for intestinal epithelial cells and erythrocytes. Microbiology 150: 1529–1538.
22. Nakamura T, Tonozuka T, Ide A, Yuzawa T, Oguma K, et al. (2008) Sugar-binding sites of the HA1 subcomponent of *Clostridium botulinum* type C progenitor toxin. J Mol Biol 376: 854–867.
23. Jin Y, Takegahara Y, Sugawara Y, Matsumura T, Fujinaga Y (2009) Disruption of the epithelial barrier by botulinum haemagglutinin (HA) proteins - differences in cell tropism and the mechanism of action between HA proteins of types A or B, and HA proteins of type C. Microbiology 155: 35–45.
24. Yamashita S, Yoshida H, Uchiyama N, Nakakita Y, Nakakita S, et al. (2012) Carbohydrate recognition mechanism of HA70 from *Clostridium botulinum* deduced from X-ray structures in complexes with sialylated oligosaccharides. FEBS Lett 586: 2404–2410.

25. Lee K, Lam KH, Kruel AM, Perry K, Rummel A, et al. (2014) High-resolution crystal structure of HA33 of botulinum neurotoxin type B progenitor toxin complex. Biochem Biophys Res Commun 446: 568–573.
26. Nakamura T, Takada N, Tonozuka T, Sakano Y, Oguma K, et al. (2007) Binding properties of *Clostridium botulinum* type C progenitor toxin to mucins. Biochim Biophys Acta 1770: 551–555.
27. Pinho SS, Seruca R, Gärtner F, Yamaguchi Y, Gu J, et al. (2011) Modulation of E-cadherin function and dysfunction by N-glycosylation. Cell Mol Life Sci 68: 1011–1020.
28. Harrison OJ, Jin X, Hong S, Bahna F, Ahlsen G, et al. (2011) The extracellular architecture of adherens junctions revealed by crystal structures of type I cadherins. Structure 19: 244–256.
29. Lommel M, Winterhalter PR, Willer T, Dahlhoff M, Schneider MR, et al. (2013) Protein O-mannosylation is crucial for E-cadherin-mediated cell adhesion. Proc Natl Acad Sci U S A 110: 21024–21029.
30. Lee K, Zhong X, Gu S, Kruel AM, Dorner MB, et al. (2014) Molecular basis for disruption of E-cadherin adhesion by botulinum neurotoxin A complex. Science 344: 1405–1410.

16

A MIV-150/Zinc Acetate Gel Inhibits SHIV-RT Infection in Macaque Vaginal Explants

Patrick Barnable[1], Giulia Calenda[1], Louise Ouattara[1], Agegnehu Gettie[2], James Blanchard[3], Ninochka Jean-Pierre[1], Larisa Kizima[1], Aixa Rodríguez[1], Ciby Abraham[1], Radhika Menon[1], Samantha Seidor[1], Michael L. Cooney[1], Kevin D. Roberts[1], Rhoda Sperling[4], Michael Piatak Jr.[5], Jeffrey D. Lifson[5], Jose A. Fernandez-Romero[1], Thomas M. Zydowsky[1], Melissa Robbiani[1], Natalia Teleshova[1]*

1 Population Council, New York, New York, United States of America, 2 Aaron Diamond AIDS Research Center, Rockefeller University, New York, New York, United States of America, 3 Tulane National Primate Research Center, Tulane University, Covington, Louisiana, United States of America, 4 Icahn School of Medicine at Mount Sinai, New York, New York, United States of America, 5 AIDS and Cancer Virus Program, SAIC-Frederick, Inc., Frederick National Laboratory for Cancer Research, Frederick, Maryland, United States of America

Abstract

To extend our observations that single or repeated application of a gel containing the NNRTI MIV-150 (M) and zinc acetate dihydrate (ZA) in carrageenan (CG) (MZC) inhibits vaginal transmission of simian/human immunodeficiency virus (SHIV)-RT in macaques, we evaluated safety and anti-SHIV-RT activity of MZC and related gel formulations ex vivo in macaque mucosal explants. In addition, safety was further evaluated in human ectocervical explants. The gels did not induce mucosal toxicity. A single ex vivo exposure to diluted MZC (1:30, 1:100) and MC (1:30, the only dilution tested), but not to ZC gel, up to 4 days prior to viral challenge, significantly inhibited SHIV-RT infection in macaque vaginal mucosa. MZC's activity was not affected by seminal plasma. The antiviral activity of unformulated MIV-150 was not enhanced in the presence of ZA, suggesting that the antiviral activity of MZC was mediated predominantly by MIV-150. In vivo administration of MZC and CG significantly inhibited ex vivo SHIV-RT infection (51–62% inhibition relative to baselines) of vaginal (but not cervical) mucosa collected 24 h post last gel exposure, indicating barrier effect of CG. Although the inhibitory effect of MZC (65–74%) did not significantly differ from CG (32–45%), it was within the range of protection (~75%) against vaginal SHIV-RT challenge 24 h after gel dosing. Overall, the data suggest that evaluation of candidate microbicides in macaque explants can inform macaque efficacy and clinical studies design. The data support advancing MZC gel for clinical evaluation.

Editor: Cheryl A. Stoddart, University of California, San Francisco, United States of America

Funding: This work was supported by the United States Agency for International Development (USAID) Cooperative Agreement GPO-A-00-04-00019-00, by the Swedish Ministry of Foreign Affairs (SMFA) and Primate Center base grant P51 OD011104-51. This research is made possible by the generous support of the American people through the USAID and supported in part with federal funds from the National Cancer Institute, National Institutes of Health (NIH), under contract HHSN261200800001E (J.D.L.). The funders had no role in study design, data collection and analysis, decision to publish, or preparation of the manuscript.

Competing Interests: The authors have declared that no competing interests exist.

* Email: nteleshova@popcouncil.org

Introduction

Considering that sexual intercourse is the main route of HIV acquisition, products that block sexual HIV-1 transmission are urgently needed. The success of the CAPRISA 004 clinical trial, in which vaginal application of a 1% tenofovir gel reduced HIV-1 acquisition by 39% overall and by 54% in high adherers [1], emphasizes the promise for anti-HIV microbicide development.

The Population Council's (PC's) lead microbicide gel (MZC) containing 50 µM of the NNRTI MIV-150 and 14 mM zinc acetate dihydrate (ZA) in carrageenan (CG) protects Depo-Provera-treated macaques against a single vaginal simian/human immunodeficiency virus (SHIV)-RT challenge completely for up to 8 h and protects ~75% of macaques when challenged 24 h after gel dosing (original and modified gels optimized for use in human) [2–4]. Although not as effective as MZC, repeated application of ZC significantly protected macaques [2,5], but this

was reduced when ZC was only administered once [4]. Importantly, the efficacy of MC was improved by the addition of ZA [2]. ZA could contribute to the antiviral activity of MZC through zinc targeting the reverse transcriptase (RT), involving sites not targeted by traditional RTIs (Mizenina et al., in preparation) [6] and induction of immune changes (e.g., cytokines/chemokines/other innate factors) [7–10] that can also influence infection.

This study was designed to determine (i) ex vivo safety of MZC in macaque and human explants, (ii) anti-SHIV-RT activity of MZC and related gel formulations ex vivo in macaque mucosal explants, and (iii) pharmacodynamics (PD) of MZC in macaque mucosal tissues after in vivo gel application to support advancing this formulation for clinical testing.

Materials and Methods

Ethics statement

Adult female Chinese and Indian rhesus macaques (*Macaca mulatta*) were housed and cared for in compliance with the regulations under the Animal Welfare Act [11], the Guide for the Care and Use of Laboratory Animals [12], at Tulane National Primate Research Center (TNPRC; Covington, LA). All studies were approved by the Animal Care and Use Committee of the TNPRC (OLAW assurance #A4499-01). Animals were monitored continuously by veterinarians to ensure their welfare. Animals in this study were socially housed unless restricted by study design. Housing restrictions were scientifically justified and approved by the IACUC as part of protocol review. Animals were housed indoors in climate controlled conditions with a 12/12 light/dark cycle. Animals in this study were fed commercially prepared monkey chow twice daily. Supplemental foods were provided in the form of fruit, vegetables, and foraging treats as part of the TNPRC environmental enrichment program. Water was available at all times through an automatic watering system. The TNPRC environmental enrichment program is reviewed and approved by the IACUC semiannually. Extensive efforts are made to find compatible pairs for every study group, with additional environmental enrichment of housing space through a variety of food supplements and physical complexity of the environment. A team of 11 behavioral scientists monitors the wellbeing of the animals and provides direct support to minimize stress during the study period. All biopsy procedures were performed by Board Certified veterinarians (American College of Laboratory Animal Medicine). Veterinarians at the TNPRC Division of Veterinary Medicine have established procedures to minimize pain and distress through several means. Anesthesia was administered prior to and during all procedures, and analgesics were provided afterwards as previously described [2,13]. Seven macaques became sick over the course of this study and were euthanized. Leftover necropsy tissues derived from 14 additional animals (8 sick and 6 healthy) were available from separate studies. These animals were euthanized using methods consistent with recommendations of the American Veterinary Medical Association (AVMA) Guidelines for Euthanasia. The animal is anesthetized with tiletamine/zolazepam (8 mg/kg IM) and given buprenorphine (0.01 mg/kg IM) followed by an overdose of pentobarbital sodium. Death is confirmed by auscultation of the heart and pupillary dilation. TNPRC is accredited by the Association for Assessment and Accreditation of Laboratory Animal Care (AAALAC#000594).

Macaques

Naïve, SIV infected, SHIV and SHIV-RT exposed/uninfected and infected recycled animals were utilized. The plasma viral load of the infected animals used for *ex vivo* gel activity testing and in PD studies ranged from undetectable (lower limit of quantification = 30 copies/ml) to 50 copies/ml by quantitative RT-PCR for SIV *gag* [14,15]. Animals used for vaginal biopsy collection in *ex vivo* gel safety and activity testing experiments ranged in age from 4–21 years old and their weights ranged from 4.4–12.15 kg. Animals used in PD studies ranged in age from 5–15 years old and their weights ranged from 5.65–10.55 kg.

Gels and active pharmaceutical ingredients (APIs)

The gel formulation attributes are summarized in Table 1. The original gels contained 95λ:5κ CG (lot PDR98-15) that is no longer available from the manufacturer, whereas the modified gels contained 60λ:40κ CG (lot RERK-4137) which is available. The design of modified gels was adapted for human use by changing 1% DMSO to 2% propylene glycol (PG) as the co-solvent and adjusting buffers [3]. Regardless of the co-solvent and other components, the gels likely contained a mixture of soluble and insoluble MIV-150. However, insoluble MIV-150 will dissolve as the soluble MIV-150 is absorbed into tissues, thereby keeping the amount of soluble MIV-150 constant. In the final manufacturing step, the pH of the gels was adjusted to 6.8 ± 0.2, so any change in pH due to the addition of ZA was neutralized. As described in [2], MIV-150 was developed by Medivir AB (Sweden) and licensed to the PC. Unformulated MIV-150 and ZA were used by diluting stock solutions of 27 mM (MIV-150; in DMSO) and 14 mM (ZA; in sodium acetate buffer, pH adjusted to 6.2 with glacial acetic acid).

Macaque vaginal tissues for testing gel safety and anti-SHIVRT activity *ex vivo*

Vaginal mucosa (n = 2 biopsies; or necropsy tissues) was obtained from untreated naïve and recycled macaques. Biopsies were collected using 3×5 mm biopsy forceps. To determine the activity of MZC in tissues from Depo-Provera (Depo) treated animals, a single 30 mg dose of Depo was given to animals by intramuscular injection and vaginal biopsies were collected 5 weeks (wks) later. Tissues were placed in complete L-15 (cL-15) medium (L-15 (HyClone Laboratories, Inc., Logan, UT) supplemented with 10% FBS (Gibco, Life Technologies, Grand Island, NY), 100 U/ml penicillin - 100 µg/ml streptomycin (Cellgro Mediatech, Manassas, VA) and transported to our laboratories at the PC on ice overnight.

Human ectocervical tissues for testing gel safety

Human ectocervical tissues without gross pathological changes from women undergoing routine hysterectomy were received from the National Disease Research Interchange (NDRI, Philadelphia, PA) and from Icahn School of Medicine at Mount Sinai (New York, NY). No patient identifiers were provided. The study was conducted under IRB approved protocol (Program for the Protection of Human Subjects, Icahn School of Medicine at Mount Sinai; #10-1213 0001 01 OB). Participants provided written informed consent. The consent procedure was approved by the IRB. Tissues were processed fresh or transported on ice overnight in RPMI medium.

In vivo treatments and sample collection for PD studies

The design of PD studies is shown in Fig. 1. Naïve and recycled animals were given Depo treatment as described above. Three wks after Depo treatment, 2 ml of the MZC gel or the corresponding CG placebo gel were applied intravaginally daily for 2 wks to mirror the conditions of the efficacy studies [2,3]. Specimens (vaginal fluids (VF); vaginal or vaginal and ectocervical tissues (two biopsies each)) were collected at baseline (4 wks prior to Depo treatment) and 24 h post last gel administration. VF were collected using foam swabs [16]. Both original (n = 4 and n = 3 animals in MZC and CG groups, respectively) and modified (n = 5 animals in MZC and CG groups each) gels were tested. Swabs were transported to our laboratories at the PC on ice overnight and processed as previously described [16,17]. Tissues were placed in cL-15 medium and immediately processed for infection (below).

Viral stocks

SHIV-RT was generated from the original stock provided by Disa Böttiger (Medivir AB, Sweden) [18] using PHA/IL-2-activated macaque PBMCs and titered in CEMx174 cells before

Table 1. Formulation attributes.

	Gels	CG (% w/w)	λ:κ CG (% of total CG)	ZA (% w/w, mM)	MIV-150 (% w/w, μM)	Buffer	Preservative	API solvent	Gel lots
Original	CG	3.0	95:5	-	-	PBS	0.2% MP	-	120118A515MR
	MZC	3.0	95:5	0.3,14	0.00184, 50	-	0.2% MP	DMSO	120119A1005MR
								-	110204A525ML
Modified	CG	3.4	60:40	-	-	$C_2H_3NaO_2$	0.2% MP	PG	110512A525MR
								PG	120111525ML
								PG	110920A525
								PG	110218A1005ML
	MZC	3.1	60:40	0.3, 14	0.00184, 50	$C_2H_3NaO_2$	0.2% MP	PG	110516A1005MR
								PG	120926A1005ML
								PG	110921A1005ML
	ZC	3.1	60:40	0.3, 14	-	$C_2H_3NaO_2$	0.2% MP	-	110211A707ML
								PG	110609A707MR
	MC	3.4	60:40	-	0.00184, 50	$C_2H_3NaO_2$	0.2% MP	PG	110513A815ML

Abbreviations: CG, carrageenan; ZA, zinc acetate dihydrate; DMSO, dimethyl sulfoxide; PG, propylene glycol; $C_2H_3NaO_2$, sodium acetate; MP, methylparaben.

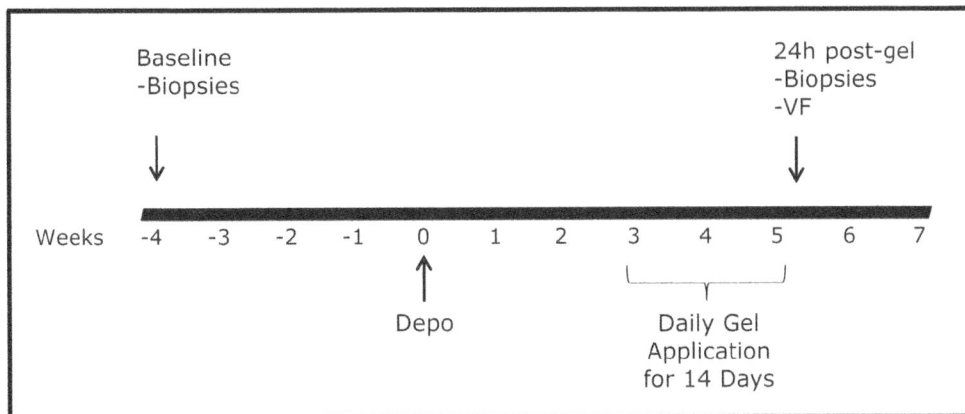

Figure 1. Study design schematic. Macaques were treated with 2 mL of MZC (n = 9) or CG gel (n = 8) intravaginally daily for 14d. Vaginal biopsies were collected from all macaques at the baseline (4 wks before Depo treatment) and 24 h after the last gel exposure to determine PD of MZC in tissues. Ectocervical biopsies were collected from 5 macaques in both groups. VF was collected just before collection of biopsies to assess PK of MIV-150.

use as previously described [2,17]. The same viral stock was used for *ex vivo* tissue challenge at the baseline and post gel exposure in PD studies.

Histology

Histological analysis was performed on macaque vaginal and human ectocervical mucosae. Tissues were washed in PBS and excess submucosa was trimmed. Then tissues were processed, adapting the protocol described for polarized explant culture [19]. Briefly, $5 \times 5 \times 3$ mm explants cut using 5 mm diameter Acu-Punches (Acuderm, Fort Lauderdale, FL) were inserted in a 3 mm hole in a 3 µm pore polyester membrane transwell insert (Costar, Corning, NY) with the mucosal side facing upwards. The positioned tissues were sealed with Matrigel basement membrane matrix (BD Biosciences, Franklin Lakes, NJ), and cDMEM (DMEM (Cellgro Mediatech) containing 10% FBS (for use in macaque tissue) (Gibco), or 10% AB serum (for use in human tissue) (Sigma Aldrich, St. Louis, MO), 100 U/ml penicillin - 100 µg/ml streptomycin (Cellgro Mediatech), 100 µM MEM non-essential amino acids (Irvine Scientific, Santa Ana, CA)) was placed in the basolateral compartment. The mucosal side of the culture was exposed overnight (~18 h) to neat or 1:10 diluted gel formulations. A single explant was set up for each condition. After washing in PBS, tissues were placed in 10% buffered formalin (Richard-Allan Scientific, Kalamazoo, MI) for at least 24 h and submitted to the Laboratory of Comparative Pathology (Cornell University, New York, NY) for paraffin embedding, cutting and Hematoxilin and Eosin (H&E) staining. Tissue sections (5 µm) were examined at the Bioimaging Resource Center of The Rockefeller University using a Zeiss Axioplan 2 widefield microscope (Carl Zeiss Microscopy, Thornwood, NY). Brightfield pictures were taken at a 10× magnification.

Tissue viability

$5 \times 5 \times 3$ mm or $3 \times 3 \times 3$ mm polarized tissue cultures of macaque vaginal and human ectocervical explants (inserted in 3 mm or 2 mm holes in the transwell inserts, respectively; single explant/condition) were set up as described in the "Histology" section. To test potential gel toxicity in an immersion culture model, $3 \times 3 \times 3$ mm explants were set up (2–4 explants/condition) in a 96-well plate. Tissues were incubated with the gels (neat or

diluted) or medium control overnight (~18 h) and toxicity measured as described [17,20].

Activity of APIs and gels against *ex vivo* SHIV-RT challenge

Transported macaque vaginal tissue was processed with minor modifications to the published protocol [20]. All viral challenges were performed in an immersion model. Briefly, $3 \times 3 \times 3$ mm explants were stimulated with 5 µg/ml PHA (Sigma Aldrich) and 100 U/ml IL-2 (Roche Applied Science, Indianapolis, IN; or NCI BRB Preclinical Repository, Frederick, MD) in cDMEM for 48 h. Then, tissues were challenged with SHIV-RT (10^4 TCID$_{50}$ per explant) overnight (~18 h) in the presence of gels (2–4 explants/condition), washed (x4) and cultured in the presence of 100 U/ml IL-2 for 14d. Tissue culture supernatants were collected at days 0 (d0; post last wash), 3, 7, 11 and 14 of culture, and infection was monitored by RETRO-TEK SIV p27 Antigen ELISA kit (ZeptoMetrix, Buffalo, NY). To test if MZC retained its activity in the presence of seminal plasma (SP), activated explants were challenged with SHIV-RT in the presence of diluted gel and 12.5% of pooled human SP (Lee Biosolutions Inc., St. Louis, MO). Pooled semen was centrifuged at 500xg for 10 minutes (min), 4°C for the separation of SP. SP was collected, aliquoted and stored at −20°C until use.

To determine if the APIs maintained any residual activity after washout, tissues were incubated with either unformulated MIV-150, ZA, or both in combination, or with the gels overnight (~18 h) in the presence of 5 µg/ml PHA and 100 U/ml IL-2 in cDMEM (2–4 explants/condition). Then tissues were washed (x4) and cultured in the presence of PHA/IL-2 in the same tissue culture plate. 24 h or 4d post API or gel washout, tissues were challenged with SHIV-RT as described above and cultured in the presence of IL-2 for 14d. In the 4d post gel exposure challenge experiments, tissues were additionally washed (x2) after ~48 h of PHA/IL2 stimulation to eliminate PHA. In separate experiments to address whether residual MIV-150 bound to the tissue culture plate contributed to the observed activity, tissues were transferred to a new plate after gel exposure and washout. As SHIV exposed animals were included in the studies, we measured basal p27 production in cultured vaginal tissues from these animals. No basal p27 production was detected (at least 40 experiments, not shown). Based on these results and due to the limited amount of biopsy

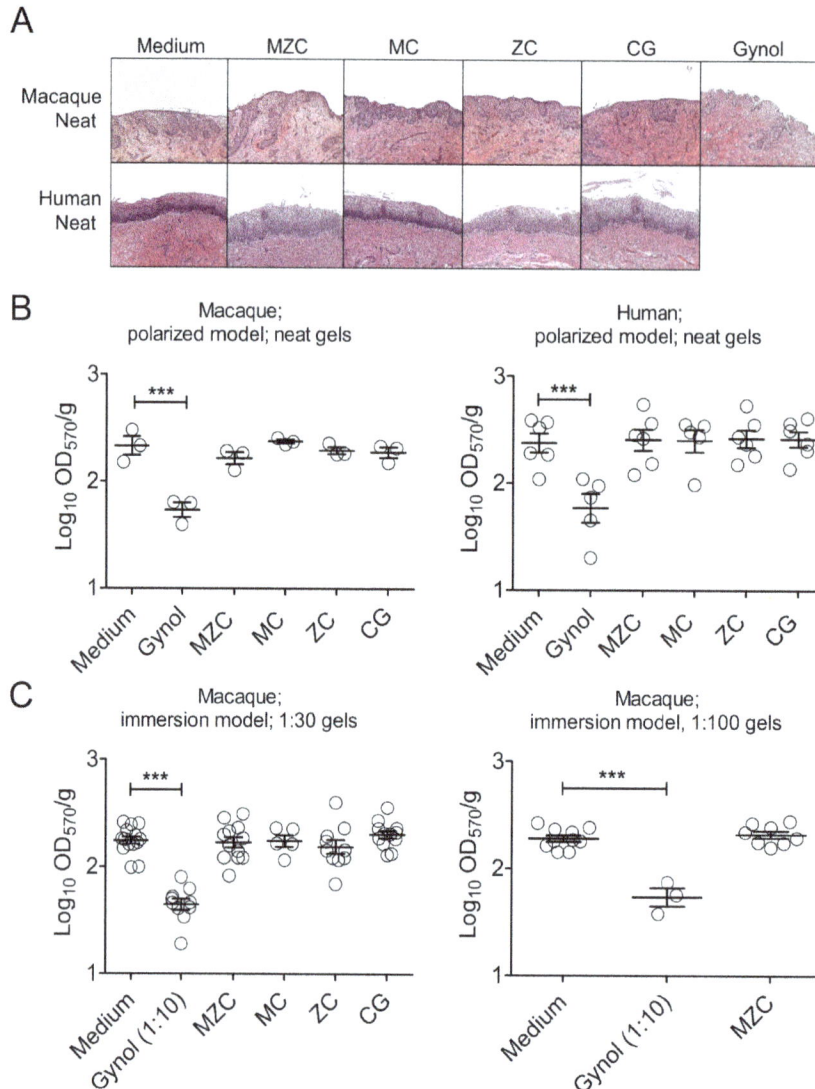

Figure 2. MZC is not toxic in macaque vaginal and human ectocervical explants. (A) Polarized macaque vaginal and human ectocervical explants were cultured for ~18 h in the presence of neat modified gels, applied on the epithelium. For histological evaluation, after exposure to MZC, MC, and ZC, or control conditions (medium, CG, Gynol), tissues were washed, paraffin-embedded, and stained with H&E. Results representative of 4–6 (macaque) and 1–2 (human) experiments are shown at $10\times$ magnification. (B) For determination of tissue viability polarized macaque vaginal and human ectocervical explants (single tissues) were incubated with neat modified MZC, MC, ZC (vs. Medium, CG and Gynol controls) applied on the epithelium for ~18 h. Each symbol indicates a donor and the Mean$\pm$SEM of the $\log_{10}(OD_{570}/g)$ for each condition is shown. (C) Macaque vaginal explants (2–4 per condition) were immersed in culture media with diluted (1:30 or 1:100) modified gels (vs. Medium and 1:10 diluted Gynol control) for ~18 h. Following incubation with the gels, tissue viability was determined using MTT assay. Each symbol indicates a donor and the Mean$\pm$SEM of the $\log_{10}(OD_{570}/g)$ for each condition is shown.

tissues, the no-SHIV-RT-challenge control condition was not included in the studies.

Tissue PD studies

Anti-SHIV-RT activity of tissue-associated MIV-150 was determined in biopsies collected 24 h post last MZC and CG placebo gel application (vs. biopsies collected at the baselines) (Fig. 1). Biopsies (n = 2 vaginal and ectocervical) were collected using 3×5 mm and 3×4.5 mm forceps for vaginal and ectocervical tissue, respectively. Tissues were processed for SHIV-RT challenge in an immersion model as described [17,20] within an hour of collection. Non-stimulated explants (3–6 vaginal explants and 2–5 ectocervical) were challenged with 10^5 $TCID_{50}$ SHIV-RT

per explant in the presence of IL-2 (100 U/ml) (no PHA) and cultured for 14d as described above. This challenge dose was found to reproducibly infect tissues that were not pre-stimulated with PHA/IL-2 [17]. ~18 h after infection, tissues were washed and cultured in cDMEM containing IL-2 for 13 or 14d. Supernatants were collected at 0 (after washes), 3, 7, 10/11 and 13/14d of culture. Infection was monitored as described above.

SIV *gag* qPCR

DNA was extracted from tissues using the DNeasy Blood and Tissue Kit (Qiagen, Germantown, MD) after 3 or 7d of culture post SHIV-RT challenge for the quantification of viral DNA. qPCR was performed using the ABsolute Blue QPCR SYBR

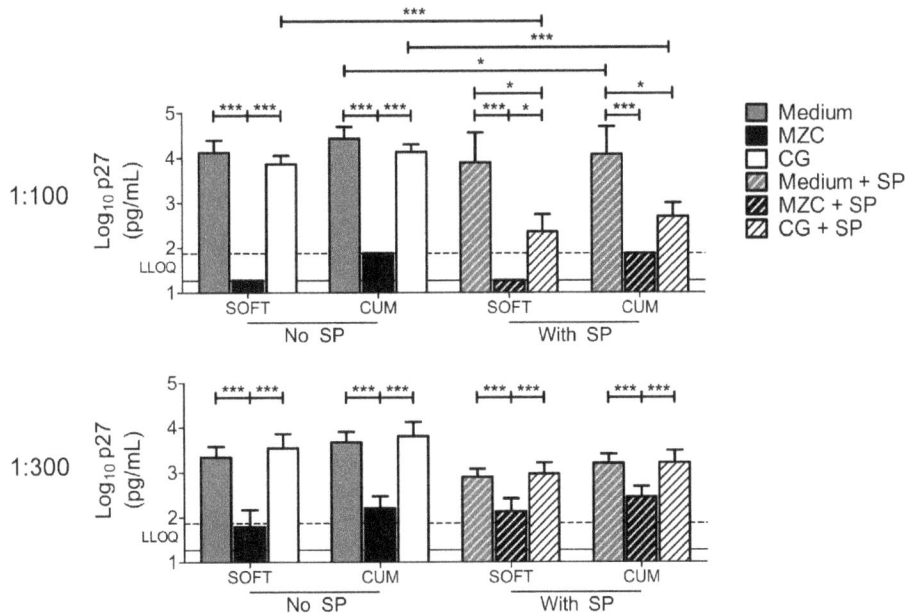

Figure 3. MZC and MC inhibit SHIV-RT infection in macaque vaginal explants. Explants stimulated with PHA/IL-2 for 48 h were challenged with SHIV-RT (10^4 $TCID_{50}$/explant; 2–4 explants/condition) in the presence of modified MZC diluted 1:100 or 1:300 with or without 12.5% human SP. Explants challenged in medium alone or in the presence of CG (± SP) served as controls. After ~18 h, the tissues were washed and cultured for 14d in the presence of IL-2. SIV p27 release was measured at 0, 3, 7, 11, 14d of culture by ELISA. Shown are log_{10}-transformed p27 SOFT and CUM analyses (Mean±SEM) (d3-14 of culture). Summary of n = 3-8 (top panel) and n = 5-8 (bottom panel) experiments are shown. The LLOQ of the assay are denoted for both SOFT (solid line) and CUM (dotted line).

Green Low ROX mix (Thermo Scientific, Waltham, MA) and the Viia 7 real time PCR system (Applied Biosystems, Carlsbad, CA). Changes in SIV *gag* expression were analyzed by the comparative crossing threshold (C_t) method ($2^{-\Delta\Delta Ct}$ method) [21] as previously described [20]. Primers for SIV were 5′-GGTTGCACCCCC-TATGACAT-3′ (SIV667gag Fwd) and 5′-TGCATAGCCGC-TTGATGGT-3′ (SIV731gag Rev); primers for the reference gene albumin were 5′-ATTTTCAGCTTCGCGTCTTT-TG-3′ (RhAlbF) and 5′-TTCTCGCTTACTGGCGTTTT-CT-3′ (RhAlbRev) [20].

RIA for MIV-150 detection

MIV-150 was detected in VF by radioimmunoassay (RIA) as previously described [2,17]. The MIV-150 concentration in the samples was calculated using a curve fitting procedure (logistic 4-parameter model).

Statistics

Tissue viability analysis. Analysis of tissue viability (MTT assay) in the presence of gels was done using a log-normal generalized linear mixed model predicting the weight-normalized OD_{570} of each replicate. A random intercept was included by animal ID. The significance was determined by the Type 3 F-test for overall treatment effect. Significant pairwise comparisons to medium were determined by t tests with Dunnett's adjustment.

Activity of gels against *ex vivo* SHIV-RT challenge. Analysis of gel activity against *ex vivo* SHIV-RT challenge was done using a log-normal generalized linear mixed model with the individual replicate data. For CUM and SOFT analyses [22,23], p27 concentrations of individual replicate values ≥ LLOQ were assumed log-normal. Any value <LLOQ (62.5 pg/ml) at 3-14d of culture was set to $62.5^{1 \vee 2} = 18.616$, a common substitution for log-normal data. CUM from 3-14d for

replicates below LLOQ corresponds to 74.46 pg/ml [17]. A random intercept was included by animal ID. For experiments with recycled animals, a random effect of biopsy date was included grouped by animal ID.

Tissue PD studies. CUM and SOFT analyses were done as described above. Comparisons of CUM and SOFT p27 endpoints were performed using a log-normal generalized linear mixed model with the individual replicate data. Because in the PD study baseline infection was monitored over 14d of culture but post gel infection was monitored over 13d of culture, a linear extrapolation was used to predict post gel p27 concentrations at day 14. The treatment (MIV-150 vs. placebo), type of gel (original vs. modified), biopsy time (baseline vs. post gel exposure), and all interactions were used a predictors. Time was also included as a random effect grouped by a random subject of animal ID. To examine vaginal vs. ectocervical tissue, the treatment, biopsy tissue type (ectocervical vs. vaginal), biopsy time, and all interactions were used a predictors. Time was also included as a random effect grouped by a random subject of animal ID. Overall significance was determined by the Type 3 F-test, and pairwise comparisons were made with Tukey-adjusted t tests. In all cases, macaque age (animal age spanned 10 years) and interactions with age were not significant and were excluded from the final models.

All analyses were performed with SAS V9.4, SAS/STAT V13.2 and $\alpha = 0.05$. Significant p-values of <0.05 (*), <0.01 (**) and < 0.001 (***) are indicated.

Results

MZC does not induce toxic changes in macaque vaginal and human ectocervical mucosa

We performed histological evaluation (H&E staining) and viability testing (MTT assay) of macaque vaginal and human

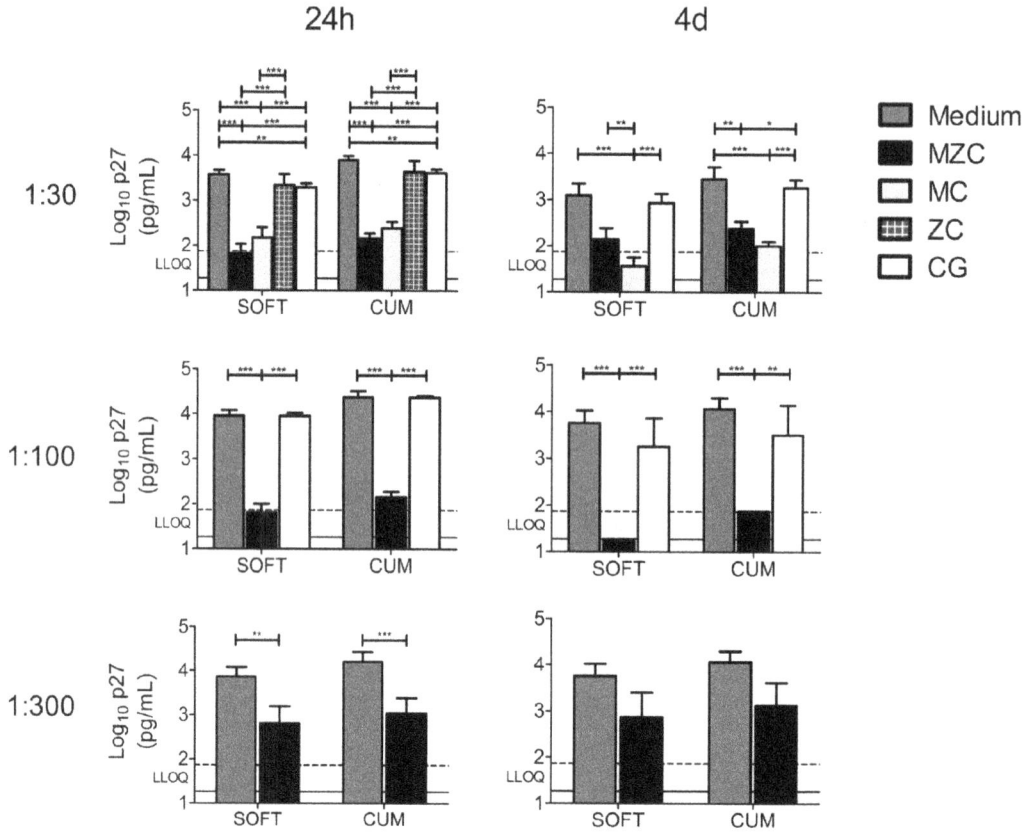

Figure 4. MIV-150 containing gels inhibit SHIV-RT infection in macaque vaginal explants up to 4d post gel exposure. Explants were challenged with SHIV-RT 24 h or 4d after exposure to diluted modified gels and cultured for 14d as described in Fig. 3. Shown are log$_{10}$-transformed p27 SOFT and CUM analyses (Mean±SEM) (d3-14 of culture). Summaries of n=6-16 (24 h post gel exposure) and n=3-6 (4d post gel exposure) experiments are shown. The LLOQ of the assay are denoted for both SOFT (solid line) and CUM (dotted line).

ectocervical tissues to assess the safety of modified MZC and related gels. No histopathological changes were detected in the macaque and human mucosa after application of neat test gels on the epithelial surface of the explants for 18 h (Fig. 2A). Similarly, no histopathological changes were observed when 1:10 diluted (to mimic gel dilution *in vivo*) MZC, MC, CG were applied on human ectocervical mucosa (3 experiments, not shown). The tissue viability (MTT assay) was not decreased after exposure to MZC, ZC, MC or CG relative to the medium control and was

significantly decreased after exposure to Gynol (Fig. 2B). The tissue viability was also determined in the immersion culture model in order to ensure that the antiviral activity testing was performed using non-toxic gel dilutions. We previously determined that ZC gel (original formulation) significantly decreases tissue viability at 1:10 (42% viability of control; p-value = 0.0001) and at 1:20 dilutions (64% viability of control; p-value = 0.0435), but not at 1:30 dilution (84% viability of control; p-value = 0.3474) (3–10 experiments; not shown). Based on this, macaque vaginal

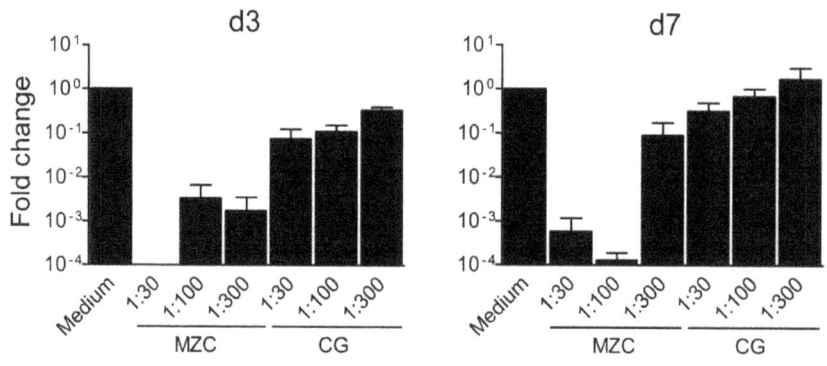

Figure 5. MZC inhibits SHIV-RT DNA burden in the tissues. Explants were challenged with SHIV-RT 24 h after exposure to diluted modified gels (Fig. 4). 3d (A) and 7d (B) after infection, tissues were collected and genomic DNA was extracted. Viral DNA was quantified by SIV *gag* qPCR. The $2^{-\Delta\Delta ct}$ method was used to calculate the fold change of SIV *gag* expression. Summary of n=2 experiments at d3 and d7 each is shown (Mean±SEM).

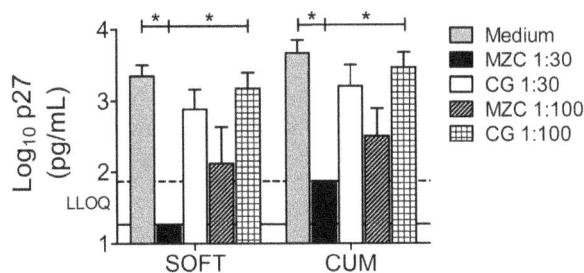

Figure 6. MZC inhibits SHIV-RT infection in macaque vaginal explants derived from Depo-treated animals. Vaginal explants from Depo-treated animals (5 wks post Depo) were challenged with SHIV-RT 24 h after exposure to diluted modified gels and cultured for 14d as described in Fig. 3. Shown are log_{10}-transformed p27 SOFT and CUM analyses (Mean±SEM) (d3-14 of culture). Summary of n = 3–5 experiments is shown. The LLOQ of the assay are denoted for both SOFT (solid line) and CUM (dotted line).

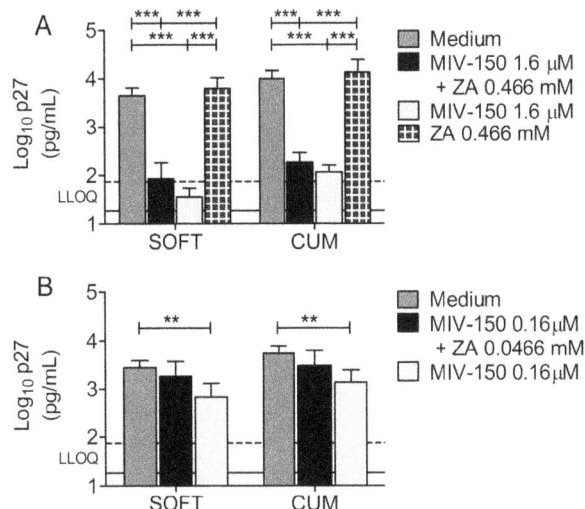

Figure 7. Unformulated MIV-150 inhibits SHIV-RT infection in macaque vaginal explants. PHA/IL-2 stimulated explants were challenged with SHIV-RT (10^4 $TCID_{50}$/explant; 2–4 explants/condition) 24 h after ~18 h exposure to unformulated MIV-150 (1.6 or 0.16 µM), ZA (0.466 mM or 0.0466 mM), or a combination of both vs. untreated control (Medium) (A, B). After challenge, the tissues were washed and cultured for 14d in the presence of IL-2. Shown are log_{10}-transformed p27 SOFT and CUM analyses (Mean±SEM) (d3-14 of culture). Summaries of n = 3–11 experiments are shown. The LLOQ of the assay are denoted for both SOFT (solid line) and CUM (dotted line).

explants were immersed in 1:30 and 1:100 diluted modified MZC and related gels for ~18 h and processed for the MTT assay. No significant changes in tissue viability were detected after immersion in diluted gels (Fig. 2C).

MZC protects against *ex vivo* SHIV-RT infection in macaque vaginal explants and this is not affected by SP

Macaque vaginal explants from untreated animals were challenged with 10^4 $TCID_{50}$ of SHIV-RT in the presence of 1:100 or 1:300 diluted modified MZC and SP (12.5%) for ~18 h. To allow reproducible infection, explants were stimulated with PHA/IL-2 prior to challenge. MZC at both 1:100 and 1:300 dilution in the absence of SP provided strong inhibition of infection relative to untreated medium and CG placebo controls (Fig. 3). Overall, tissue infection in the presence of non-toxic 12.5% SP (MTT data not shown) was lower than control infection (SOFT/CUM p-values = 0.0273/0.0142; combined Medium vs. Medium + SP data (Fig. 3)). Anti-SHIV-RT activity of MZC was not changed in the presence of SP (Fig. 3). 1:100 (but not 1:300) diluted CG + SP inhibited infection relative to medium + SP control and CG.

MZC provides post washout activity against *ex vivo* SHIV-RT infection in macaque vaginal explants

Macaque vaginal explants from untreated animals were immersed in 1:30–1:300 diluted modified gels for ~18 h in the presence of PHA/IL-2, then washed and challenged with 10^4 $TCID_{50}$ of SHIV-RT 24 h or 4d later, to quantify residual antiviral activity post gel washout. MZC strongly inhibited infection at 1:30 (significant for CUM/SOFT in 24 h experiments; significant for CUM but not for SOFT in 4d experiments) and at 1:100 (significant for CUM/SOFT) dilutions (Fig. 4). Comparable protection with 1:30 and 1:100 of MZC suggests that the inhibitory effect may be saturable. 1:30 diluted MC (the only dilution tested) also significantly inhibited tissue infection (Fig. 4). The protection provided by both MZC and MC was significant compared to untreated infection controls as well as to CG placebo controls. A 1:300 MZC dilution afforded less pronounced, but still significant, protection when tissues were challenged 24 h, but not 4d post gel exposure (Fig. 4). Exposure to CG significantly inhibited infection relative to the medium control in 24 h experiments (Fig. 4; 1:30 gel dilution). ZC demonstrated no activity against SHIV-RT (Fig. 4). Additional experiments in which explants were transferred to new plates after gel exposure

and washout determined that culture plate-bound MIV-150 does not significantly contribute to the observed MZC activity. The post washout activity of MZC against SHIV-RT challenge 24 h post gel exposure was not changed under these settings (n = 4 experiments; not shown).

SIV*gag* qPCR analysis using genomic DNA from tissues exposed to MZC gel (1:30–1:300 dilutions) and infected with SHIV-RT 24 h post gel exposure demonstrated inhibition of viral DNA (Fig. 5) and supported the ELISA results.

To determine whether Depo treatment affects the *ex vivo* activity of MZC and to further relate *ex vivo* testing results with data from efficacy studies [2–4], vaginal biopsies from Depo-treated animals were exposed to MZC, washed, and challenged with SHIV-RT 24 h later. The activity of the gel at 1:30 dilution was not changed; however, some decrease in the activity at 1:100 dilution was observed providing 90/88% (CUM/SOFT) inhibition (Fig. 6) as compared to 99/99% (CUM/SOFT) inhibition in tissues from untreated animals (Fig. 4).

Unformulated MIV-150 protects macaque vaginal explants against SHIV-RT infection

Although addition of ZA to the MC gel improved protection against SHIV-RT infection *in vivo* after repeated gel administration [2], our data suggest that the bulk of the antiviral activity measured herein is due to MIV-150. To explore this further, we tested the antiviral activity of unformulated MIV-150 and of a combination of unformulated MIV-150 and ZA (non-toxic concentrations; MTT analysis not shown). To mimic the approach used in the experiments with gels, macaque vaginal explants were incubated with 1.6 µM MIV-150 and 0.466 mM ZA (corresponding to 1:30 gel dilution) for ~18 h, then washed and challenged with SHIV-RT 24 h post APIs exposure. Exposure to MIV-150 or the combination of MIV-150 and ZA inhibited infection similarly to the corresponding MZC or MC gels (Fig. 7A). ZA did not

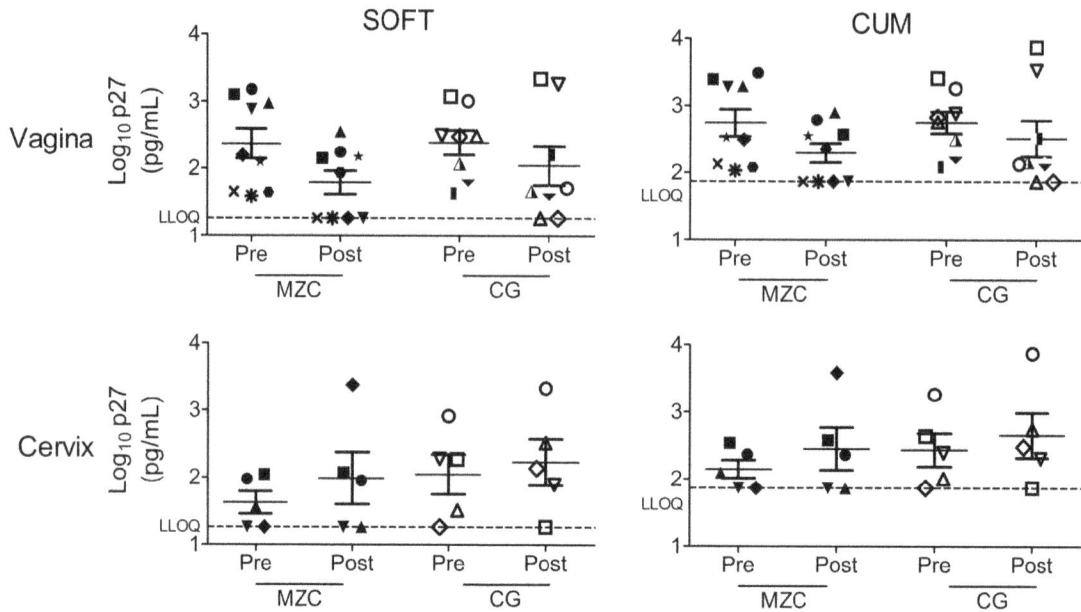

Figure 8. Repeated vaginal administration of MZC and CG reduce *ex vivo* infection of the vaginal mucosa. Depo-treated macaques were administered daily original or modified MZC vs. CG intravaginally for 2 wks. Vaginal biopsies were collected at the beginning of the study (baseline) and 24 h after the last gel application as described in Fig 1. Cervical biopsies were taken only from macaques treated with modified gels. Explants (3–6 vaginal and 2–5 cervical) were challenged with 10^5 TCID$_{50}$ SHIV-RT/explant in the presence of IL-2. After ~18 h incubation, tissues were washed and cultured for 14d in the presence of IL-2. Shown are log$_{10}$-transformed p27 SOFT and CUM analyses (Mean±SEM). Each symbol indicates an individual animal.

inhibit SHIV-RT infection *ex vivo* (Fig. 7A). Decreasing MIV-150 to 0.16 μM resulted in less significant protection (Fig. 7B).

In vivo application of MZC and CG leads to decreased *ex vivo* SHIV-RT infection in vaginal tissues

To determine if *in vivo* administration of MZC leads to inhibition of *ex vivo* SHIV-RT infection in the genital mucosa, macaques received vaginal application of MZC or CG daily for 14d, mimicking *in vivo* efficacy studies [2,3]. 24 h after the last gel application, vaginal and ectocervical biopsies were collected and challenged *ex vivo* with SHIV-RT. The VF MIV-150 concentrations in MZC exposed animals ranged from undetectable to 7.9 nM. We examined infection levels in tissues collected both prior to and after the respective gel dosing for each animal.

Infection levels in vaginal biopsies taken 24 h after the last gel were reduced (51–62% inhibition) by both MZC and CG gels relative to baselines (Fig. 8, Table 2), indicating a barrier effect of CG. Although reduced to a greater extent, the infection inhibition after MZC exposure (65–75%) was not significantly different from infection inhibition after CG exposure (32–45%). To increase the samples size, PD data on original formulations were included in these studies. There was no statistically significant difference in the inhibitory effect afforded by original and modified formulations. No inhibitory effect on *ex vivo* infection was observed in ectocervical biopsies collected 24 h after the MZC or CG exposure.

Discussion

We demonstrate that MZC is safe in the macaque and human genital mucosal models based on histological evaluation and MTT assay. These results support previous data showing that MZC is safe *in vitro* and *in vivo* (mice and macaques) [2,3].

When present at the time of *ex vivo* viral challenge, the gel provides strong protection against SHIV-RT in macaque vaginal mucosa up to 1:300 dilution, indicating that its activity is unlikely to be affected by dilution with VF or dilution associated with intercourse. If gel (2–4 ml) and VF (~ 0.5 ml) [24] mix, the net dilution effect is estimated to be 10%–30% [25,26]. In addition, there will be a range of dilutions of a gel by semen over time after ejaculation. However, given that the volume of the human ejaculate ranges from about 2 to 5 mL [24], these dilutions will be well below the 1:300 dilution that has been shown to be efficacious in the current study. A single *ex vivo* exposure to 1:30 and 1:100 diluted gel 24 h or 4d prior to virus exposure afforded significant inhibition of infection. Notably, the study design (different animals included in different experimental sets, small sample size) did not allow comparison of tissue susceptibility to SHIV-RT at varying *ex vivo* challenge time-points, to determine if changes in susceptibility to infection contributed to the observed results.

As previously reported for neat MIV-150 in cell-based assays [27], the antiviral activity of MZC in the tissues was not changed in the presence of SP. It has also been shown that VF and semen does not interfere with the activity of *in vivo* intravaginal ring (IVR)-released MIV-150 in TZMbl cells [13](Kenney et al., in preparation; Ugaonkar et al., in preparation) and in macaque vaginal explants *ex vivo* [17].

MZC was reported to be more potent than MC against SHIV-RT *in vivo* [2]. To start to address the potential mechanism of the improved antiviral activity, we compared the activity of combined unformulated MIV-150 and ZA to that of either compound alone. The combination of MIV-150 and ZA afforded no greater antiviral activity compared to MIV-150 alone. Thus, the antiviral activity measured in this system was dominated by MIV-150. Given the immunomodulatory properties of zinc [7–10] and the fact that repeated administration of ZC showed greater efficacy in macaques than a single dose [2,4,5], ZA can potentially create a

Table 2. PD study.

Predictor	SOFT			CUM		
Pairwise Comparison	Ratio	[95% CI]	p-value[1]	Ratio	[95% CI]	p-value[1]
Vaginal tissue only:						
Time: Post Gel vs. Baseline	0.38	[0.18, 0.79] *	**0.0114**	0.49	[0.26, 0.91] *	**0.0249** *
Interaction: Time – Gel (MZC, CG)		Overall	0.2929		Overall	0.2753
Interaction: Time – Gel (original, modified)		Overall	0.5219		Overall	0.5538
Cervical tissue only[2]:						
Time: Post Gel vs. Baseline	1.86	[0.50, 7.00]	0.6124	0.85	[0.43, 1.69]	0.6300
Interaction: Time – Gel (MZC, CG)		Overall	0.9355		Overall	0.9393

[1]Included are the Tukey-Kramer adjusted 95% confidence interval (95% CI) and p value for multiple pairwise comparisons.
Significant p values <0.05 (*) are indicated in boldface.
[2]Modified gels only.

non-permissive environment for viral transmission that requires more than the single exposure to ZA as utilized herein. Our observation demonstrating no antiviral activity of a single exposure to ZC in explants supports these data.

We recently demonstrated in a proof-of-principle study that tissue-associated MIV-150 released *in vivo* post MIV-150 IVR exposure inhibits *ex vivo* SHIV-RT infection [17]. Repeated *in vivo* MZC and CG dosing provided 51–62% inhibition (significant) of *ex vivo* SHIV-RT infection in vaginal mucosa 24 h post last gel exposure, indicating a barrier effect of CG. Although the effect of MZC did not significantly differ from CG, MZC afforded 65–74% inhibition of infection level vs. 32–45% inhibition afforded by CG. Because of the barrier effect of CG, we are unable to determine if there would be more antiviral activity 8 h after gel exposure, the time at which consistent complete protection against vaginal SHIV-RT challenge was observed *in vivo* [2,3]. It is plausible that MZC activity in PD studies could have been stronger at the earlier time-point as tissue MIV-150 concentrations were reported to be higher 8 h after last MZC exposure than at 24 h [2]. The barrier effect of CG in PD studies and in some experiments where tissues were immersed in diluted gel and challenged *ex vivo* is in contrast to *in vivo* efficacy data [2,3]. We cannot exclude that incomplete washout of the gel from biopsies could have contributed to this effect.

Higher protection against *ex vivo* infection was observed when tissues were immersed in diluted MZC vs. *in vivo* gel application. This could be due to greater gel/API access to mucosa and submucosa in *ex vivo* gel exposure experiments vs. exposure limited to the epithelial side during *in vivo* gel application. Some decrease in MZC *ex vivo* activity at higher gel dilution (1:100) was observed in tissues from Depo-treated animals providing 88–90% infection inhibition as compared to 99% inhibition in tissues from non-Depo-treated animals. This could have also contributed to PD results as Depo-treated animals were utilized for PD studies. Our results and published data [28,29] emphasize the need to study the impact of hormonal contraceptives on mucosal HIV infection. Furthermore, Depo treatment was reported to change biodistribution of HIV inhibitors leading to decreased tissue concentrations and increased plasma levels [30].

Although the amounts of MIV-150 in the ectocervix and vagina 24 h post last gel were reported to be similar [2], no trend of decreased infection and no barrier effect of CG in the ectocervical mucosa were detected in our study. This finding warrants further investigation. Post-Depo treatment infection level in the cervix in the placebo group was higher than the infection at the baseline, although this was not statistically significant. Changes in cervical tissue susceptibility to infection post-Depo treatment could have potentially contributed to the lack of MZC activity in the cervix. However, the sample size in the study was too small to determine the difference. We need to note that we were unable to measure tissue levels of MIV-150 in our study and had to extrapolate data from earlier study [2].

As discussed in [22,23], there is currently no consensus on the optimal virological endpoint for characterizing tissue infection. PK/PD analysis of UC-781 gel in a Phase 1 rectal safety study [22] demonstrated that SOFT and CUM measurements at the higher titer of HIV_{BaL} (10^4 $TCID_{50}$) had reliability and provided evidence of between-visit reproducibility. SOFT analysis defines an endpoint at which high rates of virus growth have been achieved and no further biologically significant increases in virus growth are apparent [22]. It is less affected by the duration of the explant assay than CUM analysis. Inclusion of SOFT analyses is important for potential comparison of data across laboratories using longer explant assay protocols [22,23]. Based on this, we performed both analyses in our study.

In conclusion, our data show that MZC is safe in macaque vaginal and human ectocervical mucosa. The gel is highly effective against *ex vivo* SHIV-RT infection in macaque vaginal explants (gel/virus co-exposure experiments and post gel washout effect). Although in our PD studies MZC activity (65–74% inhibition) 24 h post last gel administration did not significantly differ from CG (32–45% inhibition), it was comparable to the protection seen when macaques were vaginally challenged 24 h after gel administration (~75%) [2–4]. Overall, the data suggest that evaluation of safety, *ex vivo* activity and PK and PD in macaque explants can inform macaque efficacy and clinical studies design, and support advancing MZC for clinical evaluation.

Acknowledgments

We thank the veterinary staff at the TNPRC for the continued support. Disclaimer: The contents of this manuscript are the sole responsibility of the Population Council and do not necessarily reflect the views or policies of the funding agencies. None of the material in this article has been published or is under consideration elsewhere, including the Internet.

Author Contributions

Conceived and designed the experiments: NT PB GC. Performed the experiments: PB GC LO NT. Analyzed the data: PB GC MLC KDR NT. Contributed reagents/materials/analysis tools: NJP LK AR CA RM SS JAFR TMZ MP JDL. Wrote the paper: NT GC PB MR. Macaques at TNRCP: JB AG. Human tissue samples: RS.

References

1. Abdool Karim Q, Abdool Karim SS, Frohlich JA, Grobler AC, Baxter C, et al. (2010) Effectiveness and safety of tenofovir gel, an antiretroviral microbicide, for the prevention of HIV infection in women. Science 329: 1168–1174.

2. Kenney J, Aravantinou M, Singer R, Hsu M, Rodriguez A, et al. (2011) An antiretroviral/zinc combination gel provides 24 hours of complete protection against vaginal SHIV infection in macaques. PLoS One 6: e15835.

3. Kizima L, Rodríguez A, Kenney J, Derby N, Mizenina O, et al. (2014) A Potent Combination Microbicide that Targets SHIV-RT, HSV-2 and HPV. PLOS ONE

4. Kenney J, Singer R, Derby N, Aravantinou M, Abraham C, et al. (2012) A single dose of a MIV-150/zinc acetate gel provides 24 hours of protection against vaginal SHIV-RT infection, with more limited protection rectally 8–24 hours after gel use. AIDS Res Hum Retroviruses.

5. Kenney J, Rodriguez A, Kizima L, Seidor S, Menon R, et al. (2013) A Potential Non-ARV Microbicide – A Modified Zinc Acetate Gel is Safe and Effective Against SHIV-RT and HSV-2 infection In Vivo. Antimicrob Agents Chemother.

6. Fenstermacher KJ, DeStefano JJ (2011) Mechanism of HIV reverse transcriptase inhibition by zinc: formation of a highly stable enzyme-(primer-template) complex with profoundly diminished catalytic activity. J Biol Chem 286: 40433–40442.

7. Shankar AH, Prasad AS (1998) Zinc and immune function: the biological basis of altered resistance to infection. Am J Clin Nutr 68: 447S–463S.

8. Rink L, Kirchner H (2000) Zinc-altered immune function and cytokine production. J Nutr 130: 1407S–1411S.

9. Wellinghausen N, Kirchner H, Rink L (1997) The immunobiology of zinc. Immunol Today 18: 519–521.

10. Ibs KH, Rink L (2003) Zinc-altered immune function. J Nutr 133: 1452S–1456S.

11. Animal Welfare Act and Regulation of 2001. ed. Code of Federal Regulations, t., chapter 1, subchapter A: animals and animal products. US Department of Agriculture, Beltsville, MD.

12. Committee on Care and Use of Laboratory Animals of the Institute of Laboratory Animal Resources Guide for the Care and Use of Laboratory Animals (1985) US Department of Health and Human Services, National Institutes of Health Bethesda, MD 85-23: 1–83.

13. Singer R, Mawson P, Derby N, Rodriguez A, Kizima L, et al. (2012) An intravaginal ring that releases the NNRTI MIV-150 reduces SHIV transmission in macaques. Sci Transl Med 4: 150ra123.

14. Lifson JD, Rossio JL, Piatak M Jr., Parks T, Li L, et al. (2001) Role of CD8(+) lymphocytes in control of simian immunodeficiency virus infection and resistance to rechallenge after transient early antiretroviral treatment. J Virol 75: 10187–10199.

15. Cline AN, Bess JW, Piatak M Jr., Lifson JD (2005) Highly sensitive SIV plasma viral load assay: practical considerations, realistic performance expectations, and application to reverse engineering of vaccines for AIDS. J Med Primatol 34: 303–312.

16. Crostarosa F, Aravantinou M, Akpogheneta OJ, Jasny E, Shaw A, et al. (2009) A macaque model to study vaginal HSV-2/immunodeficiency virus co-infection and the impact of HSV-2 on microbicide efficacy. PLoS One 4: e8060.

17. Ouattara LA, Barnable P, Mawson P, Seidor S, Zydowsky TM, et al. (2014) MIV-150 containing intravaginal rings protect macaque vaginal explants against SHIV-RT infection. Antimicrob Agents Chemother.

18. Turville SG, Aravantinou M, Miller T, Kenney J, Teitelbaum A, et al. (2008) Efficacy of Carraguard-based microbicides in vivo despite variable in vitro activity. PLoS One 3: e3162.

19. Cummins JE Jr., Guarner J, Flowers L, Guenthner PC, Bartlett J, et al. (2007) Preclinical testing of candidate topical microbicides for anti-human immunodeficiency virus type 1 activity and tissue toxicity in a human cervical explant culture. Antimicrob Agents Chemother 51: 1770–1779.

20. Aravantinou M, Singer R, Derby N, Calenda G, Mawson P, et al. (2012) The NNRTI MIV-160 delivered from an intravaginal ring, but not from a carrageenan gel, protects against SHIV-RT infection. AIDS Res Hum Retroviruses.

21. Livak KJ, Schmittgen TD (2001) Analysis of relative gene expression data using real-time quantitative PCR and the 2(-Delta Delta C(T)) Method. Methods 25: 402–408.

22. Richardson-Harman N, Mauck C, McGowan I, Anton P (2012) Dose-response relationship between tissue concentrations of UC781 and explant infectibility with HIV type 1 in the RMP-01 rectal safety study. AIDS Res Hum Retroviruses 28: 1422–1433.

23. Richardson-Harman N, Lackman-Smith C, Fletcher PS, Anton PA, Bremer JW, et al. (2009) Multisite comparison of anti-human immunodeficiency virus microbicide activity in explant assays using a novel endpoint analysis. J Clin Microbiol 47: 3530–3539.

24. Owen DH, Katz DF (2005) A review of the physical and chemical properties of human semen and the formulation of a semen simulant. J Androl 26: 459–469.

25. Lai BE, Xie YQ, Lavine ML, Szeri AJ, Owen DH, et al. (2008) Dilution of microbicide gels with vaginal fluid and semen simulants: effect on rheological properties and coating flow. J Pharm Sci 97: 1030–1038.

26. Katz DF, Gao Y, Kang M (2011) Using modeling to help understand vaginal microbicide functionality and create better products. Drug Deliv Transl Res 1: 256–276.

27. Fernandez-Romero JA, Thorn M, Turville SG, Titchen K, Sudol K, et al. (2007) Carrageenan/MIV-150 (PC-815), a combination microbicide. Sex Transm Dis 34: 9–14.

28. Marx PA, Spira AI, Gettie A, Dailey PJ, Veazey RS, et al. (1996) Progesterone implants enhance SIV vaginal transmission and early virus load. Nat Med 2: 1084–1089.

29. Heffron R, Donnell D, Rees H, Celum C, Mugo N, et al. (2012) Use of hormonal contraceptives and risk of HIV-1 transmission: a prospective cohort study. Lancet Infect Dis 12: 19–26.

30. Malcolm RK, Veazey RS, Geer L, Lowry D, Fetherston SM, et al. (2012) Sustained release of the CCR5 inhibitors CMPD167 and maraviroc from vaginal rings in rhesus macaques. Antimicrob Agents Chemother 56: 2251–2258.

17

A Colloidal Singularity Reveals the Crucial Role of Colloidal Stability for Nanomaterials *In-Vitro* Toxicity Testing

Soledad Gonzalo[1]**ꚗ**, **Veronica Llaneza**[2]**ꚗ**, **Gerardo Pulido-Reyes**[1,3]**ꚗ**, **Francisca Fernández-Piñas**[3],
Jean Claude Bonzongo[2], **Francisco Leganes**[3], **Roberto Rosal**[1,4], **Eloy García-Calvo**[1,4],
Ismael Rodea-Palomares[3]*¤

1 Departamento de Ingeniería Química, Universidad de Alcalá, Alcalá de Henares, Madrid, Spain, **2** Department of Environmental Engineering Sciences, University of Florida, Gainesville, Florida, United States of America, **3** Departamento de Biología, Facultad de Ciencias, Universidad Autónoma de Madrid, Madrid, Spain, **4** Instituto Madrileño de Estudios Avanzados (IMDEA) Agua, Alcalá de Henares, Madrid, Spain

Abstract

Aggregation raises attention in Nanotoxicology due to its methodological implications. Aggregation is a physical symptom of a more general physicochemical condition of colloidal particles, namely, colloidal stability. Colloidal stability is a global indicator of the tendency of a system to reduce its net surface energy, which may be achieved by homo-aggregation or hetero-aggregation, including location at bio-interfaces. However, the role of colloidal stability as a driver of ENM bioactivity has received little consideration thus far. In the present work, which focuses on the toxicity of nanoscaled Fe° nanoparticles (nZVI) towards a model microalga, we demonstrate that colloidal stability is a fundamental driver of ENM bioactivity, comprehensively accounting for otherwise inexplicable differential biological effects. The present work throws light on basic aspects of Nanotoxicology, and reveals a key factor which may reconcile contradictory results on the influence of aggregation in bioactivity of ENMs.

Editor: Vishal Shah, Dowling College, United States of America

Funding: This research was supported by the Comunidad de Madrid grants S-0505/AMB/0321 and S-2009/AMB/1511 (Microambiente-CM). The Spanish Ministry of Science and Innovation grants CGL2010-15675 sub-program BOS, and the CTM2008-04239/TECNO, and the Spanish Ministry of Economy and Competitiveness (MINECO) grants CTM2013-45775-C2-1-R, CTM2013-45775-C2-2-R and the US-National Science Foundation, grant number CBET-0853347.

Competing Interests: The authors have declared that no competing interests exist.

* Email: ismael.rodea@uam.es

ꚗ These authors contributed equally to this work.

¤ Current address: Departamento de Ingeniería Química, Universidad de Alcalá, Alcalá de Henares, Madrid, Spain

Introduction

Environmental health and safety (EHS) is one of the key challenges in the field of nanotechnology [1–3]. Despite research efforts, EHS development is hampered by a series of difficulties, from misconceptions in the field, to the lack of agreement on methodological aspects of EHS execution [1,4,5]. The latter is primarily due to the peculiar and complex intrinsic characteristics of nano-sized materials [1,3,4,6,7]. Overall, more than a dozen physicochemical properties could potentially contribute to hazardous interactions at the bio-nano interface [3,8,9]. However, it is hard to find comprehensive studies where key drivers of ENM bioactivity are clearly identified [1]. Currently, the influence of aggregation on ENM bioactivity is surrounded by controversy, mainly due to its broad methodological implications and the existence of contradictory results [1,4,10]. Despite the importance of the effects of aggregation in EHS [4,5], no international consensus exists on how to handle ENM aggregation [1,4,5]. Notably, Schorus & Lison [1] stated in their recent commentary "Focusing the research effort" that "the influence of SNPs (silica nanoparticles) aggregation in the biological response remains unclear and it remains impossible to state whether or not it is necessary to have a dispersion of SNPs before testing". Although they revised SNPs toxicity data, their statement may be generalized to ENMs. Aggregation is a symptom of a more general physicochemical condition of colloidal particles, *i.e.*, colloidal stability [11]. Colloidal stability has been considered a major issue in EHS assessment of ENMs to warrant exposure and dosimetry [2,4]. However, despite the growing body of evidence regarding biophysically mediated bioactivity of ENMs in a variety of model biological systems [8,12–14], the role of colloidal stability as its driver has been essentially overlooked. In the present work, the spontaneous occurrence of an nZVI speciation phenomenon [3,15], which we have termed *colloidal singularity*, showed that taking only colloidal stability into account allowed us to reconcile otherwise inexplicable biological effects.

Results

Physicochemical characterization of nZVI

The surface area and mean particle size of pristine nZVI powder was 43.9 m^2 g^{-1} and 20 nm, respectively, as calculated by the Nova 1200 BET (Brunauer–Emmett–Teller) method. Pristine nZVI powder showed a partial degree of surface oxidation as revealed by XRD analysis (**Table S1**). This is due to the high tendency of zero valent iron to oxidize in aerobic conditions [16,17]. nZVI suspensions ranging from 2.5 to 50 mgL^{-1} were prepared in Milli-Q water and OECD TG 201 algal growth medium [18]. Their main physicochemical properties (**Table 1**) were studied to select the best option for working as a stock suspension for subsequent EHS characterization. General trends were as follows: nZVI appeared aggregated with effective diameters of approximately 100 nm in Milli-Q water, and between 200–400 nm in OECD medium. The Polydispersity index was below 0.5 in Milli-Q water, and ranged from 0.5 to 0.8 in OECD TG 201 medium, indicating a high degree of polydispersity in particle size, especially in OECD TG 201 medium. nZVI appeared with a near neutral surface charge when dispersed in Milli-Q water, however it showed a negative surface charge in OECD TG 201 medium, ranging from −17.54 mV for 2.5 mgL^{-1} to −27.5 mV for 25 mgL^{-1}. The observed tendency of nZVI to acquire negative charge in saline medium has been associated with the high affinity of ions to be adsorbed onto the oxy-hydroxide surface shell of nZVI formed under oxidant conditions [17]. Primary particle size and morphology was investigated by Transmission electron microscopy (TEM) in nZVI suspensions prepared in Milli-Q water (**Figure 1**). 50 mgL^{-1} nZVI suspensions showed a high degree of agglomeration (**Figure 1.a**), however, the visualization of diluted suspensions (5 mgL^{-1}) allowed us to observe primary sized particles and small particle aggregates (**Figure 1.b, c**). The mean size, measured by TEM, of primary aggregates were 37.5±9.7 nm, and that of individual small and big nZVI primary particles was 4.0±1.4 and 12.7±1.7 nm, respectively (**Figure S1**). XEDS (X-Ray Energy Dispersive Spectroscopy) analysis confirmed Fe as the main constituent of the visualized particles (**Figure 1.d**). 25 mgL^{-1} of nZVI in Milli-Q water had the lowest polydispersity index (PDI) and effective diameter (**Table 1**). Therefore, it was selected as stock suspension.

nZVI working suspensions show a *colloidal singularity* in a narrow dose range

The stability of the nZVI working suspensions used in EHS was evaluated under shaking conditions by measuring the residual absorbance ($\lambda = 750$ nm) at relevant experimental lapse times [4,15] (0 h, 4 h, 24 h, 48 h and 72 h) (**Figure 1. e**). Extensive sedimentation of nZVI particles was observed within the first 4 h, in agreement with the high tendency of nZVI to homo-aggregate [19–21]. However, suspensions remained stable from 4 h to 72 h (**Figure 1.e**), confirming the relevant nZVI colloidal fractions in the exposure experiments [3,15]. The actual nZVI concentrations as stable colloidal fractions in equilibrated conditions were 20–40% of nominal concentrations. They maintained a linear correlation ($R^2 = 0.95$, $p<0.05$) with the initial dosage (**Figure 1. f**). Main physicochemical properties (e.g. size, surface charge and total surface area) of the stable colloidal nZVI fractions in the test system were monitored during 72 h (**Figure 2.a, b, c**). As a general trend, nZVI appeared in the form of aggregates of sizes around 150–500 nm, with ζ-potential around −15 mV and a total surface area of near 10^{-3} m^2 L^{-1}. However, nZVI presented a substantially different physicochemical state within a small dose range (0.1–0.5 mgL^{-1}). In this narrow interval nZVI appeared suspended in its primary size (4–12 nm) along the experimental lapse time (**Figure 2.a**), with a substantially more negative surface charge (ζ-potential = −28 mV) (**Figure 2.b**), and a nearly 10 times higher total surface area (**Figure 2.c**). To our knowledge, such an extreme asymmetry in behavior of colloidal suspensions without the interplay of any external factor has never been previously reported. For simplicity, we will herein refer to this phenomenon as *colloidal singularity*.

The *colloidal singularity* prevents nZVI toxicity

EHS assessment of nZVI suspensions was evaluated using an algal growth inhibition assay based on *Pseudokirchneriella subcapitata* (*P. subcapitata*) [18]. The assay consisted of monitoring the algal growth rates based on three different biomass end-points. Besides growth inhibition, a study of alterations in intracellular levels of reactive oxygen species (ROS) and cell cycle was also performed. All biomass related end-points tested correlated ($R>0.9$, $p<0.05$) (**Figure 3.b, c**) and showed non-linear dose-effect relationships (**Figure 3.a**) which mimicked the ζ-potential and particle size profiles showed in **Figure 2.a, b**. The biological response of the test model organism showed two statistically significant ($p>0.05$) regions of toxic response, bracketing a non-toxic region. The first toxic dose region occurred within nZVI concentrations of 0.05 to 0.075 mgL^{-1}, resulting in almost 40% maximum growth inhibition. The second window of concentrations resulting in adverse biological effects ranged from 1.0 to 2.5 mgL^{-1}, resulting in almost 60% maximum growth inhibition (**Figure 3.a**). The observed toxic response was consistently accompanied by a statistically significant ($p>0.05$) increase in intracellular ROS production, and cell cycle alterations (**Figure 4.a–e**). ROS production reached nearly 180% and 250% induction with respect to control levels for 0.075 mgL^{-1} and 2.5 mgL^{-1} nZVI, respectively (**Figure 4.a**). Cell cycle alterations consisted of two levels of abnormalities in cell cycle progression of cell populations. Firstly, nearly 50% of the cell population accumulated increasing intracellular DNA quantities. This means that cells accumulated in poorly defined S and G2/M phases (Black chart in **Figure 4.e**). Secondly, almost 30% of cell population showed DNA fragmentation (Blue spots in **Figure 4.d** and F region in **Figure 4.e**) which is usually linked to apoptosis [22,23]. ROS induction, cell cycle alterations and DNA fragmentation were clearly evident and statistically significant ($p<0.05$) from early stages of nZVI exposure (24 h) (**Figure 4.e, f**). Interestingly, between the two toxic end dose regions, there was a non-toxic dose region within a narrow range of nZVI concentrations (0.1 to 0.5 mgL^{-1}), coinciding with the *colloidal singularity* described earlier (**Figure 2**).

The nZVI-algal suspension can be regarded as a whole colloidal system and its stability properties can be studied as such. Destabilization of colloidal systems is typically accompanied by an increase in sedimentation rates, and a decrease (in absolute value) in ζ- potential which typically account for aggregation and sedimentation of colloidal particles [11,24,25]. **Figure 5.a** shows that the algae-nZVI systems exhibited clear signs of colloidal destabilization with the exception of the concentration range in which the colloidal singularity occurred. Furthermore, the decrease in algal growth rates clearly correlated with the sedimentation rates of the test systems ($R = -0.689$, $p<0.001$) (**Figure 5.b**). This suggests that the toxicity of nZVI is mediated by colloidal destabilization and hetero-aggregation between algal

Table 1. Main physicochemical parameters of nZVI suspensions prepared in Milli-Q water and OECD algal growth medium.

nZVI (mgL⁻¹)	Dispersion Medium	Effective diameter (nm)[1]	Polydispersity Index (PDI)	ζ-potential (mV)	Total surface area (m² g⁻¹)[3]	pH	Conductivity(μScm⁻²)
2.5	H₂O	124.5±21.0	0.471	−8.5±5.5	7.19	6	0.011
	OECD medium	284.2±47.0	0.609	−17.5±2.3	3.15	8.2	0.214
5	H₂O	82.1±13.5	0.407	−3.4±1.3	10.91	6	0.007
	OECD medium	183.4±27.4	0.816	−23.6±2.4	4.88	8.2	0.202
25	H₂O	108.9±16.4	0.406	−5.55±2.2	8.22	6	0.0032
	OECD medium	121.3±16.1	0.522	−27.50±2.7	7.38	8.2	0.207
50	H₂O	123.0±12.5	0.451	−7.0±3.3	7.28	6	0.012
	OECD medium	369.1±56.3	0.539	−22.2±2.5	2.43	8.2	0.206

Measurements were conducted at 25 C. Effective diameter (hydrodynamic diameter and ζ-potential in milli-Q water were performed in the presence of 10 mM KCl. Data are expressed as mean ± standard deviation (SD) when applicable.
[1]: Effective diameter (hydrodynamic diameter) measured by dynamic light scattering (DLS).
[3]: Total surface area (SSA) is calculated according to equation 1.

cells and nZVI. To test this hypothesis, events occurring at the bio-interface delimiting algal cells and the external microenvironment were studied by TEM and FTIR analysis (**Figure 6.**). As shown in **Figure 6.b, d**, the presence of electron-dense particles surrounding the external cell wall of the micro algae was identified at 0.075 mgL⁻¹ and 2.5 mgL⁻¹ of nZVI. However, they were absent in control cells and 0.5 mgL⁻¹ nZVI treatment (**Figure 6.a**, **c**, respectively). XEDS analysis assigned Fe as the main constituent of those particles (**Figure 6.i, j**). Furthermore, FTIR analysis (**Figure 6.f and h**) revealed significant changes to the IR signals of algal cell where cell-nZVI attachment occurred (0.075 mgL⁻¹ and 2.5 mgL⁻¹ nZVI treatments). These changes consisted of a significant reduction of peak absorption at wavenumbers characteristic of asymmetric (2924 cm⁻¹) and symmetric (2853 cm⁻¹) sp³ C-H stretching vibrations, O-H stretching (3222 cm⁻¹) and bending (1654 cm⁻¹) vibrations in hydroxyl functional groups and C=O stretching vibration (1743 cm⁻¹) of carbonyl functional groups [26]. Interestingly, IR signals remained almost unchanged in the 0.5 mgL⁻¹ treatment (**Figure 6.g**) compared to control signal (**Figure 6.f**).

Colloidal destabilization reverts the colloidal singularity and induces nZVI toxicity

To further demonstrate the key role played by colloidal stability in the observed lack of toxicity in the colloidal singularity dose range, a destabilization experiment was performed. Destabilization of the *colloidal singularity* was induced by neutralizing the negative surface charge density of the 0.5 mgL⁻¹ nZVI suspension by using a classical hydrolyzing metal coagulant [$Al_2(SO_4)_3$] [11,24]. Al^{3+}, like other polyvalent metal cations, is specifically adsorbed on negatively charge surfaces, thereby reducing the surface charge density [11]. Based on titration experiments with $Al_2(SO_4)_3$ (**Figure 7.a**), 0.01 μM $Al_2(SO_4)_3$ was chosen, as it proved to effectively reduce the ζ-potential of the 0.5 mgL⁻¹ nZVI suspension from −28 mV to nearly −10 mV, resulting in partial colloidal destabilization. This $Al_2(SO_4)_3$ had no detrimental effect on the growth of *P. subcapitata* [27] (**Figure 7.b**), nor statistically significant effects on monitored parameters compared to control systems (**Figure 7.c,d**). The addition of $Al_2(SO_4)_3$ to the test system containing 0.5 mgL⁻¹ nZVI caused 2-fold reductions in the growth rates of *P. subcapitata* ($p > 0.05$) (**Figure 7.b**). This toxic response was accompanied by a statistically significant ($p > 0.05$) increase in sedimentation rates (**Figure 7.c**), and a statistically significant ζ-potential reduction of the nZVI-algae colloidal system ($p > 0.05$) (**Figure 7.d**).

Discussion

When suspensions of nZVI were prepared in OECD culture medium, we observed an ENM speciation phenomenon [3] as a function of dose. We denoted it as *colloidal singularity* since such a non-monotonic speciation phenomenon as a function of dose has never been previously reported. Interestingly, the ζ-Potential/particle size *vs* nZVI dosage diagrams resembled those of an inverted classical flocculation diagram [11,28]. In the classical flocculation diagram a fixed amount of suspended colloids is exposed to increasing concentrations of a hydrolyzing salt. The multivalent cations of the hydrolyzing salt (commonly Al^{3+} or Fe^{2+}) specifically adsorb on the negatively charged surface of the suspended colloids. This results in a characteristic non-linear behavior on the residual turbidity of the suspensions as a function of the resulting colloid/ions ratio. The classical flocculation diagram presents three differential stages: (1) in a first stage, the dose of coagulant is not enough to destabilize the colloids which

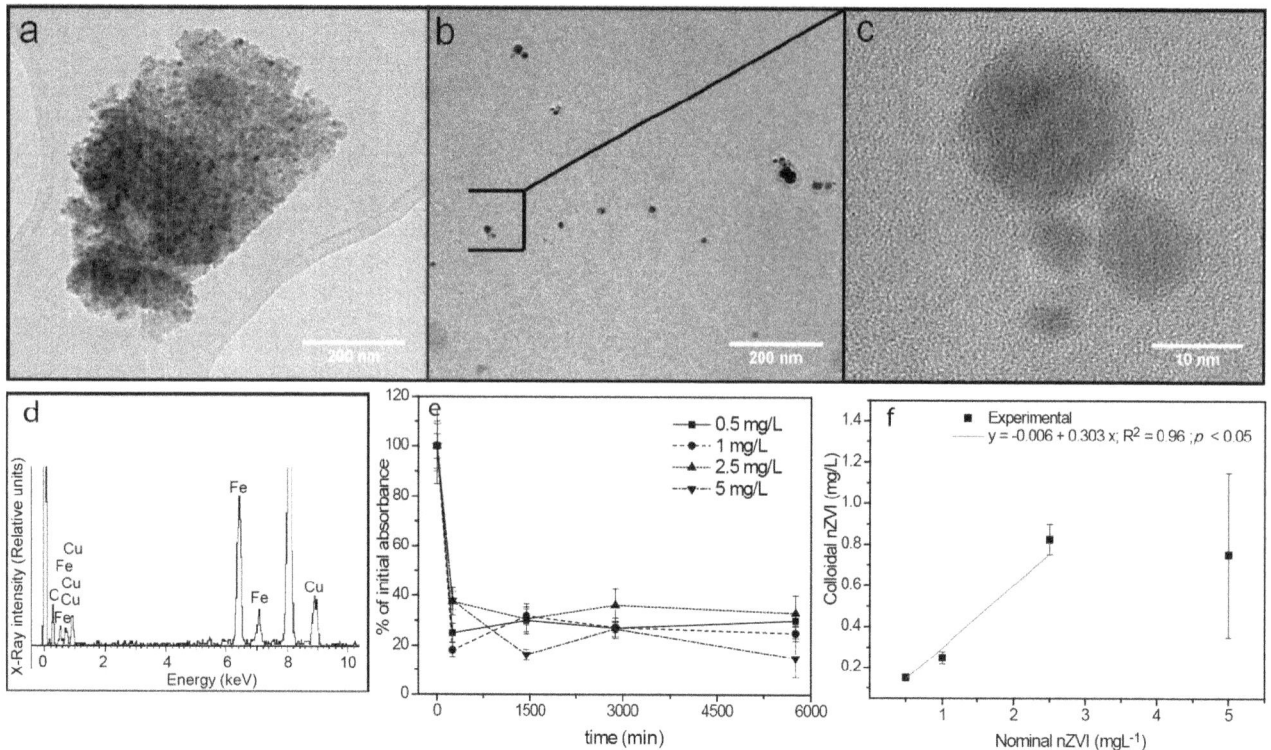

Figure 1. Characterization of nZVI working suspensions under experimental conditions. Representative TEM images of (a) nZVI aggregates in 50 mgL^{-1} nZVI stock suspensions, and (b) small nZVI aggregates and primary sized particles in diluted (5 mg/L) nZVI suspensions. (c) Detail of a small nZVI aggregate integrated by 4 agglomerated individual spherical particles. (d) XEDS spectra of (c), Fe is the main constituent. Ni and Cu intense signals in the XEDS spectra (when present), are artifacts due to the metallic grid supporting the samples, and of the sample holder of the TEM instrument, respectively. (e) Reduction of residual absorbance ($\lambda = 750$ nm) of nZVI suspensions as a result of aggregation and sedimentation at relevant experimental times (0 h, 4 h, 24 h, 48 h and 72 h). (f) Correlation between nominal nZVI concentration (mg/L) and nZVI concentration as relevant colloidal fraction (mg/L) in working suspensions at 72 h.

remained negatively charged and stable. (2) With increasing coagulant concentrations, the adsorbed cations effectively reduce the surface charge of suspended colloids, resulting in colloidal destabilization and removal of colloids. This typically happens in a very narrow coagulant dosage [11,24]. (3) A further increase in coagulant dose produces re-stabilization of colloidal particles by charge reversal resulting in a stable colloidal system. The success of the coagulation process depends entirely upon the adequate shaking [11,24]. In our case, two regions of colloidal destabilization bracketing a stable one appeared with increasing nZVI concentration. In contrast with the classical flocculation diagram, we have a fixed ion concentration (from the culture medium) and a variable number of colloidal particles. However, we also generated a variable colloid/ions ratio, which may result in a similar asymmetry in the colloidal stability of the suspensions. Interestingly, in the ζ-potential/particle size *vs* nZVI dose diagrams, we did not find charge reversal causing stabilization of nZVI in the colloidal singularity, but rather an increase in its electrophoretic mobility (ζ-potential). We hypothesized that the high affinity of anions to be adsorbed on nZVI surfaces [17] may promote an increased anion adsorption on the nZVI surface at the specific nZVI/ion ratios established in the *colloidal singularity* dose range This would result in a net increase in negative surface charge and may cause a reduction of its net surface energy, and of its tendency to interact with bio-interfaces.

The occurrence of the colloidal singularity is highly illustrative as a case study for EHS of ENMs. It clearly exemplifies the extreme relevance of ENMs speciation in their ability to interact

with living organisms, resulting in differential biological effects [1,3]. When nZVI appeared stable in its primary size (in the *colloidal singularity*), no detrimental biological effects could be observed. This was true even when its total surface area was higher than that of any aggregated nZVI suspension. This is in apparent disagreement with previous reports [29,30] and the general consensus: a higher total surface implies a higher exposure, and hence bioactivity [2,4,5]. Here, the key point to reconcile the findings is to consider colloidal stability. The concomitant increase in the negative surface charge of nZVI suspensions implies stable colloidal systems. These systems do not cause biological effects due to their lack of tendency to interact with bio-interfaces [8,25,31]. This is true regardless of considerations on their surface area or their intrinsic bioactivity. It was further confirmed by the observed correlation between toxicity and hetero-aggregation tendency, and the destabilization experiment with $Al_2(SO_4)_3$.

Nanoparticle-induced oxidative stress caused by the formation of reactive oxygen species (ROS) has been established as one of the most common paradigms involved in the toxic responses induced by engineered nanoparticles in living organisms [7,8,32,33]. According to this paradigm, when the cellular antioxidant barriers are surpassed, ROS injury can ultimately induce cell death via alterations in cell cycle [33,34] and the activation of apoptotic signaling pathways [7,33]. Our results demonstrate that the toxicity of nZVI to *P.subcapitata* is linked to the production of ROS, cell cycle alterations and pre-apoptotic DNA fragmentation; and therefore can be assigned to the oxidative stress toxicity paradigm [5,7,33]. Furthermore, toxicity, ROS and cell cycle

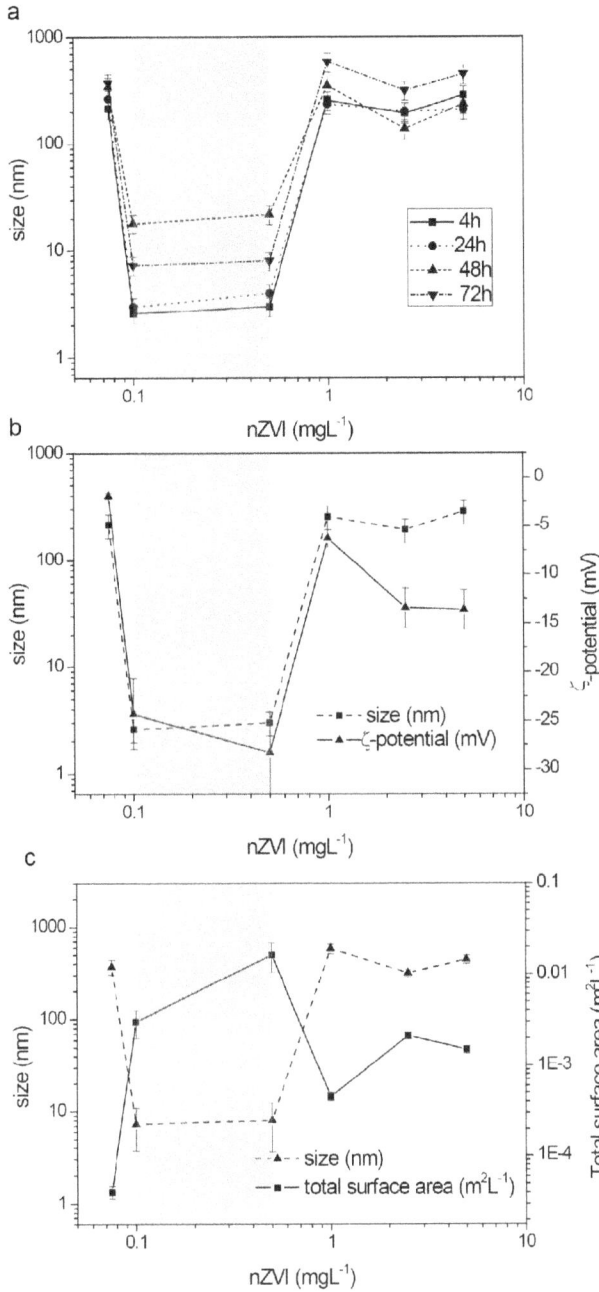

Figure 2. The occurrence of a *colloidal singularity*: Main physicochemical characteristics of colloidal nZVI fractions under experimental conditions. (a) Particle size of a range of nZVI concentrations at 4 h, 24 h, 48 h and 72 h. (b) ζ-potential and particle size for a representative exposure time (4 h). (c) Total surface area (m² L⁻¹) and particle size for a representative exposure time (4 h). The concentration range in which the *colloidal singularity* occurred is marked in blue.

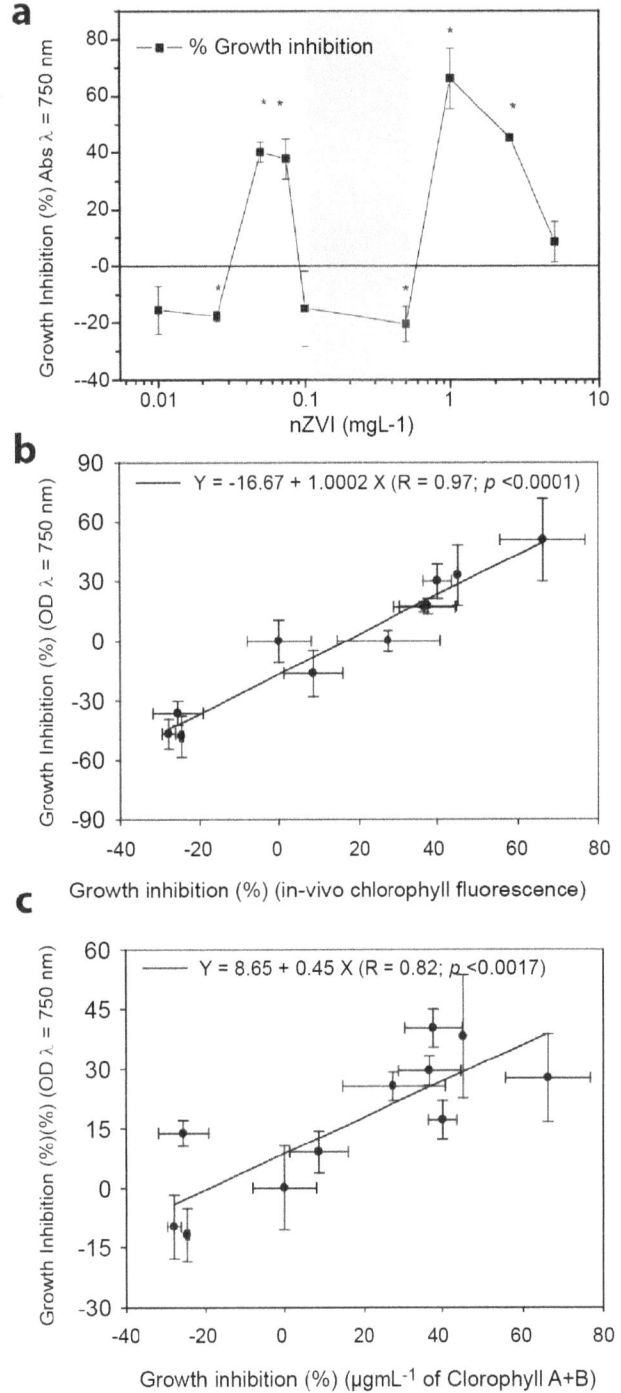

Figure 3. The *colloidal singularity* prevents nZVI toxicity. (a) Growth inhibition of *P. subcapitata* exposed to a linear nZVI dose gradient (0.025–5 mgL⁻¹) based on $OD_{\lambda = 750\,nm}$ biomass surrogate. Typically, growth rate of control replicates was 1.4 d⁻¹. The coefficient of variance for 72 h control cultures was 9%. Statistically significant differences ($p < 0.05$) are marked by an asterisk. (b,c) Correlations of biomass surrogates for growth inhibition of *P. subcapitata* exposed to an nZVI dose gradient (0.025–5 mgL⁻¹). The concentration range in which the *colloidal singularity* occurred is marked in blue.

alterations are mediated by direct attachment of nZVI particles to the outer cell envelopes of the microalgae, which was confirmed independently by TEM-XEDS and FTIR. To our knowledge, this is the first time that the mechanisms of toxicity of nZVI to a microalga have been elucidated. Keller et al. [35] described the toxicity of several commercial bare and polymer-coated nZVI to several marine and freshwater *microalgae*, including *P. subcapi-*

tata. However, the investigation of the underlying toxic mechanism was beyond the scope of their study. ROS mediated toxicity of nZVI is consistent with previous studies with other biological

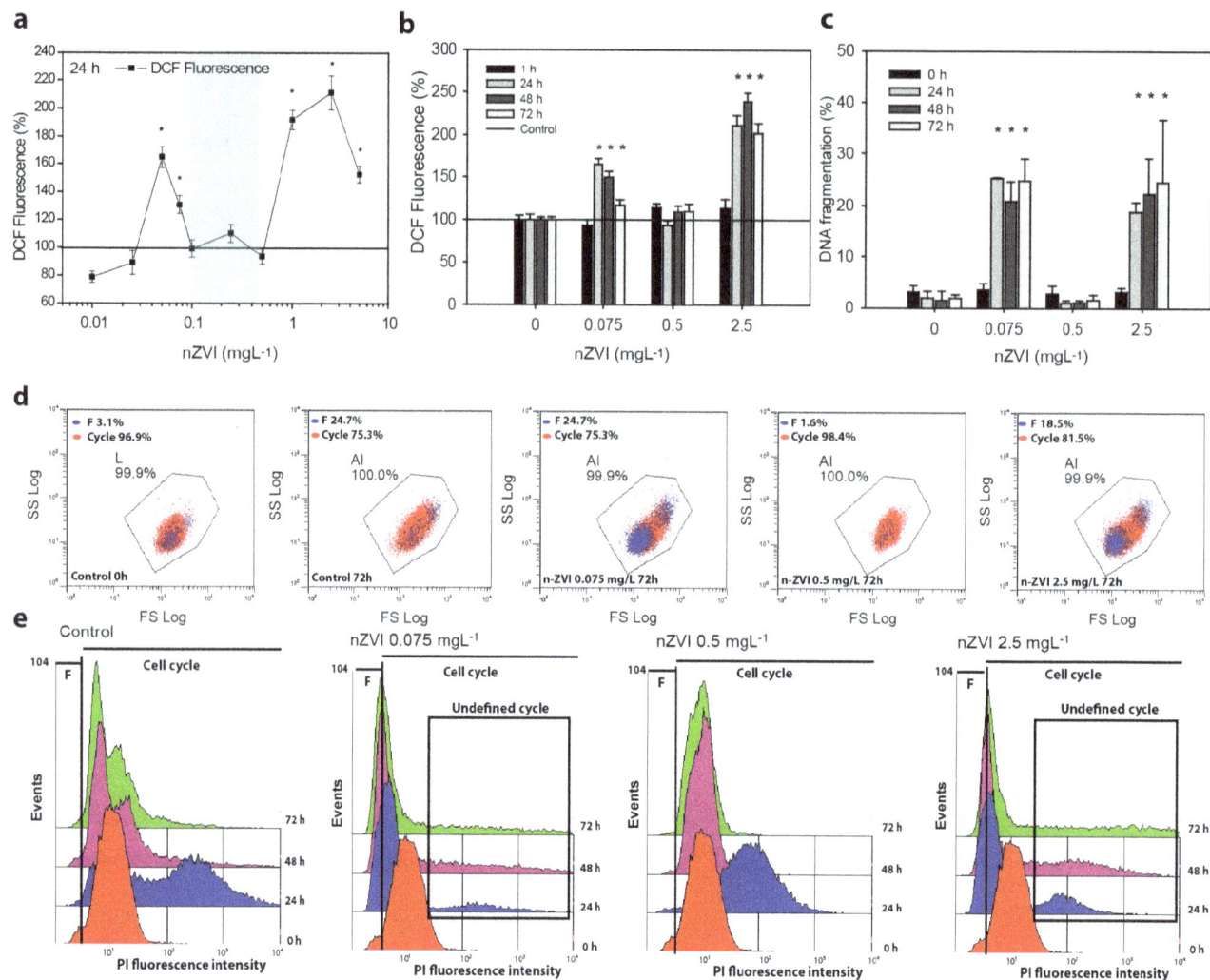

Figure 4. nZVI toxicity is consistently accompanied by ROS and cell cycle alterations. (a) Intracellular ROS formation as DCF fluorescence (% with respect to control levels) of *P. subcapitata* to an nZVI dose gradient (0.025–5 mgL^{-1}) at 24 h of exposure. The concentration range in which the *colloidal singularity* occurred is marked in blue. (b) Intracellular ROS formation as DCF fluorescence (% with respect to control levels) and (c) % of *P. subcapitata* cells showing DNA fragmentation, when exposed to representative nZVI concentrations (0, 0.075, 0.5, 2.5 mgL^{-1}) at relevant exposure times (0 h, 24 h, 48 h and 72 h). (d) Representative *FS-SS* distribution plots of *P. subcapitata* cells exposed to 0, 0.075, 0.5, 2.5 mgL^{-1} nZVI at relevant exposure times (0 h, 24 h, 48 h and 72 h). Red dots depict cells undergoing cell cycle progression, and blue dots depict cells showing DNA fragmentation (e) Representative cell cycle frequency histograms of *P. subcapitata* cells exposed to 0, 0.075, 0.5, 2.5 mgL^{-1} nZVI at relevant exposure times (0 h, 24 h, 48 h and 72 h). F: cell population containing less DNA than G0/G1 phase, which denoted a cell population showing DNA fragmentation. Black chart: cell population undergoing an abnormal cell cycle (poorly defined S and G2/M phases). Statistically significant differences ($p < 0.05$) are marked by asterisks.

model systems, including bacteria, fish and mammals [8,16,36–40]. Similarly, toxicity mediated by direct attachment of nZVI to the cell surface is also well documented for bacteria [8,16,38,39,41]. FTIR spectra found in *P. subcapitata* control cells markedly resemble those of cyclic sugars [42–44] such as those found in algal cells walls, mainly cellulose, hemicellulose and pectins. The alterations in those peak vibrations when nZVI presented toxicity may confirm a direct alteration of the main chemical bonds of *P. subcapitata* cell wall. This suggests the presence of active binding sites of nZVI to *P. subcapitata* cell wall. The latter may imply that nZVI is able to induce intracellular specific toxic responses by damaging the structure and changing the physicochemical properties of surface biomolecules [45]. Jiang et al. [45] found that oxide nanoparticles (Al$_2$O$_3$, TiO$_2$ and ZnO) induced chemical and structural changes on bacterial cell

envelopes, and that toxicity correlated with the severity of those changes. Similarly, other authors have found direct bond-interaction of oxide nanoparticles with algal and bacterial cell walls mediating toxicity and ROS [46–52]. The work by Gong et al. [46] on the toxicity of NiO nanoparticles to the microalgae *C. vulgaris* is especially interesting. They found that toxicity of NiO nanoparticles was mediated by hetero-aggregation. Furthermore, they found a decrease in the NiO signal and the appearance of a Ni° peak in the X-Ray diffraction spectra of NiO nanoparticles attached to the cell walls. This evidenced that a redox reaction was occurring between the NiO surface and the cell walls, resulting in the reduction of Ni (II) to Ni° [46]. The counterpart of the reduction is the oxidation of surface biomolecules of the organism resulting in chemical bonding. Our FTIR results and the fact that nZVI is covered by layers of iron oxy-hydroxides under oxidant

a

b

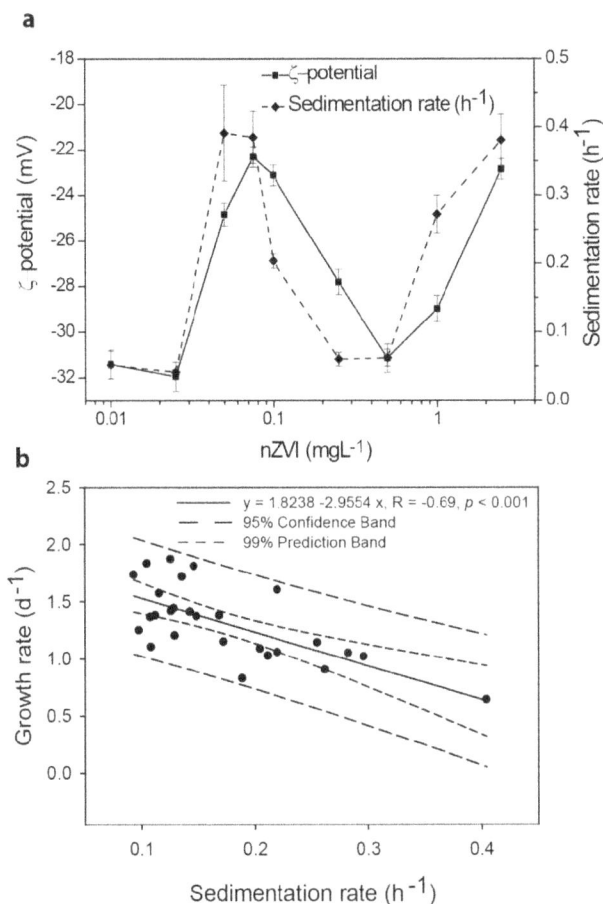

Figure 5. Toxicity of nZVI is mediated by colloidal destabilization and hetero-aggregation between algal cells and nZVI. (a) Sedimentation rates/ζ-potential of test systems exposed to increasing concentrations of nZVI. (b) Correlation of sedimentation rates *vs* algal growth rates.

conditions (Figure **S1**) [17], led us to think that the ROS production mechanism proposed for nZVI, *i.e.*, a Fenton/Haber–Weiss reactions [37–40], would not be the main mechanism of action responsible of ROS induction in *P. subcapitata*. We hypothesize that the spontaneous bio-reduction of some of the surface iron oxy-hydroxides by microalgal surface biomolecules, might be the origin of the attachment of nZVI to *P. subcapitata*. This may result in concomitant ROS induction and cell cycle alterations which is in agreement with the oxidative stress paradigm of redox cycling at the bio-interface [53]. However, this hypothesis requires further investigation.

From our results, it can be inferred that colloidal stability is a crucial regulator of the interaction of negatively charged ENMs with bio-interfaces. In this regard, there is a huge number of published research reporting toxicity of a variety of negatively charged metallic and carbonaceous ENMs on aquatic organisms [8,10,51,52]. A relevant part reported hetero-aggregation [8,10,15,51,52], despite the known negative surface charge of living cells [8,9]. This suggests a key role of negative-negative colloidal interactions in the fate and biological effects of ENMs in aquatic systems, which may be especially relevant for *microalgae* and other aquatic microorganisms in the "colloidal" size region. From our results, surface charge seems to play a major role in these negative-negative interactions because an increase (in

absolute value) of nearly 10 mV (from −15 mV to −25 mV) seems to be enough to prevent hetero-aggregation. Similarly, a reduction from −25 mV to −14 mV was enough to induce it. However, would all ENMs and organisms have the same intrinsic tendency to generate hetero-aggregation? Logic seems to indicate that probably not. However, the rules governing the interactions between negatively charged colloids and organism interfaces remain unclear [8], and at present, no systematic studies concerning this issue have been reported.

From a basic methodological view-point, our findings have some broader implications for EHS evaluation of ENMs *in-vitro*. Firstly, when performing EHS screening [5], the need to achieve stable ENM suspensions, by stabilizing agents, polymers, etc., may yield false negative bioactivity results [5] if the ENM suspensions under evaluation are excessively stabilized. Secondly, overlooking colloidal stability may have severe implications in basic aspects of EHS characterization of ENMs. For example, characteristic-dependent ENM bioactivity, when comparing different sizes, shapes, aspect ratio, redox states, etc., can be masked due to a heterogeneous degree of colloidal destabilization of the working suspensions, resulting in misleading conclusions.

Conclusions

The influence of physicochemical state on the bioactivity and toxicity of ENMs is presently under discussion. In the present work, when a linear nZVI dose-range was prepared for EHS characterization, a spontaneous speciation phenomenon occurred which resulted in a qualitatively different physicochemical state of nZVI along the dose gradient. We called it *colloidal singularity*. Basically, nZVI generally appeared in the form of aggregates. However, there was a narrow dose-range (the *colloidal singularity*), between 0.1–0.5 mgL^{-1} nZVI where it appeared highly stable (low particle size and increased ζ-potential). nZVI resulted in adverse biological effects when destabilized, but not when stable (the *colloidal singularity*). Furthermore, destabilization of those particular suspensions resulted in toxicity. Considering colloidal stability as a driver of bio-physical interactions allows us to explain our findings and previous apparently contradicting reports. In summary, our work demonstrates the role played by colloidal stability as a fundamental driver of ENM bioactivity, and opens up new perspectives on the relevant factors which may be considered in EHS and bioactivity testing of ENMs.

Experimental Section

nZVI and chemicals

All chemical reagents in the study were of reagent grade and used as purchased without further purification or pretreatment. Nano zero-valent iron (nZVI) powder was purchased from Quantum-Sphere (California, USA). The water used throughout the work was Milli-Q water. nZVI stock suspensions (25 mgL^{-1}) were prepared in Milli-Q water followed by a 10-minute sonication bath (Ultrason Selecta, Spain) to avoid particle agglomeration or aging. 25 mgL^{-1} of nZVI in Milli-Q water was selected as stock suspension since it had the lowest polydispersity index (PDI) and effective diameter (**Table 1**) and was sufficiently concentrated to be used as stock suspension for further EHS testing. Stock suspensions were always freshly prepared a few minutes prior to any physicochemical or biological experiment in order to ensure homogeneity of stock suspensions and prevent differences in aging and/or aggregation state throughout the different physicochemical and biological studies.

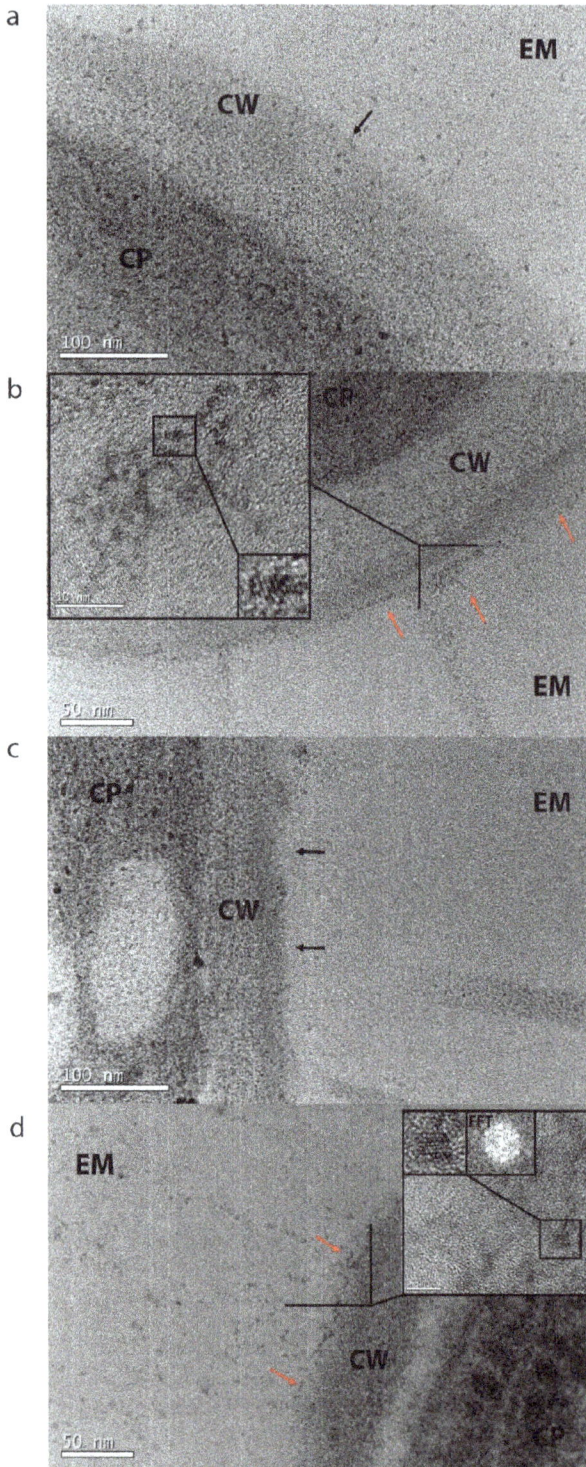

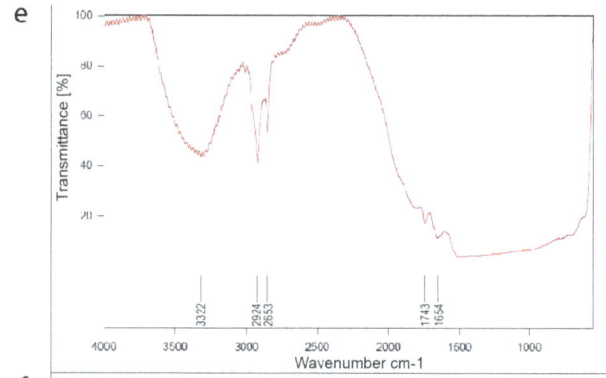

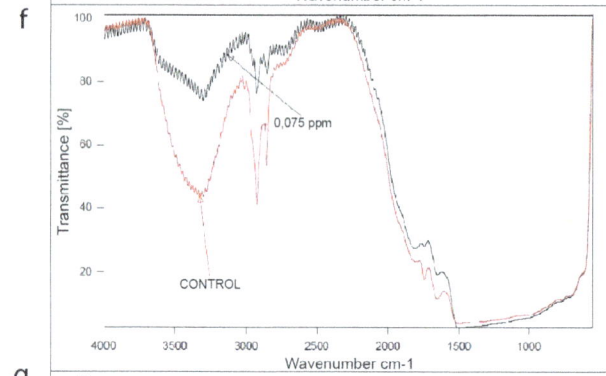

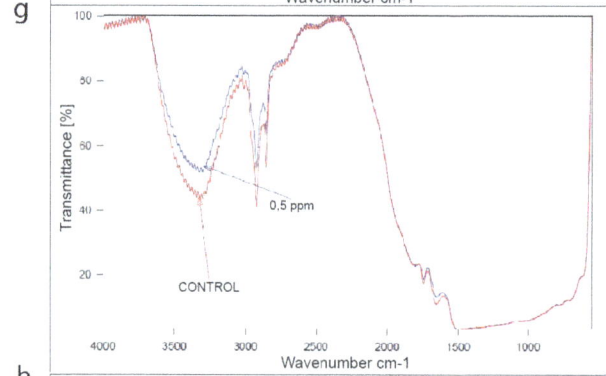

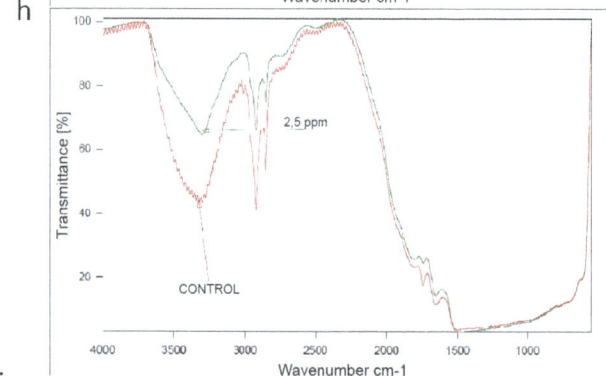

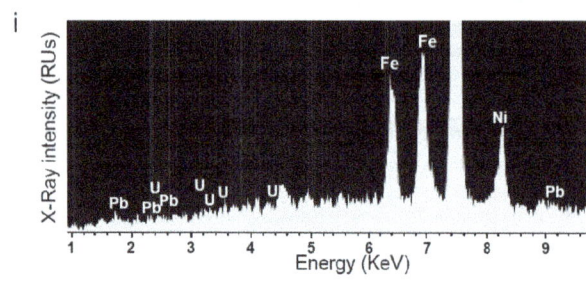

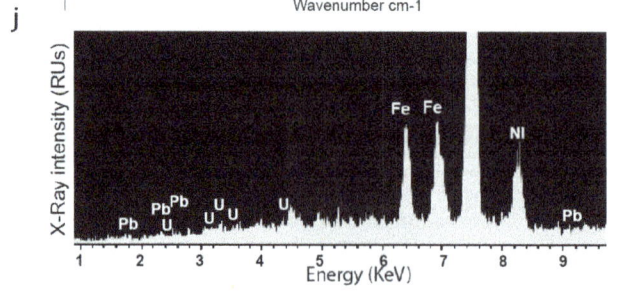

Figure 6. Tracking nZVI physicochemical interactions at the bio-nano interface. Representative TEM images showing the interfacial space delimiting algal cells and the external microenvironment of untreated *P. subcapitata* control cells (a), and *P. subcapitata* cells exposed to 0.075 mgL^{-1} (b), 0.5 mgL^{-1} (c) and 2.5 mgL^{-1} (d) of nZVI during 72 h. EM: external microenvironment, CW: Cell wall, CP: Cytoplasm. Black arrows show the CW-EM interface and red arrows the nZVI attached to the outer region of the algal CW in (b) and (d). Insets in (b) and (d) show CW-EM interfacial regions where surface-bound nZVI accumulations were observed on the CW and where nZVI particles were identified. nZVI particles showed round shapes and near squared lattices. Interplanar spaces (2.8 A° and 3.2 A°) are consistent with those of hematite and iron (III) oxide-hydroxide. Fast Fourier Transform (FFT) pattern of the nanoparticle in inset of (d) revealed a truncated octahedral tridimensional structure [56]. Representative FTIR transmission profile of untreated *P. subcapitata* control cells (e) and *P. subcapitata* cells exposed to 0.075 mgL^{-1} (f), 0.5 mgL^{-1}, (g) and 2.5 mgL^{-1} (h) of nZVI during 72 h. Wavenumber (cm^{-1}) of absorption peaks vibrations analyzed in the study are marked in (e). (i, j) XEDS spectra of the insets in b and d, respectively. Fe is the main constituent; Ni and Cu intense signals in the XEDS spectra (when present), are artifacts due to the metallic grid supporting the samples, and of the sample holder of the TEM instrument, respectively.

The current use of nZVI particles in the remediation of water resources contaminated with recalcitrant organic pollutants is based on *in situ* injection of large quantities (up to 50 g nZVI/L) into the subsurface waters [40]. Following removal from solution via sedimentation and dispersal with water flow, a nZVI concentration gradient would develop downward from the point of injection and much lower concentrations as one moves away from the point of injection. While it is still unclear what the actual nZVI concentrations are along such a gradient, in this study, we used what could reasonably be concentrations in the low end of such a gradient.

Physicochemical characterization of nZVI stock suspensions

The Nova 1200 BET (Brunauer–Emmett–Teller) was used to determine the surface area of pristine nZVI. Morphology properties were studied by Transmission Electron Microscopy (TEM) at 200 KeV operating voltage in a JEOL JEM 2100 electron microscope with 0.25 nm point resolution coupled with XEDS (X-Ray Energy Dispersive Spectroscopy). ENMs size measurements, interplanar spacing of crystalline planes and Fast Fourier Transform (FFT) power spectrums were analyzed using Digital Micrograph software, 2.30.542.0 (Gatan Inc.). The size distribution of nZVI was obtained using dynamic light scattering (DLS, Malvern Zetasizer Nano ZS). Measurements were conducted performing 500 rounds measurements and in particle number mode. ς-potential was measured via electrophoretic light scattering combined with phase analysis light scattering in the same instrument. ζ-potential was derived from electrophoretic mobility by applying the Henry equation and the Smoluchowski approximation [11]. Measurements were conducted at ambient temperature (25°C) in Milli-Q water and/or OECD TG 201 assay media without any modification. DLS measurements were performed in particle number mode. The count rates obtained in DLS and ς-potential measurements were higher than 15 kilo counts per second (kcps) for all the measurements included in analysis. The quality of data was also evaluated by visual observation of autocorrelation curves.

Physicochemical characterization of stable nZVI fractions during the toxicity experiments

The suspension stability of nZVI particles was measured over time (0, 4, 24, 48 and 72 h) by spectrophotometry (Hitachi U-2000 spectrophotometer, Japan). 250 mL of the desired nZVI concentrations were prepared in OECD TG 201 algal growth medium in 500 mL Erlenmayer flasks (chemical composition of OECD TG 201 algal growth medium can be found in **Table S2**.). Samples were maintained under identical experimental conditions as the bioassays. The temporal stability of nZVI particles was evaluated by measuring the residual absorbance at λ = 750 nm. 25 mL of samples were carefully taken from the supernatant of the test flasks to prevent catching sedimented non-colloidal nZVI fractions.

Absorbance measurements were performed in 100 mm light path spectrophotometric glass cuvettes (100-OS, Hellma, Germany). Main physicochemical properties of nZVI colloidal fractions were evaluated by measuring ζ-potential and DLS of the supernatant of the test flasks. Total surface area (SSA) was calculated accordingly to the equation 1 proposed by Macé et al. [54] assuming spherical particles:

$$SSA(m^2g^{-1}) = \frac{\pi d^2}{\rho \frac{\pi}{6} d^3} \qquad (1)$$

where ρ is the density of the solid particle (6700 Kg m^{-3} accordingly to producer) and d is the effective diameter of the particles.

Growth inhibition experiments using *P. Subcapitata*

The green alga *Pseudokirchneriella subcapitata* (*P. subcapitata*) was purchased from Microbiotests. Inc. (Denmark). *P. subcapitata* was routinely grown in 250 ml flasks at 28°C in light, ca. 65 μmol photons m^2 s^{-1} on a rotatory shaker at 135 rpm in OECD TG 201 standard algal culture medium [18] (**Table S2**.) (pH 8.2; Conductivity 0.214 μS/cm^2). Exposure experiments to nZVI suspensions were carried out in 12 mL of OECD TG 201 culture medium in 25 mL glass Erlenmeyer flasks. Before exposure to nZVI, cultures were washed once and re-suspended in fresh culture media to obtain a final optical density (OD$_{\lambda = 750 nm}$) of 0.1. nZVI was added to achieve the desired concentrations, and cultures were exposed for up to 72 h in a rotary shaker at 135 rpm and 28°C under constant illumination. Growth inhibition experiments were performed in triplicate with serial dilutions essentially as described in the standard OECD TG 201 [18]. Three biomass surrogates were measured: OD$_{\lambda = 750 nm}$, total chlorophyll content and *in-vivo* chlorophyll fluorescence. OD$_{\lambda = 750 nm}$ was determined by spectrophotometry in a Hitachi U-2000 spectrophotometer (Japan) by measuring the culture absorbance at λ = 750 nm in 10 mm light path transparent plastic cuvettes. For nZVI nominal concentrations lower than 5 mg/L no interference of nZVI was observed in the determination of OD$_{\lambda = 750 nm}$ of *P. subcapitata* (data not shown). For chlorophyll content determinations, culture aliquots (250 μL) were extracted in methanol at 4°C for 24 h in darkness. The total chlorophyll content of the extract (chlorophyll a+b) was determined by spectrophotometry as described elsewhere [52]. In-vivo fluorescence of chlorophyll was measured daily by transferring 100 μl of quadruplicate samples of cultures to an opaque black 96 well microtiter plate and by measuring fluorescence (485 nm/645 nm excitation/emission) on a Synergy HT multimode microplate reader (BioTek, USA).

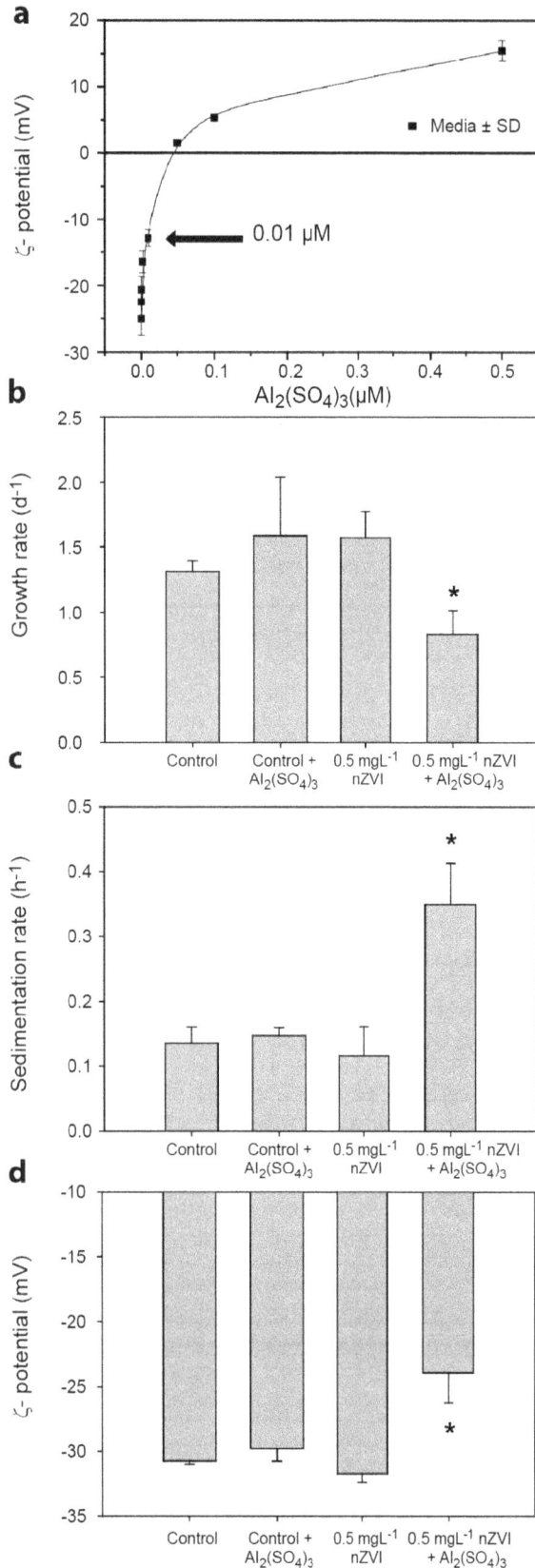

sedimentation rates and ζ-potential (respectively) of colloidal test systems of *P. subcapitata* exposed to 0.5 mgL^{-1} of nZVI in the presence or absence of 0.01 μM Al$_2$(SO$_4$)$_3$. Statistically significant differences ($p<$ 0.05) are marked by an asterisk.

Assessment of ROS generation

Intracellular reactive oxygen species (ROS) produced by *P. subcapitata* was assessed by using the fluorescent probe 2',7'-dichlorofluorescein diacetate (H$_2$DCFDA) (Invitrogen Molecular Probes; Eugene, OR, USA). The intracellular oxidation of H$_2$DCFDA generates 2,7-dichlorofluorescein (DCF), a fluorescent compound that serves as an indicator for hydrogen peroxide and other ROS, such as hydroxyl and peroxyl radicals. A 100 mM H$_2$DCFDA stock solution was freshly prepared in DMSO under dim light conditions to avoid degradation. Prior to analysis, the green alga was incubated for 30 min at room temperature (23°C) with a final concentration of 100 μM of H$_2$DCFDA. As a positive control for ROS formation, 3% H$_2$O$_2$ (v/v) was used. Fluorescence was monitored on a Synergy HT multimode microplate reader (BioTek,USA) with excitation and emission wavelengths of 488 and 530 nm, respectively. Results were normalized, for differences in cell numbers, by measuring chlorophyll content (as described previously) and expressed as arbitrary fluorescence units (AFU) per μg of chlorophyll a+b.

Cell cycle and cell population dynamics by flow cytometry

Alterations in cell cycle were assessed by propidium iodide (PI) staining and flow cytometry analysis using a Cytomix FL500 MPL flow cytometer (Beckman Coulter Inc., Fullerton, CA, USA). 1 mL of cell suspensions of the nZVI exposure experiments were collected in 1.5 mL eppendorf tubes and fixed by adding 3% paraformaldehyde (final concentration). Samples were kept in darkness at 4°C until further preparation. For PI staining, cells were gently washed with 1 mL PBS and were incubated for 3 h at 37°C in 1 mL of 0.05% Triton X-100, 0.24 mg/mL RNAase A (all from Sigma-Aldrich) in PBS buffer for membrane permeabilization. Cells were stained in 250 mL of staining buffer consisting of 0.05% Triton X-100, 0.24 mg/mL RNase A and 40 μg/mL PI and were incubated for 24 h in the dark at 4°C. Nuclear DNA content was then analysed using a Cytomics FC500 MPL flow cytometer with the MXP software (Beckman Coulter Inc., Fullerton, CA, USA) equipped with an argon-ion excitation wavelength (488 nm). The flow rate was set at 1 μL s^{-1} and at least 10,000 events (algal cells) were recorded. Light scattered by cells was collected at two angles: *FS* and *SS*, which were used in combination to distinguish between different cell populations. Non-algal particles were excluded from the analysis by setting an acquisition threshold value 1 for the FS parameter. Chlorophyll red autofluorescence was collected with a 675 nm long band pass filter (FL4), PI fluorescence was collected with a 575 nm long band pass filter (FL2), and the signal was adjusted to exclude the residual fluorescence signal coming from the chlorophyll red autofluorescence (FL4) (data not shown). Data acquisition was performed using MXP-2.2 software, and the analyses were performed using CXP-2.2 analysis software. Fluorescence was analysed in Log mode (chlorophyll fluorescence) and linear mode (PI fluorescence).

nZVI- algal interaction analyses by TEM microscopy and FTIR

For transmission electron microscopy (TEM) analysis, algal cell suspensions exposed to nZVI were collected by centrifugation using a swinging bucket rotor at low relative centrifugal forces

Figure 7. Colloidal destabilization reverts the *colloidal singularity* and induces nZVI toxicity. (a)Titration curve showing an exponential growth profile. Experimental ζ-potential of 0.5 mgL^{-1}+ nZVI Al$_2$(SO$_4$)$_3$ 0.01 μM is marked by an arrow. (b,c,d) Growth rates,

(RCFs = 1500 g) during 3 min in order to reduce the chance of artifacts [55], washed three times in phosphate buffer (0.1 M Na-phosphate, pH 7.2) to clean samples and remove unattached nZVI particles. After the final centrifugation step nearly 1 mm of cell pellet was formed. Supernatant was removed and 50 µl of 2% bacteriological ultrapure agar (CONDA, Spain) prepared in phosphate buffer was added to obtain 1 to 2 mm agar blocks. Cells were fixed in 3.1% freshly prepared glutaraldehyde in phosphate buffer (0.1 M Na-phosphate, pH 7.2) for 3 h at 4°C. Postfixation was in osmium tetroxide 1% prepared in phosphate buffer (0.1 M Na-phosphate, pH 7.2) for 2 h at 4°C. The samples were then gradually dehydrated in ethanol during 24 h and embedded in Durcupan resin (epoxy) (Fluka) sectioned (100 nm ultrathin films) in a Leica Reichert Ultracut S ultramicrotome and stained with uranyl acetate 2% in water. Samples were observed on Formvar/carbon Cu grids 300 mesh, at 200 KeV operating voltage in a JEOL JEM 2100 electron microscope with 0.25 nm point resolution coupled with XEDS (X-Ray Energy Dispersive Spectroscopy). All reagents used for TEM preparations were EM grade. Sample preparation for TEM microscopy involves centrifugation and washing of the cells, which may alter nZVI distribution on the algal cells [55]. Therefore, nZVI-cell interactions were further confirmed by IR spectroscopy. For Fourier Transformed Infrared (FTIR) analyses, 1 mL of exposed cells was centrifuged (5500 rpm, 5 min) and supernatant was removed. 5 µL of pelletized cells were transferred to 1 mm^2 glass cover-slips and were dried for 2 h at 25°C. Infrared spectra of the algal cell before and after nZVI contact experiments were obtained using a Bruker model IFs 66VFourier Transform Infrared (FTIR) spectrometer in transmission mode. This analysis allowed us to track changes in the chemical bonds and to elucidate the chemical groups involved in nZVI sorption processes.

Colloidal stability of nZVI-*algal* suspensions

Colloidal stability of nZVI-*algal* suspensions was evaluated by measuring ζ-potential, DLS and sedimentation rates of the colloidal system integrated by the algal cells and nZVI. For the determination of the sedimentation rates, a sedimentation system consisting of 15 mL falcon tubes was used. *Algae* exposed to the different nZVI concentrations were collected and $OD_{\lambda = 750\ nm}$ were determined by spectrophotometry (Hitachi U-2000 spectrophotometer, Japan). All colloidal systems were leveled to a homogeneous $OD_{\lambda = 750\ nm}$ of 0.350 with a final volume of 5 mL, therefore sedimentation rates were neither influenced by differences in initial optical densities nor test symmetry (mainly water column height). Residual turbidity ($OD_{\lambda = 750\ nm}$) of the test systems was measured spectrophotometrically as previously described by taking 250 µl of the near meniscus region of the test tubes at specific time intervals (0, 90, 180 min) and measuring the $OD_{\lambda = 750\ nm}$. Sedimentation rates were derived from the differences in residual turbidity of the test systems as a function of time according to equation 2.

$$\mu_{i-j}(h^{-1}) = \frac{\ln X_j - \ln X_i}{t_j - t_i} \qquad (2)$$

where μ_{i-j} is the average specific sedimentation rate (h^{-1}) from time i to j, X_j and X_i are the residual optical density at $\lambda = 750$ nm ($OD_{\lambda = 750\ nm}$) at times i to j; and t_j and t_i are time (h). ζ-potential of cells exposed to nZVI samples were measured at the end of the exposure experiments. For that, 2 mL of exposed cells were collected and transferred to 2.5 mL eppendorf tubes and ζ-

potential was measured directly via electrophoretic light scattering combined with phase analysis light scattering in a Malvern Zetasizer Nano ZS as describe above.

Colloidal destabilization of nZVI suspensions with Al$_2$(SO$_4$)$_3$

nZVI Destabilization experiments were performed in 15 mL Erlenmeyer flask containing 12 mL OECD TG 201 culture medium. Algal cells were exposed to 0.5 mgL^{-1} of freshly prepared nZVI as described in the toxicity exposure experiments. 0.01 µM of Al$_2$(SO$_4$)$_3$×18 H$_2$O (final concentration) was added to the test flask at the beginning of the experiments in order to induce colloidal destabilization and heteroaggregation of the nZVI-*algal* system. Growth inhibition, sedimentation rates and ζ-potential of the colloidal systems were measured as described above. The concentration of Al$_2$(SO$_4$)$_3$ used for the colloidal destabilization experiments was selected based on the results of a titration experiment of 0.5 mgL^{-1} of nZVI suspended in OECD TG 201 algal growth medium with increasing concentrations of Al$_2$(SO$_4$)$_3$ (**Figure 7.a**). 0.01 µM of Al$_2$(SO$_4$)$_3$ was selected due to its ability to partially neutralize the negative surface charges of nZVI (0.5 mgL^{-1}) without causing total neutralization or charge reversal.

Replication and Statistical Analysis

All the experimental procedures were performed at least in triplicate. The mean and standard deviations (SD) were calculated for each parameter. Statistical analyses were evaluated by R statistical program. A one-way ANOVA coupled with Student–Newman–Keuls *post-hoc* test was performed for comparing means of the toxicity effect and physicochemical characteristic among different nZVI concentrations. Statistically significant differences were considered to exist when $p < 0.05$.

Supporting Information

Figure S1　X-Ray diffraction (XRD) of pristine nZVI powder. X-ray diffraction analysis of pristine nZVI powder showed peaks of Fe$^°$ (Fe) and iron oxide formation of magnetite/maghemite (Fe$_3$O$_4$/γ-FeOOH) (M) respectively.

Table S1　Size of nZVI primary particles and aggregates by TEM microscopy.

Table S2　Chemical composition of OECD TG 201 standard algal culture medium.

Acknowledgments

Authors want to acknowledge research support services (SIDI) of Universidad Autónoma de Madrid and the Centro Nacional de Microscopía Electrónica (ICTS) for their excellent technical support in TEM, flow cytometry and FTIR analyses. Thanks to Dr. Alberto Martín Molina and Dr. Arturo Moncho Jordá for their helpful discussion in the interpretation of the colloidal stability data.

Author Contributions

Conceived and designed the experiments: IRP SG VL GPR. Performed the experiments: SG VL GPR IRP. Analyzed the data: IRP SG GPR VL. Contributed reagents/materials/analysis tools: FFP FL RR JCB. Wrote the paper: IRP FFP FL RR JCB EGC.

References

1. Schrurs F, Lison D (2012) Focusing the research efforts. Nat Nanotechnol 7: 546–548.
2. Thomas CR, George S, Horst AM, Ji Z, Miller RJ, et al. (2011) Nanomaterials in the Environment: From Materials to High-Throughput Screening to Organisms. ACS Nano 5: 13–20.
3. Westerhoff P, Nowack B (2013) Searching for Global Descriptors of Engineered Nanomaterial Fate and Transport in the Environment. Acc Chem Res 46: 844–853.
4. Handy RD, Cornelis G, Fernandes T, Tsyusko O, Decho A, et al. (2012) Ecotoxicity test methods for engineered nanomaterials: Practical experiences and recommendations from the bench. Environ Toxicol Chem 31: 15–31.
5. Nel A, Zhao Y, Madler L (2013) Environmental Health and Safety Considerations for Nanotechnology. Acc Chem Res 46: 605–606.
6. (2012) Join the dialogue. Nat Nanotechnol 7: 545.
7. Nel A, Xia T, Madler L, Li N (2006) Toxic Potential of Materials at the Nanolevel. Science 311: 622–627.
8. Ma S, Lin D (2013) The biophysicochemical interactions at the interfaces between nanoparticles and aquatic organisms: adsorption and internalization. Environ Sci Process Impacts 15: 145–160.
9. Zhu M, Nie G, Meng H, Xia T, Nel A, et al. (2013) Physicochemical Properties Determine Nanomaterial Cellular Uptake, Transport, and Fate. Acc Chem Res 46: 622–631.
10. Kahru A, Dubourguier HC (2010) From ecotoxicology to nanoecotoxicology. Toxicology 269: 105–119.
11. Gregory J (2006) Particles in Water Properties and Processes. Boca Raton (Florida): Taylor & Francis. 180 p.
12. Nel AE, Madler L, Velegol D, Xia T, Hoek EM, et al. (2009) Understanding biophysicochemical interactions at the nano-bio interface. Nat Mater 8: 543–557.
13. Verma A, Stellacci F (2010) Effect of surface properties on nanoparticle-cell interactions. Small 6: 12–21.
14. Ivask A, ElBadawy A, Kaweeteerawat C, Boren D, Fischer H, et al. (2013) Toxicity Mechanisms in Escherichia coli Vary for Silver Nanoparticles and Differ from Ionic Silver. ACS Nano 8: 374–386.
15. Hartmann NB, Engelbrekt C, Zhang J, Ulstrup J, Kusk KO, et al. (2013) The challenges of testing metal and metal oxide nanoparticles in algal bioassays: titanium dioxide and gold nanoparticles as case studies. Nanotoxicology 7: 1082–1094.
16. Lee C, Kim JY, Lee WI, Nelson KL, Yoon J, et al. (2008) Bactericidal effect of zero-valent iron nanoparticles on Escherichia coli. Environ Sci Technol 42: 4927–4933.
17. Sun Y-P, Li X-Q, Cao J, Zhang W-X, Wang HP (2006) Characterization of zero-valent iron nanoparticles. Adv Colloid Interface Sci 120: 47–56.
18. OECD (2006) OECD TG 201. Guidelines for the Testing of Chemicals. Freshwater Algal and Cyanobacteria, Growth Inhibition Test.
19. Gilbert B, Ono RK, Ching KA, Kim CS (2009) The effects of nanoparticle aggregation processes on aggregate structure and metal uptake. J Colloid Interface Sci 339: 285–295.
20. Phenrat T, Saleh N, Sirk K, Tilton RD, Lowry GV (2006) Aggregation and Sedimentation of Aqueous Nanoscale Zerovalent Iron Dispersions. Environ Sci Technol 41: 284–290.
21. Rosicka D, Sembera J (2011) Assessment of Influence of Magnetic Forces on Aggregation of Zero-valent Iron Nanoparticles. Nanoscale Res Lett 6: 10.
22. Bortner CD, Oldenburg NBE, Cidlowski JA (1995) The role of DNA fragmentation in apoptosis. Trends Cell Biol 5: 21–26.
23. Zhang JH, Xu M (2000) DNA fragmentation in apoptosis. Cell Res 10: 205–211.
24. Henderson RK, Parsons SA, Jefferson B (2010) The impact of differing cell and algogenic organic matter (AOM) characteristics on the coagulation and flotation of algae. Water Res 44: 3617–3624.
25. Long Z, Ji J, Yang K, Lin D, Wu F (2012) Systematic and quantitative investigation of the mechanism of carbon nanotubes' toxicity toward algae. Environ Sci Technol 46: 8458–8466.
26. Crews P, Rodriguez J, Jaspars M (1998) Organic Structure Analysis. University of California, Santa Cruz.
27. Pitre D, Boullemant A, Fortin C (2014) Uptake and sorption of aluminium and fluoride by four green algal species. Chem Cent J 8: 8.
28. Duan J, Gregory J (2003) Coagulation by hydrolysing metal salts. Adv Colloid Interface Sci 100–102: 475–502.
29. Rabolli V, Thomassen LC, Uwambayinema F, Martens JA, Lison D (2011) The cytotoxic activity of amorphous silica nanoparticles is mainly influenced by surface area and not by aggregation. Toxicol Lett 206: 197–203.
30. Van Hoecke K, De Schamphelaere KA, Van der Meeren P, Lucas S, Janssen CR (2008) Ecotoxicity of silica nanoparticles to the green alga Pseudokirchneriella subcapitata: importance of surface area. Environ Toxicol Chem 27: 1948–1957.
31. Lin D, Ji J, Long Z, Yang K, Wu F (2012) The influence of dissolved and surface-bound humic acid on the toxicity of TiO2 nanoparticles to Chlorella sp. Water Res 46: 4477–4487.
32. Stone V, Donaldson K (2006) Nanotoxicology: signs of stress. Nat Nanotechnol 1: 23–24.
33. Xia T, Kovochich M, Liong M, Madler L, Gilbert B, et al. (2008) Comparison of the mechanism of toxicity of zinc oxide and cerium oxide nanoparticles based on dissolution and oxidative stress properties. ACS Nano 2: 2121–2134.
34. Burhans WC, Heintz NH (2009) The cell cycle is a redox cycle: Linking phase-specific targets to cell fate. Free Radic Biol Med 47: 1282–1293.
35. Keller AA, Garner K, Miller RJ, Lenihan HS (2012) Toxicity of Nano-Zero Valent Iron to Freshwater and Marine Organisms. PloS one 7: e43983.
36. Li H, Zhou Q, Wu Y, Fu J, Wang T, et al. (2009) Effects of waterborne nano-iron on medaka (Oryzias latipes): antioxidant enzymatic activity, lipid peroxidation and histopathology. Ecotoxicol Environ Saf 72: 684–692.
37. Phenrat T, Long TC, Lowry GV, Veronesi B (2009) Partial oxidation ("aging") and surface modification decrease the toxicity of nanosized zerovalent iron. Environ Sci Technol 43: 195–200.
38. Sevcu A, El-Temsah YS, Joner EJ, Cernik M (2011) Oxidative stress induced in microorganisms by zero-valent iron nanoparticles. Microbes Environ 26: 271–281.
39. Auffan Ml, Achouak W, Rose JR, Roncato M-A, Chanéac C, et al. (2008) Relation between the Redox State of Iron-Based Nanoparticles and Their Cytotoxicity toward Escherichia coli. Environ Sci Technol 42: 6730–6735.
40. Grieger KD, Fjordoge A, Hartmann NB, Eriksson E, Bjerg PL, et al. (2010) Environmental benefits and risks of zero-valent iron nanoparticles (nZVI) for in situ remediation: Risk mitigation or trade-off? J Contam Hydrol 118: 165–183.
41. Li Z, Greden K, Alvarez PJ, Gregory KB, Lowry GV (2010) Adsorbed polymer and NOM limits adhesion and toxicity of nano scale zerovalent iron to E. coli. Environ Sci Technol 44: 3462–3467.
42. Garside P, Wyeth P (2003) Identification of Cellulosic Fibres by FTIR Spectroscopy: Thread and Single Fibre Analysis by Attenuated Total Reflectance. Stud Conserv 48: 269–275.
43. Hadjoudja S, Deluchat V, Baudu M (2010) Cell surface characterisation of Microcystis aeruginosa and Chlorella vulgaris. J Colloid Interface Sci 342: 293–299.
44. Lugo-Lugo V, Barrera-Diaz C, Urena-Nunez F, Bilyeu B, Linares-Hernandez I (2012) Biosorption of Cr(III) and Fe(III) in single and binary systems onto pretreated orange peel. J Environ Manage 112: 120–127.
45. Jiang W, Yang K, Vachet RW, Xing B (2010) Interaction between Oxide Nanoparticles and Biomolecules of the Bacterial Cell Envelope As Examined by Infrared Spectroscopy. Langmuir 26: 18071–18077.
46. Gong N, Shao K, Feng W, Lin Z, Liang C, et al. (2011) Biotoxicity of nickel oxide nanoparticles and bio-remediation by microalgae Chlorella vulgaris. Chemosphere 83: 510–516.
47. Khan SS, Mukherjee A, Chandrasekaran N (2011) Studies on interaction of colloidal silver nanoparticles (SNPs) with five different bacterial species. Colloids Surf B Biointerfaces 87: 129–138.
48. Khan SS, Srivatsan P, Vaishnavi N, Mukherjee A, Chandrasekaran N (2011) Interaction of silver nanoparticles (SNPs) with bacterial extracellular proteins (ECPs) and its adsorption isotherms and kinetics. J Hazard Mater 192: 299–306.
49. Pakrashi S, Dalai S, T CP, Trivedi S, Myneni R, et al. (2013) Cytotoxicity of aluminium oxide nanoparticles towards fresh water algal isolate at low exposure concentrations. Aquat Toxicol 132–133: 34–45.
50. Sadiq IM, Dalai S, Chandrasekaran N, Mukherjee A (2011) Ecotoxicity study of titania (TiO(2)) NPs on two microalgae species: Scenedesmus sp. and Chlorella sp. Ecotoxicol Environ Saf 74: 1180–1187.
51. Rodea-Palomares I, Boltes K, Fernandez-Pinas F, Leganes F, Garcia-Calvo E, et al. (2011) Physicochemical characterization and ecotoxicological assessment of CeO2 nanoparticles using two aquatic microorganisms. Toxicol Sci 119: 135–145.
52. Rodea-Palomares I, Gonzalo S, Santiago-Morales J, Leganes F, Garcia-Calvo E, et al. (2012) An insight into the mechanisms of nanoceria toxicity in aquatic photosynthetic organisms. Aquat Toxicol 122–123: 133–143.
53. von Moos N, Slaveykova VI (2014) Oxidative stress induced by inorganic nanoparticles in bacteria and aquatic microalgae – state of the art and knowledge gaps. Nanotoxicology 8: 605–630.
54. Macé C, Desrocher S, Gheorghiu F, Kane A, Pupeza M, et al. (2006) Nanotechnology and groundwater remediation: A step forward in technology understanding. Remediation Journal 16: 23–33.
55. Schrand AM, Schlager JJ, Dai L, Hussain SM (2010) Preparation of cells for assessing ultrastructural localization of nanoparticles with transmission electron microscopy. Nat Protoc 5: 744–757.
56. Zheng RK, Gu H, Xu B, Fung KK, Zhang XX, et al. (2006) Self-Assembly and Self-Orientation of Truncated Octahedral Magnetite Nanocrystals. Adv Mater 18: 2418–2421.

18

Cadmium-Induced Hydrogen Sulfide Synthesis Is Involved in Cadmium Tolerance in *Medicago sativa* by Reestablishment of Reduced (Homo)glutathione and Reactive Oxygen Species Homeostases

Weiti Cui[1], Huiping Chen[2], Kaikai Zhu[1], Qijiang Jin[1], Yanjie Xie[1], Jin Cui[1], Yan Xia[1], Jing Zhang[1], Wenbiao Shen[1]*

1 College of Life Sciences, Laboratory Center of Life Sciences, Nanjing Agricultural University, Jiangsu Province, Nanjing, China, **2** Key Laboratory of Protection and Development Utilization of Tropical Crop Germplasm Resources, Hainan University, Haikou, China

Abstract

Until now, physiological mechanisms and downstream targets responsible for the cadmium (Cd) tolerance mediated by endogenous hydrogen sulfide (H_2S) have been elusive. To address this gap, a combination of pharmacological, histochemical, biochemical and molecular approaches was applied. The perturbation of reduced (homo)glutathione homeostasis and increased H_2S production as well as the activation of two H_2S-synthetic enzymes activities, including L-cysteine desulfhydrase (LCD) and D-cysteine desulfhydrase (DCD), in alfalfa seedling roots were early responses to the exposure of Cd. The application of H_2S donor sodium hydrosulfide (NaHS), not only mimicked intracellular H_2S production triggered by Cd, but also alleviated Cd toxicity in a H_2S-dependent fashion. By contrast, the inhibition of H_2S production caused by the application of its synthetic inhibitor blocked NaHS-induced Cd tolerance, and destroyed reduced (homo)glutathione and reactive oxygen species (ROS) homeostases. Above mentioned inhibitory responses were further rescued by exogenously applied glutathione (GSH). Meanwhile, NaHS responses were sensitive to a (homo)glutathione synthetic inhibitor, but reversed by the cotreatment with GSH. The possible involvement of cyclic AMP (cAMP) signaling in NaHS responses was also suggested. In summary, LCD/DCD-mediated H_2S might be an important signaling molecule in the enhancement of Cd toxicity in alfalfa seedlings mainly by governing reduced (homo)glutathione and ROS homeostases.

Editor: Ji-Hong Liu, Key Laboratory of Horticultural Plant Biology (MOE), China

Funding: This research was supported by the Fundamental Research Funds for the Central Universities (KYZ201316), the National Natural Science Foundation of China (grants no. 30971711, J1210056, J1310015), and the Priority Academic Program Development of Jiangsu Higher Education Institutions. The funders had no role in study design, data collection and analysis, decision to publish, or preparation of the manuscript.

Competing Interests: The authors have declared that no competing interests exist.

* Email: wbshenh@njau.edu.cn

Introduction

Cadmium (Cd) contamination is a non-reversible accumulation process, with the estimated half-life and high plant-soil mobility, thus resulting in a serious threat to human health through food chains. Normally, Cd exposure leads to the inhibition of plant growth, decrease of crop yield, and even plant cell death [1,2]. Indirectly stimulated generation of reactive oxygen species (ROS) that modify the antioxidant defence and bring out oxidative stress is ascribed to one of the Cd toxicities in plants, and therefore lipid peroxidation is considered as a hallmark of Cd exposure [3].

In plants, there are a lot of antioxidant defence mechanisms, which could keep the normally formed ROS at a low level and prevent them from exceeding toxic thresholds [3,4]. The glutathione (GSH) and ascorbate were subsequently recognized as the heart of the redox hub [5]. In plants, GSH is synthesized by two ATP-dependent steps: γ-glutamylcysteine (γ-EC) is synthesized from L-glutamate and L-cysteine by γ-glutamyl cysteine

synthetase (γ-ECS, also called as γ-GCS); and the second step, glycine is conjunct to γ-EC by glutathione synthetase (GS) [6,7]. In soybean and alfalfa plants, GSH homolog homoglutathione (hGSH) synthesized by homoglutathione synthetase (hGS) from β-alanine and γ-EC, is more abundant than GSH [8]. The rate of glutathione reductase (GR) reaction was the same with either oxidized glutathione (GSSG) or oxidized homoglutathione (hGSSGh) as the substrate [7]. Upon Cd exposure, it was confirmed that the rapid accumulation of peroxides and depletion of GSH and hGSH causes redox imbalance in *Medicago sativa* [9]. Subsequent experiments with comparing ten pea genotypes showing that, activities of ascorbate peroxidase (APX) decreased, but concentrations of GSH increased in the less Cd-sensitive genotypes [10].

Another sulphur-containing compound, hydrogen sulfide (H_2S), previously known as a toxic gas, has been progressively recognized as a gaseous signaling molecule with multiple functions in animals [11,12]. For example, H_2S has been revealed as a cytoprotectant and a regulator in various biological processes, such as oxidative stress suppression, smooth muscle relaxation, proliferation inhibition and apoptosis triggering [13–16]. Meanwhile, although previous reports observed that many plants can emit H_2S [17–19], there have been few studies on the physiological role of H_2S in *planta* during the last century.

In mammals, the majority of endogenous H_2S was produced by two enzymes, cystathionine β-synthase (CBS, EC 4.2.1.22) and cystathionine γ-lyase (CSE, EC 4.4.1.1), from L-cysteine [20]. Cysteine-degrading enzymes such as cysteine desulfhydrases are hypothesized to be involved in H_2S release in plants [21]. Previously, two specific desulfhydrases, L-cysteine desulfhydrase (LCD, EC 4.4.1.1; also called L-CDes or L-DES) and D-cysteine desulfhydrase (DCD, EC 4.4.1.15; also called D-CDes or D-DES), have been isolated and partially analyzed from *Arabidopsis thaliana* [22–24]. The LCD, which is considered as the most important enzyme with H_2S production in plants, shares a 100% sequence homolog with CSE in mammals [25]. By using sodium hydrosulfide (NaHS) as a H_2S donor, ample evidence further suggested that H_2S can protect plants against various stress-induced damage, such as salinity stress [26], drought [27–29], heavy metal exposure [30,31], and heat shock [32]. Additionally, H_2S can act as an inducer in several developmental processes, including adventitious root formation [33] and flower senescence [34]. However, exogenously applied H_2S donor without checking the kinetics of H_2S synthesis including corresponding metabolic enzyme activities or transcripts, may not fully replicate the function of endogenous H_2S in plants.

Cyclic AMP (adenosine 3′, 5′-cyclic monophosphate, cAMP) is a well-known second messenger playing important roles in many physiological processes. The cAMP is synthesized by adenylyl cyclase and broken down by cNMP phosphodiesterase. Dedioxyadenosine (DDA) and 1,3-diazinane-2,4,5,6-tetrone (alloxan) are well characterized as the inhibitors of adenylyl cyclase. Likewise, cNMP phosphodiesterase is sensitive to the inhibitor 1-methyl-3-(2-methylpropyl)-7*H*-purine-2,6-dione (IBMX) [35,36]. In animals, there is ample evidences to show H_2S-activited cAMP level or H_2S-regulated cAMP homeostasis [37,38]. It was found that H_2S acted via cAMP-mediated PI3K/Akt/p70S6K signal pathways to inhibit hippocampal neuronal apoptosis and protect neurons from OGD/R-induced injury [39]. However, the functions of cAMP signaling in H_2S-alleviated Cd stress in plants are still poorly understood.

Thus, the aim of this study was to investigate the signaling role of endogenous H_2S in the tolerance of *Medicago sativa* seedlings to Cd stress. For this purpose, we preliminarily investigated the synthesis of endogenous H_2S under Cd stress, which has not been fully performed. Furthermore, the effects of H_2S on GSH and hGSH metabolism, as well as ROS homeostasis were checked. Our results further indicated that Cd stress triggered endogenous H_2S production catalyzed by LCD/DCD pathways, and the elevated H_2S acts as a signal improving the homeostasis of GSH pool and keeping ROS under control, both of which finally contributed to Cd tolerance. Finally, the possible involvement of cAMP signaling in NaHS responses was also suggested.

Materials and Methods

Plant material, growth condition

Commercially available alfalfa (*Medicago sativa* L. Victoria) seeds were surface-sterilized with 5% NaClO for 10 min, and rinsed extensively in distilled water before being germinated for 1 d at 25°C in the darkness. Uniform seedlings were then selected and transferred to the plastic chambers and cultured with nutrient medium (quarter-strength Hoagland's solution) in the illuminating incubator (14 h light with a light intensity of 200 $\mu mol \cdot m^{-2} \cdot s^{-1}$, 25±1°C, and 10 h dark, 23±1°C). Five-day-old seedlings were then incubated in quarter-strength Hoagland's solution with or without varying concentrations of NaHS (Sigma-Aldrich; St Louis, MO, USA) or the other indicated chemicals (2 mM DL-propargylglycine (PAG), 1 mM GSH, 1 mM L-buthionine-sulfoximine (BSO), 50 μM 8-Br-cAMP (8Br), 200 μM alloxan (All), 1 mM DDA, and 500 μM IBMX) alone, or the combination of treatments for 6 h followed by the indicated time points of incubation in 200 μM $CdCl_2$. Seedlings without chemicals were used as the control (Con). The pH for both nutrient medium and treatment solutions was adjusted to 6.0.

After various treatments, above-ground parts and root tissues of seedlings were sampled immediately or flash-frozen in liquid nitrogen, and stored at −80°C for further analysis. Among these, above-ground parts and root tissues of 240 seedlings were respectively used for the determination of Cd contents. Seedling root tissues were also used for fresh weight determination (10 seedlings), thiobarbituric acid reactive substances (TBARS) content determination (120 seedlings), and other indicated tests (30 seedlings).

Determination of H_2S content, LCD and DCD activity

Hydrogen sulfide content was determined according to the method previously reported [19,34]. 100 mg of alfalfa seedling roots from 30 seedlings were ground under liquid nitrogen and extracted by 1 ml phosphate buffered saline (50 mM, pH 6.8) containing 0.1 M EDTA and 0.2 M ascorbic acid. After centrifugation at 13000 g for 15 min at 4°C, 400 μl of the supernatant was injected to 200 μl 1% zinc acetate and 200 μl 1 N HCl. After 30 min reaction, 100 μl 5 mM dimethyl-*p*-phenylenediamine dissolved in 7 mM HCl was added to the trap followed by the injection of 100 μl 50 mM ferric ammonium sulfate in 200 mM HCl. After 15 min incubation at room temperature, the amount of H_2S was determined at 667 nm. Solutions with different concentrations of Na_2S were used in a calibration curve.

100 mg of alfalfa seedling roots from 30 seedlings were used for activity determination. The activities of LCD and DCD were determined as described by the methods previously reported [23,40]. L-cysteine desulfhydrase (LCD) activity was measured by the release of H_2S from L-cysteine in the presence of dithiothreitol (DTT). The formation of methylene blue was determined at 670 nm. To removal of the background, content of H_2S in the extracted protein solution was measured by same way with 50% trichloroacetic acid (TCA) instead of L-cysteine. The final LCD activity was calculated from the difference between the measured LCD activity and the background. D-cysteine desulfhydrase (DCD) activity was measured by the same method with following modifications: D-cysteine instead of L-cysteine, the pH of Tris-HCl was 8.0 rather than 9.0. Solutions with different concentrations of Na_2S were prepared, treated in the same way as the assay samples and were used for the quantification of enzymatically formed H_2S.

Determination of thiobarbituric acid reactive substances (TBARS), (h)GSH and (h)GSSG(h) contents

Lipid peroxidation was estimated by measuring the amount of TBARS as previously described [41]. About 400 mg of root tissues from 120 seedlings was ground in 0.25% 2-thiobarbituric acid (TBA) in 10% TCA using a mortar and pestle. After heating at 95°C for 30 min, the mixture was quickly cooled in an ice bath and centrifuged at 10,000×g for 10 min. The absorbance of the supernatant was read at 532 nm and corrected for unspecific turbidity by subtracting the absorbance at 600 nm. The concentration of lipid peroxides together with oxidatively modified proteins of plants were thus quantified in terms of TBARS amount using an extinction coefficient of 155 mM^{-1} cm^{-1} and expressed as nmol g^{-1} fresh weight (FW).

(h)GSH (GSH + hGSH) and (h)GSSG(h) (GSSG + hGSSGh) contents were measured by the 5,5′dithio-bis-(2-nitrobenzoic acid) (DTNB)-glutathione reductase (GR) recycling assay [41,42]. Frozen root tissues from 30 seedlings were homogenized in cold 5% 5-sulfosalicylic acid. The homogenate was centrifuged at 12,000×g for 20 min at 4°C and the supernatant was collected. Total glutathione ((h)GSH plus (h)GSSG(h)) was determined in the homogenates spectrophotometrically at 412 nm, using GR, DTNB, and NADPH. (h)GSSG(h) contents were determined by the same method in the presence of 2-vinylpyridine and (h)GSH contents were calculated from the difference between total glutathione and (h)GSSG(h).

Thiol analysis by reversed-phase HPLC

Low-molecular-weight thiols and their corresponding disulfides contents in root tissues from 30 seedlings were measured according to the methods previously reported [43–45], through derivatization with monobromobimane (mBBr) after reduction with DTT with or without previously blocked with N-ethylmaleimide (NEM), and separation by reversed-phase HPLC (Agilent Technologies, 1200 series Quaternary, Foster city, USA).

Histochemical analyses

Histochemical detection of lipid peroxidation and loss of plasma membrane integrity was performed with Schiff's reagent and with Evans blue described by previous reports [41,45].

Real-time quantitative RT-PCR analysis

Total RNA from root tissues of 30 seedlings was extracted using Trizol reagent (Invitrogen) according to the manufacturer's instructions. DNA-free total RNA (2 μg) from different treatments was used for first-strand cDNA synthesis in a 20-μL reaction volume containing 2.5 units of avian myeloblastosis virus reverse transcriptase XL (TaKaRa) and oligo dT primer.

Real-time quantitative RT-PCR reactions were performed with Mastercycler realplex^2 real-time PCR system (Eppendorf, Hamburg, Germany) using the SYBR *Premix Ex Taq* (TaKaRa) according to the user manual. The cDNA was amplified using primers (Table S1). The expression levels of the genes are presented as values relative to the corresponding control samples under the indicated conditions, with normalization of data to the geometic average of two internal control genes *MSC27* and *Actin2* [46].

Visualization of endogenous ROS by LSCM

Endogenous ROS was imaged using the fluorescent probe H$_2$DCFDA, and then scanned described by [45,47].

Statistical analysis

Values are means ± SD of three different experiments with three replicated measurements. Differences among treatments were analysed by one-way ANOVA, taking $P<0.05$ as significant according to Duncan's multiple range test.

Results

(h)GSH depletion and increased endogenous H$_2$S synthesis triggered by Cd stress

Considering alfalfa plants contain a thiol tripeptide homolog, hGSH, instead of or in addition to GSH [8,9], we detected the concentrations of GSH and hGSH. As shown in Table 1, the content of hGSH in alfalfa seedling roots under the control conditions, was about 8-fold higher than that of GSH. Similarly, hGSSGh is the main component of (h)GSSG(h) (total of hGSSGh and GSSG), because the GSSG content was almost negligible.

To further elucidate the correlation among GSH pool, H$_2$S and Cd tolerance, the time course of (homo)glutathione ((h)GSH; total of hGSH and GSH, and (h)GSSG(h)) contents, and H$_2$S synthesis were investigated in alfalfa seedling roots upon Cd stress. As expected, a decrease of (h)GSH content (especially hGSH) and an increase of (h)GSSH(h) (especially hGSSGh) level were progressively triggered by Cd stress within 12 h, thus leading to a decreased (h)GSH/(h)GSSH(h) ratio (12 h; Figure 1A-C), an important parameter for the intracellular redox status in *planta* upon Cd stress [3,45]. The ratio of hGSH/hGSSGh exhibited the similar tendency (Table 1). These results were consistent with the observed Cd toxicity, confirmed by the histochemical staining detecting the aggravated loss of plasma membrane integrity and lipid peroxidation with Evans blue and Schiff's reagent, increased TBARS content and growth stunt of seedling roots (Figure S1).

Because H$_2$S synthesis could be induced by oxidative stress and depletion of GSH both in animals and plants [48–50], we simultaneously investigated the production of H$_2$S in seedling roots after the exposure to Cd. Similar to the recent report [51], the production of H$_2$S was continuously increased after the exposure to Cd alone for 12 h (Figure 1D). The changes in activities of two H$_2$S synthetic enzymes LCD and DCD displayed similar tendencies (Figure 1E and F). Apparently, the reduced (homo)glutathione depletion and increased endogenous H$_2$S synthesis preceded Cd toxicity in alfalfa seedlings.

NaHS not only mimics intracellular H$_2$S content, but also alleviates Cd toxicity

Previous results revealed that the exogenously applied NaHS, a H$_2$S donor, alleviates Cd toxicity in bermudagrass seedlings [51]. Therefore, a preliminary work was carried out to compare the oxidative damage and growth performance of alfalfa seedlings upon Cd exposure with or without the indicated concentrations of NaHS pretreatment. Firstly, the results of histochemical staining and TBARS contents revealed that NaHS at 100 (in particular) and 500 μM was able to significantly decreased Cd-induced lipid peroxidation (Figure S1A and B). These beneficial roles were also supported by the changes of fresh weight of ten alfalfa seedling roots, showing that NaHS at 100 and 500 μM had the greatest effects on the alleviation of the inhibition of root growth caused by Cd stress (Figure S1C). The beneficial roles of 100 μM NaHS alone were also observed. Subsequent work confirmed that H$_2$S rather than other sulphur-containing derivatives and sodium exhibited the cytoprotective role in the improvement of Cd toxicity by using a series of sulphur- and sodium-containing chemicals including Na$_2$S, Na$_2$SO$_4$, Na$_2$SO$_3$, NaHSO$_4$, NaHSO$_3$, and NaAc, in comparison with the positive roles of NaHS (Figure S2).

Accordingly, we observed that the treatment with 100 μM NaHS for 3 h resulted in the enhancement of endogenous H$_2$S level in alfalfa seedling roots, which also mimicked a physiological response elicited by Cd alone for 12 h (Figure 2A). The addition of Cd to the NaHS-pretreated plants further strengthened the increased H$_2$S content. Therefore, 100 μM NaHS was used to mimic the physiological role of intracellular H$_2$S in the subsequent experiments.

Table 1. Concentrations of low molecular weight thiols and their disulfides, and hGSH/hGSSGh ratio in root tissues.

Treatment	cysteine (nmol g⁻¹ FW)	cysteine disulfide (nmol g⁻¹ FW)	γ-EC (nmol g⁻¹ FW)	γ-EC disulfide (nmol g⁻¹ FW)	GSH (nmol g⁻¹ FW)	GSSG (nmol g⁻¹ FW)	hGSH (nmol g⁻¹ FW)	hGSSGh (nmol g⁻¹ FW)	hGSH/hGSSGh
Con→Con	30±1 d	3.8±0.8 c	10±1 e	1.5±0.1	27±2 bc	0.2±1.9	252±16 b	28±2 c	8.86
Con→Cd	33±1 cd	5.7±0.8 b	14±2 d	1.7±0.1	21±2 c	0.2±1.4	112±13 f	33±1 bc	3.41
NaHS→Cd	40±2 c	4.0±0.6 c	18±2 bc	1.4±0.5	26±4 c	0.1±1.4	163±14 de	33±4 bc	4.89
NaHS→Con	34±2 cd	3.4±0.7 c	8±0 e	1.2±0.5	36±1 b	0.2±1.1	309±14 a	30±2 bc	10.23
NaHS + PAG→Cd	54±7 b	4.3±0.4 bc	21±3 b	1.2±0.5	29±5 bc	0.3±0.7	144±8 e	41±6 a	3.55
NaHS + PAG + GSH→Cd	65±8 a	4.7±1.6 bc	27±1 a	1.4±0.5	46±11 a	0.7±0.6	179±7 d	36±5 ab	4.91
PAG→Cd	52±6 b	7.4±0.7 a	21±1 b	1.6±0.2	29±1 bc	0.9±0.9	82±12 g	33±3 bc	2.41
PAG→Con	56±4 b	4.0±0.3 c	17±3 cd	1.2±0.5	29±2 bc	0.7±0.7	206±28 c	36±2 ab	5.67

Seedlings were pretreated with or without 100 μM NaHS, 2 mM PAG, 1 mM GSH, individual or combination for 6 h, and then exposed to 200 μM CdCl₂ for another 12 h. Values are means ± SD of three independent experiments with three replicates for each. Different letters within columns indicate significant differences ($P<0.05$) according to Duncan's multiple range test.

Changes of low molecular weight thiols and their disulfides as well as representative transcripts in response to NaHS

To determine the influence of H_2S at physiologically concentrations on (h)GSH depletion, GSH pool and corresponding metabolism associated genes were investigated. As shown in Figure 2B, the time-course analysis revealed that (h)GSH contents in seedling roots were significantly enhanced by the pretreatment with NaHS for 6 h, and remained high through 24 h of further incubation in the control solution. Meanwhile, NaHS pretreatment was able to slow down the decreased (h)GSH levels caused by Cd exposure. Changes of the (h)GSH/(h)GSSG(h) ratio also exhibited the similar tendencies (Figure 2C). Comparatively, Cd-induced cysteine and γ-EC (in particular), and cysteine disulfide contents were differentially strengthened or blocked by NaHS pretreatment, respectively (Table 1).

These results arises the question that, whether this increases in metabolites are, at least in part, duo to changes in the expression of genes involved in (h)GSH metabolism. Therefore, the expression of *ECS*, *GS*, and *GR1* genes, were analyzed by real-time RT-PCR. Results of Figure 3A and B revealed that the transcripts of *ECS*, *GS* and *GR1* (especially) in seedling roots approximately displayed a time-dependent increase during Cd stress for 24 h, while the transcriptional profiles of these genes in the control samples were relatively constant during the same period. The pretreatment with NaHS for 6 h in culture solution increased above transcripts, which were differentially strengthened by thereafter Cd stress.

NaHS-induced Cd tolerance, (h)GSH and ROS homeostases were sensitive to PAG, but rescued by GSH

To further verify the involvement of endogenous H_2S in Cd tolerance, DL-propargylglycine (PAG), an effective H_2S synthetic inhibitor [27], and GSH, applied individually and in combination, were used in the subsequent experiment. After 72 h exposure to Cd, the alfalfa seedlings displays severe growth inhibition both in roots and above ground parts, compared to control samples, both of which were improved by NaHS pretreatment (Figure 4A). By contrast, the improvement of seedling growth inhibition as well as the reestablishment of (h)GSH homeostasis triggered by NaHS were sensitive to PAG, but blocked by exogenously applied GSH (Figure 4, Figure S3A). An aggravated Cd toxicity in seedling growth inhibition was also observed when PAG was pretreated.

In an attempt to assess the potential role of endogenous H_2S in ROS homeostasis in Cd-stressed seedlings, ROS production was visualized by staining with H_2DCFDA. As expected, ROS in root tips with Cd alone were produced considerably, suggesting a perturbation in ROS homeostasis (Figure 5). However, the pretreatment with NaHS reduced the ROS abundance. Further results revealed that PAG pretreatment increased the H_2DCFDA fluorescence in Cd-stressed seedling roots, which was further blocked by the addition of GSH. The changes of TBARS content, an indictor of lipid peroxidation, exhibited the similar tendencies (Figure S3B).

Cd treatment caused the accumulation of Cd contents both in shoot and root (particularly) tissues (Figure S4). Similar to the previous reports [31], NaHS decreased Cd accumulation, which was significantly reversed by PAG, but was further blocked by the cotreatment with GSH.

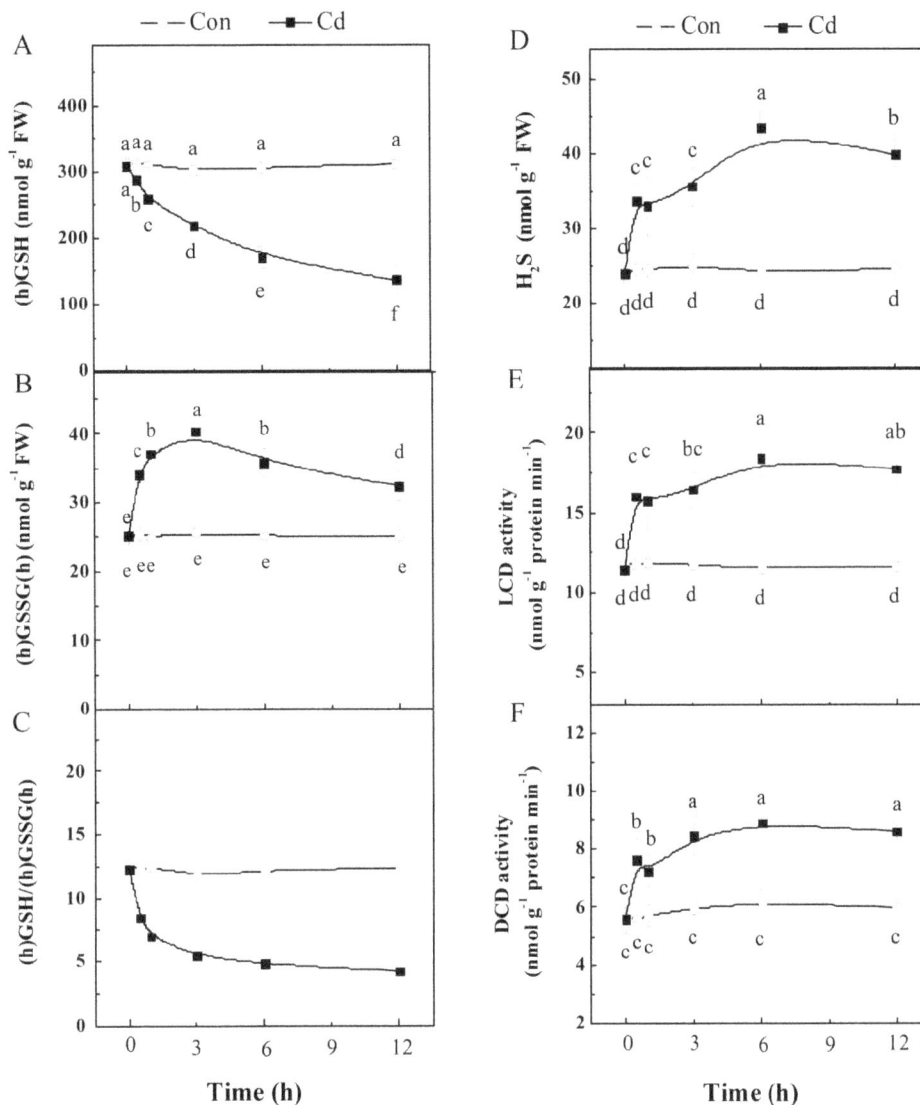

Figure 1. Time course changes of GSH pool and H₂S synthesis upon Cd stress. Upon 200 μM CdCl₂ treatment for 12 h, contents of (h)GSH (A), (h)GSSG(h) (B) and H₂S (D), the ratio of (h)GSH/(h)GSSG(h) (C), and the activities of LCD (E) and DCD (F) in root tissues were analyzed. Values are means ± SD of three independent experiments with three replicates for each. Bars denoted by the same letter did not differ significantly at $P<0.05$ according to Duncan's multiple range test.

Transcripts of representative antioxidant defense genes were sensitive to PAG, but rescued by GSH

Since ROS homeostasis was reestablished by NaHS in stressed conditions, the real-time RT-PCR test of corresponding genes involved in their metabolism, i.e. *Cu*, *Zn-SOD*, *APX1*, and *GPX* [3,5], were analysed. The results of Figure 6 revealed that in comparison with Cd alone samples, NaHS pretreatment followed by Cd exposure resulted in the enhancement in the transcript levels of *Cu*, *Zn-SOD*, *APX1*, and *GPX* in alfalfa seedling roots. The addition of PAG, however, significantly blocked the increases in the transcripts levels of these representative antioxidant enzymes induced by NaHS, all of which were reversed when GSH was added together with PAG.

NaHS responses were sensitive to a (h)GSH synthetic inhibitor, but reversed by the added GSH

The involvement of (h)GSH homeostasis in NaHS-induced cytoprotective against Cd stress were further investigated using a (h)GSH synthetic inhibitor and GSH applied exogenously. Pretreatment with NaHS, and ʟ-buthionine-sulfoximine (BSO) at 1 mM, a concentration expected to be effective [52], exhibited an aggravated Cd toxicity, which was confirmed by the severe growth stunt and TBARS overproduction, in comparison with Cd plus NaHS (Figure 7A and B). Similarly, NaHS-mediated reestablishment of (h)GSH homeostasis in Cd stressed alfalfa seedling roots was also perturbed by BSO (Figure 7C and D), which was confirmed by the significant decreased (h)GSH content and the ratio of (h)GSH/(h)GSSG(h), respect to Cd alone. By contrast, above BSO responses were sensitive to the addition of GSH when

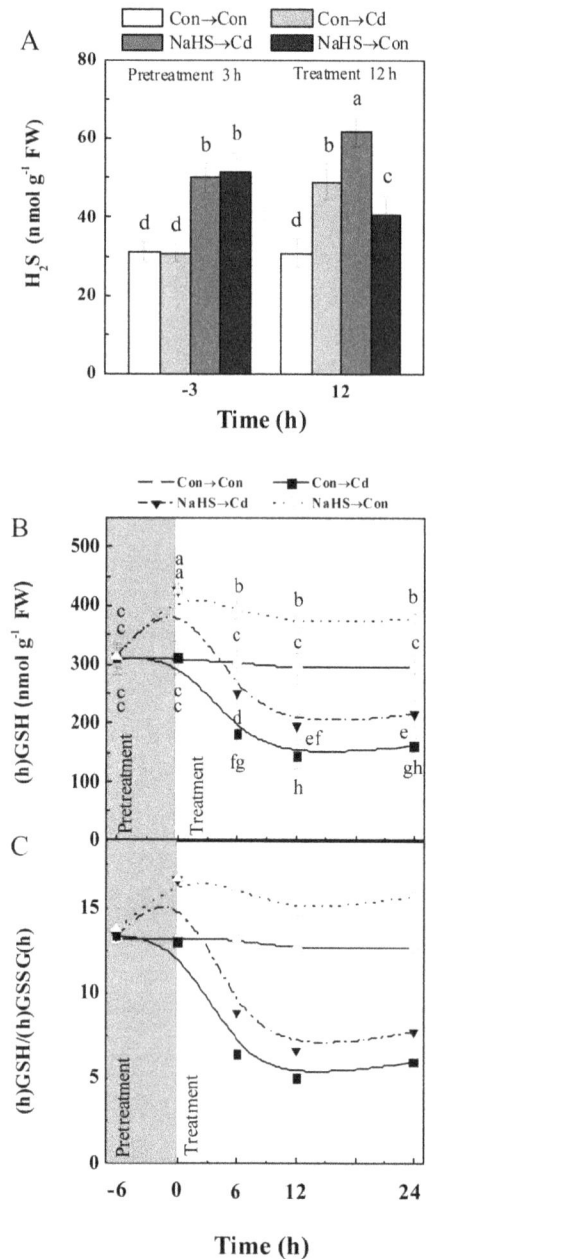

Figure 2. NaHS increased endogenous H₂S and (h)GSH contents, and the ratio of (h)GSH/(h)GSSG(h) upon Cd stress. Endogenous H_2S concentration in root tissues (A) was detected at 3 h after the beginning of 100 μM NaHS pretreatment (−3 h), and 200 μM $CdCl_2$ or chemical-free control treatments for 12 h (12 h). Meanwhile, contents of (h)GSH (B) and the ratio of (h)GSH/(h)GSSG(h) (C) in root tissues were detected at the indicated time points of treatments. Values are means ± SD of three independent experiments with three replicates for each. Bars denoted by the same letter did not differ significantly at $P<0.05$ according to Duncan's multiple range test.

Figure 3. Time course of transcripts responsible for (h)GSH metabolism regulated by NaHS and Cd. Seedlings were pretreated with or without 100 μM NaHS for 6 h and then exposed to 200 μM $CdCl_2$ for another 24 h. The expression levels of *ECS* (A), *GS* (B) and *GR1* (C) in root tissues analyzed by real-time RT-PCR are presented as values relative to the control at the beginning of pretreatment, normalized against expression of two internal reference genes in each sample. Values are means ± SD of three independent experiments with three replicates for each.

applied together. Above results clearly indicated a requirement for (h)GSH homeostasis in NaHS-mediated alleviation of Cd toxicity.

cAMP signaling might be involved in NaHS responses

To testify the hypothesis that H₂S response is associated with cAMP signaling pathway, a pharmacological approach was used to manipulate endogenous cAMP. Results presented in Figure 8A

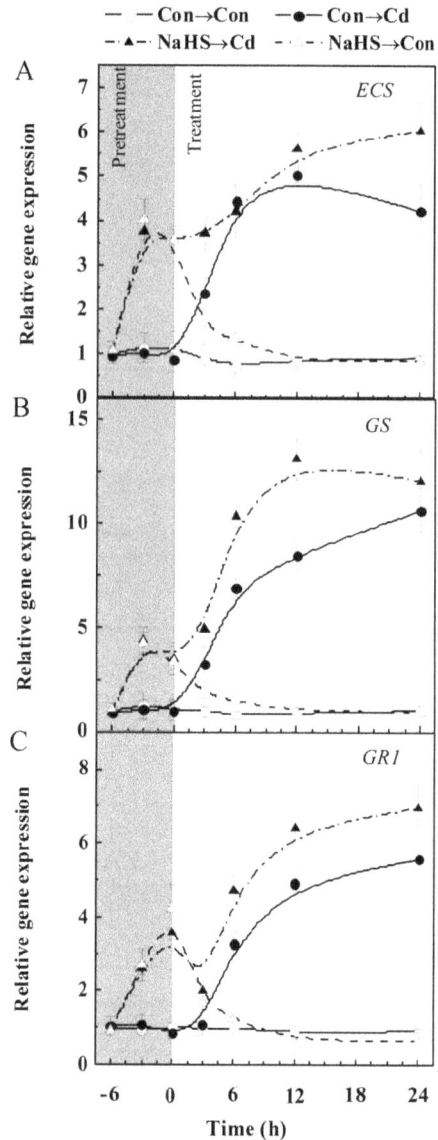

and B indicated that the pretreatment with 8-Br-cAMP, a membrane-permeable analogue of cAMP, alleviated Cd-induced decrease of fresh weight and increase of TBARS content in alfalfa seedling roots. Both of two adenylyl cyclase inhibitors, alloxan and DDA, blocked NaHS-alleviated Cd stress. Moreover, similar to the beneficial actions of 8-Br-cAMP (when was cotreated with PAG followed by Cd stress), a cNMP phosphodiesterase inhibitor IBMX also reversed the PAG responses in the aggravation of fresh weight loss and lipid peroxidation caused by Cd stress. Results from the real-time RT-PCR showed that 8-Br-cAMP and IBMX pretreatments followed by Cd stress, mimicked the effect of NaHS on *GR1* up-regulation, regardless of whether PAG was added or

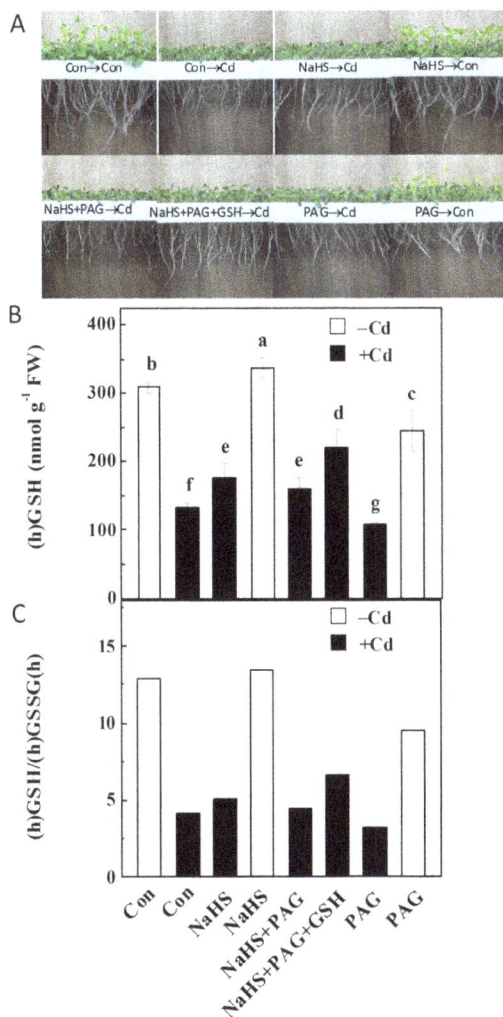

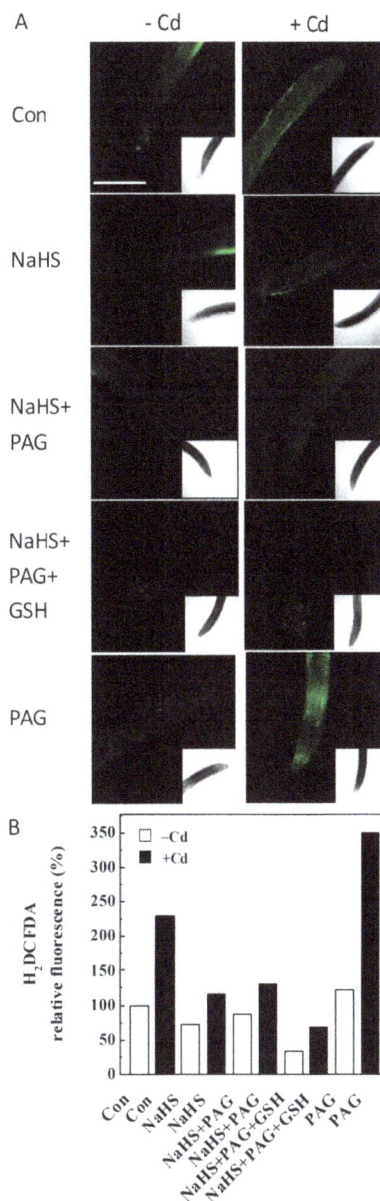

Figure 4. NaHS, PAG and GSH pretreatments differentially regulated seedling growth, (h)GSH content, and (h)GSH/(h)GSSG(h) ratio. Corresponding phenotypes were photographed after 200 μM CdCl$_2$ treatment for 72 h, with or without 100 μM NaHS, 2 mM PAG, 1 mM GSH, individual or combination pretreatments for 6 h (A). Scale bar, 2 cm. Contents of (h)GSH (B), and the ratio of (h)GSH/(h)GSSG(h) (C) in root tissues were also analyzed after 200 μM CdCl$_2$ treatment for 12 h, with or without 100 μM NaHS, 2 mM PAG, 1 mM GSH, individual or combination pretreatment for 6 h. Values are means ± SD of three independent experiments with three replicates for each. Bars denoted by the same letter did not differ significantly at $P<0.05$ according to Duncan's multiple range test.

Figure 5. NaHS and GSH pretreatments alleviated Cd-induced ROS production, but blocked by PAG. LSCM results (A). Seedlings were pretreated with or without 100 μM NaHS, 2 mM PAG, 1 mM GSH, individual or combination for 6 h, and then exposed to 200 μM CdCl$_2$ for another 6 h. After various treatments, the roots were respectively stained with H$_2$DCFDA, then washed thoroughly to removal extra dye and immediately photographed by LSCM. Scale bar, 0.5 mm. The relative DCF fluorescence intensity in the corresponding roots (B).

not (Figure 8C). Two inhibitors alloxan and DDA partially blocked NaHS plus Cd-induced *GR1* transcripts. A similar tendency was found in the changes in *GPX* transcripts (Figure 8F). Results presented in Figure 8D and E further revealed the negative effects of adenylyl cyclase inhibitors on the transcripts of *Cu, Zn-SOD* and *APX1* in NaHS-pretreated seedling roots upon Cd, in comparison with the positive responses of 8-Br-cAMP and IBMX in the presence or absence of PAG.

Discussion

Although H$_2$S is a hazardous gaseous molecule with a strong odor of rotten eggs, it has been described as an important regulator with a variety of biological roles in animals and recently

in plants [11–16,25–34,53–56]. Moreover, recent works on *Populus euphratica* cells [57] and bermudagrass seedlings [51], demonstrated that exogenously applied NaHS, a H$_2$S donor, resulted in an enhanced Cd tolerance in these species. However, possible physiological mechanisms and downstream targets responsible for the observed Cd tolerance triggered by intracellular H$_2$S remain elusive. In this report, we discovered endogenous H$_2$S production in response to Cd stress, and further provided evidence demonstrating a requirement of (h)GSH and ROS homeostases, at least partially, in the intracellular H$_2$S-mediated plant adaptation

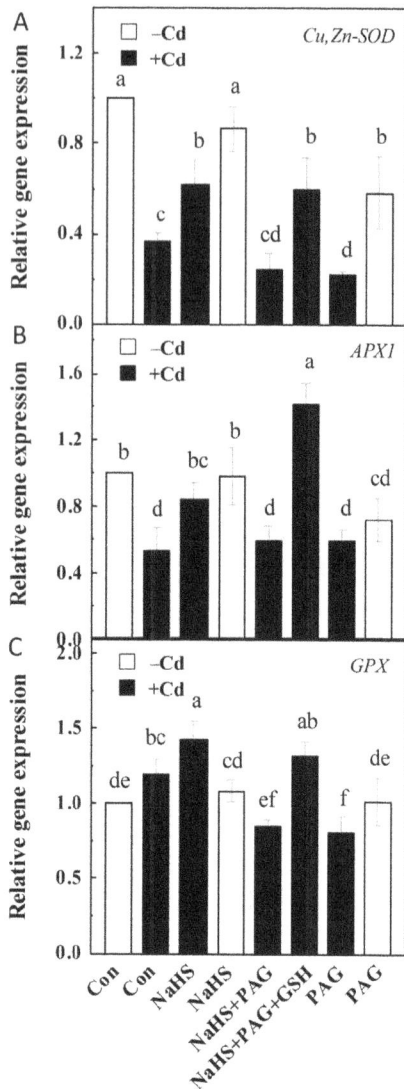

Figure 6. Transcripts of *Cu, Zn-SOD*, *APX1*, **and** *GPX* **regulated by NaHS, PAG, GSH and Cd.** Seedlings were pretreated with or without 100 μM NaHS, 2 mM PAG, 1 mM GSH, individual or combination for 6 h, and then exposed to 200 μM CdCl₂ for another 12 h. The expression levels of *Cu,Zn-SOD* (A), *APX1* (B), and *GPX* (C) transcripts in root tissues analyzed by real-time RT-PCR are presented as values relative to the control, normalized against expression of two internal reference genes in each sample. Values are means ± SD of three independent experiments with three replicates for each. Bars denoted by the same letter did not differ significantly at *P*<0.05 according to Duncan's multiple range test.

against Cd toxicity. Therefore, our results presented in this work are vital for both fundamental and applied plant biology.

Endogenous H₂S production in response to Cd stress: the possible involvement of LCD/DCD

In animals, it was previously reported that diverse stress-inducing stimuli could result in the production of H₂S, including oxidative stress [49], depletion of cysteine (or its derivatives) [58] and glutathione [50]. Recent work in Arabidopsis [25] and bermuda-grass seedlings [51] reported drought- and Cd-induced H₂S production. Because the signal compound H₂S is very reactive [53], the rapid regulation of the activity of H₂S biosynthetic

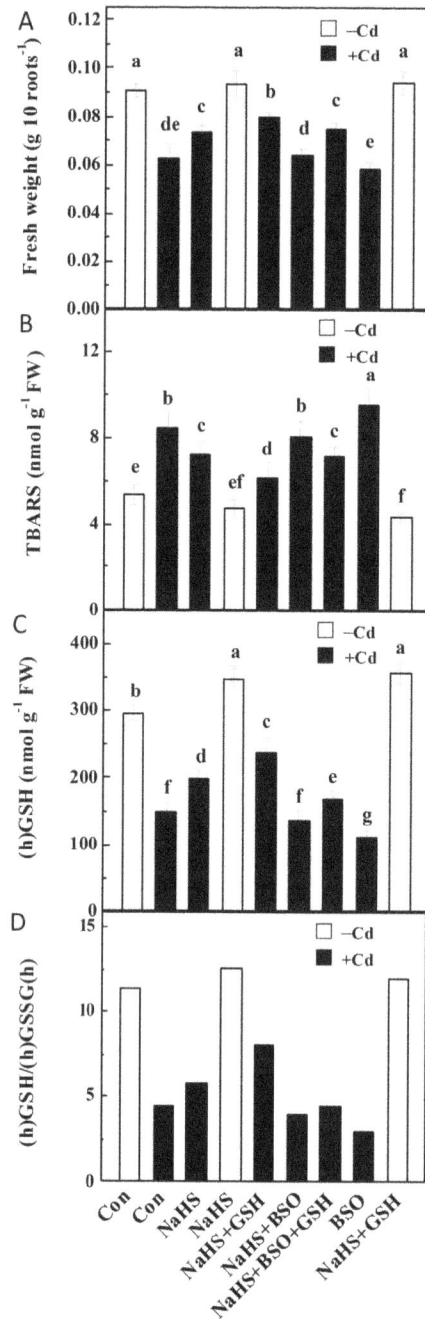

Figure 7. NaHS, GSH and BSO pretreatments differentially regulated seedling growth, TBARS accumulation, (h)GSH content, and (h)GSH/(h)GSSG(h). Fresh weight of 10 roots (A), TBARS accumulation (B), (h)GSH contents (C), and (h)GSH/(h)GSSG(h) ratio (D) in root tissues were determined after seedlings were pretreated with or without 100 μM NaHS, 1 mM GSH, 1 mM BSO, individual or combination for 6 h, and then exposed to 200 μM CdCl₂ for 72 h (A), 24 h (B) and 12 h (C and D). Values are means ± SD of three independent experiments with three replicates for each. Bars denoted by the same letter did not differ significantly at *P*<0.05 according to Duncan's multiple range test.

enzymes seems essential to fulfill H₂S-depenent functions. In this work, we further showed that Cd-triggered endogenous H₂S production might be related to LCD/DCD pathways (Figure 1D-F), since the similar increasing changes in the levels of intracellular

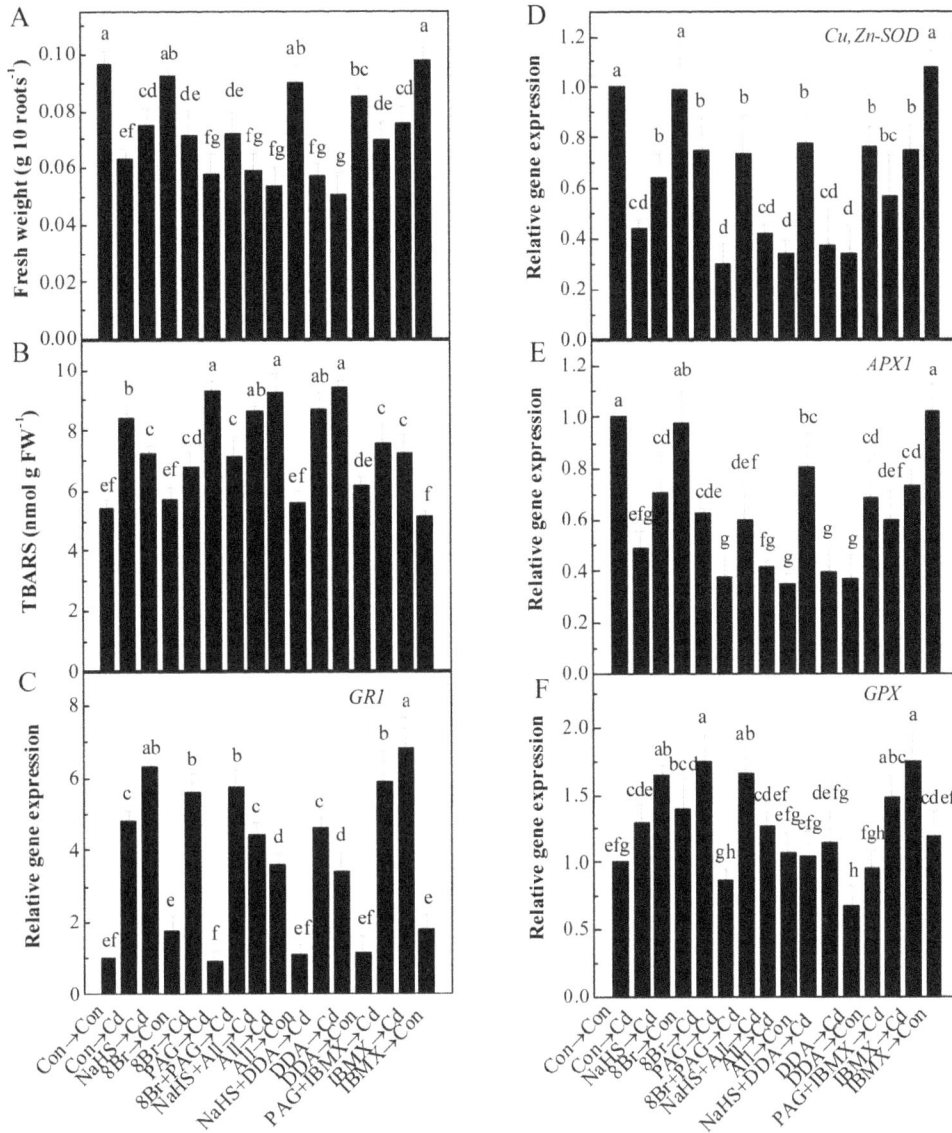

Figure 8. cAMP pathway might be involved in H₂S-alleviated Cd toxicity. Fresh weight of 10 roots (A), TBARS accumulation (B), *GR1* (C), *Cu,Zn-SOD* (D), *APX1* (E), and *GPX* (F) gene expression in alfalfa seedling roots upon Cd stress. Seedlings were pretreated with or without 100 μM NaHS, 50 μM 8-Br-cAMP (8Br), 2 mM PAG, 200 μM alloxan (All), 500 μM IBMX, 1 mM DDA alone, or the combination of treatments for 6 h, and then exposed to 200 μM CdCl₂ for 72 h (A), 24 h (B) and 12 h (C–F). The expression levels of corresponding genes analyzed by real-time RT-PCR are presented as values relative to the control, normalized against expression of two internal reference genes in each sample. Values are means ± SD of three independent experiments with three replicates for each. Bars denoted by the same letter did not differ significantly at *P*<0.05 according to Duncan's multiple range test.

H₂S as well as LCD/DCD activities were observed in the seedling roots of alfalfa challenged with Cd for 12 h. Meanwhile, similar to previous reports in wheat [30], bermudagrass [51], *Spinacia oleracea* seedlings [59], and strawberry plants [60], NaHS-induced H₂S production in alfalfa plants was also observed (Figure 2A).

In plants, both LCD and DCD are hypothesized to be involved in intracellular H₂S synthesis [21,27]. Several LCD/DCD candidates have been cloned and partially analyzed from the model plant *Arabidopsis* to *Brassica napus* [24,61]. Our above findings are consistent with those reported by Bloem et al. [40], in which they found that *Brassica napus* was able to react to *Pyrenopeziza brassicae* infection with a greater potential to release H₂S, which was reflected by an increasing LCD activity with fungal infection. More recently, auxin-induced DES-mediated

H₂S generation was also found to be involved in lateral root formation in tomato seedlings [62]. In view of the fact that all H₂S synthetic enzymes are not fully elucidated, our results suggested that LCD/DCD pathways might be, at least partially, related to Cd-induced H₂S production in alfalfa seedlings. In a future study, the role of other enzymatic and non-enzymatic pathways-mediated induction of H₂S synthesis in alfalfa seedlings upon Cd stress need be further elucidated.

The mechanism underlying the role of intracellular H₂S in the alleviation of Cd toxicity: reestablishment of reduced (homo)glutathione and ROS homeostases

Ample evidence revealed a clear relationship between metal stress and redox homeostasis and antioxidant capacity

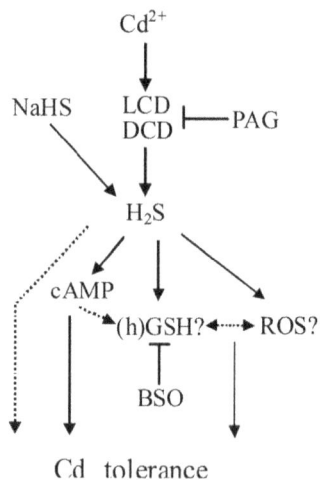

Figure 9. Simplified scheme of mechanisms involved in Cd tolerance by LCD/DCD-produced H₂S-modulated (h)GSH and ROS homeostases. Abbreviations: NaHS, sodium hydrosulfide; PAG, DL-propargylglycine; LCD, L-cysteine desulfhydrase; DCD, D-cysteine desulfhydrase; H₂S, hydrogen sulfide; ROS, reactive oxygen species; (h)GSH, reduced (homo)glutathione; BSO, L-buthionine-sulfoximine; cAMP, cyclic AMP. The dashed line denotes possible signaling cascade. T bars, inhibition.

[3,9,63–66]. Also, GSH could function as a heavy metal-ligand and an antioxidant [5,67]. In plants, H₂S serves as a signal as well as a novel antioxidant in hormonal and defense responses against abiotic stress [53,60]. Genetic evidence further revealed that the GSH deficiency mutant *pad2-1* shows the more oxidized redox state in contrast to wild type [68]. Arabidopsis mutants deficient in phytochelatins (PCs) and GSH biosynthesis respectively, *cad1* and *cad2*, are consequently more sensitive to Cd [6,69,70], that showed the crucial role of PCs, especially their precursor GSH in responding to Cd challenge. In the assays described here, as expected, when alfalfa seedling plants were upon Cd exposure, (h)GSH homeostasis is altered, which is reflected by the fact that the concentrations of reduced GSH and hGSH dropped (Table 1, Figure 1A–C), possible as a consequence of initiated PCs biosynthesis [71,72]. Similarly, a low ratio of (h)GSH/(h)GSSH(h), an important redox index related to Cd tolerance in alfalfa plants [45], was also observed. These changes thereafter cause redox imbalance and in turn Cd toxicity (Figures S1A, S3 and S4; Figures 4A and 5).

Our further experiments provide strong evidence to support the existence of a causal relationship between the endogenous H₂S signal and the alleviation of Cd toxicity in alfalfa seedlings partly by reestablishment of (h)GSH and ROS homeostases, which might be associated with the cAMP pathway. This conclusion is based on several pieces of evidence: (i) increased H₂S metabolism as well as the perturbation of (h)GSH homeostasis in alfalfa seedling roots are two early responses to the exposure of Cd (Figure 1, Table 1). These changes were consistent with the phenotypes of Cd toxicity (Figure 4A, Figure S1C). (ii) Application of a H₂S-releasing compound NaHS (also called as H₂S donor), not only mimics intracellular H₂S content triggered by Cd, but also alleviates Cd toxicity (Figures 2 and 4). Consistently, we also detected reestablishment of (h)GSH homeostasis, which was reflected by a higher (h)GSH content and ratio of (h)GSH/(h)GSSG(h) upon Cd stress. The observed Cd tolerance might be due to the available (h)GSH by the up-regulation of (h)GSH synthesis related genes, *ECS* and *GS* (Figure 3A and B), as well as *GR1* (Figure 3C), because besides

the synthesis of PCs, availability of GSH and concerted activity of GR seem to play a important role for plants to combat oxidative stress and Cd toxicity [7,72,73]. While, the inhibition of H₂S production caused by its synthetic inhibitor PAG blocked NaHS-induced Cd tolerance and reestablishment of (h)GSH and ROS homeostases, the latter of which was confirmed by the histochemical staining detecting the alleviation of plasma membrane integrity and lipid peroxidation, decreased ROS content and up-regulation of *Cu,Zn-SOD*, *APX1* and *GPX* transcripts, as well as declined TBARS level (Table 1, Figures 2–6, and Figures S1, S3 and S4). (iii) Above mentioned PAG responses were further rescued by exogenously applied GSH (Table 1, Figures 4–6). (iv) NaHS responses were sensitive to a (h)GSH synthetic inhibitor, but reversed by the added GSH (Figure 7), both of which suggesting a requirement of (h)GSH homeostasis for NaHS cytoprotective roles; and (v) Previous reports in animals showed H₂S-activited cAMP level or H₂S-regulated cAMP homeostasis [37,38]. Here, we found that two adenylyl cyclase inhibitors, alloxan and DDA, blocked the beneficial responses conferred by NaHS in alfalfa seedlings subjected to Cd stress (Figure 8). On the contrary, an analogue of cAMP 8-Br-cAMP and a cNMP phosphodiesterase inhibitor IBMX mimicked the effects of NaHS on the alleviation of Cd toxicity as well as the regulation of (h)GSH homeostasis and ROS metabolism (*GR1*, *Cu,Zn-SOD*, *APX1*, and *GPX*, etc). Above pharmacological evidence indicated the involvement of cAMP signaling in NaHS responses. Additionally, NaHS-triggered cytoprotective roles were confirmed to act as a H₂S-dependent fashion (Figure S2). Above results clearly established a casual link between intracellular H₂S in the alleviation of Cd toxicity and reestablishment of (h)GSH and ROS homeostases.

Conclusions

In summary, our pharmacological, histochemical, biochemical and molecular evidence suggested that the intracellular H₂S was able to ameliorate Cd toxicity in alfalfa seedlings at least partly by reestablishment of (h)GSH and ROS homeostases. Figure 9 illustrates a simplified scheme of mechanisms involved in Cd tolerance by LCD/DCD-produced H₂S-modulated (h)GSH and ROS homeostases, since 1) LCD/DCD-produced H₂S acts as a signal triggered by Cd to regulated (h)GSH metabolisms; 2) both (h)GSH and ROS homeostases could be reestablished by H₂S and further linked to Cd tolerance; 3) cAMP signaling pathway might be related to NaHS-triggered Cd tolerance, partially through the regulation of GSH homeostasis and ROS metabolism. Taking into account that H₂S participates in stressful responses and developmental process, our study therefore may extend our understanding of the complex system integrating environmental and developmental signals.

Supporting Information

Figure S1   NaHS pretreatment alleviates Cd toxicity.

Figure S2   H₂S or HS⁻, but not other compounds derived from NaHS contribute to NaHS responses.

Figure S3   Effects of NaHS, PAG and GSH pretreatments on the fresh weight (A) and TBARS concentrations (B) in alfalfa seedling roots upon Cd stress.

Figure S4   Effects of NaHS, PAG and GSH pretreatments on Cd concentrations in alfalfa seedlings upon Cd stress.

Table S1   The sequences of primers for real-time RT-PCR.

Author Contributions

Conceived and designed the experiments: WS. Performed the experiments: WC KZ QJ. Analyzed the data: WC Y. Xie WS. Contributed reagents/materials/analysis tools: HC JC Y. Xia JZ. Wrote the paper: WC HC WS.

References

1. Gao J, Sun L, Yang X, Liu JX (2013) Transcriptomic analysis of cadmium stress response in the heavy metal hyperaccumulator *Sedum alfredii* hance. PLoS ONE 8(6): e64643.
2. Ye Y, Li Z, Xing D (2013) Nitric oxide promotes MPK6-mediated caspase-3-like activation in cadmium-induced *Arabidopsis thaliana* programmed cell death. Plant Cell Environ 36: 1–15.
3. Sharma SS, Dietz KJ (2009) The relationship between metal toxicity and cellular redox imbalance. Trends Plant Sci 14: 43–50.
4. Tkalec M, Štefanić PP, Cvjetko P, Šikić S, Pavlica M, et al. (2012) The effects of cadmium-zinc interactions on biochemical responses in tobacco seedlings and adult plants. PLoS ONE 9(1): e87582.
5. Foyer CH, Noctor G (2011) Ascorbate and glutathione: the heart of the redox hub. Plant Physiol 155: 2–18.
6. Cobbett CS, May MJ, Howden R, Rolls B (1998) The glutathione-deficient, cadmium-sensitive mutant, *cad2-1*, of *Arabidopsis thaliana* is deficient in γ-glutamylcysteine synthetase. Plant J 16: 73–78.
7. Cruz de Carvalho MH, Brunet J, Bazin J, Kranner I, d' Arcy-Lameta A, et al. (2010) Homoglutathione synthetase and glutathione synthetase in drought-stressed cowpea leaves: expression patterns and accumulation of low-molecular-weight thiols. J Plant Physiol 167: 480–487.
8. Matamoros MA, Moran JF, Iturbe-Ormaetxe I, Rubio MC, Becana M (1999) Glutathione and homoglutathione synthesis in legume root nodules. Plant Physiol 121: 879–888.
9. Ortega-Villasante C, Rellán-Álvarez R, Del Campo FF, Carpena-Ruiz RO, Hernández LE (2005) Cellular damage induced by cadmium and mercury in *Medicago sativa*. J Exp Bot 56: 2239–2251.
10. Metwally A, Safronova VI, Belimov AA, Dietz KJ (2005) Genotypic variation of the response to cadmium toxicity in *Pisum sativum* L. J Exp Bot 56: 167–178.
11. Abe K, Kimura H (1996) The possible role of hydrogen sulfide as an endogenous neuromodulator. J Neurosci 16: 1066–1071.
12. Li L, Rose P, Moore PK (2011) Hydrogen sulfide and cell signaling. Annu Rev Pharmacol Toxicol 51: 169–187.
13. Hosoki R, Matsuki N, Kimura H (1997) The possible role of hydrogen sulfide as an endogenous smooth muscle relaxant in synergy with nitric oxide. Biochem Biophys Res Commun 237: 527–531.
14. Du J, Hui Y, Cheung Y, Bin G, Jiang H, et al. (2004) The possible role of hydrogen sulfide as a smooth muscle cell proliferation inhibitor in rat cultured cells. Heart Vessels 19: 75–80.
15. Yang G, Sun X, Wang R (2004) Hydrogen sulfide-induced apoptosis of human aorta smooth muscle cells via the activation of mitogen-activated protein kinases and caspase-3. FASEB J 18: 1782–1784.
16. Kimura Y, Goto Y, Kimura H (2010) Hydrogen sulfide increases glutathione production and suppresses oxidative stress in mitochondria. Antioxid Redox Signal 12: 1–13.
17. Wilson LG, Bressan RA, Filner P (1978) Light-dependent emission of hydrogen sulfide from plants. Plant Physiol 61: 184–189.
18. Hällgren JE, Fredriksson SÅ (1982) Emission of hydrogen sulfide from sulfur dioxide-fumigated pine trees. Plant Physiol 70: 456–459.
19. Sekiya J, Schmidt A, Wilson LG, Filner P (1982) Emission of hydrogen sulfide by leaf tissue in response to L-cysteine. Plant Physiol 70: 430–436.
20. Wang R (2002) Two's company, three's a crowd: can H$_2$S be the third endogenous gaseous transmitter? FASEB J 16: 1792–1798.
21. Papenbrock J, Riemenschneider A, Kamp A, Schulz-Vogt HN, Schmidt A (2007) Characterization of cysteine-degrading and H$_2$S-releasing enzymes of higher plants – from the field to the test tube and back. Plant Biol 9: 582–588.
22. Léon S, Touraine B, Briat JF, Lobréaux S (2002) The *AtNFS2* gene from *Arabidopsis thaliana* encodes a NifS-like plastidial cysteine desulphurase. Biochem J 366: 557–564.
23. Riemenschneider A, Wegele R, Schmidt A, Papenbrock J (2005) Isolation and characterization of a D-cysteine desulfhydrase protein from *Arabidopsis thaliana*. FEBS J 272: 1291–1304.
24. Álvarez C, Calo L, Romero LC, García I, Gotor C (2010) An O-acetylserine(thiol)lyase homolog with L-cysteine desulfhydrase activity regulates cysteine homeostasis in Arabidopsis. Plant Physiol 152: 656–669.
25. Jin Z, Shen J, Qiao Z, Yang G, Wang R, et al. (2011) Hydrogen sulfide improves drought resistance in *Arabidopsis thaliana*. Biochem Biophys Res Commun 414: 481–486.
26. Wang Y, Li L, Cui W, Xu S, Shen W, et al. (2012) Hydrogen sulfide enhances alfalfa (*Medicago sativa*) tolerance against salinity during seed germination by nitric oxide pathway. Plant Soil 351: 107–119.
27. Garcia-Mata C, Lamattina L (2010) Hydrogen sulphide, a novel gasotransmitter involved in guard cell signalling. New Phytol 188: 977–984.
28. Lisjak M, Srivastava N, Teklic T, Civale L, Lewandowski K, et al. (2010) A novel hydrogen sulfide donor causes stomatal opening and reduces nitric oxide accumulation. Plant Physiol Biochem 48: 931–935.
29. Zhang H, Jiao H, Jiang CX, Wang SH, Wei ZJ, et al. (2010) Hydrogen sulfide protects soybean seedlings against drought-induced oxidative stress. Acta Physiol Plant 32: 849–857.
30. Zhang H, Hu LY, Hu KD, He YD, Wang SH, et al. (2008) Hydrogen sulfide promotes wheat seed germination and alleviates oxidative damage against copper stress. J Integr Plant Biol 50: 1518–1529.
31. Li L, Wang Y, Shen W (2012) Roles of hydrogen sulfide and nitric oxide in the alleviation of cadmium-induced oxidative damage in alfalfa seedling roots. Biometals 25: 617–631.
32. Li ZG, Gong M, Xie H, Yang L, Li J (2012) Hydrogen sulfide donor sodium hydrosulfide-induced heat tolerance in tobacco (*Nicotiana tabacum* L) suspension cultured cells and involvement of Ca^{2+} and calmodulin. Plant Sci 185–186: 185–189.
33. Lin YT, Li MY, Cui WT, Lu W, Shen WB (2012) Haem oxygenase-1 is involved in hydrogen sulfide-induced cucumber adventitious root formation. J Plant Growth Regul 31: 519–528.
34. Zhang H, Hu SL, Zhang ZJ, Hu LY, Jiang CX, et al. (2011) Hydrogen sulfide acts as a regulator of flower senescence in plants. Postharvest Biol Tec 60: 251–257.
35. Ma W, Qi Z, Smigel A, Walker RK, Verma R, et al. (2009) Ca^{2+}, cAMP, and transduction of non-self perception during plant immune responses. Proc Natl Acad Sci U S A 106: 20995–21000.
36. Jin XC, Wu WH (1998) Involvement of cyclic AMP in ABA- and Ca^{2+}-mediated signal transduction of stomatal regulation in *Vicia faba*. Plant Cell Physiol 40: 1127–1133.
37. Kimura H (2000) Hydrogen sulfide induces cyclic AMP and modulates the NMDA receptor. Biochem Biophys Res Commun 267: 129–133.
38. Lu M, Liu YH, Ho CY, Tiong CX, Bian JS (2012) Hydrogen sulfide regulates cAMP homeostasis and renin degranulation in As4.1 and rat renin-rich kidney cells. Am J Physiol Cell Physiol 302: C59–C66.
39. Shao JL, Wan XH, Chen Y, Bi C, Chen HM, et al. (2011) H$_2$S protects hippocampal neurons from anoxia-reoxygenation through cAMP-mediated PI3K/Akt/p70S6K cell-survival signaling pathways. J Mol Neurosci 43: 453–460.
40. Bloem E, Riemenschneider A, Volker J, Papenbrock J, Schmidt A, et al. (2004) Sulphur supply and infection with *Pyrenopeziza brassicae* influence L-cysteine desulphydrase activity in *Brassica napus* L. J Exp Bot 55: 2305–2312.
41. Cui W, Gao C, Fang P, Lin G, Shen W (2008) Alleviation of cadmium toxicity in *Medicago sativa* by hydrogen-rich water. J Hazard Mater 260: 715–724.
42. Smith IK (1985) Stimulation of glutathione synthesis in photorespiring plants by catalase inhibitors. Plant Physiol 79: 1044–1047.
43. Herschbach C, Pilch B, Tausz M, Rennenberg H, Grill D (2002) Metabolism of reduced and inorganic sulphur in pea cotyledons and distribution into developing seedlings. New Phytol 153: 73–80.
44. Meyer AJ, Brach T, Marty L, Kreye S, Rouhier N, et al. (2007) Redox-sensitive GFP in *Arabidopsis thaliana* is a quantitative biosensor for the redox potential of the cellular glutathione redox buffer. Plant J 52: 973–986.
45. Cui W, Li L, Gao Z, Wu H, Xie Y, et al. (2012) Haem oxygenase-1 is involved in salicylic acid-induced alleviation of oxidative stress duo to cadmium stress in *Medicago sativa*. J Exp Bot 63: 5521–5534.
46. Vandesompele J, De Preter K, Pattyn F, Poppe B, Van Roy N, et al. (2002) Accurate normalization of real-time quantitative RT-PCR data by geometric averaging of multiple internal control genes. Genome Biol 3: research0034.
47. Kováčik J, Babula P, Klejdus B, Hedbavny J, Jarošová M (2014) Unexpected behavior of some nitric oxide modulators under cadmium excess in plant tissue. PLos ONE 9(3): e91685.
48. Rennenberg H, Filner P (1982) Stimulation of H$_2$S emission from pumpkin leaves by inhibition of glutathione synthesis. Plant Physiol 69: 766–770.
49. Kwak WJ, Kwon GS, Jin I, Kuriyama H, Sohn HY (2003) Involvement of oxidative stress in the regulation of H$_2$S production during ultradian metabolic oscillation of *Saccharomyces cerevisiae*. FEMS Microbiol Lett 219: 99–104.
50. Sohn HY, Kum EJ, Kwon GS, Jin I, Adams CA, et al. (2005) *GLR1* plays an essential role in the homeodynamics of glutathione and the regulation of H$_2$S production during respiratory oscillation of *Saccharomyces cerevisiae*. Biosci Biotechnol Biochem 69: 2450–2454.
51. Shi H, Ye T, Chan Z (2014) Nitric oxide-activated hydrogen sulfide is essential for cadmium stress response in bermudagrass (*Cynodon dactylon* (L). Pers.). Plant Physiol Biochem 74: 99–107.
52. Rüegsegger A, Schmutz D, Brunold C (1990) Regulation of glutathione synthesis by cadmium in *Pisum sativum* L. Plant Physiol 93: 1579–1584.
53. Lisjak M, Teklic T, Wilson ID, Whiteman M, Hancock JT (2013) Hydrogen sulfide: environmental factor or signalling molecule? Plant Cell Environ 36: 1607–1616.

54. García-Mata C, Lamattina L (2013) Gasotransmitters are emerging as new guard cell signaling molecules and regulators of leaf gas exchange. Plant Sci 201–202: 66–73.

55. Kimura Y, Kimura H (2004) Hydrogen sulfide protects neurons from oxidative stress. FASEB J 18: 1165–1167.

56. Li ZG, Gong M, Liu P (2012) Hydrogen sulfide is a mediator in H_2O_2-induced seed germination in *Jatropha Curcas*. Acta Physiol Plant 34: 2207–2213.

57. Sun J, Wang R, Zhang X, Yu Y, Zhao R, et al. (2013) Hydrogen sulfide alleviates cadmium toxicity through regulations of cadmium transport across the plasma and vacuolar membranes in *Populus euphratica* cells. Plant Physiol Biochem 65: 67–74.

58. Sohn HY, Kuriyama H (2001) The role of amino acids in the regulation of hydrogen sulfide production during ultradian respiratory oscillation of *Saccharomyces cerevisiae*. Arch Microbiol 176: 69–78.

59. Chen J, Wu FH, Wang WH, Zheng CJ, Lin GH, et al. (2011) Hydrogen sulphide enhances photosynthesis through promoting chloroplast biogenesis, photosynthetic enzyme expression, and thiol redox modification in *Spinacia oleracea* seedlings. J Exp Bot 62: 4481–4493.

60. Christou A, Manganaris GA, Papadopoulos I, Fotopoulos V (2013) Hydrogen sulfide induces systemic tolerance to salinity and non-ionic osmotic stress in strawberry plants through modification of reactive species biosynthesis and transcriptional regulation of multiple defence pathways. J Exp Bot 64: 1953–1966.

61. Xie Y, Lai D, Mao Y, Zhang W, Shen W, et al. (2013) Molecular cloning, characterization, and expression analysis of a novel gene encoding L-cysteine desulfhydrase from *Brassica napus*. Mol Biotechnol 54: 737–746.

62. Fang T, Cao Z, Li J, Shen W, Huang L (2014) Auxin-induced hydrogen sulfide generation is involved in lateral root formation in tomato. Plant Physiol Biochem 76: 44–51.

63. Dawood M, Cao F, Jahangir MM, Zhang G, Wu F (2012) Alleviation of aluminum toxicity by hydrogen sulfide is related to elevated ATPase, and suppressed aluminum uptake and oxidative stress in barley. J Hazard Mater 209–210: 121–128.

64. Jin CW, Mao QQ, Luo BF, Lin XY, Du ST (2013) Mutation of *mpk6* enhances cadmium tolerance in *Arabidopsis* plants by alleviating oxidative stress. Plant Soil 371: 387–396.

65. Lagorce A, Fourçans A, Dutertre M, Bouyssiere B, Zivanovic Y, et al. (2012) Genome-wide transcriptional response of the archaeon *Thermococcus gammatolerans* to cadmium. PLoS ONE 7(7): e41935.

66. Thapa G, Sadhukhan A, Panda SK, Sahoo L (2012) Molecular mechanistic model of plant heavy metal tolerance. Biometals 25: 489–505.

67. Dixit P, Mukherjee PK, Ramachandran V, Eapen S (2013) Glutathione transferase from *Trichoderma virens* enhances cadmium tolerance without enhancing its accumulation in transgenic *Nicotiana tabacum*. PLoS ONE 6(1): e16360.

68. Dubreuil-Maurizi C, Vitecek J, Marty L, Branciard L, Frettinger P, et al. (2011) Glutathione deficiency of the Arabidopsis mutant *pad2-1* affects oxidative stress-related events, defense gene expression, and the hypersensitive response. Plant Physiol 157: 2000–2012.

69. Howden R, Andersen CR, Goldsbrough PB, Cobbett CS (1995) A cadmium-sensitive, glutathione-deficient mutant of *Arabidopsis thaliana*. Plant Physiol 107: 1067–1073.

70. Howden R, Goldsbrough PB, Andersen CR, Cobbett CS (1995) Cadmium-sensitive, *cad1* mutants of *Arabidopsis thaliana* are phytochelatin deficient. Plant Physiol 107: 1059–1066.

71. Grill E, Löffler S, Winnacker EL, Zenk MH (1989) Phytochelatins, the heavy-metal-binding peptides of plants, are synthesized from glutathione by a specific γ-glutamylcysteine dipeptidyl transpeptidase (phytochelatin synthase). Proc Natl Acad Sci USA 86: 6838–6842.

72. Mishra S, Srivastava S, Tripathi RD, Govindarajan R, Kuriakose SV, et al. (2006) Phytochelatin synthesis and response of antioxidants during cadmium stress in *Bacopa monnieri* L. Plant Physiol Biochem 44: 25–37.

73. Verbruggen N, Juraniec M, Baliardini C, Meyer CL (2013) Tolerance to cadmium in plants: the special case of hyperaccumulators. Biometals 26: 633–638.

19

A Pattern of Early Radiation-Induced Inflammatory Cytokine Expression Is Associated with Lung Toxicity in Patients with Non-Small Cell Lung Cancer

Shankar Siva[1*], **Michael MacManus**[1,2], **Tomas Kron**[2,3], **Nickala Best**[4], **Jai Smith**[4], **Pavel Lobachevsky**[2,4], **David Ball**[1,2], **Olga Martin**[1,2,4]

1 Division of Radiation Oncology and Cancer Imaging, Peter MacCallum Cancer Centre, Melbourne, VIC, Australia, **2** Sir Peter MacCallum Department of Oncology, the University of Melbourne, Melbourne, VIC, Australia, **3** Physical Sciences, Peter MacCallum Cancer Centre, Melbourne, VIC, Australia, **4** Molecular Radiation Biology Laboratory, Peter MacCallum Cancer Centre, VIC, Australia

Abstract

Purpose: Lung inflammation leading to pulmonary toxicity after radiotherapy (RT) can occur in patients with non-small cell lung cancer (NSCLC). We investigated the kinetics of RT induced plasma inflammatory cytokines in these patients in order to identify clinical predictors of toxicity.

Experimental Design: In 12 NSCLC patients, RT to 60 Gy (30 fractions over 6 weeks) was delivered; 6 received concurrent chemoradiation (chemoRT) and 6 received RT alone. Blood samples were taken before therapy, at 1 and 24 hours after delivery of the 1st fraction, 4 weeks into RT, and 12 weeks after completion of treatment, for analysis of a panel of 22 plasma cytokines. The severity of respiratory toxicities were recorded using common terminology criteria for adverse events (CTCAE) v4.0.

Results: Twelve cytokines were detected in response to RT, of which ten demonstrated significant temporal changes in plasma concentration. For Eotaxin, IL-33, IL-6, MDC, MIP-1α and VEGF, plasma concentrations were dependent upon treatment group (chemoRT vs RT alone, all p-values <0.05), whilst concentrations of MCP-1, IP-10, MCP-3, MIP-1β, TIMP-1 and TNF-α were not. Mean lung radiation dose correlated with a reduction at 1 hour in plasma levels of IP-10 ($r^2 = 0.858$, $p < 0.01$), MCP-1 ($r^2 = 0.653$, $p<0.01$), MCP-3 ($r^2 = 0.721$, $p<0.01$), and IL-6 ($r^2 = 0.531$, $p = 0.02$). Patients who sustained pulmonary toxicity demonstrated significantly different levels of IP-10 and MCP-1 at 1 hour, and Eotaxin, IL-6 and TIMP-1 concentration at 24 hours (all p-values <0.05) when compared to patients without respiratory toxicity.

Conclusions: Inflammatory cytokines were induced in NSCLC patients during and after RT. Early changes in levels of IP-10, MCP-1, Eotaxin, IL-6 and TIMP-1 were associated with higher grade toxicity. Measurement of cytokine concentrations during RT could help predict lung toxicity and lead to new therapeutic strategies.

Editor: Yong J. LEE, University of Pittsburgh School of Medicine, United States of America

Funding: Dr Shankar Siva has received National Health and Medical Research Council scholarship funding for this research, APP1038399. http://www.nhmrc.gov.au/grants/apply-funding/postgraduate-scholarships. The funders had no role in study design, data collection and analysis, decision to publish, or preparation of the manuscript.

Competing Interests: The authors have declared that no competing interests exist.

* Email: Shankar.Siva@petermac.org

Introduction

Lung cancer is the leading cause of cancer-related death in both sexes [1]. Non-small cell lung cancer (NSCLC) accounts for 85% of cases. Radiotherapy (RT), alone or in combination with chemotherapy, is a standard definitive treatment approach for patients with locally advanced NSCLC or inoperable patients with early stage disease [2,3]. Over half of NSCLC patients are currently treated with RT. This rate may increase in the future with the optimal RT utilization rate being estimated to be 76% [4]. However, local failures are a major cause for the relatively poor survival reported for patients treated with RT. A recent meta-analysis suggests that local failures still occur in up to 38% of patients [5]. Efforts to intensify RT, however, are severely limited by the need to constrain dose to the surrounding normal lung in order to preserve function [6]. Lung toxicity caused by RT (termed *pneumonitis*) is a real and potentially debilitating toxicity, sometimes leading to patient death [7]. In the modern era symptomatic pneumonitis still occurs in 29.8% of patients and fatal pneumonitis in 1.9% [8]. Currently used RT planning constraints that were designed to limit the risk of pneumonitis are based on evidence over a decade old [9]. These constraints apply to populations and give no indication of an individual patient's susceptibility to lethal toxicity, beyond the fact that on average higher RT doses to larger volumes are more likely to be toxic. It is therefore imperative to establish *in vivo* biomarkers for prediction or early assessment of pneumonitis that will ultimately assist in avoiding RT induced lung dysfunction by individualizing treatment.

The pathophysiology of radiation-induced lung toxicity is incompletely understood at present. A large body of evidence from animal models, molecular biology and clinical observations suggests that normal tissue injury is a dynamic and progressive process [10,11]. A complex interaction between radiation-induced damage to parenchymal cells, supporting vasculature and associated fibrotic reactions results in acute and late radiation toxicities. In the lung, these changes can manifest themselves as reduced pulmonary function and in a chronic inflammatory cascade known as pneumonitis [12]. There are many factors that influence the likelihood of severe respiratory toxicity including the volume of irradiated parenchyma, pre-existing lung disease and the use of radiosensitizing chemotherapy [13]. However, the exact biological mechanisms of inflammatory cascade and eventual pulmonary fibrosis are not fully elucidated.

Cytokine release in response to ionizing radiation is a documented phenomenon and may play a major role in subsequent radiation induced lung toxicity (reviewed in [14–18]. A non-specific acute reaction, or "cytokine storm" usually resolves within 24 hours [19]. Fractionated radiation, however, creates a constant complex stress response and a cytokine profile is different to that induced by a single radiation dose [20]. RT-related plasma concentrations of one or more cytokines in humans have correlated with lung toxicity. Transforming growth factor (TGF)-β1 [21–24], interleukin (IL)-6 and IL-10 [25,26] during RT have been suggested as possible risk markers in these studies. However, other studies have reported contradictory or negative findings [27,28].

The rationale for the composition of our panel of 22 potential biomarkers for lung tissue toxicity was based on several published reports dissecting inflammatory and radiation response. The plasma levels of a range of cytokines have been previously investigated in context of both murine [29] and cell models [20]. A range of pro-inflammatory cytokines are expressed as acute phase reactants, including tumour necrosis factor (TNF)-α, i IL-1 and IL-6 [14,18]. Chemokines act as chemoattractants for leukocytes which potentiate the inflammatory response, such as interferon-inducible protein-10 (IP-10) which attracts predominantly neutrophils, macrophage inflammatory protein (MIP)-1α, and macrophage chemoattractant protein (MCP)-3 which attracts predominantly monocytes, and MIP-1β and MIP-3α which attract predominantly lymphocytes [16,17]. Induction of MIP-3β results in chemoattraction of dendritic cells and antigen engaged B-cells [30]. MCP-1 is a cytokine that has been associated with many inflammation-related diseases and has been implicated in the progression and prognosis of several cancers [29,31]. Upregulation of MCP-3 gene expression has been shown to be maximal at 1-hour in response to radiation in rat liver [32]. Excessive release of interferon-gamma (IFNγ) has been associated with the pathogenesis of chronic inflammatory and autoimmune diseases [17]. Macrophage-derived chemokine (MDC), is involved in chronic inflammation and dendritic cell and lymphocyte homing [17]. Eotaxin is a chemoattractant for eosinophils and is implicated in acute inflammatory lung injury responses, particularly in emphysema and asthma [33,34]. IL-3, IL-11, IL-22 and IL-33 are all acute phase reactants that potentiate cellular immune signalling and inflammatory responses [35–37]. The induction of all these inflammatory cytokines in response to radiation stimulate the subsequent expression of fibrotic cytokines such as the TGF-β

family and vascular endothelial growth factor (VEGF). These in turn facilitate the progression from pneumonitis to lung fibrosis [38,39]. Helping to balance this process, both IL-22 and IL-10 can act to down-regulate the pneumonitic response by blocking pro-inflammatory cytokines and function of antigen-presenting cells [25,37]. Additionally, tissue inhibitors of metalloproteinase (TIMP)-1 acts to down-regulate the profibrotic response and is elevated in chronic inflammatory disease states [29,40].

In this study, we report the modulation of plasma concentrations of these cytokines in patients receiving RT alone or RT with concurrent radiosensitising chemotherapy. In contrast to many previous studies, we consider the differential patterns of response in patients receiving radiosensitizing chemotherapy compared to those receiving RT alone. We assess a homogenous cohort of patients receiving identical dose/fractionation schedules, and employ a large panel of candidate cytokines. Additionally, we report the effect of treatment volume and dose to normal lung tissue on plasma cytokine concentrations, suggesting that these cytokines could be used as *in-vivo* 'biodosimeters' of individual radiation dose. Finally, we identified five cytokines that that could be predictive of pulmonary lung toxicity and should be validated in a larger cohort as early predictive markers for clinical radiation pneumonitis.

Materials and Methods

This research was the translational component of an institutional ethics committee approved prospective clinical trial at the Peter MacCallum Cancer Centre (Universal Trials Number U1111-1138-4421). All patients provided written consent to participate in this study. Consecutive patients undergoing definitive RT with or without concurrent chemotherapy underwent serial venipuncture and blood collection for inflammatory cytokine testing. Patients were followed up at three monthly intervals after treatment. Toxicity scoring was performed prospectively at each clinical visit using Common Terminology Criteria for Adverse Events (CTCAE) version 4.0.

Radiotherapy

All patients were planned to receive 60 Gy in 30 fractions of RT delivered over 6 weeks using 3D conformal techniques. Respiratory-sorted four-dimensional computed tomography (4DCT) was used for RT planning. Target delineation was performed on an Elekta FocalSim workstation (Stockholm, Sweden). An internal target volume (ITV) was delineated from the maximal intensity projection (MIP) series, and a further isotropic expansion of 5 mm expansion was used to generate the clinical target volume, and a further 10 mm isotropic expansion was used to create the planning target volume (PTV). The lung organ at risk volume was defined as the volume of both lungs minus the volume of the ITV. Typically a 3–4 field RT technique using 6MV photons was used with effort made to avoid the contralateral unaffected lung and spare spinal cord whilst ensuring the PTV was within −5% and +7% of the prescribed dose, as per ICRU 62 recommendations. Dose constraints to organ at risks dose were as follows: spinal canal ≤45 Gy, mean lung dose ≤20 Gy, the volume of lung receiving 5 Gy (V5) ≤60%, V20≤35%, V30≤30%. In patients receiving concurrent chemotherapy, this was delivered using platinum doublets. This consisted of either 2×3 weekly cycles of 50 mg/m^2 cisplatin days 1 and 8, with 50 mg/m^2 etoposide days 1–5, or 6x weekly cycles of carboplatin AUC 2 day 1 with 45 mg/m^2 paclitaxel day 1. The first cycle of concurrent chemotherapy was commenced immediately prior to the first fraction of radiotherapy. No patient received adjuvant chemotherapy after the concurrent chemotherapy delivery. All patients in our institution are planned for concurrent chemotherapy unless precluded by cardiovascular comorbidities or renal insufficiency.

Blood Sample Processing

Patient blood samples were collected and processed at five time points in this study. Baseline blood samples were collected before therapy, and 4 consecutive samples were collected at 1 hour after the first fraction of RT, 24 hours after the first fraction of RT, 4 weeks into the course of RT, and 12 weeks after the completion of RT. The early time points were chosen for pragmatic purposes to reflect clinical practicality; patients are routinely within the department within 1 hour of the first and second fraction of RT. These times allow for the possibility of early adaptation of the RT plan based on cytokine response. The 4 week time point was chosen as this typically coincides with approximately 40 Gy of delivered dose, which allows an ideal opportunity for adaptive radiation planning prior to completion of RT [41]. The final time point, at 12 weeks after completion of RT, coincides with the period in which fibrosing alveolitis and the clinical manifestation of subsequent pneumonitis can occur [42]. Blood was collected in 9 mL ethylenediaminetetraacetic acid (EDTA) tubes, and centrifuged twice at 2000 rpm for 10 minutes and then at 4000 rpm for 10 minutes. The upper 90% of the plasma was transferred into 2 mL aliquots into cryovials and stored at –80°C. These were subsequently processed in batch using a commercial flow cytometry system.

Cytokine Analysis

Each patient sample was run in duplicate using 100 µl of plasma diluted by a factor of 2. A commercial multiplexed sandwich ELSIA-based array was used (Quantibody custom array, RayBiotech Inc., Norcross, GA, USA). All of the samples were tested using a panel of 22 cytokines: Eotaxin, IFNγ, IL-6, IL-10, IL-11, IL-22, IL-3, IL-33, IP-10, MCP-1, MCP-3, MDC, MIP-1α, MIP-1β, MIP-3α, MIP-3β, TGF-β1, TGF-β2, TGF-β3, TIMP-1, TNF-α, VEGF. The antibody array is a glass-chip-based multiplexed sandwich ELISA system designed to determine the concentrations all 22 cytokines simultaneously. One standard glass slide was spotted with 16 wells of identical biomarker antibody arrays. Each antibody, together with the positive and negative control, was arrayed in quadruplicate. The samples and standards were added to the wells of the chip array and incubated for 3 h at 4 °C. This was followed by three to four washing steps and the addition of primary antibody and HRP-conjugated streptavidin to the wells. The signals (Cy3 wavelengths: 555 nm excitation, 655 nm emission) were scanned and extracted with a Genepix laser scanner (Axon Instruments, Foster City, CA), and quantified using Quantibody Analyzer software (Ray Biotech Inc). Each signal was identified by its spot location. The scanner software calculated background signals automatically. Concentration levels, expressed in picograms per milliliter (pg/ml), were calculated against a standard curve set for each biomarker from the positive and negative controls.

Statistical Methods

Patients were grouped into those receiving concurrent chemotherapy (chemoRT) and those receiving RT alone. Two-way ANOVA assuming repeated measures testing for different time points was used to assess differences in cytokine concentrations between chemoRT and RT groups and across sampled time points. These changes from baseline cytokine concentration were measured at an individual patient level. Subsequently, 95% confidence intervals were calculated and corrections for multiple comparisons were performed using Dunnett's method and an alpha of 0.05. Clinical toxicities secondary to treatment were assessed using CTCAE v4.0 at baseline, 4 weeks into treatment and 12 weeks after treatment completion. The PTV and mean lung dose (MLD) of RT were recorded for each patient and correlated with the change in cytokine concentrations from baseline at one hour after the first fraction, four weeks into treatment and 12 weeks after treatment completion using a linear regression model. Patients were dichotomized into those experiencing severe respiratory toxicity (grade 2+) and those who did not. Unpaired two-tailed t-tests were used to compare the mean cytokine concentrations at these timepoints between the patient toxicity groups. All statistical analyses were performed using PRISM v6.0 software.

Results

Twelve patients were enrolled into this study with a median age of 67 years (range 46–89 years). All patients received 60 Gy in 30 fractions of RT. Six patients received concurrent chemotherapy and six received RT alone (due to comorbidities). Six patients had stage III disease, three had stage II disease and three had stage I disease. Patient population characteristics are listed in **Table 1.** Individual patient characteristics are further given in **Table S1**.

Effect of Treatment Group and Sample Time Point

Of the 22 cytokines analysed, results from 12 cytokines were above the limit of detection. These were Eotaxin, IL-33, IL-6, MCP-1, MDC, MIP-1α, VEGF, IP-10, MCP-3, MIP-1β, TIMP-1 and TNF-α. Of these 12 cytokines, all except for IL-33 and TNF-α demonstrated significant variation in concentrations across the different time points (all p-values ≤0.02, **Table S2**). The absolute changes in concentrations for each of the 12 plasma cytokines are depicted in **Figure 1.** Levels of Eotaxin, IL-33, IL-6, MDC, MIP-1α and VEGF were different in those patients receiving chemoRT as compared to those receiving RT alone (all p-values <0.01), whilst concentrations of IP-10, MCP-1, MCP-3, MIP-1β, TIMP-1 and TNF-α were not dependent upon the treatment group. In the RT alone group, the peak change in cytokine levels (depression or elevation) was seen at 4 weeks during treatment for IL-33, IP-10, MCP-1, & MIP-1α. The peak change in cytokine level was seen at 12 weeks after treatment completion for Eotaxin, IL-6, MCP-3, TIMP-1 and VEGF. By comparison, in the chemoRT group, the peak change in cytokine levels (depression or elevation) was seen at 4 weeks during treatment for MDC and MIP-1α only. The peak change in cytokine level was seen at 12 weeks after treatment completion for Eotaxin, MCP-3, MIP-1β, and VEGF. There was significant interaction between treatment group and sample time point (p-values <0.01) for concentrations of IL-6, IP-10, MCP-1, MDC, MIP-1α, MIP-1β and VEGF, indicating that the variation of plasma cytokine concentrations over time was not the same for the RT group as for the chemoRT group. Conversely, there was no significant interaction between treatment groups and sample time points for Eotaxin, MCP-3 and TIMP-1 indicating that the variation of cytokine concentrations across time was the same for both treatment groups. A summary of the cytokine levels that varied by time and those that varied by treatment group are depicted in **Figure 2**.

Effect of Mean Lung Dose and PTV volume

The median (range) PTV volume in all patients was 320 cm^3 (87 cm^3–1138 cm^3). Plasma concentrations decreased from baseline at 1-hour post irradiation in all patients in a linear volume dependent manner for IL-6 ($r = 0.887$, $p<0.01$), MCP-1

Table 1. Patient Characteristics.

Characteristic	Number (%)
Sex	
Male	8 (67%)
Female	4 (33%)
Age	
Median	67 years
Range	46–89 years
Planning Target Volume (PTV)	
Median	320 cm³
Range	87 cm³–1138 cm³
Mean Dose to Normal Lung	
Median	11.7 Gy
Range	6.0 Gy–19.1 Gy
Total Radiation Dose	60 Gy (100%)
Clinical Stage	
I	3 (25%)
II	3 (25%)
III	6 (50%)
Histology	
Squamous Cell Carcinoma	6 (50%)
Adenocarcinoma	3 (25%)
Non-Small Cell [NOS*]	2 (17%)
Large Cell Neuroendocrine	1 (8%)
Treatment:	
RT alone	6 (50%)
Concurrent Cisplatinum/Etoposide	1 (8%)
Concurrent Carboplatinum/Paclitaxel	5 (42%)

*NOS – Not Otherwise Specified.

($r = 0.664$, $p = 0.03$), and IP-10 ($r = 0.819$, $p < 0.01$), which is depicted in **Figure 3**. The extent of the reduction in these plasma cytokine concentrations correlated with irradiated target volume. The strongest correlation was observed for IL-6 (**Figure 3A**). Change in plasma concentration at 1-hour in the 9 remaining cytokines did not correlate with PTV volume. The changes in plasma concentration from baseline for all 12 cytokines did not correlate with irradiated target volume at either 4 weeks into treatment or 12 weeks after treatment.

The median (range) MLD in all patients was 11.7 Gy (5.97 Gy–19.14 Gy). Similar to the effect seen with PTV volume, plasma concentrations decreased from baseline at 1-hour post irradiation in a linear dose dependent manner for IL-6 ($r = 0.729$, $p = 0.02$), MCP-1 ($r = 0.808$, $p < 0.01$), and IP-10 ($r = 0.926$, $p < 0.01$), depicted in **Figure 4**. In addition, MCP-3 also demonstrated linear reduction in plasma concentration at 1-hour for a given MLD ($r = 0.849$, $p < 0.01$). The MLD was proportional to a reduction in these plasma cytokine concentrations at 1-hour. Change in plasma concentration at 1-hour in the 8 remaining cytokines did not correlate with MLD. Again, the change in plasma concentration from baseline for all 12 cytokines did not correlate with mean normal lung dose at either 4 weeks into treatment or 12 weeks after treatment.

Association of Plasma Cytokine Concentrations with Likelihood of Severe Respiratory Toxicity

Five of the twelve patients had severe lung toxicity (CTCAE grade 2 or higher) in this study. Three of these patients were in the chemoRT group and two of these patients received RT alone. Overall, patients with a greater depression in concentrations of MCP-1 and IP-10 levels at 1-hour post the first fraction of radiation subsequently sustained severe lung toxicity (**Figure 5A, 5B**). For those patients with severe toxicities, the mean (+/− standard deviation) reduction of plasma concentration of MCP-1 and IP-10 was 167.0 pg/ml (+/−119.0 pg/ml) and 233.0 pg/ml (+/−232.0 pg/ml), respectively. These levels were significantly more reduced than those in patients who subsequently did not have severe lung toxicity, with corresponding mean reductions of 38.6 pg/ml (+/−62.2 pg/ml), and 4.0 pg/ml (+/−76.7 pg/ml), respectively ($p = 0.05$). At 24-hours post the first fraction of radiation, patients with a reduction in concentrations of Eotaxin and IL-6 levels subsequently sustained respiratory toxicity (**Figure 6A, 6B**). For those patients with severe toxicities, the mean (+/− standard deviation) decrease in Eotaxin and IL-6 from pre-treatment levels was 6.8 pg/ml (+/−36.6 pg/ml) and 8.9 pg/ml (+/−8.8 pg/ml), respectively. By comparison, those patients without toxicity had increased Eotaxin and IL-6 levels by a mean (+/− standard deviation) of 31.9 pg/ml (+/−20.4 pg/ml), $p = 0.03$ and 1.4 (+/−2.5 pg/ml) $p = 0.04$, respectively. In

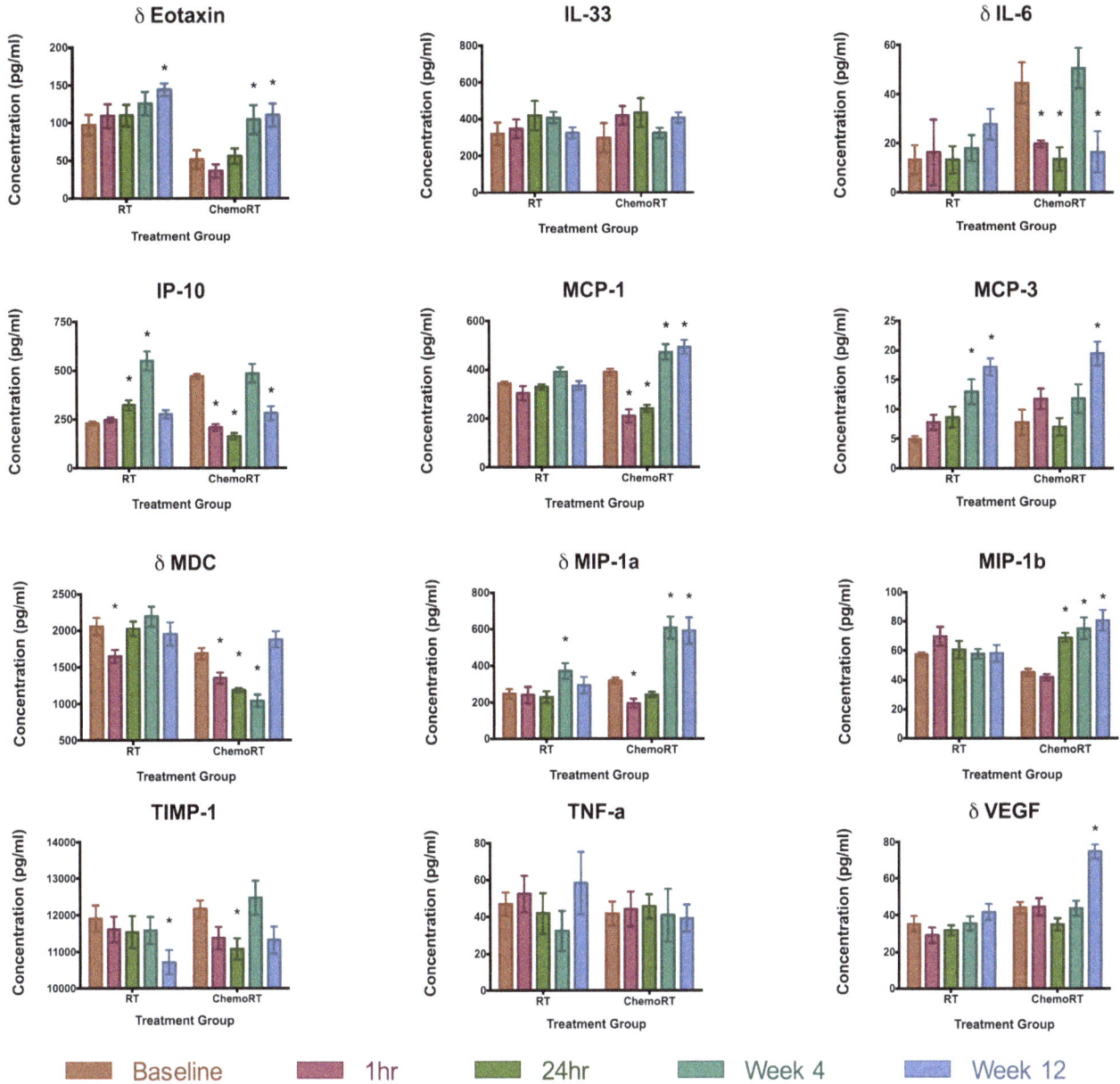

Figure 1. Mean plasma (+/− standard error) cytokine levels at each time point during radiotherapy (1-hour, 24-hours, 4-weeks) and after radiotherapy (12-weeks). Levels are grouped into treatment type (ChemoRT vs RT alone). Cytokines in which the variation is different dependent upon treatment groups are marked with a *delta* (δ). Within each cytokine graph, an *asterisk* (*) denotes at which time the level of plasma cytokines are significantly different from the baseline level, with corrections for multiple comparisons performed using Dunnett's method.

contrast, elevated concentrations of TIMP-1 at 24-hours were associated with more severe toxicity (**Figure 6C**). The mean increase (+/− standard deviation) of TIMP-1 in those with severe lung toxicity was 337.0 pg/ml (+/−867.0 pg/ml), versus a decrease of 762.0 pg/ml (+/−292.0 pg/ml) for those without, $p = 0.02$. None of the cytokines tested were significantly different in those with or without severe lung toxicities at 4 weeks into treatment or 12 weeks after completion of treatment.

Discussion

In this study, we discovered that severe lung toxicity (CTCAE grade 2 or higher) is associated with significant change in some cytokine plasma concentrations from pre-treatment levels in

patients receiving definitive RT for NSCLC. Specifically, severe toxicity was associated with the detection of depressed levels of IP-10 and MCP-1 at 1-hour post irradiation, as well as lower levels of Eotaxin and IL-6 at 24 hours post-irradiation as compared to patients who did not subsequently develop severe lung toxicity (**Figure 5 and 6**). Levels of TIMP-1 at 24-hours were significantly elevated in patients with severe lung toxicity compared to those without. These early prognostic variations in cytokine levels may represent an individual's lung sensitivity and subsequent risk for lung toxicity and fibrosis after repetitive exposure to a radiation insult. The detection of a prognostic signal after the first fraction of a 30-fraction course of RT is particular clinical significance, as this may allow for an early intervention during the treatment course. From a clinical perspective, this may

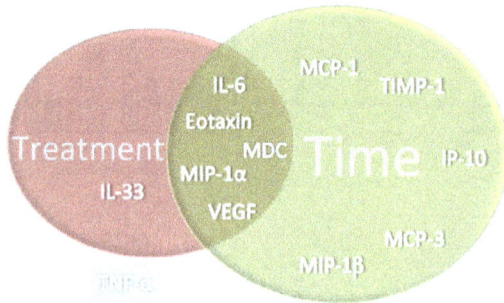

Figure 2. Venn diagram depicting the relationship between variation of plasma cytokine levels and the variables of time and treatment. MCP-1, TIMP-1, IP-10, MCP-3 and MIP-1β vary differently across the different timepoints sampled, but are have similar variance between treatment groups (RT alone vs ChemoRT). IL-6, Eotaxin, MDC, MIP-1α and VEGF levels varied across both time and treatment group. TNF-α levels were no different across time points sampled or treatment delivered.

manifest as more intensive inter-fractional toxicity assessment and early supportive intervention, potentially with corticosteroids. Alternatively, early prognostic cytokine signature may allow for personalised biological adaptation of the treatment course through increasing or decreasing the intensity of treatment. These cytokine signals were generally less apparent at 4 weeks into therapy. We hypothesize that during a long 6-week course radiation therapy that patients may adjust to the repeated exposures and manifest a less brisk acute inflammatory cytokine response than the initial exposures at the start of therapy. Interpretation of the prognostic significance of cytokines 12 weeks post-therapy is challenging, as this is a complex timepoint due to variability introduced by a broad spectrum of individual patient clinical outcome. At this time some patients will have complete tumoral responses to therapy, whilst others will have stable or progressive disease. In addition, this time may also be influenced by nutritional deficits induced by treatment related esophagitis and dysphagia. The results of this study indicate that early changes in these cytokines should be further investigated as prognostic markers of likelihood of toxicity within patients receiving definitive irradiation for NSCLC.

From a mechanistic perspective, these putative biomarkers of lung injury appear to be reasonable prognostic candidates due to their pro-inflammatory role in various disease states. MCP-1 causes cellular activation of specific functions related to host defence and inflammation, including monocyte, granulocytes and lymphocyte migration [43]. IP-10 also selectively stimulates directional migration of T-cells and monocytes, as well as participating in T cell adhesion [16]. TIMP-1 acts to down-regulate the profibrotic response and is associated with the degree of inflammation in the mucosa of patients with chronic inflammatory states (such as inflammatory bowel disease) [40]. In previous studies, early reduction (within 3–6 hours) of serum levels of IP-10, MCP-1 and TIMP-1 have been previously demonstrated in murine strains more sensitive to fibrosis [C57BL/6] in comparison to more tolerant strains [C3H/HeN] [29,43,44]. Later after irradiation, induction of IP-10 and MCP-1 mRNA gene expression up to 6 months post RT in murine models

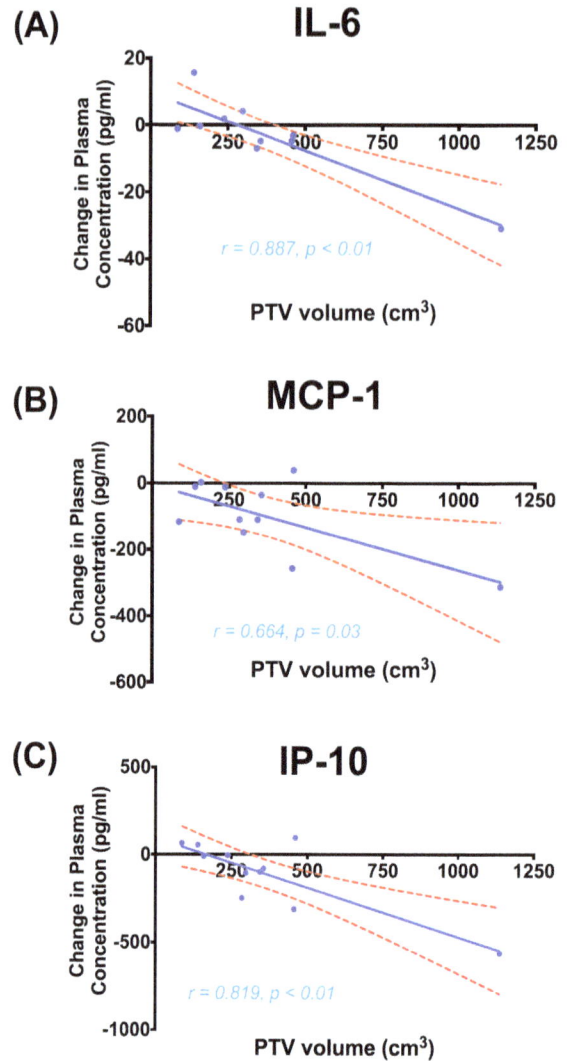

Figure 3. Correlation between irradiated PTV volume and the change in plasma cytokine concentration from baseline to 1-hour post irradiation for IL-6 (A), MCP-1 (B), and IP-10 (C). Linear-regression line (blue) is displayed with 95% confidence intervals (red broken line).

is thought subsequently lead to late tissue fibrosis and subsequent lung damage [16,45]. IL-6 is a pleiotropic cytokine secreted by T-lymphocytes and involved in maturation of B-lymphocytes, and is thought to mediate clinical fever and regulates inflammation and fibrosis through immune cells [38]. Chen et al. [26] observed early reductions in IL-6 cytokine in patients who sustained pneumonitis, similar to findings in the present study. Eotaxin is a primary mediator of IgE-related allergic inflammatory reactions in lung, which are characteristically associated with an early, transient accumulation of neutrophils and subsequent neutrophil-dependent acute lung inflammatory injury [34]. Plasma levels of Eotaxin have been previously documented in a murine model to initially decrease then subsequently increase over days, similar to the temporal pattern observed in the present study [29].

The concentrations of several cytokines were also demonstrated to be dependent on the dose to normal lung tissue and the irradiated tumour volume (**Figures 2 and 3**). In 3D conformal radiotherapy, the dose to the irradiated normal lung tissue (the MLD) is influenced by the number of beams selected, beam angles

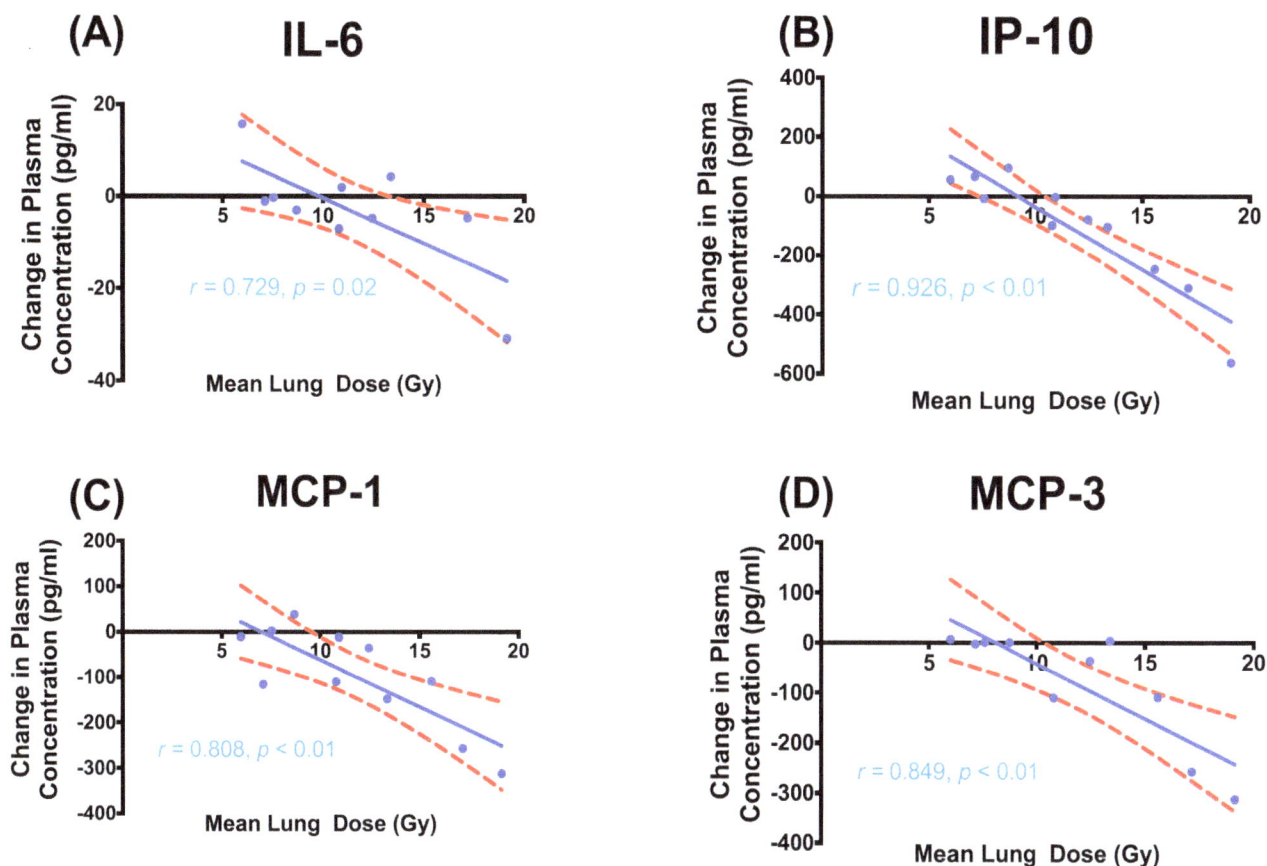

Figure 4. Correlation between mean lung dose (MLD) and the change in plasma cytokine concentration from baseline to 1-hour post irradiation for IL-6 (A), IP-10 (B), MCP-1 (C) and MCP-3 (D). Linear-regression line (blue) is displayed with 95% confidence intervals (red broken line).

employed, the location of the tumour, and the degree of sparing of the uninvolved lung. In contrast, the target volume (PTV) is a direct function of the tumour volume and geometric margins applied to account for microscopic disease and delivery error. In this study, both measurements were related to cytokine concentrations at 1-hour post irradiation. Cytokine levels were linearly correlated to the dose to the irradiated normal lung tissue (the MLD), with depression of IP-10, MCP-1 and MCP-3 being the most strongly correlated. Reduction in circulating levels of IL-6, MCP-1 and IP-10 at 1-hour were also correlated with the PTV, although this association was less robust. In particular, this relationship was influenced in particular by the response in one specific patient with a large PTV volume of 1138 cm³. Despite this potential confounder, a plausible explanation for a differential relationship between cytokine levels and PTV/MLD is that larger radiation fields (with a larger PTV) which cover mediastinal nodal involvement may not necessarily traverse larger volumes of normal lung parenchyma than smaller tumour volumes located deeper within the lung tissue (which may have a larger MLD). By the same token, this may indicate that IL-6, which significantly correlated with PTV volume but not MLD, may be influenced more by a more generic response in irradiated tissues rather than by response specifically in the bronchoalveolar environment. Conversely, MCP-3, which was correlated to MLD but not PTV volume, may be more specific to a response in pulmonary tissue. IL-6 functions to stimulate the growth and differentiation of B and T lymphocytes and is synthesised by a variety of cells in the lung

parenchyma, including alveolar macrophages, lung fibroblasts and type II pneumocytes [38,44]. The MCP family, including MCP-3 (CCL7), attract cells through activation of their cognate receptor, CC-chemokine receptor (CCR)-2 [32]. A target volume/cytokine response relationship with partial lung irradiation has yet to be described in the context of these cytokines. In addition, a previous study by Arpin et al [25] has previously documented a relationship between serum levels of IL-6 and MLD, however, to our knowledge our findings of a correlation between MLD and IP-10, MCP-1 and MCP-3 are novel discoveries. Further investigation is required to assess whether patients with cytokine responses beyond that of the levels predicted by irradiated target volume or dose to normal lung are more prone to severe toxicity. Again, the ability to detect these dose and volume dependent cytokines at such an early time point during treatment suggests a potential for these biomarkers to be of great clinical utility.

We were able to demonstrate differential plasma cytokine concentrations in patients undergoing RT alone compared to those treated with chemoRT for NSCLC. Whilst previous clinical studies assessing cytokine concentration during lung irradiation have tested for associations between use of chemotherapy and risk of pneumonitis, these have not assessed the differential patterns in plasma cytokine levels for patients receiving RT alone versus chemoRT [24,25]. In the present study, the overall concentration of cytokines were different dependent upon treatment group for Eotaxin, IL-33, IL-6, MDC, MIP-1α and VEGF. In both groups, the peak elevation in plasma cytokine concentration for MIP-1α

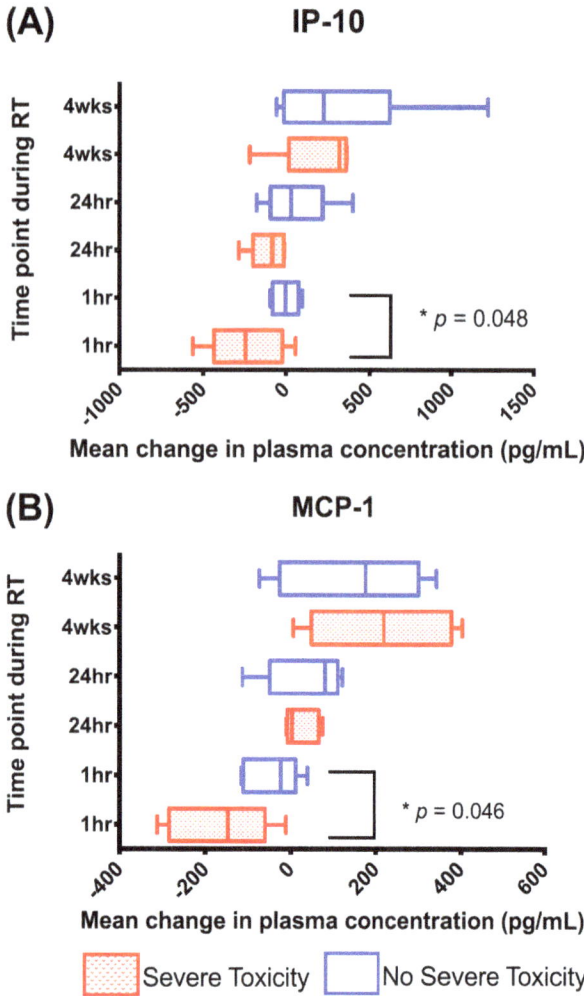

Figure 5. Box and Whisker plot demonstrating median population change of cytokine levels from baseline for IP-10 and MCP-1 during treatment delivery, with 5th–95th centiles. Statistically significant differences between patients with severe respiratory toxicities [shaded boxes] and those without [open boxes] are highlighted with associated p - values.

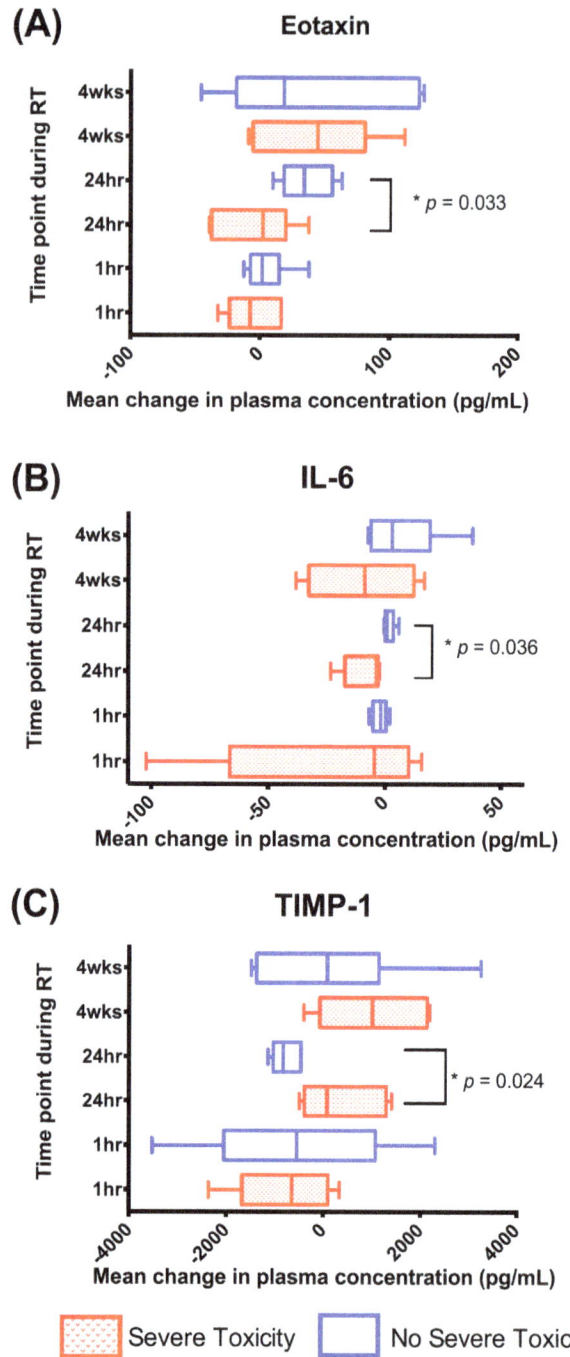

Figure 6. Box and Whisker plot demonstrating median population change of cytokine levels from baseline for Eotaxin, IL-6 and TIMP-1 during treatment delivery, with 5th–95th centiles. Statistically significant differences between patients with severe respiratory toxicities [shaded boxes] and those without [open boxes] are highlighted with associated p - values.

occurred at 4 weeks into treatment, whilst peak elevation in Eotaxin and VEGF occurred at 12 weeks after treatment completion. Changes in plasma concentrations of cytokines varied considerably between treatment groups at other time points. Our findings suggest that future studies investigating the kinetics of these plasma levels should not uniformly group patients receiving RT alone and chemoRT together.

In several previous clinical studies of patients treated for NSCLC there have been contradictory reports of associations between RT induced blood cytokine levels and clinical toxicity. A study by Arpin et al. [25] also found changes in IL-6 to be prognostic for radiation pneumonitis, along with combined covariations of IL-6 and IL-10. Similar to our study, TNF-α was not correlated to toxicity. In a study by Crohns et al. [27], VEGF, TNF-α, IL-1β, IL-6 and IL-8 levels in the serum were analysed in patients receiving various regimes of RT with mean (range) dose of 46.9 Gy (30–60 Gy). These investigators were not able to demonstrate any significant changes from the baseline levels of these cytokines (including IL-6) at two weeks or 3 months after the commencement of RT. Similarly, a study by Rűbe et al. [28]

measured levels of TGF-β and IL-6 weekly during RT and could not find correlation with symptomatic pneumonitis and plasma level cytokines levels. In their study, patients received a range of treatments including definitive RT alone, definitive chemoRT, low dose twice weekly palliative accelerated RT. This is in contrast to several reports from Anscher's group [21–23,46] and a study from Zhao et al. [24] indicating that elevation of TGF-β late during RT is associated with risk of pulmonary toxicity. In a study by Chen et al. [26], levels of IL-1α and IL-6 but not TNFα were consistently elevated prior to and throughout treatment in patients whom developed radiation pneumonitis. In this same patient cohort, levels of e-selectin, l-selectin, TGF-β1 and bFGF varied but were not correlated with radiation pneumonitis. Again this cohort had considerable treatment heterogeneity, with 4 of the total 24 patients not having NSCLC, with an average delivered radiation dose of 60–64 Gy, and 3 of 15 patients having had their chemotherapy delivered neoadjuvantly prior to the radiation course. In the context of the previous literature, the strength of our study lies in the standardised treatment characteristics in our patient cohort and the large panel of cytokines tested. We discovered that in addition to IL-6, early changes in plasma levels of four previously unreported cytokines (IP-10, MCP-1, Eotaxin and TIMP-1) were associated with the risk of pulmonary toxicity.

It is important to recognise the limitations of this study. In this study we were not able to control for potential confounding effects of patient stage and volume of irradiated amongst the RT and chemoRT cohorts. Patients treated with RT alone generally had a lower disease stage, smaller PTV and MLD than those patients treated with chemoRT.

Conclusions

We describe the association between plasma cytokine concentrations and clinical toxicities in a prospective cohort of patients with NSCLC treated with a standardised radiation dose and technique. In our cohort of patients, we observed that inflammatory cytokines were induced in patients with NSCLC both during and after RT. We were able to demonstrate that the concentrations of IL-6, MCP-1, MCP-3 and IP-10 correlated with the mean lung dose. This suggests that these cytokines have potential as *in-vivo* 'biodosimeters' of individual radiation dose. In our cohort we also observed the levels of cytokines after RT were different in patients who received concurrent chemotherapy in Eotaxin, IL-33,

IL-6, MDC, MIP-1α and VEGF. The change in plasma concentration of IP-10 and MCP-1 at 1-hour post irradiation, and Eotaxin, IL-6 and TIMP-1 at 24-hour post irradiation, were significantly different in those patients who sustained severe lung toxicities. Early decreases in IP-10 and MCP-1 at 1-hour, Eotaxin and IL-6 at 24-hours, and increases in TIMP-1 at 24-hours into RT should be externally validated in other cohorts as potential future biomarkers of severe lung toxicity.

Supporting Information

Table S1   Individual patient characteristics and respiratory toxicity.

Table S2   Two-way ANOVA testing the effect of treatment (chemoRT vs RT alone) and sample time point. All cytokines in which the plasma concentrations varied significantly dependent upon the treatment group are highlighted in bold and by an asterisk (*). The Interaction between treatment group and sample time point is tested, and residual errors are given.

Acknowledgments

The authors gratefully acknowledge the editorial review of this manuscript by Dr Carl Sprung and Dr Helen Forrester from Monash University, and Dr Nicole Haynes from the Peter MacCallum Cancer Centre.

Author Contributions

Conceived and designed the experiments: SS MPM TK PL DB OM. Performed the experiments: SS NB JS OM PL. Analyzed the data: SS OM PL DB MPM TK. Contributed reagents/materials/analysis tools: PL NB JS OM. Contributed to the writing of the manuscript: SS MPM TK NB JS PL DB OM.

References

1. Siegel R, Naishadham D, Jemal A (2012) Cancer statistics, 2012. CA: a cancer journal for clinicians.
2. Ettinger DS, Akerley W, Bepler G, Blum MG, Chang A, et al. (2010) Non small cell lung cancer. Journal of the National Comprehensive Cancer Network 8: 740–801.
3. Australian Cancer Network Colorectal Cancer Guidelines Revision Committee (2004) Clinical Practice Guidelines for the Prevention, Diagnosis and Management of Lung Cancer. The Cancer Council Australia and Australian Cancer Network, National Health and Medical Research Council Canberra.
4. Delaney G, Jacob S, Featherstone C, Barton M (2005) The role of radiotherapy in cancer treatment: estimating optimal utilization from a review of evidence-based clinical guidelines. Cancer 104: 1129–1137.
5. Mauguen A, Le Péchoux C, Saunders MI, Schild SE, Turrisi AT, et al. (2012) Hyperfractionated or Accelerated Radiotherapy in Lung Cancer: An Individual Patient Data Meta-Analysis. Journal of Clinical Oncology 30: 2788–2797.
6. Fay M, Tan A, Fisher R, MacManus M, Wirth A, et al. (2005) Dose-volume histogram analysis as predictor of radiation pneumonitis in primary lung cancer patients treated with radiotherapy. International Journal of Radiation Oncology* Biology* Physics 61: 1355–1363.
7. Emami B, Lyman J, Brown A, Cola L, Goitein M, et al. (1991) Tolerance of normal tissue to therapeutic irradiation. International Journal of Radiation Oncology* Biology* Physics 21: 109–122.
8. Palma DA, Senan S, Tsujino K, Barriger RB, Rengan R, et al. (2013) Predicting radiation pneumonitis after chemoradiation therapy for lung cancer: an international individual patient data meta-analysis. Int J Radiat Oncol Biol Phys 85: 444–450.
9. Graham MV, Purdy JA, Emami B, Harms W, Bosch W, et al. (1999) Clinical dose-volume histogram analysis for pneumonitis after 3D treatment for non-small cell lung cancer (NSCLC). Int J Radiat Oncol Biol Phys 45: 323–329.
10. Rodemann HP, Blaese MA. Responses of normal cells to ionizing radiation; 2007. Elsevier. 81–88.
11. Brush J, Lipnick SL, Phillips T, Sitko J, McDonald JT, et al. (2007) Molecular mechanisms of late normal tissue injury. Elsevier. 121–130.

12. Vujaskovic Z, Marks LB, Anscher MS (2000) The physical parameters and molecular events associated with radiation-induced lung toxicity. Elsevier. 296–307.
13. Mehta V (2005) Radiation pneumonitis and pulmonary fibrosis in non small-cell lung cancer: Pulmonary function, prediction, and prevention. International Journal of Radiation Oncology* Biology* Physics 63: 5–24.
14. Provatopoulou X, Athanasiou E, Gounaris A (2008) Predictive markers of radiation pneumonitis. Anticancer research 28: 2421–2432.
15. McBride WH, Chiang C-S, Olson JL, Wang C-C, Hong J-H, et al. (2004) A Sense of Danger from Radiation 1. Radiation research 162: 1–19.
16. Johnston CJ, Williams JP, Okunieff P, Finkelstein JN (2009) Radiation-induced pulmonary fibrosis: examination of chemokine and chemokine receptor families.
17. Thomson AW, Lotze MT (2003) The Cytokine Handbook, Two-Volume Set: Gulf Professional Publishing.
18. Ding N-H, Li JJ, Sun L-Q (2013) Molecular Mechanisms and Treatment of Radiation-Induced Lung Fibrosis. Current drug targets 14: 1247.
19. Talas M, Szolgay E, Varteresz V, Koczkas G (1972) Influence of acute and fractional X-irradiation on induction of interferon in vivo. Arch Gesamte Virusforsch 38: 143–148.
20. Desai S, Kumar A, Laskar S, Pandey B (2013) Cytokine profile of conditioned medium from human tumor cell lines after acute and fractionated doses of gamma radiation and its effect on survival of bystander tumor cells. Cytokine 61: 54–62.
21. Fu X-L, Huang H, Bentel G, Clough R, Jirtle RL, et al. (2001) Predicting the risk of symptomatic radiation-induced lung injury using both the physical and biologic parameters V$_{30}$ and transforming growth factor β. International Journal of Radiation Oncology* Biology* Physics 50: 899–908.
22. Anscher MS, Murase T, Prescott DM, Marks LB, Reisenbichler H, et al. (1994) Changes in plasma TGF [beta] levels during pulmonary radiotherapy as a predictor of the risk of developing radiation pneumonitis. International Journal of Radiation Oncology* Biology* Physics 30: 671–676.
23. Anscher MS, Kong F-M, Andrews K, Clough R, Marks LB, et al. (1998) Plasma transforming growth factor β1 as a predictor of radiation pneumonitis. International Journal of Radiation Oncology* Biology* Physics 41: 1029–1035.

24. Zhao L, Wang L, Ji W, Wang X, Zhu X, et al. (2009) Elevation of Plasma TGF-β1 During Radiation Therapy Predicts Radiation-Induced Lung Toxicity in Patients With Non-Small-Cell Lung Cancer: A Combined Analysis From Beijing and Michigan. International Journal of Radiation Oncology*Biology*Physics 74: 1385–1390.

25. Arpin D, Perol D, Blay J-Y, Falchero L, Claude L, et al. (2005) Early variations of circulating interleukin-6 and interleukin-10 levels during thoracic radiotherapy are predictive for radiation pneumonitis. Journal of clinical oncology 23: 8748–8756.

26. Chen Y, Rubin P, Williams J, Hernady E, Smudzin T, et al. (2001) Circulating IL-6 as a predictor of radiation pneumonitis. International Journal of Radiation Oncology* Biology* Physics 49: 641–648.

27. Crohns M, Saarelainen S, Laine S, Poussa T, Alho H, et al. (2010) Cytokines in bronchoalveolar lavage fluid and serum of lung cancer patients during radiotherapy-association of interleukin-8 and VEGF with survival. Cytokine 50: 30–36.

28. Rübe CE, Palm J, Erren M, Fleckenstein J, König J, et al. (2008) Cytokine plasma levels: reliable predictors for radiation pneumonitis? PLoS One 3: e2898.

29. Zhang M, Yin L, Zhang K, Sun W, Yang S, et al. (2012) Response patterns of cytokines/chemokines in two murine strains after irradiation. Cytokine 58: 169–177.

30. Rossi DL, Vicari AP, Franz-Bacon K, McClanahan TK, Zlotnik A (1997) Identification through bioinformatics of two new macrophage proinflammatory human chemokines: MIP-3alpha and MIP-3beta. J Immunol 158: 1033–1036.

31. Conti I, Rollins BJ (2004) CCL2 (monocyte chemoattractant protein-1) and cancer. Elsevier. 149–154.

32. Malik IA, Moriconi F, Sheikh N, Naz N, Khan S, et al. (2010) Single-dose gamma-irradiation induces up-regulation of chemokine gene expression and recruitment of granulocytes into the portal area but not into other regions of rat hepatic tissue. Am J Pathol 176: 1801–1815.

33. Lilly C, Nakamura H, Kesselman H, Nagler-Anderson C, Asano K, et al. (1997) Expression of eotaxin by human lung epithelial cells: induction by cytokines and inhibition by glucocorticoids. Journal of Clinical Investigation 99: 1767.

34. Guo R-F, Lentsch AB, Warner RL, Huber-Lang M, Sarma JV, et al. (2001) Regulatory effects of eotaxin on acute lung inflammatory injury. The Journal of Immunology 166: 5208–5218.

35. Yagami A, Orihara K, Morita H, Futamura K, Hashimoto N, et al. (2010) IL-33 mediates inflammatory responses in human lung tissue cells. J Immunol 185: 5743–5750.

36. Neta R (1997) Modulation with cytokines of radiation injury: suggested mechanisms of action. Environmental health perspectives 105: 1463.

37. Sonnenberg GF, Nair MG, Kirn TJ, Zaph C, Fouser LA, et al. (2010) Pathological versus protective functions of IL-22 in airway inflammation are regulated by IL-17A. J Exp Med 207: 1293–1305.

38. Chen Y, Williams J, Ding I, Hernady E, Liu W, et al. (2002) Radiation pneumonitis and early circulatory cytokine markers. Seminars in Radiation Oncology 12: 26–33.

39. Gallet P, Phulpin B, Merlin J-L, Leroux A, Bravetti P, et al. (2011) Long-term alterations of cytokines and growth factors expression in irradiated tissues and relation with histological severity scoring. PloS one 6: e29399.

40. Naito Y, Yoshikawa T (2005) Role of matrix metalloproteinases in inflammatory bowel disease. Molecular aspects of medicine 26: 379–390.

41. Feng M, Kong F-M, Gross M, Fernando S, Hayman JA, et al. (2009) Using Fluorodeoxyglucose Positron Emission Tomography to Assess Tumor Volume During Radiotherapy for Non-Small-Cell Lung Cancer and Its Potential Impact on Adaptive Dose Escalation and Normal Tissue Sparing. International Journal of Radiation Oncology* Biology* Physics 73: 1228–1234.

42. Rübe CE, Wilfert F, Palm J, König J, Burdak-Rothkamm S, et al. (2004) Irradiation induces a biphasic expression of pro-inflammatory cytokines in the lung. Strahlentherapie und Onkologie 180: 442–448.

43. Cappuccini F, Eldh T, Bruder D, Gereke M, Jastrow H, et al. (2011) New insights into the molecular pathology of radiation-induced pneumopathy. Radiotherapy and Oncology 101: 86–92.

44. Ao X, Zhao L, Davis MA, Lubman DM, Lawrence TS, et al. (2009) Radiation produces differential changes in cytokine profiles in radiation lung fibrosis sensitive and resistant mice. J Hematol Oncol 2.

45. Johnston CJ, Wright TW, Rubin P, Finkelstein JN (1998) Alterations in the expression of chemokine mRNA levels in fibrosis-resistant and-sensitive mice after thoracic irradiation. Experimental lung research 24: 321–337.

46. Anscher MS, Kong F-M, Marks LB, Bentel GC, Jirtle RL (1997) Changes in plasma transforming growth factor beta during radiotherapy and the risk of symptomatic radiation-induced pneumonitis. International Journal of Radiation Oncology* Biology* Physics 37: 253–258.

20

Overexpression of Glutathione Transferase E7 in Drosophila Differentially Impacts Toxicity of Organic Isothiocyanates in Males and Females

Aslam M. A. Mazari[1✎]**, Olle Dahlberg**[2✎]**, Bengt Mannervik**[1*]**, Mattias Mannervik**[2*]

1 Department of Neurochemistry, The Wenner-Gren Institute, Stockholm University, Stockholm, Sweden, **2** Department of Molecular Biosciences, The Wenner-Gren Institute, Stockholm University, Stockholm, Sweden

Abstract

Organic isothiocyanates (ITCs) are allelochemicals produced by plants in order to combat insects and other herbivores. The compounds are toxic electrophiles that can be inactivated and conjugated with intracellular glutathione in reactions catalyzed by glutathione transferases (GSTs). The *Drosophila melanogaster* GSTE7 was heterologously expressed in *Escherichia coli* and purified for functional studies. The enzyme showed high catalytic activity with various isothiocyanates including phenethyl isothiocyanate (PEITC) and allyl isothiocyanate (AITC), which in millimolar dietary concentrations conferred toxicity to adult *D. melanogaster* leading to death or a shortened life-span of the flies. *In situ* hybridization revealed a maternal contribution of GSTE7 transcripts to embryos, and strongest zygotic expression in the digestive tract. Transgenesis involving the GSTE7 gene controlled by an actin promoter produced viable flies expressing the GSTE7 transcript ubiquitously. Transgenic females show a significantly increased survival when subjected to the same PEITC treatment as the wild-type flies. By contrast, transgenic male flies show a significantly lower survival rate. Oviposition activity was enhanced in transgenic flies. The effect was significant in transgenic females reared in the absence of ITCs as well as in the presence of 0.15 mM PEITC or 1 mM AITC. Thus the GSTE7 transgene elicits responses to exposure to ITC allelochemicals which differentially affect life-span and fecundity of male and female flies.

Editor: Efthimios M. C. Skoulakis, Alexander Fleming Biomedical Sciences Research Center, Greece

Funding: This work was supported by grants from the Swedish Research Council and the Swedish Cancer Society. The funders had no role in study design, data collection and analysis, decision to publish, or preparation of the manuscript.

Competing Interests: The Drosophila GSTE7 open reading frame was kindly provided by DNA 2.0, Inc., Menlo Park, CA.

* Email: bengt.mannervik@neurochem.su.se (BM); mattias.mannervik@su.se (MM)

✎ These authors contributed equally to this work.

Introduction

Organic isothiocyanates (ITCs) are reactive biomolecules that occur in plants, particularly abundantly in cruciferous species. The compounds are predominantly stored as unreactive glucosinolates from which they are released by hydrolysis catalyzed by the enzyme myrosinase [1]. ITCs are regarded as important contributors to the protection of plants from attacks by insects and microorganisms, and the release of the compounds is triggered by insults to the plant tissue. The interplay between plants and the offending insects have obviously evolved such that the emergence of insecticidal compounds has been countered by the evolution of detoxication enzymes in insects. For example, glucosinolates of brassicaceous plants stimulate oviposition of specialist insects such as *Pieridae* butterflies, and the caterpillars feeding on the plants resist the ITC toxicity [2] Studies of *Arabidopsis thaliana*, a plant displaying glucosinolates as their primary defensive trait, demonstrate that six different ITC chemotypes are present in different proportions in separate geographical populations [3]. The *A. thaliana* ITC chemotype was altered in five generations in response to experimental exposure to different herbivorous aphids feeding on the plants.

ITCs are strong electrophiles that exert toxicity by reacting with sulfhydryl groups and other nucleophilic chemical residues in biological tissues. The most abundant low-molecular cellular thiol nucleophile is glutathione, which inactivates ITCs by the formation of dithiocarbamates [4]. This reaction is efficiently catalyzed by many glutathione transferases (GSTs) [5], and it has been proposed that feeding on mustard plants and ITC activity of GSTs has coevolved [6]. It would appear likely that insect GSTs provide protection against ITC toxicity, but this notion has not been experimentally tested. In the present study we have created a transgenic *Drosophila melanogaster* overexpressing the enzyme GSTE7, which is shown to be highly active with ITC substrates *in vitro*, and studied the effect of allyl isothiocyanate (AITC) and phenethyl isothiocyanate (PEITC) on the transgenic flies.

Materials and Methods

Synthesis of DNA encoding GSTE7 and plasmid for heterologous expression of the enzyme

The GSTE7 open reading frame of *Drosophila melanogaster* gene CG17531 was custom synthesized by DNA 2.0, Inc., Menlo Park, CA, and provided in the pJ201 plasmid as kind gift by Dr.

Table 1. Assay conditions for the specific activity determination with alternative substrates.

Substrate	GSH (mM)	Substrate (mM)	Wavelength (nm)	$\Delta\epsilon$ (mM^{-1} cm)	pH
Allyl isothiocyanate	1	0.4	274	7.45	6.5
Benzyl isothiocyanate	1	0.4	274	9.25	6.5
Propyl isothiocyanate	1	0.4	274	8.35	6.5
Phenethyl isothiocyanate	1	0.4	274	8.89	6.5
Sulforaphane	1	0.4	274	8.00	6.5
1-chloro-2,4-dinitrobenzene	1	1	340	9.6	6.5

All measurements were performed in 0.1 M sodium phosphate buffer, pH 6.5 at 30°C. The stock solutions for Isothiocyanates were prepared in acetonitrile (2% final concentration in the assay), and that for CDNB in ethanol (5% final concentration in the assay).

Claes Gustafsson. For heterologous expression of the GSTE7 protein a synonomous nucleotide sequence with a codon usage optimized for *Escherichia coli* was synthesized and a His$_6$-tag introduced at the N-terminus. The optimized nucleotide sequence was provided in the pJexpress401 expression vector and used for transformation of electroporation-competent *E. coli* XL-1 Blue cells by electroporation. Briefly, 2 µl of plasmid DNA (17 ng/µl) and 48 µl of *E. coli* cells were gently mixed and placed in a Gene pulser cuvette with 0.1 cm electrode gap (Bio-Rad). After an electric pulse of 4.5 s at a voltage of 1.25 kV, 960 µl of LB medium was added followed by gentle mixing by repeated pipetting. Finally, the cells were incubated at 37°C in a 10 ml tube containg LB medium and agitated at 200 rpm for 45 min. After incubation, the cells were spread on LB agar plates containing 50 µg/ml of kanamycin and incubated for 16 h at 37°C.

Protein expression and purification

A starter culture of 50 ml of 2×YT medium containing 50 µg/ml of kanamycin was inoculated with a single colony of freshly transformed cells and incubated for 16 h with shaking at 200 rpm at 37°C. After incubation, 5 ml starter culture was taken to inoculate 500 ml of 2×YT medium containing 50 µg/ml kanamycin. The inoculated cells were incubated at 30°C in an incubate-shaker at 200 rpm till the required OD$_{600}$ 0.46 was obtained. Subsequently the GST expression was induced with 0.1 mM isopropyl-β-D-thiogalactopyranoside (IPTG). After incubation for 16 h at 37°C, the cells were harvested at 7000 rpm for 10 min at 4°C (Avanti J-20 XP Beckman Coulter USA). The supernatant was discarded and the pellet containing bacteria was resuspended in 30 ml of ice-cold buffer A (20 mM sodium phosphate pH 7.4, containing 85 mM imidazole, 500 mM NaCl,

10 mM β-mercaptoethanol and 0.02% sodium azide) and 0.2 mg/ml of lysozyme and one tablet of EDTA-free protease inhibitor (Roche Germany). After incubation for 30 min on an ice bath, the cells were lysed by sonication (Vibra cell USA) for 5×20 s at an output control of 7.5 with an interval of 1 min on an ice bath to avoid heating the sample. The resultant suspension was centrifuged at 27200 g for 1 h at 4°C. The pellet was discarded and the supernatant containing the protein was incubated with pre-equilibrated Ni-IMAC gel on an ice bath for 30 min. After incubation, the gel was washed extensively with milli-Q water and packed into a column. The unbound proteins were further washed away by buffer A and the bound protein eluted with buffer B (20 mM sodium phosphate pH 7.4, containing 500 mM imidazole, 500 mM NaCl, 10 mM β-mercaptoethanol and 0.02% sodium azide). The eluted fractions were pooled and dialyzed overnight against 10 mM Tris HCl buffer pH 7.8, containing 0.2 mM DTT and 1 mM EDTA. The protein was concentrated and the concentration was measured by Bradford Standard Assay (Bio-Rad USA). The homogeneity and the purity of the enzyme was confirmed by SDS-PAGE analysis, using a 12.5% (w/v) polyacrylamide resolving gel [7]. The purified enzyme was stored in aliquots at −80°C.

Enzyme activity assays of GSTE7

1-Chloro-2,4-dinitrobenzene (CDNB) was used as standard substrate for monitoring the enzyme purification. The reaction was measured at 30°C by the increase in absorbance ($\Delta\epsilon_{340nm} = 9600$ M^{-1}cm^{-1}) accompanying GSH conjugation (Table 1). For screening assays allyl isothiocyanate (AITC) and phenethyl isothiocyanate (PEITC) were also included (Table 1). Reaction with the ITC substrates was detected at 274 nm with a

Table 2. Specific activities of purified GSTE7 with alternative substrates.

Substrate	Specific activity (µmol/min per mg of protein)
1-chloro-2,4-dinitrobenzene (CDNB)	36.5±1.12
Allyl isothiocyanate	11.2±0.48
Benzyl isothiocyanate	16.0±0.76
Propyl isothiocyanate	7.9±0.49
Phenethyl isothiocyanate	7.5±1.37
Sulforaphane	23.8±1.08

The data are means ± SD of 3 replicate measurements and the background activities were corrected by using the same concentration of solvent without enzyme.

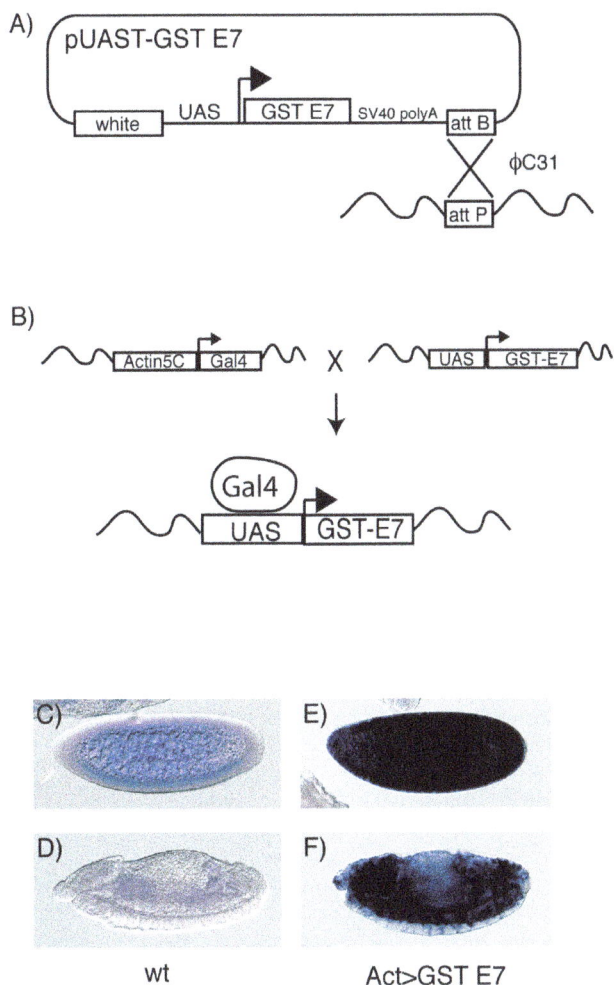

Figure 1. Overexpression of the GSTE7 protein in fruit flies (*Drosophila melanogaster*). (A) Scheme outlining the binary Gal4-UAS system [10] used to induce ubiquitous overexpression. The coding sequence of the *Drosophila* GSTE7 gene was inserted in the pUAST attB vector [9] containing an upstream activating sequence (UAS) and a downstream attB crossover site. The plasmid was introduced by injection into fly embryos carrying an attP landing site on chromosome 3 and inserted into the chromosome by way of ΦC31-mediated integration. (B) Flies overexpressing the GSTE7 transgene were obtained by crossing the UAS-GSTE7 flies with a strain bearing the Actin-Gal4 driver, which induces ubiquitous expression. (C–F) Embryonal expression of *D. melanogaster* GSTE7 detected by *in situ* mRNA hybridization. Fly embryos were stained with a digoxigenin-labeled probe recognizing GSTE7 by whole-mount *in situ* hybridization, and are oriented with anterior to the left and dorsal up. (C) Noninduced pre-cellular embryo showing endogenous GSTE7 mRNA, which is maternally contributed and therefore distributed ubiquitously in the embryo. (D) Stage 13 embryo showing GSTE7 expression lower than in the early embryo, but with pronounced staining in the midgut. In the presence of the Actin-Gal4 driver, the transgenic tissues displayed high ubiquitous expression of GSTE7 in both pre-cellular (E) and stage 13 (F) embryos.

$\Delta\epsilon_{274nm}$ at 7450 $M^{-1}cm^{-1}$ and 8890 $M^{-1}cm^{-1}$ for AITC and PEITC, respectively [5].

In situ hybridization

The GSTE7 coding sequence was PCR amplified from the pJ201 plasmid with the forward and reverse primers CTCGGGATCCATGCCCAAATTGAT and ATCGAAGCT-TATTCGATGCGAAAGTG, and TA-cloned into pGEM-T Easy (Promega). The resulting plasmid was linearized with Apa I and a digoxigenin-labeled RNA probe transcribed with SP6 RNA polymerase. Whole mount *in situ* hybridization of *Drosophila* embryos with the GSTE7 probe was performed as previously described [8].

Construction of transgenic flies

The coding sequence of GSTE7 in the pJ201 plasmid was inserted into the pUAST attB vector [9] with Not I and Xba I. Embryos from $y^1 \ sc^1 \ v^1 \ P\{nos-phiC31\backslash int.NLS\}X; \ P\{Cary-P\}attP2$ flies containing the attP2 landing site on chromosome 3L (68A4) were injected with the plasmid by Rainbow Transgenic Flies, Inc., Camarillo, CA, USA.

Survival and egg-laying assays

Approximately 100 *Actin-Gal4/CyO* virgin females were crossed to 40 w^{1118} (control) or to 40 *UAS-GSTE7* homozygous males and kept in food bottles. Offspring from the crosses were allowed to mate with each other for 3–5 days after hatching in the bottles. Male and female 3–5 days old non-*Cy* flies were then divided into separate vials at the first day of the assays. Flies were transferred to new vials 3 times per week during the survival assay. Oviposition was measured on seven consecutive days. The flies were transferred to fresh vials with or without ITC each day, and the number of eggs deposited in the vials counted.

One liter of food was prepared by adding 12.9 g of yeast, 40 g instant mashed potatoes, and 10 g of agar in 1.1 liter of boiling water, and 50 ml of syrup added when the mixture was homogeneous. After boiling for 20 min, the mixture was cooled to 55°C before being supplemented with 0.875 g ascorbic acid, 8.5 ml 10% Nipagin in ethanol, 6.25 ml propionic acid, and 10 ml of 99% ethanol (control) or isothiocyanate compound dissolved in 10 ml of 99% ethanol. Ten males or 10 females were placed in each vial and kept at 25°C.

Furthermore, we tested for any food avoidance behavior induced by the addition of PEITC. For this, we used xylene cyanol in fly food prepared as above. We monitored the uptake of food by means of flies turning blue. Males and females were separated and put in food vials (5 flies/vial). After three hours in food vials, all flies of both sexes had a blue colored abdomen, and no differences between flies given control food and drug supplemented food could be seen.

Results

Enzymatic activities of Drosophila melanogaster GSTE7

The purified GSTE7 was functionally characterized by enzymatic assays with alternative substrates (Table 2). In order to test general GST activity, the standard substrate 1-chloro-2,4-dinitrobenzene (CDNB) was used. The specific activity of 36.5 $\mu mol \cdot min^{-1} \cdot mg^{-1}$ determined is within the common range of efficient GSTs, demonstrating that GSTE7 is catalytically highly competent. Further, a series of organic isothiocyanates (ITCs) of biological significance were investigated as substrates. These compounds are plant allelochemicals and include ITCs that feature both aliphatic and aromatic chemical substituents. The specific ITC activities were of similar magnitude and ranged between 7.5 and 23.8 $\mu mol \cdot min^{-1} \cdot mg^{-1}$ (Table 2). These values are approaching the specific ITC activities of the most active human GSTs [5] and suggest that *D. melanogaster* expressing GSTE7 are highly competent to conjugate and inactivate ITCs when the flies are exposed to these electrophilic toxicants. For our *in vivo* experiments the demonstration of high enzyme activities

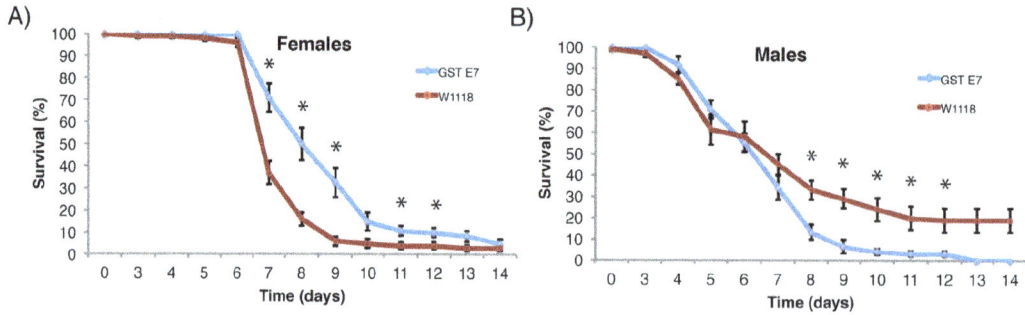

Figure 2. Survival of adult flies on 0.25 mM PEITC added to standard fly food. Actin-Gal4 flies crossed to wild-type control (w^{1118}) or to UAS-GSTE7 flies (GSTE7) were allowed to mate and then transferred to fly food supplemented with PEITC. Twelve vials with 10 flies each (GSTE7) or 10 vials with 10 flies each (w^{1118}) were used and the number of surviving flies in each vial scored. Females (A) and males (B) were kept in separate vials. Error bars represent standard error of the mean. * indicates $P<0.05$, two-tailed unpaired Student's t-test.

with allyl-ITC (AITC) and phenethyl-ITC (PEITC) are particularly relevant.

Overexpression of D. melanogaster GSTE7

To investigate the *in vivo* effects of elevated GST levels, we took advantage of the binary Gal4-UAS system to induce overexpression of GSTE7 protein in *Drosophila* [10]. We cloned the coding sequence of the *Drosophila* GSTE7 gene together with an upstream activating sequence (UAS) and introduced the construct on chromosome 3 carrying an attP landing site by way of phiC31-mediated integration (Fig. 1A) [9].

The ubiquitous Actin-Gal4 driver was crossed to UAS-GSTE7 flies to induce expression from the transgene (Fig. 1B). To confirm the overexpression, embryos were stained with a probe recognizing GSTE7 mRNA by whole-mount *in situ* hybridization (Fig. 1C–F). As shown in Fig. 1C, endogenous GSTE7 mRNA is generally distributed throughout pre-cellular embryos prior to zygotic transcription, demonstrating a maternal contribution. In stage 13 embryos, GSTE7 expression is considerably weaker, but most pronounced in the midgut (Fig. 1D). In the presence of the Actin-Gal4 driver, we observed high-level expression of GSTE7 throughout both pre-cellular and stage 13 embryos (Fig. 1E and F). This shows that transgenic GSTE7 is overproduced both in the female germline and in somatic tissues. Expression of the transgene did not diminish survival, suggesting that the GSTE7 transgene by itself did not cause any harmful physiological effects.

Impact of transgenic GSTE7 expression on PEITC exposure

We tested the *in vivo* effects of PEITC, an efficient substrate of GSTE7, in our *in vitro* assay. A concentration of 0.25 mM PEITC in standard fly food was shown to be toxic and significantly shortened the lifespan of wild-type flies, and higher PEITC concentrations were fatally toxic. We noticed that overexpression of GSTE7 could protect females from the toxic effects of 0.25 mM PEITC during days 7–12 of exposure, but had no positive effect on long-term survival (Fig. 2A). By contrast, the effect on males was the opposite to that on females, such that a significantly higher mortality was seen in fly males overexpressing GSTE7 after one week of exposure (Fig. 3B). A concentration of 4 mM AITC in the standard fly food had a lethal effect on both wild-type and transgenic flies, resulting in death in a few hours after first exposure, but subjected to 1 mM AITC in the diet, flies and larvae were viable and showed no obvious weakness or other phenotype.

Effect of ITCs on oviposition

As a corollary to the differential effects of the transgene on males and females, the *in vivo* effects of AITC and PEITC, demonstrated to be effective GSTE7 substrates in our *in vitro* studies (Table 2), were investigated with respect to oviposition. The number of eggs produced by 3–5 days old flies were counted for seven consecutive days, which demonstrated a significant influence of the transgene on the rate of oviposition (Figure 3). Unexpectedly, the effect was demonstrated both in the presence and the absence of ITCs. The number of eggs was increased about twofold in the transgenic flies on day 1 as compared to the wild-type flies, and consistently showed a higher oviposition rate than the wild-type flies both in the absence and presence of ITCs (Fig. 3). This effect of the transgene was statistically significant on most days of the assay. Exposure to AITC or PEITC in the diet demonstrated an approximately 5- to 2-fold increase in egg-laying ability in the transgenic flies expressing GSTE7 in comparison with that of wild-type flies.

Discussion

For half a century a chemical warfare of plants against herbivorous insects has been viewed as a major driver of co-evolution of the combatants [11]. In *Brassica* plants the chemical weapons are largely based on glucosinolates that decompose into ITCs and other toxic products when plant tissues are crushed or when they are digested by the feeding insects. The detoxication of the released ITCs has not been studied in great detail, but it is generally assumed that GSTs play a pivotal role in the biotransformation of ITCs.

Differences among insects may occur, but recent studies of lepidopteran caterpillars feeding on glucosinolate-containing plants show that a major fraction of the corresponding ITC metabolites are excreted in the feces as products of glutathione conjugation [12]. Two strains of the whitefly *Bemisia tabaci*, biotype B and biotype Q, feeding on *Arabidopsis thaliana* plants showed different responses to their host plants suggesting evolutionary divergence [13]. *A. thaliana* is known to contain at least three dozen glucosinolates distinguishable by their sidechains that can be aliphatic or aromatic, and the whiteflies were exposed to transgenic *A. thaliana* plants that overproduced either aliphatic or aromatic (indole-containing) glucosinolates. Significant differences in the behavioral and biochemical responses were noted between whiteflies of biotypes B and Q, but both biotypes responded by a reduced number of oviposited eggs when reared on transgenic plants overproducing glucosinolates. Biotype Q

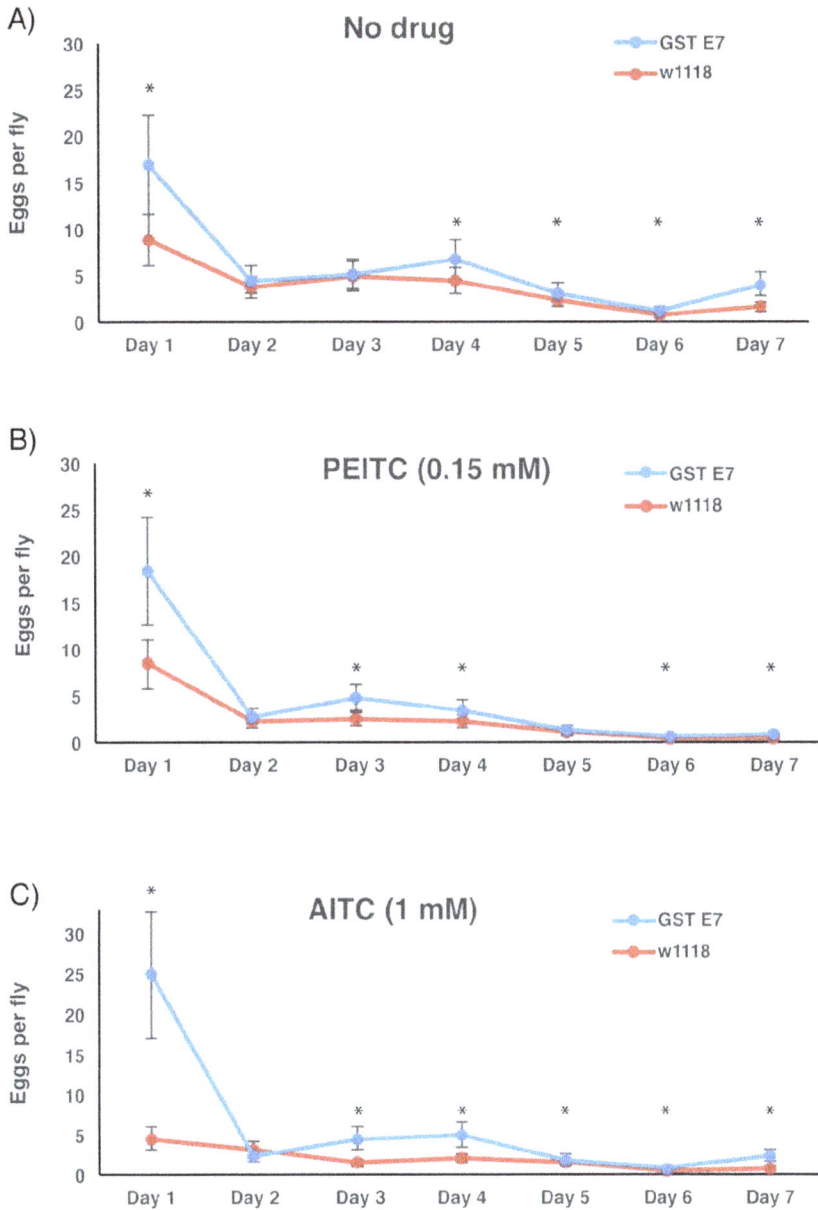

Figure 3. Effect of ITCs on egg laying of fruit flies. Actin-Gal4 flies crossed to wild-type control (w^{1118}) or to UAS-GSTE7 flies (GSTE7) were mated for 3–5 days and the females separated and then kept for 7 days on standard control, 0.15 mM PEITC, or 1 mM allyl-ITC supplemented food. They were transferred to fresh vials each day and the number of eggs in the vials counted. Ten vials (approximately 10 flies/vial) with control (w^{1118}) or GSTE7 flies were used and the number of eggs divided by the number of flies. The graph shows the number of eggs laid per fly, and error bars represent standard error of the mean. * indicates $P<0.05$, two-tailed unpaired Student's t-test.

displayed higher constitutive levels than biotype B of several detoxication enzymes and two GSTs were significantly induced during exposure to *A. thaliana* featuring overproduction of indole-containing glucosinolates, suggesting a role of GSTs in detoxication and an explanation of the good performance on the indole-glucosinolate accumulating plants [13].

Our experiments establish overexpression of GSTE7 in *D. melanogaster* without detectable behavioral or morphological alterations of the phenotype. However, GSTE7 overexpression had a significant effect on oviposition (Fig. 3). Since elevated levels were present both in the germline, as shown by enhanced transcript levels in embryos prior to zygotic transcription (Fig. 1),

and in somatic tissues, GSTE7 could be influencing oogenesis either directly or indirectly.

AITC and PEITC are produced in plants and are known toxins that may have influenced the evolutionary interplay between *Drosophila* species and *Brassica* plants [14,15]. GSTs are catalyzing the inactivation of these compounds as well as other ITCs that have been investigated [16]. GSTE7 overexpression provided significant protection of female *D. melanogaster* against ITCs represented by PEITC in Fig. 2. For unclear reasons, no significant protection of males was noted. Instead, the transgene significantly reduced the survival rate of males. The explanation of this deleterious influence is not clear, but excessive ITC conjugation may lead to glutathione depletion and toxicity or apoptosis [17]. Glutathione is involved in a multitude of reactions and conjugation may shift the tissue redox balance, which is known to influence longevity.

ITCs are not only directly toxic but also act as repellants of certain insects. Studies of *Drosophila* demonstrate that isothiocyanates in wasabi (including AITC) lead to a food avoidance behavior involving the transient receptor potential A1 gene *painless* in gustatory receptor neurons [18,19]. However, the detrimental effects of ITCs in our experiments cannot be explained primarily by food avoidance and starvation, since the GSTE7 transgene provided protection. Furthermore, to check for food avoidance behavior xylene cyanol was added to the diet, which results in a blue abdomen when consumed, but no difference in coloring was noted between flies fed ITC-containing or control food.

Canonical GSTs occur in *D. melanogaster* in multiple forms, the GSTome, encoded by 36 different genes in six distinct classes and the enzymes have previously been subjected to preliminary enzymological studies [20,21]. GSTs from the Delta (D) and Epsilon (E) classes have the largest number of members and are those primarily associated with the defense against toxicants. GSTE7 is one of a limited number of GSTs that are overexpressed in long-lived *D. melanogaster* [22], suggesting that it may provide survival value to flies. This enzyme was therefore selected for biochemical characterization and transgenesis studies. Individual GSTs often have overlapping substrate acceptance, and gene disruption studies may consequently not reveal the importance of individual genes, whereas overexpression of a particular GST from a corresponding transgene could be more informative.

Examination of embryos of different ages demonstrates that the GSTE7 transcript has a general distribution with particularly high concentration in the digestive tract (Fig. 1). Furthermore, the gut of larvae and adult flies also show high expression of GSTE7 according to the FlyAtlas Anatomical Expression Data [23], suggesting that the normal function of GSTE7 could involve metabolism of ITCs and other toxic electrophiles originating from food and liquid uptake. In summary, it would appear that flies exposed to ITCs at the first level of protection rely on avoidance mechanisms based on gustatory receptors [18,19]. However, any ITCs ingested will at the second level undergo chemical inactivation catalyzed by GSTs leading to conjugates suitable for excretion. Even though reactions catalyzed by GSTs are instrumental in the biotransformation of ITCs and other toxic electrophiles, excessive conjugation activity may lead to glutathione depletion and be detrimental, as suggested by the data obtained with the male flies (Fig. 2B). Obviously, the cellular processes have to be properly regulated for optimal adaptation to ambient life conditions.

Acknowledgments

Dr. Claes Gustafsson, DNA2.0 Inc., Menlo Park, CA, generously provided the synthesized GSTE7 DNA in the plasmid pJ201 as a gift.

Author Contributions

Conceived and designed the experiments: AM OD BM MM. Performed the experiments: AM OD. Analyzed the data: AM OD BM MM. Wrote the paper: BM MM.

References

1. Halkier BA, Gershenzon J (2006) Biology and biochemistry of glucosinolates. Annual Review of Plant Biology 57: 303–333.
2. Hopkins RJ, van Dam NM, van Loon JJ (2009) Role of glucosinolates in insect-plant relationships and multitrophic interactions. Annual Review of Entomology 54: 57–83.
3. Zust T, Heichinger C, Grossniklaus U, Harrington R, Kliebenstein DJ, et al. (2012) Natural enemies drive geographic variation in plant defenses. Science 338: 116–119.
4. Josephy PD, Mannervik B (2006) Molecular Toxicology, 2nd Edn., Oxford University Press, N.Y. 303–332.
5. Kolm RH, Danielson UH, Zhang Y, Talalay P, Mannervik B (1995) Isothiocyanates as substrates for human glutathione transferases: structure-activity studies. The Biochemical Journal 311 (Pt 2): 453–459.
6. Gloss AD, Vassao DG, Hailey AL, Nelson Dittrich AC, Schramm K, et al. (2014) Evolution in an ancient detoxification pathway is coupled with a transition to herbivory in the Drosophilidae. Molecular Biology and Evolution 9: 2441–2456.
7. Laemmli UK (1970) Cleavage of structural proteins during the assembly of the head of bacteriophage T4. Nature 227: 680–685.
8. Jiang J, Kosman D, Ip YT, Levine M (1991) The dorsal morphogen gradient regulates the mesoderm determinant twist in early Drosophila embryos. Genes & Development 5: 1881–1891.
9. Bischof J, Maeda RK, Hediger M, Karch F, Basler K (2007) An optimized transgenesis system for Drosophila using germ-line-specific phiC31 integrases. Proceedings of the National Academy of Sciences of the United States of America 104: 3312–3317.
10. Brand AH, Perrimon N (1993) Targeted gene expression as a means of altering cell fates and generating dominant phenotypes. Development 118: 401–415.
11. Ehrlich PR, Raven PH (1964) Butterflies and Plants: A Study in Coevolution. Evolution, 18: 586–608.
12. Schramm K, Vassao DG, Reichelt M, Gershenzon J, Wittstock U (2012) Metabolism of glucosinolate-derived isothiocyanates to glutathione conjugates in generalist lepidopteran herbivores. Insect Biochemistry and Molecular Biology 42: 174–182.
13. Elbaz M, Halon E, Malka O, Malitsky S, Blum E, et al. (2012) Asymmetric adaptation to indolic and aliphatic glucosinolates in the B and Q sibling species of Bemisia tabaci (Hemiptera: Aleyrodidae). Molecular Ecology 21: 4533–4546.
14. Whiteman NK, Gloss AD, Sackton TB, Groen SC, Humphrey PT, et al. (2012) Genes involved in the evolution of herbivory by a leaf-mining, Drosophilid fly. Genome Biology and Evolution 4: 900–916.
15. Whiteman NK, Groen SC, Chevasco D, Bear A, Beckwith N, et al. (2011) Mining the plant-herbivore interface with a leafmining Drosophila of Arabidopsis. Molecular Ecology 20: 995–1014.
16. Zhang Y, Kolm RH, Mannervik B, Talalay P (1995) Reversible conjugation of isothiocyanates with glutathione catalyzed by human glutathione transferases. Biochemical and Biophysical Research Communications 206: 748–755.
17. Orr WC, Radyuk SN, Sohal RS (2013) Involvement of redox state in the aging of Drosophila melanogaster. Antioxidants & Redox Signaling 19: 788–803.
18. Al-Anzi B, Tracey WD, Jr., Benzer S (2006) Response of Drosophila to wasabi is mediated by painless, the fly homolog of mammalian TRPA1/ANKTM1. Current Biology 16: 1034–1040.
19. Kang K, Pulver SR, Panzano VC, Chang EC, Griffith LC, et al. (2010) Analysis of Drosophila TRPA1 reveals an ancient origin for human chemical nociception. Nature 464: 597–600.
20. Saisawang C, Wongsantichon J, Ketterman AJ (2012) A preliminary characterization of the cytosolic glutathione transferase proteome from Drosophila melanogaster. The Biochemical Journal 442: 181–190.
21. Tu CP, Akgul B (2005) Drosophila glutathione S-transferases. Methods in Enzymology 401: 204–226.
22. McElwee JJ, Schuster E, Blanc E, Piper MD, Thomas JH, et al. (2007) Evolutionary conservation of regulated longevity assurance mechanisms. Genome Biology 8: R132.
23. Chintapalli VR, Wang J, Dow JA (2007) Using FlyAtlas to identify better Drosophila melanogaster models of human disease. Nature Genetics 39: 715–720.

21

Toxicity of TiO$_2$ Nanoparticles to *Escherichia coli*: Effects of Particle Size, Crystal Phase and Water Chemistry

Xiuchun Lin[1,2], Jingyi Li[2], Si Ma[2], Gesheng Liu[2], Kun Yang[2,3], Meiping Tong[4], Daohui Lin[2,3]*

1 College of Environmental and Biological Engineering, Putian University, Fujian, China, **2** Department of Environmental Science, Zhejiang University, Hangzhou, China, **3** Zhejiang Provincial Key Laboratory of Organic Pollution Process and Control, Zhejiang University, Hangzhou, China, **4** College of Environmental Sciences and Engineering, Peking University, Beijing, P. R. China

Abstract

Controversial and inconsistent results on the eco-toxicity of TiO$_2$ nanoparticles (NPs) are commonly found in recorded studies and more experimental works are therefore warranted to elucidate the nanotoxicity and its underlying precise mechanisms. Toxicities of five types of TiO$_2$ NPs with different particle sizes (10~50 nm) and crystal phases were investigated using *Escherichia coli* as a test organism. The effect of water chemistry on the nanotoxicity was also examined. The antibacterial effects of TiO$_2$ NPs as revealed by dose-effect experiments decreased with increasing particle size and rutile content of the TiO$_2$ NPs. More bacteria could survive at higher solution pH (5.0–10.0) and ionic strength (50–200 mg L^{-1} NaCl) as affected by the anatase TiO$_2$ NPs. The TiO$_2$ NPs with anatase crystal structure and smaller particle size produced higher content of intracellular reactive oxygen species and malondialdehyde, in line with their greater antibacterial effect. Transmission electron microscopic observations showed the concentration buildup of the anatase TiO$_2$ NPs especially those with smaller particle sizes on the cell surfaces, leading to membrane damage and internalization. These research results will shed new light on the understanding of ecological effects of TiO$_2$ NPs.

Editor: Elena A. Rozhkova, Argonne National Laboratory, United States of America

Funding: This work was supported by the 973 Program of China (2014CB441104), National Natural Science Foundation of China (21337004, 21477107), Natural Science Foundations of Zhejiang Province (LR12B07001) and Fujian Province (2014J01053) of China. The funders had no role in study design, data collection and analysis, decision to publish, or preparation of the manuscript.

Competing Interests: The authors have declared that no competing interests exist.

* Email: lindaohui@zju.edu.cn

Introduction

Due to their unique chemical and physical properties, titanium dioxide (TiO$_2$) nanoparticles (NPs) are produced at a large scale for industrial applications to meet with ever-increasing market demands [1]. The annual production of TiO$_2$ NPs is predicted to reach 2.5 million tons by 2025 [2]. The widely used TiO$_2$ NPs would find their way into aquatic environments [3–6] and interact with aquatic organisms [7]. Eco-toxicity of TiO$_2$ NPs is therefore received worldwide research attentions [8–16].

Bacteria, e.g., *Escherichia coli* (*E.coli*), as single cell organisms and ubiquitous in aquatic environments, are good model organisms for studying the eco-toxicity of NPs and the cell/organism-NP interaction. Many research works [8] have investigated the toxicity of various TiO$_2$ NPs toward *E.coli*, with a focus on the influencing factors such as: (1) Size. Many studies attributed the toxicity of TiO$_2$ NPs to their small particle size [17–22]. (2) Crystal structure. It is generally concluded that anatase TiO$_2$ NPs are more toxic than rutile NPs by inducing greater oxidative stress [15,23,24]. (3) Experimental matrix. Changes in water chemistry (e.g., pH and ionic strength) may influence the agglomeration and sedimentation characteristics of NPs and then their toxicity [12,21,25–28]. (4) Solar radiation, especially those in the UVA region, is also considered as a critical factor of aquatic nanotoxicity [13,29–35]. These researches substantially increased our knowledge on the eco-toxicity of TiO$_2$ NPs.

However, controversial and inconsistent results on the toxicity of TiO$_2$ NPs are commonly found in recorded studies and precise mechanisms of the nanotoxicity warrant more specific researches. For example, Adams et al. (2006) reported 44% reduction in the growth of *E.coli* by 1 g L^{-1} and 72% reduction by 5 g L^{-1} TiO$_2$ NPs (66 nm, crystal structure not determined) [36]; Tong et al. (2013) [21] reported 70% reduction in the growth of *E.coli* by 10 mg L^{-1} TiO$_2$ NPs while 30% reduction was observed by Planchon et al. (2013) [15] with the same TiO$_2$ NPs at 10 mg L^{-1} (P25, consisting of an 80:20 ratio of anatase:rutile). So an acute lack of emphasis on the environmental and nanoparticle parameters prevents a meaningful comparative assessment from the hitherto available nanotoxicity data, and it highlights the necessity to provide additional eco-toxicological studies and physicochemical characterization of TiO$_2$ NPs to ensure consistency of research results.

This study is aimed to elucidate the roles of particle size and crystal structure in the toxicity of TiO$_2$ NPs using *E.coli* as a model organism. Five types of well characterized TiO$_2$ NPs with different particle sizes and crystal phases were examined. The toxicity assays were conducted at different concentrations of the NPs and various solution pHs and ionic strengths. In addition, the interactions between TiO$_2$ NPs and bacteria and the cell reactive responses were examined with transmission electron microscopy (TEM) and measurements of intracellular reactive oxygen species

(ROS) and malondialdehyde (MDA) to address the toxicity mechanism. The results are believed to increase our understanding of the nanotoxicology.

Materials and Methods

1. Nanoparticles and characterizations

Five types of TiO_2 NPs were purchased and used in this study. They were anatase TiO_2 with particle sizes measured to be around 10 nm (TiO_2-NP 10A), 25 nm (TiO_2-NP 25A), and 50 nm (TiO_2-NP 50A) and rutile TiO_2 of 50 nm (TiO_2-NP 50R) and mixed anatase and rutile TiO_2 of 25 nm (TiO_2-NP25 AR). TiO_2-NP 10A and TiO_2-NP 50R were from Hongsheng Material Sci & Tech Co., Zhejiang, China and the other three TiO_2 NPs from Wangjing New Material Sci & Tech Co., Zhejiang, China.

Morphologies of the NPs were examined using TEM (JEM-1230, JEOL Ltd., Tokyo, Japan). Powered X-ray diffraction analysis (XRD, X'Pert Pro, Holland) was carried out to characterize the crystal structure of the NPs. Elemental compositions of the NPs were determined by using an X-ray energy dispersion spectroscope (EDS, GEN-ESIS 4000, EDAX Inc. America). Hydrodynamic diameters and zeta potentials of the NPs (50 mg L^{-1}) were measured with a Zetasizer (Nano ZS90, Malvern, UK) after being sonicated (100 W, 40 kHz, 30 min) into 100 mg L^{-1} NaCl solution at 25°C and various pH values. Points of zero charge (pH_{pzc}) of the NPs were obtained from the zeta potential versus pH curves. Specific surface areas of the NPs were determined using the multi-point Brunauer-Emmett-Teller (BET) method (Quantachrome NOVA 2000e, America).

2. Dose-effect experiments

E.coli O111 (Genbank access no. GU237022.1) isolated from a sewage water was used as the test organism, as reported in our previous studies [37,38]. The bacteria were maintained in Luria Bertani (LB) solid plates at 4°C and inoculated in LB broth (pH 7.2~7.4) at 37°C overnight (12~16 h) at 150 rpm. The bacteria were separated from the broth by centrifugation at 3000 g for 15 min and washed twice with 0.85% physiology salt-water. The bacterium stock suspension was prepared by resuspending the bacterial pellets in 0.85% NaCl physiology salt-water with the cell concentration determined by the absorbance at 600 nm (OD_{600}) being adjusted to 1.0.

The stock NP suspensions (500 mg L^{-1}) were obtained by sonicating (100 W, 40 kHz, 30 min) 50 mg of the TiO_2 NPs into 100 mL of ultra-pure water. The stock NP suspensions after sonication were diluted using ultra-pure water to the target test concentrations (10–500 mg L^{-1}).

One mL of the *E.coli* stock suspension was added into 100 mL of the test NP suspensions. The mixtures were placed on a shaker at 37°C and 150 rpm for 3 h with natural light. The bacteria in the resultant mixtures were spread on LB agar plates and incubated at 37°C for 24 h, and the colonies were counted. The percentage viabilities of the bacteria in the NP suspensions were calculated by dividing their colony forming units (CFU) mL^{-1} by that in the NP-free control. All treatments including the control were repeated in triplicate.

3. TEM Observations

TEM was used to observe the direct contact between the NPs and the bacterial cells. A drop of the bacteria exposed to the NPs (50 mg L^{-1}, 3 h) and the NP-free control was air-dried onto a

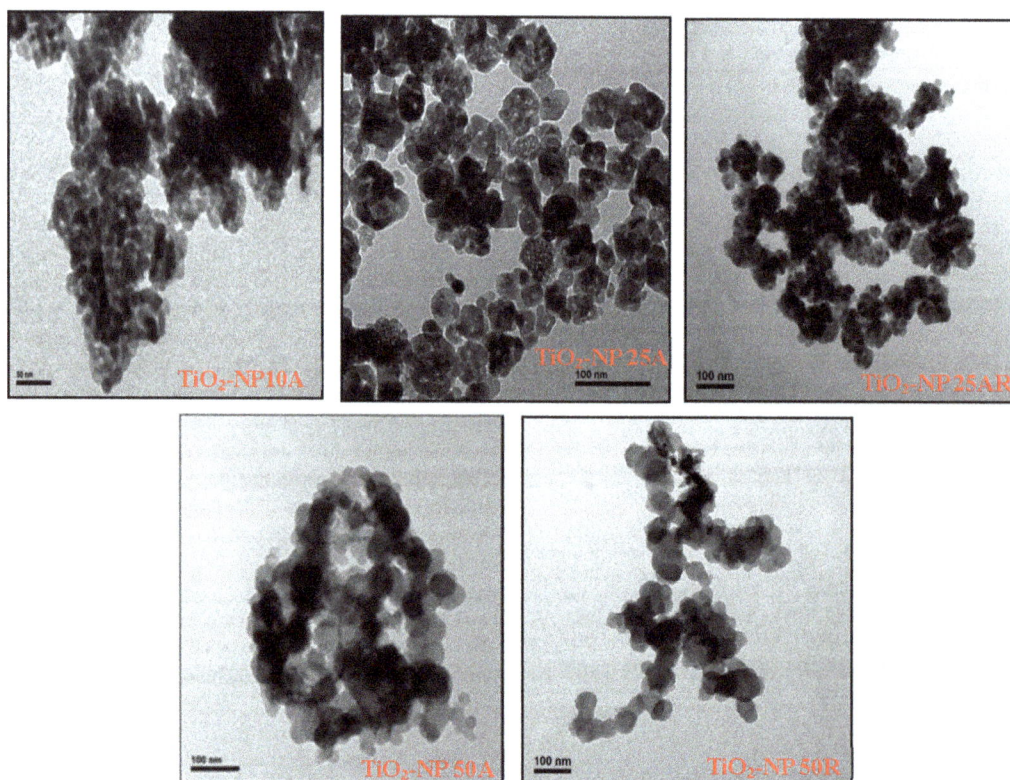

Figure 1. TEM images of the as-received TiO_2 NPs.

Table 1. Characteristics of the nanoparticles.

Sample	Crystal phase, %	Zeta potential, mV	pH_{pzc}	hydrodynamic size, nm	S_{BET}, $m^2 g^{-1}$	TEM size, nm
TiO$_2$-NP10A	Anatase, 100	−21.6	6.2	314±8	324	11.0±3.4
TiO$_2$-NP25A	Anatase, 99.2	−5.48	5.6	251±37	77	26.2±6.1
TiO$_2$-NP25AR	Anatase, 93.0	−13.6	5.2	202±57	66	26.7±5.0
TiO$_2$-NP50A	Anatase, 98.8	−9.34	6.0	486±12	105	57.1±14.0
TiO$_2$-NP50R	Rutile, 100	−33.8	3.6	260±10	30	57.2±17.8

Note: the crystal phase was determined by the XRD; zeta potential and hydrodynamic size were measured in the toxicity test medium at pH 6.5 by the Zetasizer; pH_{pzc} was calculated from the zeta potential versus pH curves shown in Figure 2; S_{BET} (specific surface area) measured using the BET (Brunauer-Emmett-Teller) method; TEM size shows the NP size measured with the TEM images.

copper grid and was then imaged by the TEM. To observe the internalization and localization of the NPs in the cells and the changes in cellular structure as affected by the NPs, the NPs-treated and untreated bacteria were fixed in 2.5% glutaraldehyde, dehydrated in graded concentrations of ethanol, embedded in Epon resin, and stained with OsO$_4$ [12,39]. Ultrathin sections were then cut and counterstained with Reynold's and uranyl acetate for the TEM observation.

4. Reactive oxygen species (ROS) and lipid peroxidation measurements

The fluorescence probe 2'7'-dichlorodihydrofluorescein diacetate (H$_2$DCFDA) was used to quantify the formation of intracellular ROS as described in our previous papers [12,39] with minor modifications. The bacterial cells after exposure to the test media were collected by centrifugation (8000 g, 5 min). The pellet was resuspended in 0.85% physiology salt-water containing 10 μM H$_2$DCFDA and incubated on the shaker (150 rpm) for 30 min at 37°C. The bacteria were further pelleted and resuspended in 300 μL of 0.85% physiology salt-water. The fluorescence values were measured in a 96-well plate using a

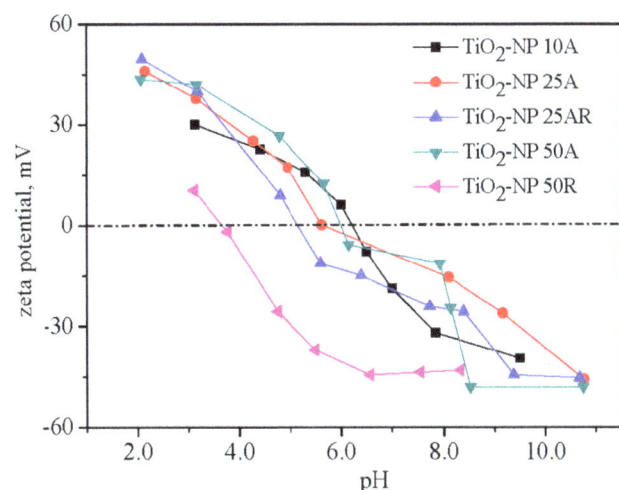

Figure 2. Changes of zeta potentials of the TiO$_2$ NPs against a solution pH.

multifunctional microplate reader (M200 PRO, Ltd., Austria) with the excitation and emission wavelengths of 485 nm and 528 nm, respectively. Relative ROS levels were calculated by the fluorescence ratio of the treatments to the control.

The concentration of malondialdehyde (MDA) was determined as an indicator of lipid peroxidation as described in previous works [12,39]. Briefly, the exposed bacteria were mixed into 1 mL of 10% (wt/vol) trichloroacetic acid and left at room temperature for 10 min; the supernatant of the mixture was collected after centrifugation at 11,000 g for 40 min and then mixed with 1.5 mL of a freshly prepared 0.67% (wt/vol) thiobarbituric acid solution; the resultant mixture was incubated for 20 min in a boiling water bath and after cooling the absorbance was measured at 532 nm; and the MDA concentration was calculated using the Hodges' equations.

5. Effects of pH and ionic strength

TiO$_2$-NP 10A with a concentration fixed at 10 mg L^{-1} was selected as a type of representative NPs in examining the effect of water chemistry on the nanotoxicity. In the pH effect experiment, the suspension pHs were adjusted to 5.0, 7.0, 8.0 and 10.0 using 0.1 M HCl and NaOH; NaCl was added to maintain a constant background ionic strength (100 mg L^{-1}). For the ionic strength effect experiment, difference concentrations of NaCl (0, 50, 100, 150 and 200 mg L^{-1}) were added into the NP suspensions; the final suspension pH remained at about neutral without further adjustment. The bacterial exposure method in the pH and ionic strength effect experiments was the same as the above dose-effect experiments. Zeta potential of 1 mL of the stock bacterial suspension after being mixed into 100 mL of ultra-pure water at pH 5.0, 7.0, 8.0 and 10.0 was also measured by the Zetasizer.

6. Statistical analysis

Each treatment including the blank control was conducted in triplicate and the results were presented as mean ± SD (standard deviation). For each datum point, data were normalized by the reference without NPs. This representation thus strictly reports the incremental impacts of TiO$_2$ NPs, excluding the medium stress. The Student's t test was performed to analyze the significance of difference between two groups of data. Origin 8.0 was used to make graphs. The concentration resulting in 50% mortality (LC$_{50}$) was calculated with SPSS 20.0.

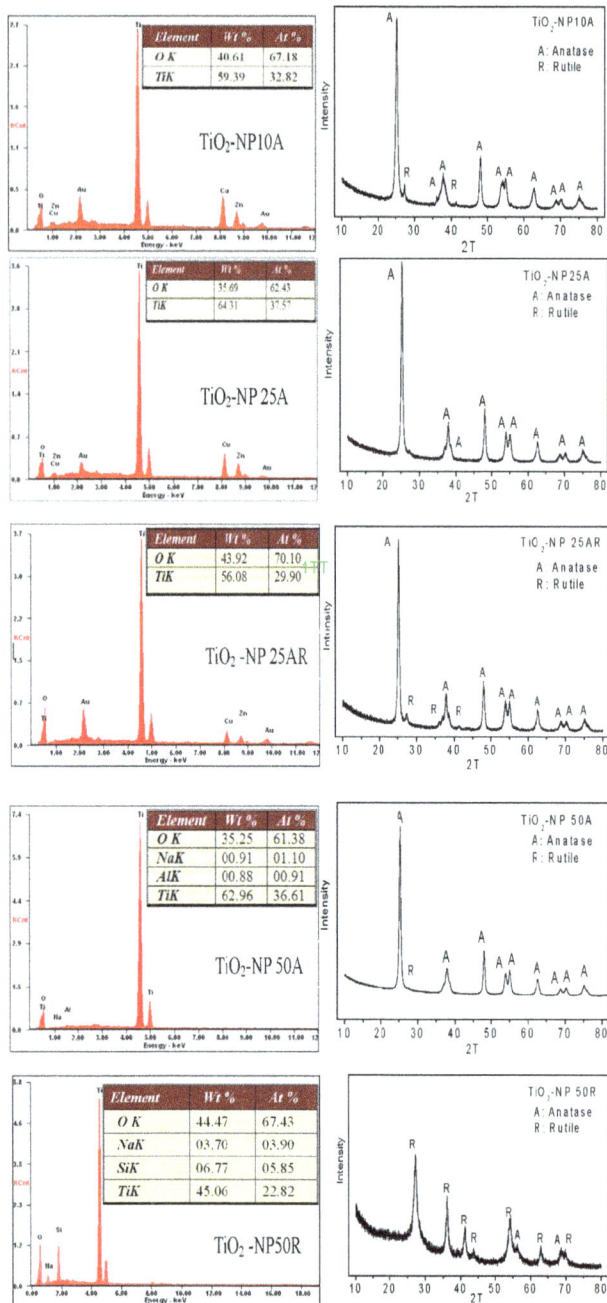

Figure 3. EDS (left column) and XRD (right column) figures of the as-received TiO₂ NPs.

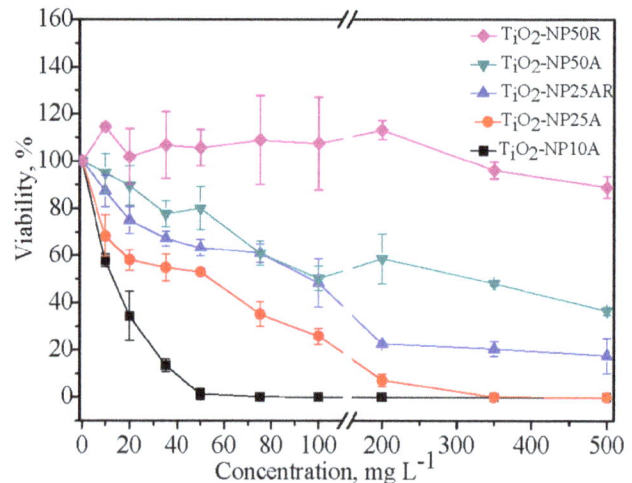

Figure 4. Variations of the bacteria viability with concentrations of the TiO₂ NPs. The viability was the ratio of bacterial cell number under the NP treatment to the blank control.

Results and Discussion

1. Characteristics of TiO₂ NPs

Selected properties of the TiO₂ NPs are listed in Table 1 with their TEM images shown in Figure 1. Big NP aggregates present in the TEM images and the measured large hydrodynamic diameters indicate the aggregation of the NP suspensions even after the sonication. TiO₂-NP 10A, having the smallest particle size (TEM size of 11.0 ± 3.4 nm) among the five TiO₂ NPs, owned relative greater hydrodynamic diameter of 314 ± 8 nm. Changes in zeta potentials of the TiO₂ NPs against a solution pH are shown in Figure 2. The calculated pH_{pzc} of the NPs varied from pH 3.6 to

6.2, which could account for their negative zeta potentials (-33.8– -5.48 mV) in the neutral toxicity test medium. Specific surface areas of the TiO₂ NPs ranged from 30 m² g⁻¹ (TiO₂-NP 50R) to 324 m² g⁻¹ (TiO₂-NP 10A). Purities of the NPs determined by the EDS were all above 98.0% (Figure 3). XRD patterns of the TiO₂ NPs (Figure 3) confirmed the predominance of anatase phase of TiO₂-NP 10A, TiO₂-NP 25A, and TiO₂-NP 50A and the rutile nature of TiO₂-NP 50R; and TiO₂-NP 25AR was a mixture of anatase (93%) and rutile (7%).

2. Cell viability assessment

The particle dose, size and phase dependent reductions in the cell viability of *E.coli* upon exposure to the TiO₂ NPs for 3 h were observed through the plate count assay (Figure 4). The four anatase NPs were more or less toxic to *E.coli* and the viability of the bacteria exhibited a pronounced concentration-dependent decrease. The calculated 3 h LC_{50} of the four anatase NPs had an order of TiO₂-NP 10A (17.0 mg L⁻¹)<TiO₂-NP 25A (59.2 mg L⁻¹)<TiO₂-NP 25AR (163 mg L⁻¹)<TiO₂-NP 50A (304 mg L⁻¹). The enhancement of bactericidal effect of the NPs with decreasing particle size was observed throughout the various particle concentrations. TiO₂-NP 10A in the anatase phase with the minimum particle size and the largest BET surface area was determined to be the most toxic to *E.coli*. The presence of rutile phase in the NPs lowered the bactericidal activity in comparison to the pure anatase NPs. As shown in Figure 4, although similar in particle size, the toxicity of TiO₂-NP 25AR was much lower than that of TiO₂-NP 25A; the pure rutile TiO₂-NP 50R was nontoxic to the bacteria with concentration up to 500 mg L⁻¹, while the anatase TiO₂-NP 50A could inactivate half of the bacteria at 304 mg L⁻¹.

3. TEM observations of the direct NP-cell interactions

Nanoparticle-type-dependent bacterial cell membrane localizations of the TiO₂ NPs as well as morphological changes of the NPs-exposed cells were captured by the TEM images (Figure 5). The stronger NP-cell interaction was observed for the TiO₂ NPs with anatase crystal structure and smaller particle size. Numerous TiO₂-NP 10A aggregates with various sizes were observed tightly attached to the bacterial cell surfaces (Figure 5B). The big and tight NP-cell aggregate in Figure 5C indicates the strong

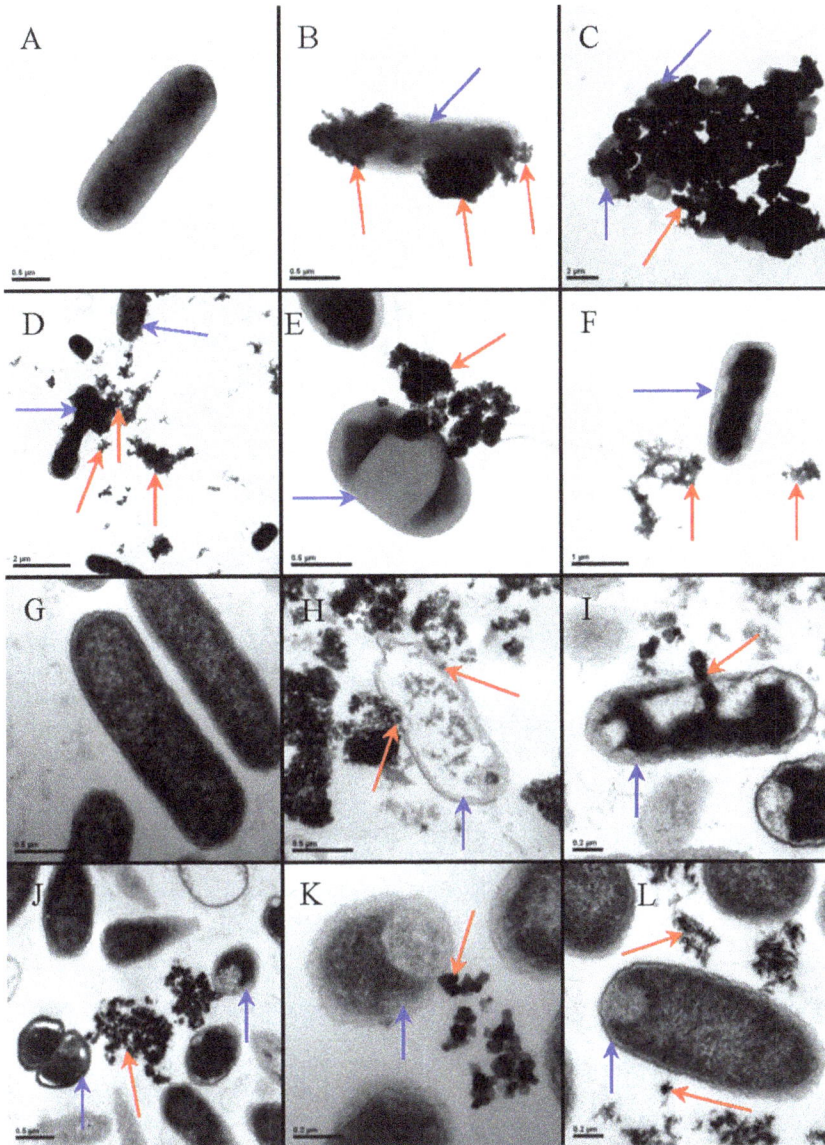

Figure 5. Selected TEM images of the unsliced (A to F) and sliced (G to L) *E.coli* cells without (A and G) and with the treatments of TiO₂-NP 10A (B and H), TiO₂-NP 25A (C and I), TiO₂-NP 25AR (D and J), TO₂-NP 50A (E and K) and TiO₂-NP 50R (F and L). The blue arrows point to the cells and the red arrows direct to the NP aggregates.

interaction between TiO₂-NP 25A and the cells. Some of the TiO₂-NP 25AR aggregates were also observed attaching to the bacterial cells but some present away from the cells (Figure 5D), implying the relatively weaker NP-cell interaction as compared with the pure anatase TiO₂-NP 25A of the same size. A few big aggregates of TiO₂-NP 50A were observed loosely attached to the bacterial cell (Figure 5E), which suggests the much weaker interaction of TiO₂-NP50A than the smaller sized TiO₂-NP 25A and TiO₂-NP 10A with the cells. No obvious attachment between the TiO₂-NP 50R aggregates and the bacterial cells was observed (Figure 5F).

Figures 5G to 5L show TEM images of the sliced bacterial cells untreated or treated with 50 mg L^{-1} of the TiO₂ NPs. The untreated (Figure 5G) and TiO₂-NP 50R-treated (Figure 5L) cells remained intact with unimpaired cell morphology and structure, indicating the nontoxicity of TiO₂-NP 50R. However, the NPs with smaller size and anatase phase were observed sticking to the

cell surfaces (Figure 5H to 5K), which apparently induced cell distortion, plasmolysis and cell wall and membrane damage; penetration and internalization of the nanoparticles into the bacterial cells were also observed (Fig. 5H and 5I).

From the above TEM observations, it can be concluded that anatase TiO₂ NPs are more prone to attaching on the bacterial surfaces than rutile NPs, and the larger NPs interact weaker with cells compared to the smaller NPs. As particle size decreases, the ratio of surface area to mass increases and changes in the physicochemical properties (e.g., surface atom reactivity, electronic and optical properties) of the nanoparticles occur, consequently, the smaller particles tend to agglomerate to a greater extent, which can further influence their reactivity and binding characteristics [40]. The NP-cell attachment may inhibit the movement of substances in and out of bacterial cells, thereby causing homeostatic imbalance, cellular metabolic disturbance and even cell death [41]. Moreover, the NP-cell attachment would facilitate

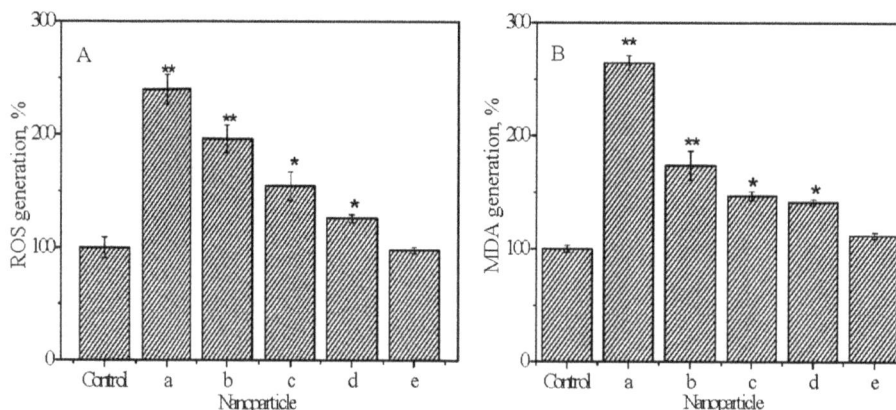

Figure 6. Relative contents of intracellular ROS (A) and MDA (B) in the bacterial cells after 3 h exposure to the TiO$_2$ NPs (50 mg L^{-1}). a–e stand for TiO$_2$-NP 10A, TiO$_2$-NP 25A, TiO$_2$-NP 25AR, TiO$_2$-NP 50A, and TiO$_2$-NP 50R, respectively. Asterisk indicates a significant difference relative to the control (*, $p<0.05$; **, $p<0.01$) based on the Student's t test. Error bars represent standard deviation (n = 3).

the cell internalization of NPs and the intracellular ROS production [12]. If TiO$_2$ NPs are sufficiently small, they can penetrate in the cells, and then induce the potential photocatalytic process inside and adsorb and deactivate biomolecules such as proteins [22,42]. Therefore, the physical NP-cell attachment and interaction could substantially contribute to the observed nanotoxicity. Many studies [13,43–44] suggest that the antibacterial mechanism of NPs includes the disruption of bacterial cellular membrane. However, we do not know for sure yet why the anatase NPs had higher affinity to the cell surfaces than the rutile NPs, which could be possibly due to their different surface properties. It is indicated that the coordination and surface properties allow anatase but not rutile NPs after dispersion induce the generation of ROS [24].

4. Oxidative stress and lipid peroxidation induced by the NPs

Relative intracellular ROS productions following the exposures to the TiO$_2$ NPs (50 mg L^{-1}) are shown in Figure 6A. The produced ROS in the bacterial cells exposed to the anatase TiO$_2$ NPs was significantly ($p<0.05$) higher than that in the blank control cells and increased with decreasing particle size; whereas the pure rutile TiO$_2$-NP 50R had insignificant effect on the intracellular ROS production and TiO$_2$-NP 25AR containing 7% rutile induced significantly lower intracellular ROS production

compared with the pure anatase TiO$_2$-NP 25A of the same particle size. The enhanced intracellular ROS would affect protein expression and function in the bacteria by interrupting translation and post-translational modification [45]. It has been indicated that TiO$_2$ NPs in anatase phase are capable of inducing generation of more ROS than that in the rutile phase [24,46] and thereby may cause higher cytotoxicity [47] including toward *E.coli* [13,48–52]. The increased ROS generation in *E.coli* exposed to the smaller and anatase TiO$_2$ NPs coincided with their enhanced bactericidal effects as shown in Figure 4, which suggests that the size and crystal phase of TiO$_2$ NPs played a critical role in the nanotoxicity and the nanotoxicity could be caused mainly by the elevated oxidative stress.

MDA productions in the *E coli* cells upon the exposures to the 50 mg L^{-1} TiO$_2$ NPs are shown in Figure 6B. Significantly higher MDA contents were observed in the NPs-treated cells compared with the blank control, indicating the cell membrane lipid peroxidation induced by the NPs. The MDA content was the highest in the TiO$_2$-NP 10A treated cells and overall decreased with decreasing size of the anatase NPs, which was in the same order of the ROS production by anatase NPs. This implies that the lipid peroxidation could be mainly caused by the increased ROS.

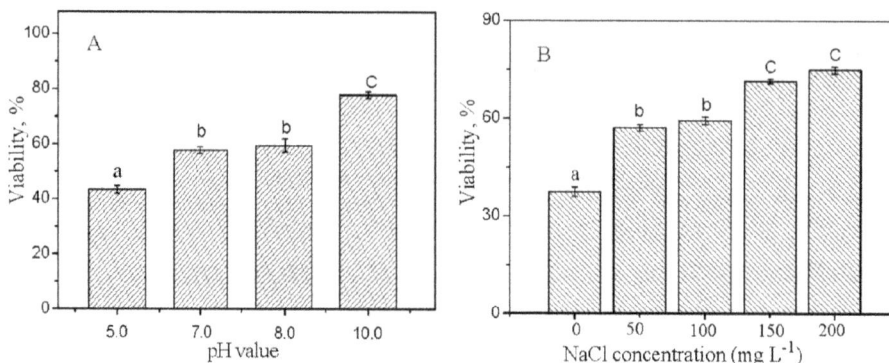

Figure 7. The effects of pH (A) and NaCl (B) on the relative viability of *E.coli* exposed to 10 mg L^{-1} TiO$_2$-NP 10A for 3 h. Significant difference ($p<0.05$) between two treatments is presented by different lowercase letters a, b and c. Error bars represent standard deviation (n = 3).

5. Effects of pH and ionic strength on the nanotoxicity

The bactericidal effect of the 10 mg L^{-1} TiO$_2$-NP 10A exhibited a significant dependence on the solution chemistry (Figure 7). No significant difference in the nanotoxicity was observed between pH 7.0 and 8.0, while the exposed bacteria presented significantly lower and higher viability at pH 5.0 and 10.0 as compared with that at pH 7.0, respectively (Figure 7A). Increasing the suspension pH from 5.0 to 10.0, the bacterial viability increased from 43.3% to 77.9%, indicating the decreasing nanotoxicity with increasing pH. It is generally considered that direct contact and adherence of NPs with the organism cell surfaces plays a critical role in the nanotoxicity [12]. Zeta potentials of the bacterial cells were all negative at the four pHs, being -58.4, -56.7, -56.7 and -52.5 mV at pH 5.0, 7.0, 8.0 and 10.0, respectively; whereas zeta potential of TiO$_2$-NP 10A decreased from about 20 mV at pH 5.0 to lower than -40 mV at pH 10.0 (Figure 2). The positively-charged NPs at pH 5.0 could have a higher potential of contact and hetero-agglomeration with the negatively-charged bacterial cells through the electrostatic attraction and therefore had a higher antibacterial effect compared with the negatively-charged NPs at the three higher pHs.

It has been indicated that the antibacterial effect of TiO$_2$ NPs (25 nm, P25) on *E.coli* was stronger at pH 5.5 versus 7.0 and 9.5 and the stronger antibacterial effect at the lower pH was attributed to the stronger accumulation of the NPs on the cell surfaces [53,54]. However, contradictory research result has also been reported. Planchon et al. (2013) found a stronger adsorption of TiO$_2$ NPs (25 nm, P25) on *E.coli* but a slightly lower toxicity at pH 5.0 versus 8.0, which was attributed to a better physiological state of *E.coli* bacteria at pH 5.0 (artificial water) versus 8.0 (surface water sample) [15]. Rincon and Pulgarin (2004) did not observe any difference in the *E.coli* deactivation rate by the TiO$_2$ NPs (25 nm, P25) in the pH range of 4.0–9.0 [55]. The contradictory results on the pH effect may partly come from the difference in the crystal structure of the used TiO$_2$ NPs, but the exact mechanisms remain to be studied.

It is observed that the addition of NaCl ($50 \sim 200 \text{ mg L}^{-1}$) reduced the toxicity of TiO$_2$-NP 10A toward *E.coli* to an extremely significant extent (Figure 7B). Some studies suggest that NaCl introduced to the medium can decrease the toxicity effect on the bacteria by providing a barrier of steric hindrance between NPs and cells [15]. Li et al. (2013) found that saline ions promoted NP aggregation and reduced surface charge, and then inhibited the adsorption of NPs on bacterial surfaces, so higher saline ions could lead to higher cell viability [38]. Furthermore, ionic strength can also influence the tolerance of bacteria to toxicants [37,38]. The ionic strength of physiology salt-water (8.5 g L^{-1} NaCl) is isotonic and favorable for the bacterium survival. Hence, the bacteria were more tolerant to the NP suspensions at the higher ionic strengths (closer to the physiology salt-water).

Conclusions

The present study investigated the antibacterial effect of five types of TiO$_2$ NPs with various crystal phase and particle size. A marked particle size and crystal phase dependent nanotoxicity was observed. Water chemistry, i.e. pH and ionic strength, could also significantly influence the bactericidal activity of the anatase TiO$_2$ NPs. In conclusion, the TiO$_2$ NPs with anatase phase and smaller particle size had higher affinity to the cell surfaces and induced heavier oxidative damage and toxicity to the bacterial cells, and the toxicity decreased with increasing pH (5.0–10.0) and ionic strength (50–200 mg L^{-1} NaCl). These findings substantiate the need to correlate the NP characterization and behavior in environmental matrices with the toxicological endpoints and to develop a common test strategy for the eco-toxicity study of NPs taking into consideration of various confounding factors relating to the NPs, bacterial cells, and the test environment in the near future.

Author Contributions

Conceived and designed the experiments: DHL XCL. Performed the experiments: XCL JYL SM GSL. Analyzed the data: XCL. Contributed reagents/materials/analysis tools: DHL KY MPT. Contributed to the writing of the manuscript: XCL DHL.

References

1. Chen XB, Mao SS (2007) Titanium dioxide nanomaterials: synthesis, properties, modifications, and applications. Chem Rev 107: 2891–2959.

2. Robichaud CO, Uyar AE, Darby MR, Zucker LG, Wiesner MR (2009) Estimates of upper bounds and trends in nano-TiO$_2$ production as a basis for exposure assessment. Environ Sci Technol 43: 4227–4233.

3. Kaegi R, Ulrich A, Sinnet B, Vonbank R, Wichser A, et al. (2008) Synthetic TiO$_2$ nanoparticle emission from exterior facades into the aquatic environment. Environ Pollut 156: 233–239.

4. Lin DH, Tian XL, Wu FC, Xing BS (2010) Fate and transport of engineered nanomaterials in the environment. J Environ Qual 39: 1896–1908.

5. Johnson AC, Bowes MJ, Crossley A, Jarvica HP, Jurkschat K, et al. (2011) An assessment of the fate, behaviour and environmental risk associated with sunscreen TiO$_2$ nanoparticles in UK field scenarios. Sci Total Environ 409: 2503–2510.

6. Batley GE, Kirby JK, Mclaughlin MJ (2013) Fate and risks of nanomaterials in aquatic and terrestrial environments. Accounts Chem Res 46: 854–862.

7. Ma S, Lin DH (2013) The biophysicochemical interactions at the interfaces between nanoparticles and aquatic organisms: adsorption and internalization. Environ Sci: Processes Impacts 15: 145–160.

8. Menard A, Drobne D, Jemec A (2011) Ecotoxicity of nanosized TiO$_2$: review of in vivo data. Environ Pollut 159: 677–684.

9. Griffitt RJ, Luo J, Gao J, Bonzongo JC, Barber DS (2008) Effects of particle composition and species on toxicity of metallic nanomaterials in aquatic organisms. Environ Toxic Chem 27: 1972–1978.

10. Battin TJ, Kammer FVD, Weilhartner A, Ottofuelling S, Hofmann T (2009) Nanostructured TiO$_2$: transport behavior and effects on aquatic microbial communities under environmental conditions. Environ Sci Technol 43: 8098–8104.

11. Ji J, Long ZF, Lin DH (2011) Toxicity of oxide nanoparticles to the green algae *Chlorella* sp. Chem Eng J 170: 525–530.

12. Lin DH, Ji J, Long ZF, Yang K, Wu FC (2012) The influence of dissolved and surface-bound humic acid on the toxicity of TiO$_2$ nanoparticles to *Chlorella* sp. Water Res 46: 4477–4487.

13. Dalai S, Pakrashi S, Kumar RSS, Chandrasekaran N, Mukherjee A (2012) A comparative cytotoxicity study of TiO$_2$ nanoparticles under light and dark conditions at low exposure concentrations. Toxicol Res 1: 116–130.

14. Clément L, Hurel C, Marmier N (2013) Toxicity of TiO$_2$ nanoparticles to cladocerans, algae, rotifers and plants – effects of size and crystalline structure. Chemosphere 90: 1083–1090.

15. Planchon M, Ferrari R, Guyot F, Gélabertb A, Menguy N, et al. (2013) Interaction between *Escherichia coli* and TiO$_2$ nanoparticles in natural and artificial waters. Colloid Surface B 102: 158–164.

16. Kim J, Lee S, Kim C, Seo J, Park Y, et al. (2014) Non-monotonic concentration–response relationship of TiO$_2$ nanoparticles in freshwater cladocerans under environmentally relevant UV-A light. Ecotox Environ Safe 101: 240–247.

17. Jiang JK, Oberdörster G, Biswas P (2009) Characterization of size, surface charge, and agglomeration state of nanoparticle dispersions for toxicological studies. J Nanopart Res 11: 77–89.

18. Kim DS, Kwak SY (2009) Photocatalytic inactivation of *E.coli* with a mesoporous TiO$_2$ coated film using the film adhesion method. Environ Sci Technol 43: 148–151.

19. Deckers AS, Loo S, L'hermite MM, Boime NH, Menguy N, et al. (2009) Size-, composition- and shape-dependent toxicological impact of metal oxide nanoparticles and carbon nanotubes toward bacteria. Environ Sci Technol 43: 8423–8429.

20. Park S, Lee S, Kim B, Lee S, Lee J, et al. (2012) Toxic effects of titanium dioxide nanoparticles on microbial activity and metabolic flux. Biotechnol Bioproc E 17: 276–282.

21. Tong TZ, Binh CTT, Kelly JJ, Gaillard JF, Gray KA (2013) Cytotoxicity of commercial nano-TiO$_2$ to *Escherichia coli* assessed by high-throughput screening: effects of environmental factors. Water Res 47: 2352–2362.

22. Xiong SJ, George SJ, Ji ZX, Lin SJ, Yu HY, et al. (2013) Size of TiO$_2$ nanoparticles influences their phototoxicity: an in vitro investigation. Arch Toxicol 87: 99–109.

23. Nel A, Xia T, Mädler L, Li N (2006) Toxic potential of materials at the nanolevel. Science 311: 622–627.

24. Jin C, Tang Y, Yang FG, Li XL, Xu S, et al. (2011) Cellular Toxicity of TiO$_2$ nanoparticles in anatase and rutile crystal phase. Biol Trace Elem Res 141: 3–15.

25. Whirter MJM, Quillan AJM, Bremer PJ (2002) Influence of ionic strength and pH on the first 60 min of *Pseudomonas aeruginosa* attachment to ZnSe and to TiO$_2$ monitored by ATR-IR spectroscopy. Colloid Surface B 26: 365–372.

26. French RA, Jacobson AR, Kim B, Isley SL, Penn RL, et al. (2009) Influence of ionic strength, pH, and cation valence on aggregation kinetics of titanium dioxide nanoparticles. Environ Sci Technol 43: 1354–1359.

27. Chowdhury I, Cwiertny DM, Walker SL (2012) Combined factors influencing the aggregation and deposition of nano-TiO$_2$ in the presence of humic acid and bacteria. Environ Sci Technol 46: 6968–6976.

28. Ng AMC, Chan CMN, Guo MY, Leung YH, Djurišić AB, et al. (2013) Antibacterial and photocatalytic activity of TiO$_2$ and ZnO nanomaterials in phosphate buffer and saline solution. App Microbiol Biot 97: 5565–5573.

29. Brunet L, Lyon DY, Hotze EM, Alvarez PJJ, Wiesner MR, et al. (2009) Comparative photoactivity and antibacterial properties of C$_{60}$ fullerenes and titanium dioxide nanoparticles. Environ Sci Technol 43: 4355–4360.

30. Jiang W, Mashayekhi H, Xing BX (2009) Bacterial toxicity comparison between nano- and micro-scaled oxide particles. Environ Pollut 157: 1619–1625.

31. Kim SW, An YJ (2012) Effect of ZnO and TiO$_2$ nanoparticles preilluminated with UVA and UVB light on *Escherichia coli* and *Bacillus subtilis*. App Microbiol Biot 95: 243–253.

32. Bokare A, Sanap A, Pai M, Sabharwal S, Athawale AA (2013) Antibacterial activities of Nd doped and Ag coated TiO$_2$ nanoparticles under solar light irradiation. Colloid Surface B 102: 273–280.

33. Li S, Wallis LK, Ma H, Diamond SA (2014) Phototoxicity of TiO$_2$ nanoparticles to a freshwater benthic amphipod: Are benthic systems at risk? Sci Total Environ 466–467: 800–808.

34. Li S, Wallis LK, Diamond SA, Ma H, Hoff DJ (2014) Species sensitivity and dependence on exposure conditions impacting phototoxicity of TiO$_2$ nanoparticles to benthic organisms. Environ Toxicol Chem 33: 1563–1569.

35. Li S, Pan X, Wallis LK, Fan ZY, Chen ZL, et al. (2014) Comparison of TiO$_2$ nanoparticle and graphene-TiO$_2$ nanoparticle composite phototoxicity to *Daphnia magna* and *Oryzias latipes*. Chemosphere 112: 62–69.

36. Adams LK, Lyon DY, Alvarez PJJ (2006) Comparative eco-toxicity of nanoscale TiO$_2$, SiO$_2$, and ZnO water suspensions. Water Res 40: 3527–3532.

37. Li M, Zhu LZ, Lin DH (2011) Toxicity of ZnO nanoparticles to *Escherichia coli*: Mechanism and the influence of medium components. Environ Sci Technol 45: 1977–1983.

38. Li M, Lin DH, Zhu LZ (2013) Effects of water chemistry on the dissolution of ZnO nanoparticles and their toxicity to *Escherichia coli*. Environ Pollut 173: 97–102.

39. Long ZF, Ji J, Yang K, Lin DH, Wu FC (2012) Systematic and quantitative investigation of the mechanism of carbon nanotubes' toxicity toward algae. Environ Sci Technol 46: 8458–8466.

40. Suresh AK, Pelletier DA, Doktycz MJ (2013) Relating nanomaterial properties and microbial toxicity. Nanoscale 5: 463–474.

41. Wang Z, Lee YH, Wu B, Horst A, Kang Y, et al. (2010) Anti-microbial activities of aerosolized transition metal oxide nanoparticles. Chemosphere 80: 525–529.

42. Szczupak AM, Ulfig K, Morawski AW (2011) The application of titanium dioxide for deactivation of bioparticulates: An overview. Catalysis Today 169: 249–257.

43. Adams CP, Walker KA, Obare SO, Docherty KM (2014) Size-dependent antimicrobial effects of novel palladium nanoparticles. Plos One 9 (1), e89581: 1–12.

44. Musee N, Thwalaa M, Nota N (2011) The antibacterial effects of engineered nanomaterials: implications for wastewater treatment plants. J Environ Monitor 13: 1164–1183.

45. Jiang GX, Shen ZY, Niu JF, Bao YP, Chen J, et al. (2011) Toxicological assessment of TiO$_2$ nanoparticles by recombinant *Escherichia coli* bacteria. J Environ Monitor 13: 42–48.

46. Linsebigler AL, Lu GQ, Yates JT (1995) Photocatalysis on TiO$_2$ surfaces: Principles, mechanisms, and selected results. Chem Rev 95: 735–758.

47. Kelly K, Havrilla C, Brady T, Abramo K, Levin E (1998) Oxidative stress in toxicology: established mammalian and emerging piscine model systems. Environ Health Persp 106: 375–384.

48. Neal AL (2008) What can be inferred from bacteria–nanoparticle interactions about the potential consequences of environmental exposure to nanoparticles? Ecotoxicology 17: 362–371.

49. Dastjerdi R, Montazer M (2010) A review on the applications of inorganic nano-structured materials in the modification of textiles: focus on anti-microbial properties. Colloid Surface B 79: 5–18.

50. Foster HA, Ditta IB, Varghese S, Steele A (2011) Photocatalytic disinfection using titanium dioxide: spectrum and mechanism of antimicrobial activity. App Microbiol Biot 90: 1847–1868.

51. Kumar A, Pandey AK, Singh SS, Shanker R, Dhawan A (2011) Engineered ZnO and TiO$_2$ nanoparticles induce oxidative stress and DNA damage leading to reduced viability of *Escherichia coli*. Free Radical Bio Med 51: 1872–1881.

52. Barnes RJ, Molina R, Xu JB, Dobson PJ, Thompson IP (2013) Comparison of TiO$_2$ and ZnO nanoparticles for photocatalytic degradation of methylene blue and the correlated inactivation of gram-positive and gram-negative bacteria. J Nanopart Res 15: 1432.

53. Pagnout C, Jomini S, Dadhwal M, Caillet C, Thomasc F, et al. (2012) Role of electrostatic interactions in the toxicity of titanium dioxide nanoparticles toward *Escherichia coli*. Colloid Surface B 92: 315–321.

54. Schwegmann H, Ruppert J, Frimmel FH (2013) Influence of the pH-value on the photocatalytic disinfection of bacteria with TiO$_2$ - explanation by DLVO and XDLVO theory. Water Res 47: 1503–1511.

55. Rincon AG, Pulgarin C (2004) Effect of pH, inorganic ions, organic matter and H$_2$O$_2$ on *E.coli* K12 photocatalytic inactivation by TiO$_2$ implications in solar water disinfection. Appl Catal B: Environ 51: 283–302.

Permissions

The contributors of this book come from diverse backgrounds, making this book a truly international effort. This book will bring forth new frontiers with its revolutionizing research information and detailed analysis of the nascent developments around the world.

We would like to thank all the contributing authors for lending their expertise to make the book truly unique. They have played a crucial role in the development of this book. Without their invaluable contributions this book wouldn't have been possible. They have made vital efforts to compile up to date information on the varied aspects of this subject to make this book a valuable addition to the collection of many professionals and students.

This book was conceptualized with the vision of imparting up-to-date information and advanced data in this field. To ensure the same, a matchless editorial board was set up. Every individual on the board went through rigorous rounds of assessment to prove their worth. After which they invested a large part of their time researching and compiling the most relevant data for our readers.

The editorial board has been involved in producing this book since its inception. They have spent rigorous hours researching and exploring the diverse topics which have resulted in the successful publishing of this book. They have passed on their knowledge of decades through this book. To expedite this challenging task, the publisher supported the team at every step. A small team of assistant editors was also appointed to further simplify the editing procedure and attain best results for the readers.

Apart from the editorial board, the designing team has also invested a significant amount of their time in understanding the subject and creating the most relevant covers. They scrutinized every image to scout for the most suitable representation of the subject and create an appropriate cover for the book.

The publishing team has been an ardent support to the editorial, designing and production team. Their endless efforts to recruit the best for this project, has resulted in the accomplishment of this book. They are a veteran in the field of academics and their pool of knowledge is as vast as their experience in printing. Their expertise and guidance has proved useful at every step. Their uncompromising quality standards have made this book an exceptional effort. Their encouragement from time to time has been an inspiration for everyone.

The publisher and the editorial board hope that this book will prove to be a valuable piece of knowledge for researchers, students, practitioners and scholars across the globe.

List of Contributors

Ana Luisa Caetano, Catarina R. Marques and Fernando Gonçalves
Department of Biology, University of Aveiro, Campus Universitário de Santiago, Aveiro, Portugal
CESAM, University of Aveiro, Campus Universitário de Santiago, Aveiro, Portugal

Fernando Carvalho
Nuclear and Technological Institute (ITN) Department of Radiological Protection and Nuclear Safety, Sacavém, Portugal

Eduardo Ferreira da Silva
Department of Geosciences, University of Aveiro, GeoBioTec Research Center, Campus Universitário de Santiago, Aveiro, Portugal

Ruth Pereira
Department of Biology, Faculty of Sciences of the University of Porto, Porto, Portugal
Interdisciplinary Centre of Marine and Environmental Research (CIIMAR/CIMAR), University of Porto, Porto, Portugal

Ana Gavina
CESAM, University of Aveiro, Campus Universitário de Santiago, Aveiro, Portugal
Interdisciplinary Centre of Marine and Environmental Research (CIIMAR/CIMAR), University of Porto, Porto, Portugal

Vítor Seabra
CESPU, Institute of Research and Advanced Training in Health Sciences and Technologies, Department of Pharmaceutical Sciences, Higher Institute of Health

Aurea Lima
CESPU, Institute of Research and Advanced Training in Health Sciences and Technologies, Department of Pharmaceutical Sciences, Higher Institute of Health Sciences-North (ISCS-N), Gandra PRD, Portugal
Molecular Oncology Group CI, Portuguese Institute of Oncology of Porto (IPO-Porto), Porto, Portugal
Abel Salazar Institute for the Biomedical Sciences (ICBAS) of University of Porto, Porto, Portugal

Rita Azevedo
Molecular Oncology Group CI, Portuguese Institute of Oncology of Porto (IPO-Porto), Porto, Portugal
Faculty of Medicine of University of Porto (FMUP), Porto, Portugal

Hugo Sousa
Molecular Oncology Group CI, Portuguese Institute of Oncology of Porto (IPO-Porto), Porto, Portugal
Faculty of Medicine of University of Porto (FMUP), Porto, Portugal
Virology Service, Portuguese Institute of Oncology of Porto (IPO-Porto), Porto, Portugal

Miguel Bernardes
Faculty of Medicine of University of Porto (FMUP), Porto, Portugal
Rheumatology Department of São João Hospital Center, Porto, Portugal

Rui Medeiros
Molecular Oncology Group CI, Portuguese Institute of Oncology of Porto (IPO-Porto),Porto, Portugal
Abel Salazar Institute for the Biomedical Sciences (ICBAS) of University of Porto, Porto, Portugal
Virology Service, Portuguese Institute of Oncology of Porto (IPO-Porto), Porto, Portugal
Research Department-Portuguese League Against Cancer (LPCC-NRNorte), Porto, Portugal

John A. Zebala, Aaron D. Schuler and Stuart J. Kahn
Syntrix Biosystems, Inc., Auburn, Washington, United States of America

Alan Mundell
Animal Dermatology Service, Edmonds, Washington, United States of America

Linda Messinger
Veterinary Referral Center of Colorado, Englewood, Colorado, United States of America

Craig E. Griffin
Animal Dermatology Clinic, San Diego, California, United States of America

Yves Brand, Vesna Radojevic, Michael Sung, Eric Wei, Cristian Setz, Andrea Glutz, Katharina Leitmeyer, Daniel Bodmer
Department of Biomedicine, University Hospital Basel, Basel, Switzerland and Clinic for Otolaryngology, Head and Neck Surgery, University Hospital Basel, Basel, Switzerland

Han Sung Kim and Yeong-Min Yoo
Department of Biomedical Engineering, College of Health Science, Yonsei University, Wonju, Gangwon-do, Republic of Korea

Seung-Il Choi
Cornea Dystrophy Research Institute and Department of Ophthalmology, Yonsei University College of Medicine, Seoul, Republic of Korea

Eui-Bae Jeung
Laboratory of Veterinary Biochemistry and Molecular Biology, College of Veterinary Medicine, Chungbuk National University, Cheongju, Republic of Korea

Chao Sun, Xin Wang, Padraig D'Arcy and Stig Linder
Cancer Center Karolinska, Department of Oncology and Pathology, Karolinska Institute, Stockholm, Sweden

Peristera Roboti, Marjo-Riitta Puumalainen, Eileithyia Swanton and Stephen High
Faculty of Life Sciences, University of Manchester, Manchester, United Kingdom

Mårten Fryknäs
Department of Medical Sciences, Division of Clinical Pharmacology, Uppsala University, Uppsala, Sweden

Malin Hult
Center for Inherited Metabolic Diseases, Karolinska University Hospital, Stockholm, Sweden

Aaron D. Schimmer and Carolyn Goard
Princess Margaret Cancer Centre, Toronto, Ontario, Canada

Azra Raza
Columbia University Medical Center, New York, New York, United States of America

Thomas H. Carter
The University of Iowa, Iowa City, Iowa, United States of America

David Claxton
Penn State, Hershey, Pennsylvania, United States of America

Harry Erba
University of Alabama at Birmingham, Birmingham, Alabama, United States of America

Daniel J. DeAngelo
Dana-Farber Cancer Institute, Boston, Massachusetts, United States of America

Martin S. Tallman
Leukemia Service, Memorial Sloan-Kettering Cancer Center, Weill Cornell Medical College, New York, New York, United States of America

Gautam Borthakur
The University of Texas MD Anderson Cancer Center, Houston, Texas, United States of America

Xiaoe Yang and Tariq Rafiq
MOE Key Lab of Environmental Remediation and Ecosystem Health, College of Environmental and Resource Sciences, Zhejiang University, Zijingang Campus, Hangzhou, P.R. China

Shufeng Wang
MOE Key Lab of Environmental Remediation and Ecosystem Health, College of Environmental and Resource Sciences, Zhejiang University, Zijingang Campus, Hangzhou, P.R. China
Research Institute of Subtropical Forestry, Chinese Academy of Forestry, Fuyang, Hangzhou, P.R. China

Xiang Shi, Haijing Sun, Yitai Chen and Hongwei Pan
Research Institute of Subtropical Forestry, Chinese Academy of Forestry, Fuyang, Hangzhou, P.R. China

Francesco Ballestriero, Malak Daim, Jadranka Nappi and Suhelen Egan
School of Biotechnology and Biomolecular Sciences and Centre for Marine Bio-Innovation, University of New South Wales, Sydney, New South Wales, Australia

Anahit Penesyan
Department of Chemistry and Biomolecular Sciences, Macquarie University, Sydney, New South Wales, Australia

David Schleheck
Biology Department, University of Konstanz,Konstanz, Germany

Paolo Bazzicalupo and Elia Di Schiavi
Institute of Genetics and Biophysics "Adriano Buzzati Traverso", National Research Council, Naples, Italy

Meiyan Jiang, Qi Wang, Takatoshi Karasawa, Hongzhe Li and Peter S. Steyger
Oregon Hearing Research Center, Oregon Health & Science University, Portland, Oregon, United States of America

Ja-Won Koo
Oregon Hearing Research Center, Oregon Health & Science University, Portland, Oregon, United States of America
Department of Otorhinolaryngology, Seoul National University College of Medicine, Bundang Hospital, Seongnam, Gyeonggi, Republic of Korea

Sang Y. Lee, Becky Slagle-Webb, Elias Rizk, Akshal Patel and James R. Connor
Department of Neurosurgery, Pennsylvania State University College of Medicine, Penn State M.S. Hershey Medical Center, Hershey, Pennsylvania, United States of America

Patti A. Miller
Department of Radiology, Pennsylvania State University College of Medicine, Penn State M.S. Hershey Medical Center, Hershey, Pennsylvania, United States of America

Shen-Shu Sung
Department of Pharmacology, Pennsylvania State University College of Medicine, Penn State M.S. Hershey Medical Center, Hershey, Pennsylvania, United States of America

Yong Zhang, Xiaoen Wang and Ming Zhao
AntiCancer, Inc., San Diego, California, United States of America

Chengyu Wu
Department of Traditional Chinese Medicine Diagnostics, Nanjing University of Chinese Medicine, Nanjing, China

Robert M. Hoffman and Lei Zhang
AntiCancer, Inc, San Diego, California, United States of America
Department of Surgery, University of California San Diego, San Diego, California, United States of America

Fang Liu
AntiCancer, Inc., San Diego, California, United States of America
Department of Anatomy, Second Military Medical University, Shanghai, China

Alenka Smid, Natasa Karas-Kuzelicki, Miha Milek and Irena Mlinaric-Rascan
Faculty of Pharmacy, University of Ljubljana, Ljubljana, Slovenia

Janez Jazbec
University Children's Hospital, University Medical Centre Ljubljana, Ljubljana, Slovenia

Daleya Abdulaziz Bardi, Mohammed Farouq Halabi, Mahmood Ameen Abdulla and Nahla Saeed Al-Wajeeh
Department of Biomedical Science, Faculty of Medicine, University of Malaya, Kuala Lumpur, Malaysia

Pouya Hassandarvish
Department of Medical Microbiology, Faculty of Medicine, University of Malaya, Kuala Lumpur, Malaysia

Elham Rouhollahi, Mohammadjavad Paydar and Nor Azizan Abdullah
Department of Pharmacology, Faculty of Medicine, University of Malaya, Kuala Lumpur, Malaysia

Soheil Zorofchian Moghadamtousi
Biomolecular Research Group, Biochemistry Program, Institute of Biological Sciences, Faculty of Science, University of Malaya, Kuala Lumpur, Malaysia

Abdulwali Ablat
Institute of Biological Science, Faculty of Science, University of Malaya, Kuala Lumpur, Malaysia

Yo Sugawara, Masahiro Yutani, Sho Amatsu, Takuhiro Matsumura, Yukako Fujinaga
Laboratory of Infection Cell Biology, International Research Center for Infectious Diseases, Research Institute for Microbial Diseases, Osaka University, Yamada-oka, Suita, Osaka, Japan

Patrick Barnable, Giulia Calenda, Louise Ouattara, Ninochka Jean-Pierre, Larisa Kizima, Aixa Rodríguez, Ciby Abraham, Radhika Menon, Samantha Seidor, Michael L. Cooney, Kevin D. Roberts, Jose A. Fernandez-Romero, Thomas M. Zydowsky, Melissa Robbiani and Natalia Teleshova
Population Council, New York, New York, United States of America

Agegnehu Gettie
Aaron Diamond AIDS Research Center, Rockefeller University, New York, New York, United States of America

James Blanchard
Tulane National Primate Research Center, Tulane University, Covington, Louisiana, United States of America

Rhoda Sperling
Icahn School of Medicine at Mount Sinai, New York, New York, United States of America

Michael Piatak Jr. and Jeffrey D. Lifson
AIDS and Cancer Virus Program, SAIC-Frederick, Inc., Frederick National Laboratory for Cancer Research, Frederick, Maryland, United States of America

Soledad Gonzalo
Departamento de Ingeniería Química, Universidad de Alcalá, Alcalá de Henares, Madrid, Spain

Gerardo Pulido-Reyes
Departamento de Ingeniería Química, Universidad de Alcalá, Alcalá de Henares, Madrid, pain
Departamento de Biología, Facultad de Ciencias, Universidad Autónoma de Madrid, Madrid, Spain

Veronica Llaneza and Jean Claude Bonzongo
Department of Environmental Engineering Sciences, University of Florida, Gainesville, Florida, United States of America

Francisca Fernández-Piñas, Francisco Leganes and Ismael Rodea-Palomares
Departamento de Biología, Facultad de Ciencias, Universidad Autónoma de Madrid, Madrid, Spain

Roberto Rosal and Eloy García-Calvo
Departamento de Ingeniería Química, Universidad de Alcalá, Alcalá de Henares, Madrid, Spain
Instituto Madrileño de Estudios Avanzados (IMDEA) Agua, Alcalá de Henares, Madrid, Spain

Weiti Cui, Kaikai Zhu, Qijiang Jin, Yanjie Xie, Jin Cui, Yan Xia, Jing Zhang and Wenbiao Shen
College of Life Sciences, Laboratory Center of Life Sciences, Nanjing Agricultural University, Jiangsu Province, Nanjing, China

Huiping Chen
Key Laboratory of Protection and Development Utilization of Tropical Crop Germplasm Resources, Hainan University, Haikou, China

Shankar Siva
Division of Radiation Oncology and Cancer Imaging, Peter MacCallum Cancer Centre, Melbourne, VIC, Australia

Michael MacManus and David Ball
Division of Radiation Oncology and Cancer Imaging, Peter MacCallum Cancer Centre, Melbourne, VIC, Australia
Sir Peter MacCallum Department of Oncology, the University of Melbourne, Melbourne, VIC, Australia

Tomas Kron
Sir Peter MacCallum Department of Oncology, the University of Melbourne, Melbourne, VIC, Australia
Physical Sciences, Peter MacCallum Cancer Centre, Melbourne, VIC, Australia

Nickala Best and Jai Smith
Molecular Radiation Biology Laboratory, Peter MacCallum Cancer Centre, VIC, Australia

Pavel Lobachevsky
Sir Peter MacCallum Department of Oncology, the University of Melbourne, Melbourne, VIC, Australia
Molecular Radiation Biology Laboratory, Peter MacCallum Cancer Centre, VIC, Australia

Olga Martin
Division of Radiation Oncology and Cancer Imaging, Peter MacCallum Cancer Centre, Melbourne, VIC, Australia
Sir Peter MacCallum Department of Oncology, the University of Melbourne, Melbourne, VIC, Australia
Molecular Radiation Biology Laboratory, Peter MacCallum Cancer Centre, VIC, Australia

Aslam M. A. Mazari and Bengt Mannervik
Department of Neurochemistry, The Wenner-Gren Institute, Stockholm University, Stockholm, Sweden

Olle Dahlberg and Mattias Mannervik
Department of Molecular Biosciences, The Wenner-Gren Institute, Stockholm University, Stockholm, Sweden

Xiuchun Lin and Jingyi Li
College of Environmental and Biological Engineering, Putian University, Fujian, China
Department of Environmental Science, Zhejiang University, Hangzhou, China

Si Ma and Gesheng Liu
Department of Environmental Science, Zhejiang University, Hangzhou, China

Kun Yang and Daohui Lin
Department of Environmental Science, Zhejiang University, Hangzhou, China
Zhejiang Provincial Key Laboratory of Organic Pollution Process and Control, Zhejiang University, Hangzhou, China

Meiping Tong
College of Environmental Sciences and Engineering, Peking University, Beijing, P. R. China

Index